AF327292

Dedicated

To my wife
for putting up with me

and

To my parents
for their love, support and encouragement

Bioscience and Bioengineering of Titanium Materials

Yoshiki Oshida

Indiana University, Professor Emeritus
Syracuse University, Distinguished Research Professor

ELSEVIER

Amsterdam • Boston • Heidelberg • London • New York • Oxford
Paris • San Diego • San Francisco • Singapore • Sydney • Tokyo

Elsevier
The Boulevard, Langford Lane, Kidlington, Oxford OX5 1GB, UK
Radarweg 29, PO Box 211, 1000 AE Amsterdam, The Netherlands

First edition 2007

Notice
No responsibility is assumed by the publisher for any injury and/or damage to persons
or property as a matter of products liability, negligence or otherwise, or from any use
or operation of any methods, products, instructions or ideas contained in the material
herein. Because of rapid advances in the medical sciences, in particular, independent
verification of diagnoses and drug dosages should be made

British Library Cataloguing in Publication Data
A catalogue record for this book is available from the British Library

Library of Congress Cataloging-in-Publication Data
A catalog record for this book is availabe from the Library of Congress

ISBN–13: 978-0-08-045142-8
ISBN–10: 0-08-045142-X

For information on all elsevier publications
visit our website at www.books.elsevier.com

Printed and bound in Great Britain

07 08 09 10 11 10 9 8 7 6 5 4 3 2 1

Working together to grow
libraries in developing countries
www.elsevier.com | www.bookaid.org | www.sabre.org

ELSEVIER   BOOK AID International   Sabre Foundation

Contents

Prologue   xiii

CHAPTER 1
Introduction
References   8

CHAPTER 2
Materials Classification
2.1. General   11
2.2. Medical/Dental Titanium and its Alloys   13
    2.2.1   Commercially Pure Titanium (CpTi)   13
    2.2.2   Ti-6Al-4V   16
    2.2.3   Ti-6Al-7Nb   16
    2.2.4   Ti-3Al-2.5V   16
    2.2.5   Ti-5Al-3Mo-4Zr   16
    2.2.6   Ti-5Al-2.5Fe   17
    2.2.7   Ti-Ni   17
    2.2.8   Ti-Cu   18
    2.2.9   Ti-Mo   18
    2.2.10   Ti-5Al-2Mo-2Fe (SP700)   18
    2.2.11   Other Ti-Based Alloys   19
    2.2.12   Intermetallic Alloys   19
References   20

CHAPTER 3
Chemical and Electrochemical Reactions
3.1. Discoloration   26
3.2. Corrosion in Media Containing Fluorine Ion Bleaching Agents   28
3.3. Corrosion Resistance, Effects of Environmental and Mechanical-Assisted Actions   31
3.4. Metal Ion Release and Dissolution   42
3.5. Galvanic Corrosion   51
3.6. Microbiology-Induced Corrosion   53

3.7.   Reaction with Hydrogen ... 61
References ... 64

CHAPTER 4
Oxidation and Oxides
4.1.   Formation of Titanium Oxides ... 81
    4.1.1   Air-Formed Oxide ... 82
    4.1.2   Passivation ... 84
    4.1.3   Oxidation at Elevated Temperatures ... 85
4.2.   Influences on Biological Process ... 86
4.3.   Crystal Structures of Ti Oxides ... 87
4.4.   Characterization of Oxides ... 90
4.5.   Unique Applications of Titanium Oxide ... 91
4.6.   Oxide Growth, Stability, and Breakdown ... 91
4.7.   Reaction with Hydrogen Peroxide ... 94
References ... 97

CHAPTER 5
Mechanical and Tribological Behaviors
5.1.   Fatigue ... 107
5.2.   Fracture and Fracture Toughness ... 111
5.3.   Biotribological Actions ... 113
    5.3.1   General ... 113
    5.3.2   Friction ... 114
    5.3.3   Wear and Wear Debris Toxicity ... 116
References ... 120

CHAPTER 6
Biological Reaction
6.1.   Toxicity ... 127
6.2.   Cytocompatibility (Toxicity to Cells) ... 130
6.3.   Allergic Reaction ... 133
6.4.   Metabolism ... 136
6.5.   Biocompatibility ... 138
6.6.   Biomechanical Compatibility ... 143
References ... 148

CHAPTER 7
Implant-Related Biological Reactions

7.1. Bone Healing   162
7.2. Hemocompatibility   164
7.3. Cell Adhesion, Adsorption, Spreading, and Proliferation   166
7.4. Roughness and Cellular Response to Biomaterials   179
7.5. Cell Growth   182
7.6. Tissue Reaction and Bone Ingrowth   185
7.7. Osseointegration and Bone/Implant Interface   193
7.8. Some Adverse Factors for Loosening Implants   199
References   200

CHAPTER 8
Implant Application

8.1. General   217
8.2. Clinical Reports   222
8.3. Surface and Interface Characterization   227
8.4. Reactions in Chemical and Mechanical Environments   230
8.5. Reaction in Biological Environment   235
References   246

CHAPTER 9
Other Applications

9.1.   Denture Bases   257
9.2.   Crowns and Bridges   259
9.3.   Clasps   265
9.4.   Posts and Cores   266
9.5.   Titanium Fiber and Titanium Oxide Powder as Reinforcement   267
       for Bone Cement
9.6.   Sealer Material   268
9.7.   Shape-Memory Dental Implant   268
9.8.   Orthodontic Appliances   269
9.9.   Endodontic Files and Reamers   271
9.10. Clamps and Staples   273
References   274

x   *Contents*

CHAPTER 10
Fabrication Technologies

10.1.	Casting	285
10.2.	Machining	288
10.3.	Electro Discharge Machining (EDM) and CAD/CAM	289
10.4.	Isothermal Forming	291
10.5.	Superplastic Forming (SPF)	291
10.6.	Diffusion Bonding (DB)	294
10.7.	Powder Metallurgy	295
10.8.	Metal Injection Molding (MIM)	297
10.9.	Laser Welding/Forming	299
10.10.	Soldering	301
10.11.	Heat Treatment (HT)	302
References		303

CHAPTER 11
Surface Modifications

11.1.	Sandblasting and Surface Texturing		314
11.2.	Shot-Peening and Laser-Peening		317
11.3.	Chemical, Electrochemical, and Thermal Modifications		319
11.4.	Coating		322
	11.4.1	Carbon, Glass, and Ceramic Coating	322
	11.4.2	Hydroxyapatite Coating	324
	11.4.3	Ca-P Coating	335
	11.4.4	Composite Coating	342
	11.4.5	TiN Coating	347
	11.4.6	Ti Coating	353
	11.4.7	Titania Film and Coating	357
11.5.	Porosity Controlled Surface and Texturing		360
11.6.	Foamed Metal		363
11.7.	Coloring		364
References			364

CHAPTER 12
Future Perspectives

12.1.	Titanium Industry and New Materials Development	385
12.2.	Amorphous Material	388
12.3.	Gradient Functional Material System	389
12.4.	Coating	390

12.5.	Fluoride Treatment	396
12.6.	Surface Texturing and Porous/Foamed Material	397
12.7.	Laser Application	398
12.8.	Near-Net Shape (NNS) Forming and Nanotechnology	400
12.9.	Tissue Engineering and Scaffold Structure and Materials	403
12.10.	Bioengineering and Biomaterial-Integrated Implant System	406
References		411
Epilogue		419

Prologue

Writing a book like this one, which is based on reviewing published articles, is a typical example of exercising an evidence-based learning (EBL). The evidence-based literature review can provide fundamental skills in research-oriented bibliographic inquiry, with an emphasis on the review and synthesis of applicable literature. All information is obtained from numerous sources provided by scholars and researchers who are involved in various disciplines. Available sources possess a variety of features, especially the degree of reliability. There is a known rank of reliability for evidence sources, which is particularly important in the medical and dental fields: (from the most to least reliable evidence source) (1) clinical reports using placebo and double blind studies, (2) clinical reports not using placebo, but conducted according to well-prepared statistical test plans, (3) study reports on time-effect on one group of patients, (4) study reports, at one limited time, on many groups of patients, (5) case reports on a new technique and/or idea, and (6) retrospective reports on clinical evidence. Unfortunately, the number of published articles increases in this descending reliability order.

The most remarkable advantage of the EBL-type reviewing of numerous published articles is to encounter an unexpected idea and issue which might provide a fundamental basis persisting in all related but separately and scattered manner being presented in different sources. Accordingly, it is actually an inductive practice.

Comparing articles seen in medical/dental journals and those published in engineering journals, there is a dilemma of a big difference between them. Obtained results present in the medical/dental journals are normally further analyzed statistically; while the data presented in the engineering journals are normally not subjected to statistical analysis, rather characterization of data is considered more important to interpret results. Even though a statistical analysis is conducted for medical/dental results, there are some cases which do not reach a conclusive point. For example, although the mercury issue (which is used for condensation of dental amalgam alloy to form Ag_2Hg_3 and $Sn_{7-8}Hg$) has been debated for more than 20 years, we still have not come to one solid consensus on its safety issue yet. Actually mercury amalgam has been completely abandoned in Scandinavian counties and Japan, whereas it is still used in the United States and other countries, although tooth-colored restorations are becoming more popular for aesthetic reasons.

There is another challenge in the medical and dental field. It is a controversial issue on the relationship between *in vivo* test results and those obtained by *in vitro* tests. To make this situation more complicated and confusing, there could be differences even among various *in vivo* tests, due to different models (different types of animals versus humans). Moreover, it is almost impossible to comply and summarize to extract any conclusiveness from various *in vitro* test results, since it is very rare to find articles where identical test methods were employed, unless some standardized (specified) methodology was followed. For example, it is easy to list at least ten electrolytes for electrochemical evaluation of the corrosion behavior and resistance of a dental metallic material. In addition to this, wide variations in pH value, concentration, temperature, and other intraoral environmental factors can easily alter the experimental outcomes. This is the reason why in the materials science and engineering area, a phenomenon is characterized to find the basic mechanisms as well as kinetics. Once the basics are elucidated, if any environmental factor is changed, the possible results would not be hard to anticipate. Although a statistical test on obtained data is considered as a powerful tool to analyze results, it should be based on postulated assumptions (which should be statistically tested once data are collected), and the results are not applicable to entire patient populations. Suppose a newly developed medicine was statistically evaluated, *in vivo* test results show that it is effective to 95% of the patient population, which indicates, on the other side of the coin that it definitely does not work for the remaining 5% of human patients.

In this book, such differentiation and comparison in analyzing results between medical/dental and engineering fields will be emphasized and an attempt has been made to fill the gaps where possible. The materials and most of the technologies employed in medical and dental fields originally developed in the engineering/technology field. If they prove their safety and can be miniaturized, they can be accommodated for human uses. Such successfully performed technology transfers from engineering to medical/dental applications are also introduced in this book. Hence, this book has a two-fold uniqueness; (1) the EBL concept has been applied and practiced throughout the entire course of preparation, writing and editing of this book, and (2) a bridging of the gap between medical/dental fields and engineering/technology areas, owing to the author's unique experience in both fields for about an equal length of 20 years each.

Chapter 1
Introduction

References 8

Chapter 1
Introduction

In Greek mythology, the Giant divine being, Titan, a son of Uranos (Father Heaven) and Gaia (Mother Earth), had lost several wars against the Olympic Gods, which resulted in his being confined in the underground dark world. The element titanium, confined within an ore (called rutile ore), was discovered by a German chemist Martin Heinrich Klaproth. Klaproth confirmed it as a new element, and in 1795 he named it for the Latin word for Earth (also the name for the Titans of Greek myth) [1-1, 1-2].

Because of their lightweight, high specific strength (which is referred to a high strength-to-weight ratio), low modulus of elasticity, and excellent corrosion resistance, titanium materials (both unalloyed and alloyed) have become important materials for the aerospace industry since the early 1950s [1-3]. It was very hard to predict, at that time, that titanium materials would currently receive their attention, interest and importance not only for industrial applications but also equally for dental and medical applications. Titanium is one material which receives equal attention and interest from both the engineering and medical/ dental fields.

Research and development in materials design, manufacturing technologies, characterization, and evaluation methods involved in titanium materials are some of the best examples of interdisciplinary science and technology; such disciplines might include physics, chemistry, metallurgy, mechanics, and surface and interface sciences, as well as biological science, engineering, and technology (see Figure 12-1). One of the typical examples can be seen in dental and orthopedic implant systems, which have been developed by integrating industrial engineering, bioengineering, as well as supportive advanced techniques.

Although empiricism and trial and error played a major role in the early stage of research and development on titanium materials, in the last 20 years progress has been made toward a multidisciplinary scientific approach to the study of titanium materials and the development of materials with better performance and properties [1-4]. The research and development of titanium materials has been heavily dependent on the aerospace industries, and this area will continue to be a significant percentage of total consumption of titanium materials for coming years [1-5]. Titanium usage on Boeing aircraft has continuously increased from 1% on the 727 (1963), to 3% on the 747 (1969), 5% on the 757 (1983), and 9% on the 777 (1994). Over the last decade, the focus of titanium alloy development has shifted from aerospace [1-6] to industrial application [1-5]. Different industrial sectors have been looking for different types of development in new titanium alloys; they include Ti-5.8Al-4Sn-3.5Zr-0.7Nb-0.5Mo-0.35Si-0.06C, Ti-6Al-2Sn-4Zr-6Mo, and Ti-4Al-4Mo-2Sn-0.5Si for gas turbine engine materials, Ti-10V-2Fe-3Al, Ti-15V-3Cr-3Sn-3Al, and Ti-15Mo-2.8Al-3Nb-0.2Si for airframe materials, Ti-6Al-1.8Fe-0.2Si for ballistic armor, Ti-6.8Mo-4.5Fe-1.5Al for geothermal and offshore tubular materials, Ti-15V-3Cr-3Sn-3Al (having higher-strength and lower-modulus) for sporting goods [1-7], V-free Ti-6Al-4V equivalent alloys for medical and dental applications, and NiTi-Cu alloys for medical orthopedic devices. Not only materials but also technologies developed and successfully used in industry can be transferred to medical and dental areas. The computer-aided design and machining (CAD/CAM) process is an excellent example to design and fabricate dental restorations. Investigators are also combining CAD and AI (artificial intelligence) to design complex prosthetic devices such as partial dentures. Computers have also been used to analyze the stress levels and stress distribution on dental restorations, dental implants and orthopedic implants. They can simulate internal stress distributions under different conditions and loading situations, and the results can be used optimize the computer-aided design process (e.g., two- or three-dimensional finite element modeling and stress analyses). The application of lasers to dentistry is another good example of technology transfer [1-8]. Various ways of laser application for surface modifications of metals have also been introduced, and include cutting, welding, surface hardening, laser surface alloying, and forming [1-9]. Superplastic forming (SPF) of denture base, and superplastic forming with diffusion bonding (SPF/DB) for modification and roughening of implant surfaces are another excellent examples of the technical transfer. These transferred technologies should not be limited to technology *per se*, but it should also include technical concepts such as composites, as well as gradually functional structure, which can be frequently seen in the surface modification of implant systems to promote the bone ingrowth.

The popularity of titanium and its alloys in the dental and medical fields can be recognized by counting the manuscripts published in different journals, most of which will be cited in this review. Referring to Figure 1-1, there are three straight lines in semi-log plot for demonstrating the total accumulated number of published articles for every 5 years. The first top line represents accumulated numbers of all articles listed in Chemical Abstracts. Therefore, the counted articles are scattered in wider areas in refining, metallurgy, medicine, dentistry, bioengineering, engineering, industries, pure chemistry as well as chemical engineering. The second straight line is obtained when the search area is limited to materials and science in both the engineering and medical/dentistry fields. If our search is furthermore defined within only medicine/dentistry, we still have an exponential increase in publications.

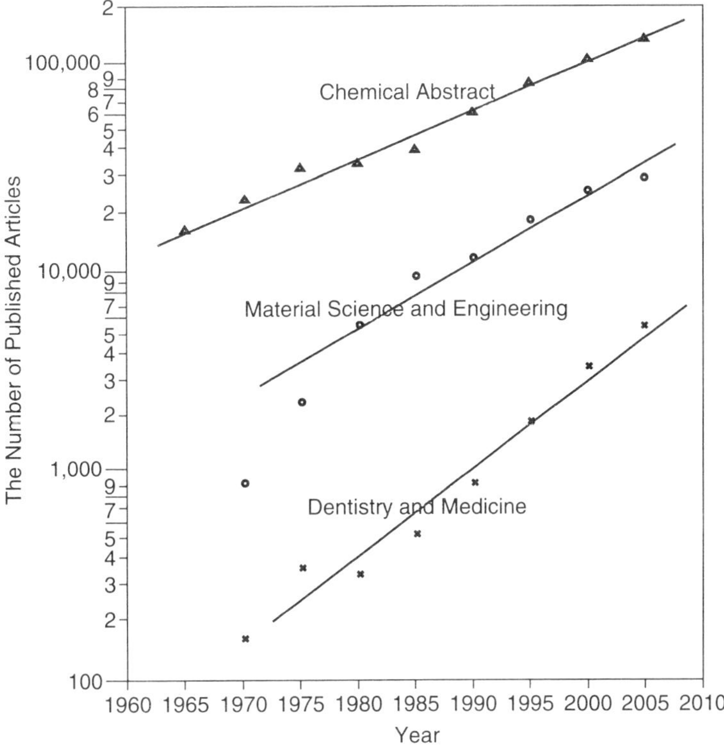

Figure 1-1. Accumulated number of published articles on titanium materials for every 5 years.

From the third straight line, only eight papers (not shown in the figure) were published in the time period from 1960 to 1965, followed by 156 papers in 1966–1970, 352 in 1971–1975, 309 in 1976–1980, 493 in 1981–1985, 916 in 1986–1990, and it jumped to 1914 papers in the 5-year period from 1991 to 1995. In 1996–2000, we already reached more than 3000 articles, followed by 5500 articles which have been published during a 5-year period from 2001 to 2005. These numbers should be considered as conservative, since articles not written in English have not been accounted for.

The exponentially increasing trend of published papers on medical and dental titanium materials may be attributed to different reasons that might include (1) needs from medical and dental sectors, (2) increased researchers and scientists involved in these medical titanium materials, and (3) expanded industrial scale being associated with the above.

About 10 years later from the previous forecasting on titanium industry and materials developments [1-5–1-7], another remarkable change has been seen. In addition to the ever-growing aerospace applications, precise forming technologies have been developed such as near-net shape (NNS) manufacturing, nanotechnology, metal-injection molding (MIM) powder metallurgy techniques, etc. [1-10–1-13], resulting in increasing application areas of titanium materials. As for biomaterials, reflecting to a natural demand for allergy-free metallic materials composed of non-toxic element(s), research and development in metallic biomaterials, dental, and healthcare materials have been advanced remarkably [1-14, 1-15].

Owing to various characteristic properties associated with titanium and titanium-based alloys, different types of dental and medical prostheses have been developed and are currently being utilized. They include cardiac (especially, mechanical heart valves, pacemakers), operational devices and equipment, and orthopedic implants. Furthermore, in the dental field, the list can expand to specifically orthodontic brackets and wires, dental implants, prosthetic appliances, and endodontic files.

It is well recognized that the first reaction of a vital hard/soft tissue (i.e., *host tissue*) to any type of biomaterial (ceramics, polymers, metals and alloys, composites) is a rejection; accordingly, biomaterial is normally recognized as a *"foreign material"* by the host tissue. The biological acceptance of these foreign materials by the living tissues is essentially controlled by the surface and interfacial reaction between the organic substance and inorganic substrate. The surface is not just a free end of a substance, but it is a contact and boundary zone with other substances (either in gaseous, liquid, or solid). A physical system which comprises of a homogeneous component such as solid, liquid, or gas and is clearly distinguishable from each other is called a phase, and a boundary at which two or three of these individual phases are in contact is called an interface. Surface and interface reactions include reactions with organic or inorganic materials, vital or non-vital species, hostile or friendly environments, etc. Surface activities may vary from mechanical actions (fatigue crack initiation and propagation, stress intensification, etc.), chemical action (discoloration, tarnishing, contamination, corrosion, oxidation, etc.), mechano-chemical action (corrosion fatigue, stress-corrosion cracking, etc.), thermo-mechanical action (thermal fatigue), tribological and biotribological actions (wear and wear debris toxicity, friction, etc.) to physical and biophysical actions (surface contact and adhesion, adsorption, absorption, diffusion, cellular attachment, etc.). Accordingly, the longevity, safety, reliability, and structural integrity of dental and medical materials are greatly governed by these surface phenomena, which can be detected, observed, characterized, and analyzed by virtue of various means of devices and technologies [1-5, 1-16].

Interfaces are as important as surfaces. Such interfaces can be found in various combinations, such as dentin/resin bonding, post/lute/tooth system, implant/hard tissue system, metal coping/porcelain system, etc. Common phenomenon underlying these interfacing couples includes the fact that there is always an intermediate layer between two distinct bodies. Accordingly, strictly speaking, there are two bodies with one (intermediate) interface. For example, when the porcelain is fired on the metal coping, the resultant interface between them is normally considered as an inter-diffusion layer in which both elements from porcelain and metal coping are involved. The stability, bonding or adherent strength, as well as structural integrity of these systems are strongly reliant on the interfacial behaviors under various mechanical and/or chemical environments and actions.

To determine the compatibilities (biological, mechanical, and morphological) of a biomaterial to receiving human hard/soft tissue, it is also important to understand the interfacial phenomena between the biomaterial and the biological system into which the material is to be implanted [1-16]. At such an interface, the molecular constituents of the biological system meet and interact with the molecular constituents of the surface of the biomaterials [1-17]. Since these interactions occur primarily on the molecular level and in a very narrow interface zone having a width of less than 1 nm (1 nanometer=0.001 μm=0.1 Å), surface properties on an atomic scale, and in particular the composition and structure of surface layers of the biomaterial, may play an important role in interfacial phenomena [1-16]. Hence, surface and interface characterizations and their biochemical and biomechanical roles in biological environments are one of the most important portion of this book. Of interest and will be seen in later (particularly, in Chapters 6 and 7), scientific approaches to, characterization on, and interpretation such interfacial layers differ between materials scientists, and bio-scientists and clinicians; the former tends to investigate the interface layers from material's side, whereas the latter observes them from the living tissue side. In most cases, these results do not corporate each other. Actually, there is no single report which compiles results obtained from both sides.

It has been noted that hazardous substances are used with relative safety in dentistry when their identity is known, such as mercury in dental amalgam and strong acids in cements and etching agents [1-17]. The following species and elements can be added to this category; beryllium in cements and alloys, asbestos in periodontal packs and drug solutions, lead in impression materials, and cadmium in silver solders and temporary crowns. Alloys containing Cr, V, and Ni (which are recognized to exhibit the heavy metal toxicity) are known to be carcinogenic in animals and man, with local and systemic dissemination of released metallic ions in the tissues. Such examples indicate a need for full scientific consideration of the biological effects of materials at an early stage of development when the implications of use of certain ingredients or techniques may be recognized [1-18]. It can be clearly recognized, currently, that most of the materials or elements mentioned in the above are totally or partially eliminated from their original contents.

In this evidence-based overview book, accordingly, a general characterization of titanium materials, various actions (mechanical, tribological, biotribological, chemical, biochemical, electrochemical, thermal, mechano-chemical, biological, etc.), types of application, fabrication technologies, the surface/interface characterization of implants, and new and advanced materials and technologies for dental/medical titanium will be covered.

REFERENCES

[1-1] http://www.scescye.net/~woods/elements/titanium.

[1-2] http://en.wikipedia.org/wiki/Titanium.

[1-3] Lautenschlager EP, Monaghan P. Titanium and titanium alloys as dental materials. Int Dent J 1993;43:245–253.

[1-4] Moore BK, Oshida Y. Materials science and technology in dentistry. In: Encyclopedic handbook of biomaterials and bioengineering. Wise DL, Trantolo DJ, Altobelli DE, Yaszemski MJ, Gresser JD, Schwartz ER, editors. New York: Marcel Dekker. 1995. pp. 1325–1430 (Chapter 48).

[1-5] Allen P. Titanium alloy development. Adv Mater Process 1996;154:35–37.

[1-6] Tuominen S, Wojcik C. Alloys for aerospace. Adv Mater Process 1995; 153:3–26.

[1-7] Ratner BD, Johnston AB, Lenk TJ. Biomaterials surfaces. J Biomed Mater Res 1987;21:59–90.

[1-8] Loiacono C, Shuman D, Darby M, Luton JG. Lasers in dentistry. Gen Dent 1993;Sep/Oct:378–381.

[1-9] Singh J. The constitution and microstructure of laser surface-modified metals. J Metals 1992;34:551–562.

[1-10] Froes FH. Titanium – Is the time now? J Metals 2004;56:30–31.

[1-11] Faller K, Froes FH. The use of titanium in family automobiles: current trend. J Metals 2001;53:27–28.

[1-12] Kosaka Y, Fox SP, Faller K. Newly developed titanium alloy sheets for the exhaust systems of motorcycles and automobiles. J Metals 2004;56:32–34.

[1-13] Kearns M. Titanium: alive, well and booming! Adv Mater Process 2005;163:63–64.

[1-14] Roncone K. A conversation with titanium suppliers and end users. J Metals 2005;57:11–13.

[1-15] Niionomi M, Hanawa T, Narushima T. Japanese research and development on metallic biomedical, dental, and healthcare materials. J Metals 2005;57: 18–24.

[1-16] Oshida Y, Hashem A, Nishihara T, Yapchulay MV. Fractal dimension analysis of mandibular bones – toward a morphological compatibility of implants. J Bio-Med Mater Eng 1994;4:397–407.

[1-17] Kasemo B, Lausmaa J. Surface science aspects on inorganic biomaterials. CRC Crit Rev Biocompat 1986;2:335–380.

[1-18] Smith DC. The biocompatibility of dental materials. In: Biocompatibility of dental materials, Vol. 1: Characteristics of dental tissues and their response to dental materials. Smith DC, Williams DF, editors. Boca Raton: CRC Press Inc. 1982. pp. 1–37.

Chapter 2
Materials Classification

2.1.	General	11
2.2.	Medical/Dental Titanium and its Alloys	13
	2.2.1 Commercially Pure Titanium (CpTi)	13
	2.2.2 Ti-6Al-4V	16
	2.2.3 Ti-6Al-7Nb	16
	2.2.4 Ti-3Al-2.5V	16
	2.2.5 Ti-5Al-3Mo-4Zr	16
	2.2.6 Ti-5Al-2.5Fe	17
	2.2.7 Ti-Ni	17
	2.2.8 Ti-Cu	18
	2.2.9 Ti-Mo	18
	2.2.10 Ti-5Al-2Mo-2Fe (SP700)	18
	2.2.11 Other Ti-Based Alloys	19
	2.2.12 Intermetallic Alloys	19
References		20

Chapter 2
Materials Classification

Materials research, development, and application should include needs-oriented and seeds-oriented approaches. A good example of needs-oriented materials R&D is the search for a light-weight structural material (particularly demanded by the aerospace industries). This has accelerated the development of titanium alloys. On the other hand, NiTi provides a typical example of seeds-oriented material. Opposing its initial specific aim for development (which was a strategic search for submarine structural materials with a relatively low damping capacity), its unique shape-memory effect (SME), as well as superelasticity (SE) properties, needed a successful application. This was realized when used for medical applications (orthopedic implants, stunt, Harrington bar, and others) and dental applications (blade-type implants, orthodontic wires, splints, as well as endodontic files).

Recently owing to advanced quantitative metallography, as well as molecular/atomic dynamics and designing concepts, these two needs-oriented and seeds-oriented concepts have been integrated systematically to develop nano-scale materials designs and manufacturing technologies.

2.1. GENERAL

Basically, titanium and titanium-based alloys can be classified into α type (HCP: hexagonal closed-packed crystalline structure), near α type, $(\alpha+\beta)$ type, and β type (BCC: body centered-cubic crystalline structure) alloy groups. Alloying elements added to titanium are divided into two groups: alpha (α) stabilizers and beta (β) stabilizers. Elements, such as Al, Sn, Ga, Zr, and interstitial elements (either singly of C, O, and N or in combination), dissolve into the titanium matrix, and are strong solid solution strengtheners which produce little change at the transformation temperature (β-transus: 885°C for pure Ti) from the HCP (α) to the BCC (β) structure of pure titanium when heating, and from BCC to HCP when cooling. Hence they are known as α-stabilizers and exhibit good high-temperature performance. Alloying elements which decrease this phase transformation temperature are referred to as β-stabilizers. Generally, β-stabilizing elements are the transition metals, such as V, Mo, Nb, Ta, and Cr, providing much friability [2-1]. Besides these alloying elements, Fe, Cu, Ni, Si, and B are frequently added to Ti-based alloys for improving mechanical strength, chemical stability, castability, and/or grain refining. By increasing the α-phase portion, it is generally recognized that (1) β-transus temperature increases, (2) creep strength as well as high temperature strengths enhance, (3) flow stress increases, and (4) weldability improves. By increasing the β-phase portion, it is known that (1) room temperature strength increases, (2) heat treatment and forming capabilities enhance, and (3) strain-rate sensitivity increases, so that superplastic forming is more favorably applicable.

When any titanium material possesses either one of the aforementioned types as a major constitutional phase, such titanium material is named after the type of phase.

Alpha alloys generally have creep resistance superior to that of beta alloys, and are preferred for high-temperature applications. The absence of a ductile–brittle transition (a feature of beta alloys) makes alpha alloys suitable for cryogenic applications, too. Unlike beta alloys, alpha alloys cannot be strengthened by heat treatment because the alpha structure is a stable phase.

Alpha+beta alloys have compositions of a mixture of α and β phases and may contain between 10 and 50% β phase at room temperature. The most common type within this group alloy is Ti-6Al-4V[1]. Within the $\alpha+\beta$ type class, an alloy containing much more alpha than beta is often called a near-alpha alloy. Generally, when strengthening is needed, the alloys are rapidly cooled (i.e., quenched) from a temperature high in the alpha-beta range above the β-transus. This solution treatment is followed by an intermediate-temperature treatment (aging) to produce an appropriate mixture of alpha and transformed beta products.

Beta alloys. Beta titanium has a wider solubility for alloying elements without precipitating any intermetallic compounds. Beta alloys contain transition elements, such as V, Mo, Nb, Ta, and Cr, which tend to reduce the temperature of the $\alpha \leftrightarrow \beta$ phase transformation (or simply β-transus). They have excellent forgeability over a wider range of forging temperature than α alloys, and β-alloy sheet is cold formable in the solution-treated condition. Beta alloys have excellent work-hardening and heat-treatment capabilities. A common thermal treatment involves solution treatment followed by aging at temperatures ranging from 450 to 650°C [2-2]. Beta alloys are one of the most promising groups of titanium alloys in terms

---

[1] Ti-6Al-4V indicates the Ti-based alloy added with 6 wt.% (or w/o) of Al and 4 wt.% of V. The text uses this alloy description, otherwise if atomic % (a/o) is used, both a/o and w/o will be indicated. For conversion between w/o and a/o, let W_x and A_x be wt.% and atomic % of element X in the X–Y binary alloy, where X and Y are atomic weights of respective elements. For each of the respective elements, conversion from w/o to a/o and from a/o to w/o is calculated as follows:

$$A_x = 100/\{1+(X/Y)[(100/W_x) - 1]\} \quad \text{and} \quad A_y = 100 - A_x$$
$$W_x = 100/\{1+(Y/X)[(100/A_x) - 1]\} \quad \text{and} \quad W_y = 100 - W_x$$

of processing, properties, and potential applications. This group of alloys (including beta, metastable beta, and beta-rich alpha/beta compositions) represents the highest range of strength, fatigue resistance, and environmental resistance among all titanium materials. Of course, in these alloys, the single α-phase and single β-phase regions are separated by a two-phase $(\alpha+\beta)$ region. As a result, the alloys utilize multi-component elements and are composed of mixtures of α and β stabilizers. Depending on the ratio of α and β phases, they can be furthermore sub-grouped into near (α) and near (β) alloys. Typical beta group Ti alloys include (1) beta type (Ti-35V-15Cr, Ti-40Mo, Ti-13V-11Cr-3Al, Ti-3Al-8V-6Cr-4Mo-4Zr, Ti-30Mo), metastable beta type (Ti-6V-5.7Fe-2.7Al, Ti-12V-11Cr-3Al, Ti-1Al-8V-5Fe, Ti-12Mo-6Zr-2Fe,Ti-4.5Fe-6.8Mo-1.5Al,Ti-15V-1Mo-0.5Nb-3Al-3Sn-0.5Zr, Ti-3Al-8V-4Mo-4Zr, Ti-15Mo, Ti-8V-8Mo-2Fe-3Al, Ti-15Mo-2.6Nb-3Al-0.2Si, Ti-15V-3Cr-3Sn-3Al, Ti-11.5Mo-6Zr-4.5Sn, Ti-10V-2Fe-3Al, Ti-5V-5Mo-1Cr-1Fe-5Al, Ti-5Al-2Sn-4Mo-4Cr, Ti-4.5Al-3V-2Mo-2Fe), and beta-rich group (Ti-5Al-2Sn-2Cr-4Mo-4Zr-1Fe, Ti-13Nb-13Zr, Ti-4.5Al-3V-2Mo-2Fe) [2-3–2-6]. Table 2.1 summarizes the above discussion.

2.2. MEDICAL/DENTAL TITANIUM AND ITS ALLOYS

Currently, pure titanium and $\alpha+\beta$ type Ti-6Al-4V ELI (Extra Low level of Interstitial content) alloys are widely used as structural and/or functional biomaterials for the replacement of hard tissues in devices such as artificial total hip or knee replacements and dental implants, since they exhibit excellent specific strengths and corrosion resistance, and the best biocompatibility characteristics among metallic biomaterials. They are used more than any other titanium biomaterials; however, other new titanium alloys for biomedical applications have now been included in ASTM standardizations [2-7, 2-8]. For example, β-type Ti-15Mo [2-9] has been registered in ASTM standardizations, and β- type Ti-35Nb-7Zr-5Ta [2-10] and $\alpha+\beta$ type Ti-3Al-2.5V [2-11] are in the process of being registered.

Among several dozen commercially available alloys, the following are typical titanium materials which are utilized or experimentally and clinically tried in both the medical and dental fields.

2.2.1 Commercially pure titanium (CpTi)

Under the category of "unalloyed grades" of ASTM specification, there are five materials classified in this group; they include ASTM grade 1 (99.5%Ti), grade 2 (99.3%Ti), grade 3 (99.2%Ti), grade 4 (99.0%Ti), and grade 7 (99.4%Ti). Although each material contains slightly different levels of N, Fe, and O, C is specified <0.10 wt.% (wt.% or w/o) and H is also specified <0.015 wt.%. ASTM CpTi

Table 2.1. Three major types of titanium materials and influencing effects of major alloying elements.

Type/material property	α and near α	$\alpha + \beta$	β and near β
α-Stabilizing elements	Al, Sn, Ga, Zr, C, O, N		
β-Stabilizing elements			V, Mo, Nb, Ta, Cr
Typical materials	Commercially pure Ti	Ti-5Al-2.5Fe	Ti-3Al-8V-6Cr-4Mo-4Zr
	Ti-5Al-2.5Sn	Ti-5Al-2Mo-2Fe	Ti-4.5Al-3V-2Mo-2Fe
	Ti-5Al-6Sn-2Zr-1Mo	Ti-5Al-3Mo-4Zr	Ti-5Al-2Sn-2Zr-4Mo-4Cr
	Ti-6Al-2Sn-4Zr-2Mo	Ti-5Al-2.5Fe	Ti-6Al-6Fe-3Al
	Ti-8Al-1Mo-1V	Ti-6Al-7Nb	Ti-10V-2Fe-3Al
		Ti-6Al-4V	Ti-13V-11Cr-3Al
		Ti-6Al-6V-2Sn	Ti-15V-3Cr-3Al-3Sn
		Ti-6Al-2Sn-4Zr-6Mo	Ti-35V-15Cr
			Ti-8Mo-8V-2Fe-3Sn
			Ti-11.5Mo-6Zr-4.5Sn
			Ti-30Mo, Ti-40Mo
			Ti-13Nb-13Zr
			Ti-25Pd-5Cr
			Ti-20Cr-0.2Sn
			Ti-30Ta
β-Transus temperature	Higher ←——————————————————— Lower		
Specific density	Lower ——————————————————→ Higher		
Room temperature strength	——————————————————→		
Room temperature toughness	——————————————————→		
Modulus of elasticity	←———————————————		
Machinability	←———————————————		
Age hardenability	——————————————————→		
Heat resistance	←———————————————		
Weldability	←———————————————		
High-temperature strength	←———————————————		
Heat-treatability	——————————————————→		
Plastic formability	——————————————————→		
Strain-rate sensitivity	——————————————————→		
Superplastic formability	——————————————————→		
Creep resistance	←———————————————		

Note: This chart does not include TiNi, Ti$_3$Al (α2), and TiAl (γ) intermetallic type alloys.

grades 1–4 (unalloyed titanium) allow a hydrogen content up to 0.015 wt.% (i.e., 15 ppm). It was reported that if CpTi contains more than 250 ppm of hydrogen, the material would be susceptible to stress corrosion cracking and hydrogen embrittlement [2-12].

Grade 1: Grade 1 CpTi is the lowest strength unalloyed titanium with a slightly lower residual content. Both oxygen and iron residuals improve the impact strength. Oxygen acts as an interstitial strengthener, maintaining a single α-phase hexagonal closed-packed microstructure. Iron acts as a second β phase BCC grain refiner, offering moderate strengthening capabilities. The lower residual content makes grade 1 the lowest strength CpTi grade, but it has the highest ductility, with an excellent cold formability.

Grade 2: The grade 2 is the most frequently selected titanium grade in industrial service, having well-balanced properties of both strength and ductility. The strength levels are very similar to those of common stainless steel and its ductility allows for good cold formability.

Grade 3: CpTi grade 3 possesses a slightly higher strength due to its slightly higher residual content (primarily oxygen, and also nitrogen) with slightly lower ductility.

Grade 4: The grade 4 is the highest strength grade of the CpTi series, so that grade 4 serves mainly in the aerospace/aircraft industry.

For all CpTi grades 1–4, the 0.2% off-set equivalent-yield strength ($\sigma_{0.2YS}$) and the ultimate tensile strength (σ_{UTS}) appear well correlated to oxygen contents [2-13]. Through linear regression analysis, with the oxygen content expressed by [O], it was found that $\sigma_{0.2YS} = 1336.2 \times$ [O] -66.7 in MPa with r (correlation coefficient) of 0.9865, and $\sigma_{UTS} = 1351.5 \times$ [O] -3.7 in MPa with r of 0.9946.

By far the most widely used of the CpTi grades is grade 2 [2-13]. From this base, the other grades have been developed for better formability or higher strength levels, significantly increasing corrosion resistance at higher temperatures, and/or improving corrosion resistance at lower pH (or higher acidity) levels [2-14]. There is a distinct difference in using CpTi between industry and medicine. Among various applications of CpTi materials in dental and medical fields, the dental CpTi implant is the most frequently and widely employed. Information on CpTi grade selection for dental application indicates that, despite the popularity of grade 2 in industry [2-11], grades 3 and 4 are equally selected in most of the 20 major dental implant systems. Only two companies use grades 1 and 2 [2-15–2-18]. More interesting, however, is their availabilities. In preparing CpTi grades 1–4 for corrosion resistance comparisons, Hernandez [2-15] searched 14 material suppliers, but found that only 2 out of the 14 suppliers were able to provide all four grades. Grades 2 and 4 were available without any problems, whereas CpTi grade 3 is the least popular and difficult to obtain; 2 out of the 14 suppliers were able to provide CpTi grade 3. After conducting electrochemical corrosion testing, using 37°C Ringer's solution as an electrolyte, it was found that CpTi grade 3 exhibited the least corrosion resistance. Regardless of material availability [2-15, 2-16], this supported the unpopularity of CpTi grade 3. Hence, despite the popularity and availability of grade 2 in industries,

CpTi grade 2 is the least popular, although its corrosion resistance is somewhat superior to grade 3, which is the worst grade in terms of corrosion resistance among the four grades.

2.2.2 Ti-6Al-4V

This alloy belongs to the $\alpha + \beta$ phase alloy group and is particularly popular because of its high corrosion resistance and the reputed low toxicity of ions released from the surface due to dense and protective passive oxide (which is mainly TiO_2) film formation. Ti-6Al-4V (which is, in some articles, marked as Ti-6/4) exhibits good mechanical and excellent tissue compatibility properties, which make it well suited for biomedical applications where a bone anchorage is required, particularly for implant applications [2-18]. Ti-6Al-4V ELI is also available and employed in the medical area [2-19].

2.2.3 Ti-6Al-7Nb

As a result of searches for vanadium-free Ti-6Al-4V equivalent alloys, this alloy was developed to enhance the wear resistance [2-20] and castability [2-21]. The optimal composition was found to be Ti-6Al-7Nb (or simply Ti-6/7). This custom-made alloy designed for implants shows the same alpha/beta structures as Ti-6Al-4V and exhibits equally good mechanical properties. The corrosion resistance of Ti-6Al-7Nb in sodium chloride solution was evaluated to be equivalent to that of pure titanium and Ti-6Al-4V, due to formation of a very dense and stable passive layer. Highly stressed anchorage stems of different medical prostheses (including hip, knee, and wrist joints) have been made from hot-forged Ti-6Al-7Nb. The surfaces of these Ti-6Al-7Nb prostheses were further hardened by means of a very hard 3–5 μm thick titanium nitride coating or by oxygen diffusion hardening to a depth of 30 μm, in order to enhance the biotribological properties [2-22].

2.2.4 Ti-3Al-2.5V

Ti-3Al-2.5V alloy possesses an excellent ductility and cold formability, allowing it to be cold-worked by standard tube-making processes and bent for installation. It is easily welded, and may be heat-treatable to a wide range of strengths and ductility [2-23].

2.2.5 Ti-5Al-3Mo-4Zr

A newly developed titanium for surgical implant application, Ti-5Al-3Mo-4Zr (or simply referred to Ti-5/3/4), was evaluated, and its properties were compared with conventional biomaterials. Mechanical properties and metallic ion elution were also examined. It was reported that the new alloy is advantageous over Ti-6Al-4V

ELI because it does not contain bio-hazardous alloying elements (e.g., vanadium), and has superior mechanical properties to stainless steel. Corrosion abrasive wear resistance is also improved in Ti-5Al-3Mo-4Zr. As a result, newly designed artificial hip joints were fabricated with this alloy [2-23, 2-24].

2.2.6 Ti-5Al-2.5Fe

In the course of developing a vanadium-free titanium-based alloy, the another new alloy Ti-5Al-2.5Fe (which is an $\alpha + \beta$ phase alloy) has been developed to avoid the presence of toxic vanadium which is an alloying element in the Ti-6Al-4V alloy and may increase to more than 15% in the β phase. Hip prostheses and hip prosthesis heads were fabricated from the Ti-5Al-2.5Fe alloy with a grain size of 20 μm or less. The frictional biotribological behavior of a hip prosthesis head in contact with an ultrahigh molecular weight polyethylene (UHMWPE) cup was investigated. It was shown that (i) the frictional behavior of a head coated with an oxide layer about 1–3 μm thick produced by induction heating of the surface equal to that of a head fabricated from alumina ceramics, and (ii) the oxide layer was still present after 10^7 cycles in 0.9% NaCl solution, indicating excellent corrosion resistance of this alloy [2-25].

2.2.7 Ti-Ni

Historically, when the US Navy searched for a new submarine material exhibiting good damping capacity, NiTi alloy was developed. Buehler, who conducted the bending tests on NiTi alloys, discovered that a large deformation was completely recovered by a slight heating to return to its original shape. This phenomenon is later referred to as a SME. Since the test was performed and the material was developed at the US Naval Ordnance Laboratory, this unique material is called as NITINOL (NIckel-TItanium Naval Ordnance Laboratory) [2-26]. This material is also often called and marked as NiTi, TiNi or Nitinol. In this book, since the Ni element contains more than 50 wt.% (weight percentage), the term "NiTi" is used according to the ordinal way to describe alloy systems in the conventional metallurgy.

It is well known that this alloy exhibits unique phenomena such as SME as well as SE. By SME, a material first undergoes a martensitic transformation. After deformation in the martensitic condition, the apparently permanent strain is recovered when the specimen is heated to cause the reverse martensitic transformation. Upon cooling, it does not return to its deformed shape. When SME alloys are deformed in the temperature regime a little above the temperature at which martensite normally forms during cooling, a stress-induced martensite is formed. This martensite disappears when the stress is released, giving rise to a superelastic stress–strain loop with some stress hysteresis. This property applies to the parent phase undergoing a stress-induced martensitic transformation. The other unique property associated with NiTi is SE. When SME alloys are deformed, a superelastic alloy deforms reversibly to very high strains up to 10% by the creation of stress-induced phase. When the load is removed, the new phase becomes unstable and the material regains its original shape. Unlike SME alloy, no change in temperature is needed for the alloy to recover its initial shape. There is a term called pseudoelasticity, which is a more generic term that encompasses both superelastic and rubber-like behavior [2-27].

Because of the uniqueness of SME and the SE associated with NiTi, there have been many applications in both the industrial and dental/medical fields, including dental and orthopedic implants, artificial heart valve, stunt, endodontic files, orthodontic archwires, etc. [2-28–2-33]. Superelastic devices take advantages of their large, reversible deformation and their applications are not limited to medical/dental fields, but can be found in parabola antenna used at the space shuttle, frames for eyeglasses, and sporting goods such as fishing line.

2.2.8 Ti-Cu

Aiming for developing an alloy for dental casting with better mechanical properties than unalloyed CpTi, Ti-Cu (Cu: 0.5–10 wt.%) was cast in an argon-arc melting furnace. It was found that (i) the mean tensile strength was significantly higher than for cast CpTi, and (ii) increases of 30% in the tensile strength and yield strengths of 40% over CpTi were obtained for the Ti-5Cu alloy [2-34].

2.2.9 Ti-Mo

TiMo alloy is widely used as an orthodontic archwire for performing an orthodontic mechanotherapy [2-35]. When NiTi wire, TMA (β-phase Ti-Mo) wire, and austenitic stainless steel wire were compared, it was found that (i) TMA showed the greatest plastic strain and springback, followed by NiTi and stainless steel, (ii) NiTi wire showed the highest stored energy value in bending-torsion, followed by TMA and stainless steel, and (iii) the highest spring ratio (stiffness) in bending-torsion was found in stainless steel, followed by TMA and NiTi [2-36]. Lin *et al.* [2-37] studied alloying effect of Fe on the Ti–Mo alloy, and found that Ti–Mo with iron contents in the range of 2–5 wt.% appeared to have a great potential for use as an implant material.

2.2.10 Ti-5Al-2Mo-2Fe (SP700)

Ti-6Al-4V is the most widely used titanium alloy. However, Ti-6Al-4V still suffers from limited applications in non-aerospace fields compared to CpTi, aluminum alloys, and steels. The reasons for this include (1) poor hot-workability, with a normal hot-working temperature range that is too high and too narrow, (2) poor cold-workability that virtually prohibits cold-working, (3) poor hardenability that is frequently inconsistent, and (4) a high superplastic forming temperature of around 900°C that results in rapid die wear. To overcome these disadvantages associated with Ti-6Al-4V, a new beta-rich alpha-beta duplex phase titanium alloy has been developed (called SP700, since the newly developed Ti alloy remarkably exhibits high superplastic capability at operating temperatures around 700°C). It was reported that SP700 provides excellent hot- and cold-workability, as well as greatly improved superplastic forming characteristics (i.e., the strain rate sensitivity exponent, m-value, is about 1.0, indicating that flow is close to that of a Newtonian viscous flow), high strength, and high toughness. Dental denture bases have been fabricated using this SP700 alloy by superplastic forming technique [2-38].

2.2.11 Other Ti-based alloys

An interesting study was done by Brème *et al.* [2-39], using isoelastic porous sintered Ti-30Ta alloy and titanium wire loop, to accomplish the mechanical compatibility in endosseous dental implant systems. Their mechanical properties were optimized by the production parameter such as sintering and diffusion bonding. The functionality was tested after insertion into an artificial jaw, which had properties corresponding to the natural mandibular. It was reported that the elastic properties of both implants are similar to the properties of the bone, and the implant has a safe anchorage bone ingrowth [2-39].

Qazi *et al.* [2-6] developed metastable β-Ti alloy (mainly Ti-35Nb-7Zr-5Ta) and solution-treated, followed by aging at 482°C. It was reported that (i) the heat-treated Ti-35Nb-7Zr-5Ta alloys exhibited 0.2% off-set yield strength of 1300 MPa with 8% elongation, due to $\omega+\beta$ and $\omega+\alpha+\beta$-phase precipitations, and (ii) this enhanced strength makes the alloy candidates for bone plates and screws.

2.2.12 Intermetallic Alloys

Although not ordinary alloys, titanium aluminide-based alloys, including TiAl (γ-phase) and Ti_3Al ($\alpha2$ phase), are used for high temperature applications. The titanium aluminide-based alloys provide an inherently low-density material and exhibit excellent creep-rupture properties [2-13, 2-40]. TiAl amorphous alloy provides high strength, linear elastic behavior, and the infinite fatigue life necessary for high device reliability. Tregilgas [2-41] developed amorphatized TiAl alloys for a material for the digital micromirror device chip. In order to enhance the wear resistance of γ-TiAl (Ti-47Al-2Nb) alloy, the surface of this alloy was modified in a nitrogen-ion plasma atmosphere [2-42].

REFERENCES

[2-1]   Collings EW. The physical metallurgy of titanium alloys. Columbus, OH: The American Society for Metals, 1984. pp. 3–5.

[2-2]   Ivansishin OM, Markovsky PE. Enhancing the mechanical properties of titanium alloys with rapid heat treatment. J Metals 1996;48:48–52.

[2-3]   Eylon D, Vassel A, Combres Y, Boyer RR, Bania PJ, Schutz RW. Issues in the development of beta titanium alloys. J Metals 1994;46:14–15.

[2-4]   Niiomi M. Recent metallic materials for biomedical applications. TMS Metallurgical and Materials Transactions 2002;33A;477–486.

[2-5]   Jha SK, Ravichandran KS. High-cycle fatigue resistance in beta-titanium alloys. J Metals 2000;52:30–35.

[2-6]   Qazi JI, Marquardt B, Rack HJ. High-strength metastable beta-titanium alloys for biomedical applications. J Metals 2004;56:49–51.

[2-7]   Niiomi M. Fatigue performance and cyto-toxicity of low rigidity titanium alloy, Ti-29Nb-13Ta-4.6Zr. Biomaterials 2003;24:2673–2683.

[2-8]   Niiomi M, Kuroda D, Fukunaga K, Morinaga M, Kato Y, Yashiro T, Suzuki A. Corrosion wear fracture of new β type biomedical titanium alloys. Mat Sci Eng A 1999;A263:193–199.

[2-9]   ASTM designation F2066-01. Standard specification for wrought titanium-15 molybdenum alloy for surgical implant applications. Philadelphia PA: American Society for Testing and Materials. 2001. pp. 1605–1608.

[2-10] ASTM designation draft #3. Standard specification for wrought titanium-35Niobium-7Zirconium-5Tantalum alloy for surgical implant applications (UNS R58350). Philadelphia, PA: American Society for Testing and Materials. 2000.

[2-11] American Society for Testing and Materials designation draft #6. Standard specification for wrought titanium-3Aluminum-2.5Vanadium alloy for surgical implant applications (UNS R58350). Philadelphia PA, ASTM, 2000.

[2-12] Paton NE, Williams JC. Effect of hydrogen on titanium and its alloys. In: American society for metals. Hydrogen in metals. 1974. pp. 409–432.

[2-13] Titanium and Titanium Alloys – Source Book. Donachie MJ, editor. Metals Park, OH: American Society for Metals. 1982.

[2-14] Mountford JA Jr. Titanium alloys for the CPI (Chemical Process Industry). Adv Mater Processes 2001;159:55–58.

[2-15] Hernandez FE. Corrosion behavior for commercially pure titanium with different grades and from different material suppliers. Indiana University Master Degree Thesis, 2003.

[2-16] Lim WC. Comparison of electrochemical corrosion behavior among four grades of commercially pure titanium. Indiana University Master Degree Thesis, 2001.

[2-17] Binon PP. Implants and components: entering the new millennium. Int J Oral Maxillofac Implants 2000;15:76–94.

[2-18] McCraken M. Dental implant materials: commercially pure titanium and titanium alloys. J Prosthodont 1999;8:40–43.

[2-19] Moore BK, Oshida Y. Materials science and technology in dentistry. In: Encyclopedic handbook of biomaterials and bioengineering. Wise DL, Trantolo DJ, Altobelli DE, Yaszemski MJ, Gresser JD, Schwartz ER, editors. New York: Marcel Dekker. 1995. pp. 1325–1430, Chapter 48.

[2-20] Iijima D, Yoneyama T, Doi H, Hamanaka N, Kurosaki N. Wear properties of Ti and Ti-6Al-7Nb castings for dental prostheses. Biomaterials 2003; 24: 1519–1524.

[2-21] Soma H. Clinical application of Ti-6Al-7Nb alloy "Ti-Alloy Tough". Quintessence Dent Technol 1998;23:98–104.

[2-22] Semlitsch MF, Weber H, Streicher RM, Schoen R. Joint replacement components made of hot-forged and surface-treated Ti-6Al-7Nb alloy. Biomaterials 1992;13:781–788.

[2-23] Tuominen S, Wojcik C. Alloys for aerospace. Adv Mater Processes 1993; 147:23–26.

[2-24] Higo Y, Ouchi C, Tomita Y, Murota K, Sugiyama H. Properties evaluation and its application to artificial hip joint of newly developed titanium alloy for biomaterials. Proc. 3rd Japan International SAMPE Symposium 1993. pp. 1621–1625.

[2-25] Zwicker U, Brème J. Investigations of the friction behavior of oxidized Ti-5%Al-2.5%Fe surface layers on implant material. J Less-Common Metals 1984;100:371–375.

[2-26] Buehler WJ, Gilfrich JV, Wiley KC. Superelasticity in TiNi Alloy. J Appl Phys 1963;34:1467–1469.

[2-27] Duerig TW, Melton KN, Stöckel D, Wayman CM, editors. Engineering aspects of shape memory alloys.London: Butterworth-Heinemann. 1990.

[2-28] Civjan S, Huget EF, DeSimon LB. Potential applications of certain nickel-titanium (Nitinol) alloys. J Dent Res 1975;54:89–96.

[2-29] Bensmann G, Baumgart F, Haasters J. Osteosyntheseklammern aus NiTi. Herstellung, Vorversuche und klinischer Ensatz. Tech Mitt Krupp, Forsch-Ber 1982;40:123–134.

[2-30] Miura F, Mogi M, Ohura Y, Hamanaka H. The superelastic property of the Japanese NiTi alloy wire for use in orthodontics. Am J Orthod Dentofacial Orthop 1986;90:1–10.

[2-31] Yoneyama T, Doi H, Hamanaka H, Okamoto Y, Mogi M, Miura F. Superelasticity and thermal behavior of Ni-Ti alloy orthodontic arch wires. Dent Mater J 1992;11:1–10.

[2-32] Yoneyama T, Doi H, Hamanaka H. Bending properties and transformation temperatures of heat treated Ni-Ti alloy wire for orthodontic appliances. J Biomed Mater Res 1993;27:399–402.

[2-33] Iijima M, Endo K, Ohno H, Mizoguchi I. Effect of Cr and Cu addition on corrosion behavior of Ni-Ti alloys. Dent Mater J 1998;17:31–40.

[2-34] Kikuchi M, Takada Y, Kiyosue S, Yoda M, Woldu M, Cai Z, Okuno O, Okabe T. Mechanical properties and microstructures of cast Ti-Cu alloy. Biomaterials 2003;19:174–181.

[2-35] Ho WF, Ju CP, Chern Lin, JH. Structure and properties of cast binary Ti-Mo alloys. Biomaterials 1999;20:2115–2122.

[2-36] Drake SR, Wayne DM, Powers JM, Asgar K. Mechanical properties of orthodontic wires in tension, bending, and torsion. Am J Orthod 1982;82: 206–210.

[2-37] Lin DJ, Chern Lin JH, Ju CP. Structure and properties of Ti-7.5Mo-xFe alloys. Biomaterials 2002;23:1723–1730.

[2-38] Ouchi C, Minakawa K, Takahashi K, Ogawa A, Ishikawa M. Development of β-rich α-β titanium alloy: SP-700. NKK Technical Review 1992;65:61–67.

[2-39] Brème J, Biehl V, Schulte W, D'Hoedt B, Donath K. Development and functionality of isoelastic dental implants of titanium alloys. Biomaterials 1993;14:887–892.

[2-40] Jha SK, Larsen JM, Rosenberger AH. The role of competing mechanisms in the fatigue-life variability of a titanium and gamma-TiAl alloy. J Metals 2005;57:50–54.

[2-41] Tregilgas J. Amorphous hinge material. Adv Mater Processes 2005;163: 46–49.

[2-42] Thongtem S, McNallan M, Thongtem T. Surface modifications of γ-TiAl alloys by nitrogen plasma deposition. Adv Technol Mater Mater Processing 2004;6:170–175.

Chapter 3
Chemical and Electrochemical Reactions

3.1.	Discoloration	26
3.2.	Corrosion in Media Containing Fluorine Ion and Bleaching Agents	28
3.3.	Corrosion Resistance, Under Influences of Environmental and Mechanical-Assisted Actions	31
3.4.	Metal Ion Release and Dissolution	42
3.5.	Galvanic Corrosion	51
3.6.	Microbiology-Induced Corrosion	53
3.7.	Reaction with Hydrogen	61
References		64

Chapter 3
Chemical and Electrochemical Reactions

The intraoral environment is hostile due to corrosive and mechanical actions. It is continuously full of saliva, an aerated aqueous solution of chloride with varying amounts of Na, K, Ca, PO_4, CO_2, sulfur compounds, and mucin. The pH value is normally in the range of $6.5-7.5$ but under plaque deposits it may be as low as 2.0. Temperatures can vary $\pm 36.5^{\circ}C$, and a variety of food and drink concentrations (with pH values ranging from 2.0 to 14.0) stay inside the mouth for short periods of time. Loads can go up to 1000 N (normally 200 N as a masticatory force), sometimes in an impact manner. Trapped food debris may decompose, and sulfur compounds cause natural teeth and restorative materials to discolor. Under these chemical and mechanochemical intraoral environments, materials in service in the mouth are still expected to last for relatively long period of time [3-1].

It is well documented that Ti materials possess an excellent corrosion resistance and form stable and protective oxides, such that Ti materials are one of the best material choices for medical and dental applications. It is inevitable to note that this chapter on corrosion, as well as the following chapter on oxidation occupy a quite a large portion of this book. What it implies is that environments to which titanium materials have been exposed are still diverse and vary widely in terms of chemical, mechanical, physical, and any combinations of these parameters. In this chapter, we will be reviewing the corrosion behavior of Ti materials when they are exposed to various atmospheres.

It is known that all primitive living species from which *Homo sapiens* are originated were created from the ocean, containing about $32-35$ mg/l of chlorine ion, while the human body contains about 35 mg/l of chlorine ion and saliva has about 5 mg/l of it. Most of the information presented and reviewed in this chapter was obtained from the *in vitro* tests, which were conducted in various corrosive media mainly containing chlorine ions. Such media can be divided into four groups: (1) containing less amount of chlorine ion (*ca.*, 50 mg/dl Cl^-, i.e. about 1% NaCl concentration) to simulate saliva, and phosphate-buffered artificial saliva, (2) since internal fluids have a concentration of chlorine ions seven times (35 mg/l of chlorine ion) higher than that of oral fluids (5 mg/l of chlorine ion), simulated body fluid with higher ion concentration (350 mg/dl Cl^-), (3) lactic acid, simulating the chemistry of major constitutents of plaque, and (4) dental treatment agents, including whitening agents and fluoride treatment agents. These corrosive media cause dental materials discoloration, tarnishing, corrosion, or oxidation.

3.1. DISCOLORATION

Currently, mouth-rinses, toothpaste, dental prophylactic agents containing fluoride, and bleaching treatment agents are popular for esthetic purposes and prevention of plaque and cavity formation. Normally, orthodontic patients are referred to general dentists for fluoride treatments once in every 6 months during the course of orthodontic mechanotherapy. Among various bleaching treatments for whitening stained teeth, an overnight (normally for 8 h) bleaching agent containing 10% carbamide peroxide is popular. To achieve a satisfactory outcome, this treatment must be repeated for two consecutive weeks. The corrosive effect of these agents (normally containing fluoride and hydrogen peroxide) on dental metallic materials has not been well documented, although it was reported to decrease the corrosion resistance of titanium in solutions containing fluoride [3-2–3-6].

Titanium application in dentistry has steadily increased because of its numerous biocompatibility benefits and mechanical advantages. In orthodontics, titanium is valuable in the treatment of patients with an allergy to nickel and other specific substances [3-7]. Adell *et al.* [3-8] acknowledged the benefits of titanium and how the body responds to it while working on permanent tooth replacements. Owing to their biological advantages and excellent corrosion resistance, research and clinical evidence supports the use of unalloyed titanium and its alloys in the body, which consistently remain the first choice for implants in both medical and dental applications.

Such excellent corrosion resistance of Ti materials might be jeopardized when they are in contact with substances containing fluoride. An *in vivo* study by Harzer *et al.* [3-7] showed that orthodontic brackets made of titanium show higher plaque accumulation and more discoloration than stainless steel brackets. This was speculated to be due to the morpholical alteration of surface layers. When a surface becomes rougher with higher unevenness, the outer surface appearance becomes duller with discolored-like appearance. Nonetheless, studies have found that, in a fluoridated medium (especially in acid-fluoridated solutions), titanium is degraded [3-9, 3-10]. Watanabe and Watanabe [3-11] found that titanium orthodontic appliances with high titanium content (such as unalloyed CpTi: commercially pure titanium) have also shown lower tarnish resistance, especially when immersed in an acidulated phosphate fluoride (APF) solution.

Teeth that have been exposed to a long-term coffee and/or cigarette usage are normally stained. Viral bleaching of such teeth reflects patients' increasing desire to achieve an optimal esthetic appearance [3-12–3-15]. Internal discoloration caused, for example, by tobacco, dentino-genesis imperfecta, or fluorosis, may be removed. Oxygenating agents like carbamide peroxide or hydrogen peroxide, H_2O_2, are used for effective bleaching purpose. Carbamide peroxide is used as a vehicle for transporting H_2O_2. First, carbamide reacts with uric acid, ammonia, and H_2O_2. Then, as a second step, H_2O_2 reacts into H_2O and O, which takes electrons from the substance to bleach [3-13]. The bleaching efficacy is influenced by the application time and the concentration of the effective agent. Such application of the bleaching agents is performed in offices by clinicians (with relatively higher concentration of H_2O_2) or at home by patients themselves (relatively lower concentration of H_2O_2) [3-14]. Commercially available bleaching products range from 3% to 9.5% for hydrogen peroxide (which can be converted to a range from 6% to over 19% carbamide peroxide). Although the above discussion is regarding the discolored natural teeth and whitening of such stained teeth, some patients undergoing the nightguard bleaching (e.g., for 8-h treatment) are likely to have some metallic restorations (i.e., amalgam, crowns made of gold or porcelain fused to a base metal, fixed or removable prosthodontic bridge or partial denture frameworks made of base alloys, and/or titanium implant fixtures). Recently, it was found that exposure of amalgam to common bleaching agents caused an increase in mercury levels in the solutions [3-16].

The conscientious health-care provider needs to know the consequences of the effects of fluoride agents and bleaching agents on various dental metallic materials intraorally. While it is acknowledged that tarnishing of the metallic dental materials is the visible consequence of these treatments, the implications on biological effects and the structural integrity of the components are unclear. Clinicians need to understand the potentially corrosive nature of the commercially available fluoride and bleaching treatments on intraoral metals, and know when and where to use them [3-17]. The discoloration efficacy of fluoride treatment agents (2.0% NaF with pH 7.0, 0.4% SnF_2 with pH 7.0, and 1.23% APF with pH 3.5) was tested on Ti-6Al-4V and 17Cr-4Ni PH (precipitation hardening) type stainless steel, and bleaching agents (10% carbamide peroxide) were applied on CpTi (grade 2), 70Ni-15Cr-5Mo, type IV gold alloy (70Au-10Ag-15Cu), and Disperalloy amalgam. The degree of discoloration on these treated alloys was examined by a colorimeter and naked eyes. After the baseline measurements of three L^* (lightness), a^* (position on the red/green axis), and b^* (position on yellow/blue axis), comparisons were made with the Commission Internationale d'Eclairage (CIE-L^*a^*b) color system, and the value of ΔE^* (defined as the Euclidean distance) can be calculated by $[(L_i-L_f)^2 + (a_i-a_f)^2 + (b_i-b_f)^2]^{1/2}$, where subscripts "i" and "f" indicate the intial value and final value, respectively. It was found that (i) all tested metallic materials exhibit discoloration to various degrees, ranging from 10 to 18 in ΔE^*, (ii) the tooth brashing between each treatment for both fluoride and bleaching treatments indicate a remarkable reduction in the degree of discoloration (i.e., ΔE^* reduced to 2−8) of all tested materials, and (iii) results on naked eyes evaluation performed by three clinicians did not agree well with those in the lower range of ΔE^*, whereas when the discoloration advanced, both evaluations agreed well [3-17].

When the temperature effect is added to the aforementioned fluoride discoloration, the resultant will get worse. Ti implants were occasionally strongly discolored after autoclaving, which contains fluorine ions. It is a common procedure to treat endosseous titanium implants with a fluoride solution before they are placed into the bone cavity. This treatment results in a blue discoloration of the titanium surface [3-18]. The phenomenon was first observed on a titanium box used for storage of titanium implants during autoclaving and surgical procedures [3-3]. It was found that the oxide film formed on autoclaved Ti implants thickened up to 650 Å (which is about 10 times thicker than on normal implants). Through the microanalyses performed by secondary ion mass spectroscopy (SIMS), X-ray photoelectron spectroscopy (XPS) , and electron spectroscopy for chemical anaysis (ESCA), Lausmaa *et al.* [3-3] had shown that these oxide films contained considerable amounts of fluorine, alkali metals, and silicon. In the cases where discoloration was observed in clinical situations, the source of fluorine was the textile cloths in which the titanium implant storage box had been wrapped during the autoclaving procedure. The cloths contained residual Na_2SiF_6, which had been used as an additive to the rinsing water used in the last step of the cloth laundry procedure. Since the biocompatibility of titanium implants is closely related to their surface oxides, it is advisable to avoid all sources of fluorine ions in the implant preparation procedures [3-3].

3.2. CORROSION IN MEDIA CONTAINING FLUORINE ION AND BLEACHING AGENTS

While numerous studies have been carried out on the corrosion of titanium and titanium-based alloys, few have taken into account the influence of fluoride and pH on pure titanium [3-19−3-27]. Fluorine ion containing solution can be found mainly sodium fluoride, NaF, solution, as seen in the followings.

During the orthodontic mechanotherapy, practitioners recommended their patients to use a fluoride mouthwashing agent. Fluoride promotes the formation of calcium fluoride globules that adhere to the teeth and stimulate remineralization while protecting against acid attack. Fluoride mouthwashes thus help prevent the development of cavities and protect dental enamel. Corrosion behaviors of TMA (75.5Ti-14Mo-5.5Sn-5.5Zr), NbTi (52Nb-48Ti), NiTi (55Ni-45Ti), and NiTi-Cu (48Ni-46.5Ti-5.5Cu) were evaluated in the Fusayama−Meyer artificial saliva, and three commercially available mouthewashes containing NaF of 150 ppm, 125 ppm amine fluoride, and 125 ppm stanneous fluoride, SnF_2, and 65.9 ppm sodium monofluoro-phosphate (all in pH 4.2 to 4.5). It was concluded that (i) NiTi alloy showed a strong corrosion in the presence of monofluoro-phosphate, (ii) NbTi alloy exhibited the most resistance to corrosion, and (iii) TMA corroded strongly with the stanneous fluoride, SnF_2 [3-28].

Under a combination of materials (CpTi, Ti-6Al-4V, 17-4 PH-type stainless steel, 70Ni-15Cr-5Mo alloy, type IV (70Au-10Ag-15Cu) gold alloy, and Disperalloy dental amalgam) and dental treatment agents (artificial saliva solution, NaF-containing solution, SnF_2-containing solution, acidulated phosphate fluoride (APF) and carbamide peroxide solution) [3-17], electrochemical polarization tests were conducted to evaluate their corrosion resistance. It was found that (i) the V/Al elemental ratios of Ti-6Al-4V brackets were not significantly different between a sound structure (0.526±0.031) and an irregularly attacked structure (0.533±0.026) after treatment in SnF_2-contaning agent and APF agents, (ii) with the same treatments as (i), Cr/Fe ratios of 17-4 (17Cr-4Ni-79Fe) stainless steel brackets were significantly different between sound structure (0.246±0.007) and intergranularly damaged structures (0.177±0.003), indicating the 17-4 stainless steel suffered from the localized Cr depletion, (iii) no stable passivation stages were observed on both Ti-6Al-4V alloy and 17-4 stainless steels under the anodic polarization in the APF agents, and (iv) no stable passivation was established with all tested alloys when polarized in carbamide peroxide solution. Hence, it was concluded that, although fluoride and bleaching treatments are indicated for patients, it was proven that these treatments are contra-indicated for dental metallic materials [3-17, 3-19, 3-29, 3-30]. Yoon *et al.* tested CpTi, Ti-6Al-4V, and NiTi in 2% NaF electrolyte, containing $1-20$ ppm fluorine ion. It was reported that CpTi and Ti-6Al-4V showed similar current density, whereas NiTi had higher current density (meaning less corrosion resistance) than the other two alloys [3-31].

The corrosion behaviors of CpTi, Ti-6Al-4V, Ti-6Al-7Nb, and the new experimental alloys Ti-Pt ($0.1-2$ w/o) and Ti-Pd ($0.1-2$ w/o) were investigated using anodic polarization and corrosion potential measurements in an environment containing 0.2% NaF (905 ppm F) adjusted to the pH of 4 by H_3PO_4 at 37°C. It was found that, although the surfaces of the Ti-Pt and Ti-Pd alloys were not affected by an acidic environment containing fluoride, the surfaces of the CpTi, Ti-6Al-4V, and Ti-6Al-7Nb were markedly roughened by corrosion, providing artifact discolored appearance, too. It was also reported that (i) the surfaces of CpTi, Ti-6Al-4V, and Ti-6Al-7Nb were microscopically damaged by corrosion when they were immersed in the solution containing a low concentration of dissolved oxygen, even with a fluoride concentration included in the commercial dentifrices, but (ii) in this situation, the surfaces of the new Ti-Pt and Ti-Pd alloys were not affected. As a result, it was suggested that Ti-Pt and Ti-Pd alloys are expected to be of use in dental work as new titanium alloys with high corrosion resistance [3-32].

During the course of the fluoride corrosion process, hydrogen as a reaction product may cause a delayed fracture, similar to a well-known hydrogen embrittlement

(HE), in high-strength steels. Hydrogen embrittlement of a beta-titanium orthodontic wire has been examined by means of a delayed-fracture test in acid and neutral fluoride aqueous solutions and hydrogen thermal desorption analysis. It was found that (i) the time to fracture increased with decreasing applied stress in 2.0 and 0.2% APF solutions, (ii) the fracture mode changed from ductile to brittle when the applied stress was lower than 500 MPa in 2.0% APF solution, but (iii) on the other hand, the delayed fracture did not occur within 1000 h in neutral NaF solutions, although general corrosion was also observed to be similar to that in APF solutions. It was also reported that the amount of absorbed hydrogen was 5000−6500 ppm under an applied stress in 2.0% APF solution for 24 h. Based on these findings, it was concluded that immersion in fluoride solutions leads to the degradation of the mechanical properties and HE-related fracture of beta-titanium alloy associated with hydrogen absorption [3-33].

Not only plastic strain, but there is a unique study on the effect of even elastic strain on corrosion behavior. The corrosion resistance of CpTi (grade 2) in 1% NaCl + 0−1% NaF solution (pH 6) under different elastic tensile strains (0, 1, 2, and 4%) was tested by the EIS[1] (electrochemical impedance spectroscopy) method. The results showed that (i) the NaF concentration and the elastic strain had a significant influence, (ii) the R_P (polarization resistance, which is equivalent to corrosion resistance) decreased on increasing the NaF concentration and elastic tensile strain, (iii) when the NaF concentration was lower than 0.01%, the R_P value ($>3.4 \times 10^5 \ \Omega \ cm^2$) was mainly ascribed to the formation of a protective TiO_2 on the metal surface, regardless of applied strain, but (iv) when the NaF was higher than 0.1%, the protectiveness of TiO_2 was destroyed by F ions, leading to severe pitting corrosion of CpTi [3-34].

Experimental evidence obtained by EMPA (electron probe microanalyzer) analysis indicated that the detected fluorine on the Ti surface immersed in the solution with fluoride was at a high level, while the detection of sodium was relatively low. Therefore, it can be speculated that fluoride−titanium or fluoride−hydrogen compounds were deposited on the Ti surface, rather than sodium fluoride. This is supported by suggestions made by Wilhelmsen *et al.* [3-4] and Nakagawa *et al.* [3-22]. Based on the above evidence, it was postulated that hydrofluoric acid (HF)

---

[1] The EIS technique is conducted under AC current, hence the results are frequency dependent, whereas the ordinary electrochemical polarization tests are performed under DC current. The physical meaning of AC impedance can be considered equivalent to that of DC resistance. Although the EIS technique was developed initially for evaluating the performance of organic coating/metal systems, recently it has been extensively employed in corrosion science and engineering. Only a few millivolts is required for application on the test metallic sample surface, so that the sample surface is not altered, while DC corrosion tests apply, in some cases, up to 1.5 V. This extremely low level of voltage is one of the remarkable advantages over the DC method [3-35, 3-36].

responds by destroying the passive film on Ti but not the fluoride ion. At first, NaF is decomposed into a sodium ion and fluoride ion in the solution. The fluoride ion becomes HF partially depending on the pH of the solution. The HF attacks the passive films on Ti surface. By knowing the increase in pH value after immersion of Ti in the solution with fluoride, it might be that the HF decomposed to form fluoride ions and protons or water molecules on the titanium surface by following formulas: $Ti_2O_3 + 6HF \rightarrow 2TiF_3 + 3H_2O$, $TiO_2 + 4HF \rightarrow TiF_4 + 2H_2O$, or $TiO_2 + 2HF \rightarrow TiOF_2 + H_2O$. Ti fluoride compounds are formed because fluoride ions bound to the Ti or Ti oxide are decomposed and dissolved in the solution. Protons are absorbed by the titanium substrate or dissolved into the solution as water molecules, since it is reported that the titanium alloys absorb hydrogen [3-4, 3-32, 3-33].

When albumin was mixed with fluoride containing corrosion media, the corrosion behavior was noticed to change remarkably. The corrosion behavior and surface characterization of passive films on titanium immersed in a solution containing 2.0 g/l fluoride and albumin (either 0.1 or 1.0 g/l) were investigated. It was found that (i) fluorine was detected on the titanium surface immersed in the solution containing fluoride, and dissolution of the titanium was confirmed, (ii) the titanium immersed in a solution containing both fluoride and albumin had an albumin film regardless of the albumin concentration level, and (iii) in addition, the amount of dissolved titanium from the titanium immersed in the solution was less than when the solution contained no albumin [3-2]. It was therefore suggested that the formation of adsorbed albumin films on the passive film acted to not only protect the titanium from attack by the fluoride but also suppressed dissolution of the titanium−fluoride compounds.

3.3. CORROSION RESISTANCE, UNDER INFLUENCES OF ENVIRONMENTAL AND MECHANICAL-ASSISTED ACTIONS

Before the corrosion behavior of titanium and its alloys is discussed, intraoral environments needed to be reviewed. It is obvious that the mouth is full of saliva, which can be defined as an ocean of ions, non-electrolytes, amino acids, proteins, carbohydrates, lipids, which flows into dental surfaces, and contains chloride ions, and sulfurated compounds. Sulfurated compounds are easily metabolized into more corrosive agents by microorganisms. In some cases, blood is another aggressor [3-37]. Internal fluids have a concentration of chloride ions seven times higher than that of oral fluids, as mentioned previosuly. A diet rich in sodium chloride, added to large volumes of acidulated beverages (phosphoric acid), provides a continuous source of corrosive agents, despite the relatively short exposure. In addition, it has been calculated that an average urban mouth-breather inhales about a cubic meter of air every two hours, with a potential intake of between 0.11 and 2.3 mg of sulfur dioxide [3-38]. Both sulfur dioxide and hydrogen sulfide have been found to accelerate tarnishing and corrosion of metal implants.

The pH value of intraoral liquid is ever-changing, depending on age (the saliva pH tends to increase in acidity with age) and type of food and beverage. The pH of several normal beverages is as follows: milk 6.7, beer and wine 4.5, orange juice 3.3, diet coke 3.0, vinegar 2.1, lemonade 2.0, and stomach acid pH is about 2.0. Even among different animal models, there can be remarkable differences in pH values. For example, the saliva pH of fowl, swine, and dog is in a range from 6.5 to 7.5, while the pH of saliva in sheep, cattle, and horses ranges from 7.8 to 8.2. Therefore, the results using animal models possess a risk of material overestimation when they are used in human saliva, since saliva pH in these animal models is less acid than the pH in human saliva. If the plaque is built on tooth structure, it is basically due to lactic acid, so that the acidity of the plaque is very low (<3.0).

Besides chlorine ion levels and pH of intraoral fluids, there is one more important factor involved in corrosion evaluation. This is an intraoral electrochemical potential. Electrochemical corrosion studies are normally performed by means of potentiostatic or potentidynamic measurements. Interpretation of the electrochemical corrosion data requires knowledge of expected intraoral potentials. Nilner and Holland [3-39] reported that the intraoral potential ranges from -431 to -127 mV, which were measured on 407 amalgam restorations in the mouth of 28 patients. Corso *et al.* [3-40] reported that the intraoral potential is in a range from -300 to $+300$ mV. Reclaru and Meyer [3-41] reported a range from 0 to $+300$ mV. Ewers and Thornber [3-42] reported that it ranges between -380 and $+50$ mV. Hence, the range of overlapping data is a very narrow window of potential zone from 0 to $+50$ mV. It is important to estimate the passivity capability of the material concerned in such an oral environment, and such estimation can be performed without conducting any electrochemical corrosion tests. For example, referring to the Pourbaix [E-pH] diagram [3-43], one can evaluate the passivation stability as well as capability of the material concerned by knowing the reasonable range of pH in saliva (i.e., $5.5-7.5$) and the above-discussed potential (E) range (i.e., $0-50$ mV).

In addition to the above listed intraoral chemistry and potential, it is important to know how to simulate the intraoral environments when the *in vitro* chemical or electrochemical corrosion test is prepared and conducted. Many studies have been done to measure the corrosion characteristics of implant metals [3-44]. Most of these studies have been carried out using physiological isotonic electrolyte solutions such as 0.9% saline [3-45−3-50], Ringer's [3-51−3-55], Tyrode's [3-56, 3-57], Hank's [3-58] solutions, or lactic acid to simulate the accumulated plaque [3-59, 3-60].

Owing to differences in used electrolytes, the corrosion results could be varied. Alkhateeb *et al.* investigated the influence of the spontaneous surface modification of titanium by exposure to Ringer's solution at open-circuit conditions on the passive behavior [3-61]. The electrochemical behavior of Ti was compared in a simple NaCl and in Ringer's physiological solution. Potentiodynamic polarization curves showed significantly higher passive current densities when tested in Ringer's solution as compared with the simple saline solution, indicating that Ringer's solution is a more corrosive agent. Furthermore, impedance spectra measured at the open-circuit potential as a function of time showed that in saline solution a long-term exposure over some days leads to a strong increase of the protectiveness of the passive film [3-61]. Hence, the passive film can grow in mild corrosive solution. Microanalytical studies using SEM and XPS showed that the surface of Ti was modified with Ca and P species from Ringer's physiological solution [3-60].

Although the above mentioned electrolytes simulate the body fluids by reproducing the concentrations of various salts, the *in vivo* corrosion measurements have indicated lower corrosion rates than those predicted from these *in vitro* experiments. The actual *in vivo* environment is made up of salts and proteins [3-62, 3-63]. The amount of protein changes is related to the number of amino and carboxyl groups. The net charge varies with the pH of the environment; they have a neutral charge at their isoelectric point. The principal protein present in extracellualr body fluids is albumin, which has an isoelectric point of 4.5, and therefore has a negative charge at a neutral pH. It is possible that these charged molecules could in some way interact with the corrosion reactions and account for the differences between *in vitro* and *in vivo* corrosion rates. Several studies have been done to attempt to understand the effect of proteins on corrosion rates. After various metals were implanted in the body, or in simulated body conditions, corrosion products were found in the blood, sweat, urine, hair, and local tissue around the implant [3-64 – 3-66]. In *in vitro* tests, it was found that the serum proteins albumin and fibrinogen were absorbed to different extents on different metals [3-67]. In some cases, the corrosion rate was increased by the presence of these proteins, causing denaturation of the proteins [3-68]. Clark and Williams [3-69] found that pure metals which form stable passive layers, such as Ti and Al, were unaffected by the presence of proteins; however, the corrosion rate of other transition metals with variable valencies was increased, suggesting that this could be due to the ability of these metals to form stable complexes with the proteins. Speck and Fraker [3-70] looked at the effect of the amino acids cysteine and tryptophan on titanium alloys. They found that the pitting potential of NiTi was adversely affected by cysteine, but Ti-6Al-4V was not affected by both amino acids. Brown and Merritt [3-71] used weight loss and chemical analysis techniques to study the static dissolution rate of 316L (low carbon containing 18Cr-8Ni-2Mo) stainless steel cylinders to which they applied a 5 V anodic potential relative to a stainless steel counter electrode in 0.9% saline,

1% and 10% newborn calf serum. They found that the presence of serum increased the corrosion rate, and that 10% serum had a greater effect than the 1% serum.

There are normally chemical or electrochemical ways to investigate and evaluate corrosion behavior. By chemical corrosion, the corrosion rate can be calculated by weight change (in terms of weight loss due to dissolution, or weight gain due to corrosion products attached). On the other hand, methods of electrochemical thermodynamics (electode potential-pH equilibrium diagram) and electrochemical kinetics (polarization curves) may help to understand and predict the corrosion behavior of metals and alloys in chosen corrosive media. The corrorion rate can be estimated by measuring the R_p or directly corrosion current density, I_{CORR}. It is important to investigate the interface phenomena between the material and environment. To determine the biocompatibility of a biomaterial, it is important to understand the phenomena at the interface between the biomaterial and the biological system into which the material is to be implanted. At such an interface, the molecular constituents of the biological system meet and interact with the molecular constituents of the surface of the biomaterial. Since these interactions occur primarily on the molecular level and in a very narrow interface zone having a width of less than 1 nm, surface properties on the atomic scale, in particular the composition and structure of the surface layer of biomaterials, may play important roles in interfacial phenomena [3-71, 3-72].

There are several studies on corrosion of CpTi and comparison with other materials. The corrosion of pure metals Al, Co, Cu, Cr, Mo, Ni, and Ti and of a Co-Cr-Mo casting alloy has been studied in buffered saline with and without the presence of the proteins, serums albumin and fibrinogen. It was found that (i) the corrosion of Al and Ti was unaffected by the protein, (ii) the corrosion rates of Cr and Ni showed a slight increase, while Co and Cu dissolved to a much greater extent in the presence of protein, but (iii) the corrosion of the Mo element, however, was inhibited by protein [3-68].

Biocompatibility is one of the important conditions for biological materials, and since titanium is considered to have a high biological affinity, it is used as dental and/or medical materials. In fact, in patients with hypersensitive reaction to the present dental alloys, pure titanium is one of the metals that can be used for dental restoration. However, there have been several studies describing patients who did not adapt to titanium [3-73−3-75], or patients allergic to titanium [3-76−3-79]. In the oral cavity, dental plaque, which includes organic acids such as lactic and formic acid, is more likely to precipitate on the titanium surface when compared with other dental metal alloys [3-80]. The kinds and concentration of organic acids can vary depending on whether the environment is aerobic or anaerobic [3-81]. Furthermore, the pH around titanium was reduced when in contact with other metals such as amalgam [3-82], due to the electrogalvanism. The pH of dental plaque after consuming sugar is about 4.0 [3-83], but it can range from 2.0 to 14.0, depending on the foods and beverages consumed [3-84]. Based on such a complexity of the intraoral environments, Kasemo *et al.* [3-73] conducted immersion−corrosion tests on cast CpTi (grade 2), in 128 mmol/l of lactic acid and formic acid to pH of 1.0, 4.0, 5.5, 7.0, and 8.5 using 1N HCl or 1N NaOH. Tests were performed at 37°C for 3 weeks. It was concluded that (i) CpTi dissolved in all lactic acid aqueous solutions, and the amount of dissolved Ti tended to decrease with a higher pH level, (ii) in formic acid, the amount of dissolved Ti at pH 1.0 was larger than that in lactic acid at the same pH, but less than the detectable limit at pH 4.0 or higher, (iii) significant discoloration was macroscopically observed only in formic acid at pH 2.5 and 4.0, (iv) the weights of all Ti samples immersed in lactic acid decreased (indicating the dissolution of the titanium element), but were not affected by pH, and (v) in formaic acid, the weight decreased at pH 2.5–5.5 and the dissolution increased at pH 2.5−5.5 [3-73]. These results support the evidence that CpTi exhibits excellent corrosion resistance in strong oxidizing agents and in high chloride concentrations [3-85].

CpTi and nitrided CpTi (with gold-colored TiN, whereas Ti_2N reveals silver color) were tested in 1.5 M H_2SO_4 at 20°C. The CpTi, when coupled with less precious metals (like stainless steel), was easily subjected to a strong cathodic polarization. It was concluded that (i) 316L stainless steel behaves as a cathode in galvanic couples of freshly prepared samples with freshly prepared CpTi samples, but (ii) 316L stainless steel is an anode in galvanic couples when it was coupled with TiN coatings [3-86], since freshly polished Ti surfaces exhibit chemically more active than the counterelectrode 316L stainless steel, which might be covered already with spinel type oxides (i.e., Fe_3O_4, $(Fe,Ni)_2O{\cdot}(Fe_2O_3)$, or $(Fe,Ni)O{\cdot}(Fe,Cr)_2O_3$) prior to corrosion tests.

Hernandez [3-87] investigated and compared electrochemical corrosion behaviors of four different CpTi grades (grade 1−4), which are obtained from different material suppliers. The tests were performed in 37°C Ringer's solution. It was found that (i) CpTi grade 3 was the most corrosive in comparison among suppliers and grades and (ii) CpTi grade 4 exhibits the best corrosion resistance.

Kuphasuk *et al.* [3-88] tested and compared CpTi (grade 2), Ti-5Al-2.5Fe, Ti-4.5Al-3V-2Mo-2Fe, Ti-5Al-3Mo-4Zr, Ti-6Al-4V, and NiTi for their corrosion resistances and passivation capabilities in 37°C Ringer's solution (8.6 g NaCl, 0.3 g KCl, 0.33 g $CaCl_2$ to make 1000 cc with distilled water). It was found that (i) all samples showed good resistance to electrochemical corrosion over the potential of revelance to intraoral conditions, (ii) CpTi and Ti-5Al-2.5Fe showed significantly different corrosion properties than Ti-6Al-4V and NiTi; the former pair exhibited the lowest corrosion rate; the latter pair exhibited the highest corrosion rate, (iii) but, NiTi alloys revealed transpassive (in other words, breakdown of stable passivation) behavior at the potential between 0.5 and 0.75 V (vs. SCE, saturalated calomel electrode as a standard electrode, having 0.2444 V vs. hydrogen electrode) accompanied with pitting corrosion, and (iv) all samples tested were covered mainly with the rutile type of TiO_2, which was identified by electron transmission diffraction [3-88].

Güleryüz and Çimenoğlu [3-89] investigated the thermal oxidation for Ti-6Al-4V to determine the optimum oxidation conditions and to evaluate corrosion-wear performance. Corrosion test was conducted in 5-M HCl solution. The examined Ti-6Al-4V exhibited excellent resistance to corrosion after oxidation at 600°C for 60 h. This oxidation condition achieved 25 times higher wear resistance than the unoxidized alloy during reciprotating wear test in 0.9% NaCl solution [3-89]. Although it was reported, in this study, that pre-oxidation of Ti-6Al-4V at 600°C improved the wear-corrosion resistance, the oxidation duration of 60 h is questionable. According to the author's experiments, such prolonging oxidation of Ti-6Al-4V should result in too thick of an oxide layer (not film) of grayish-white TiO_2. Such a thick oxide layer does not adhere firmly to the substrate surface, but rather tends to peel off from it.

As seen in the following, there are numerous studies on corrosion behavior of NiTi materials. One reason for this is the fact that almost half of atomic percentage of this alloy is Ni, which is considered as one of the three heavy toxic elements along with Cr and V, and therefore the safety and biocompatibility should be examined thoroughly. Over the last decade, due to its unique SME (shape memory effect) and SE (superelasticity) characteristics, NiTi alloys have been increasingly considered for use in external and internal biomedical devices, e.g., orthodontic wires, endodontic files, blade-type dental implants, self-expanding cardiovascular and urological stents, bone fracture fixation plates and nails, etc. For application in the human body, the corrosion resistance of NiTi becomes extremely important, as the amount and toxicity of corrosion products control the alloy biocompatibility. NiTi (with 55.6 wt.% of Ni) was coated with the powder immersion reaction assisted nitrization (TiN), followed by annealing at 900 and 1000°C. Samples were corrosion-tested in a solution (9.00 g/l NaCl, 0.20 g/l $NaHCO_3$, 0.25 g/l $CaCl_2·6H_2O$, 0.4 g/l KCl) at 37°C. It was concluded that (i) the modified NiTi surface consisted of thin outer layer of gold-colored titanium nitride (TiN), and thicker Ti_2Ni layer underneath, (ii) untreated NiTi was found susceptible to pitting corrosion, but (iii) the nitriding treatment improved the corrosion resistance of NiTi [3-90].

Huang evaluated the corrosion resistance of stressed NiTi and stainless steel orthodontic wires using cyclic potentiodynamic and potentiostatic tests in acid artificial saliva at 37°C [3-91]. An atomic force microscope was used to measure the 3-D surface topography of as-received wires. It was found that (i) the cyclic potentiodynamic test results showed that the pH had a significant influence on the corrosion parameters of the stressed NiTi and stainless steel wires, (ii) the pitting potential, protection potential, and passive range of stressed NiTi and stainless steel wires decreased by increasing acidity, whereas the passive current density increased by decreasing pH, (iii) the load had no significant influence on the above corrosion parameters, and (iv) for all pH and load conditions, the stainless steel wire showed higher pitting potential and wider passive range than the NiTi wire, whereas the NiTi wire had lower passive current density than the stainless steel wire [3-91]. This suggests that NiTi can be passivated easier than stainless steel. As seen previously, it was reported that even elastic strain caused titanium materials to be more corrosive under the presence of fluorine ion [3-34], suggesting that even this study [3-91] and a previous study [3-34] do not agree well with each other. The fluorine ion must be very corrosive substance.

The breakdown potentials (for passivation) were measured for unpolished and mechanically polished NiTi wires in simulated body fluids. It was reported that (i) significantly higher passivation breakdown potentials (which indicate a higher stability of passive film formed on NiTi surface) were observed for cross-section wire samples, and (ii) some wires were tested in human blood and the passivation breakdown values were higher than that measured in Ringer and 0.9% NaCl solutions. It was also reported that the oxide film formed on NiTi was predominantly made up of TiO_2 with a very thin layer of NiO at the outer surface. Carroll and Kelly [3-92] furthermore conducted the galvanic corrosion tests on NiTi wires, which were coupled with gold, elgiloy/phynox, and stainless steel. NiTi was found to be anodic in all combinations. In tests in which NiTi−gold couples were immersed in 0.9% NaCl for periods of up to 12 months, only very small amounts of nickel (in the parts per billion range) were detected [3-92], indicating that TiO_2 with trace amounts of NiO protects the substrate material well.

Similar studies, which were performed using a cyclic potentiodynamic test in artificial saliva with various acidities, were done on as-received commercial NiTi dental orthodontic archwires from different manufacturers by Huang [3-93]. The results showed that the surface structure of the passive film on the tested NiTi wires were identical, containing mainly TiO_2 with small amounts of NiO. The corrosion tests showed that both the wire manufacturer and solution pH had a statistically significant influence on the corrosion potential, corrosion rate, passive current, passivation breakdown potential, and crevice−corrosion susceptibility [3-93]. It is well known that different orthodontic wire manufacturers employ their own proprietary techniques for wire drawing into the final specified dimensions, resulting in various surface cleaness as well as roughness.

Hence, from the above, the good corrosion resistance of NiTi may result from the formation of stable, conntinuous, highly adherent, and protective oxide films on its surface. In fact, due to the high chemical activity of the Ti surface, a damaged oxide film can generally reheal itself if at least traces of oxygen or water (moisture) are present in the environment [3-94, 3-95]. In addition, calcium−phosphate surface films can be naturally formed on Ti alloys in a biological environment [3-60, 3-96, 3-97], which can act as a further barrier against ion diffusion from the subsurface alloy. Indeed any surface treatments of NiTi devices have a critical influence on biocompatibility. Two methods have been considered for improving the corrosion resistance and performance of an additional surface-barrier layer against ion diffusion, such as TiN, TiC, and TiB_2 by implantation with diffusion species [3-98−3-100], and the other relies on increasing the thickness of the oxide layer by anodizing, thermal oxidation, or implantation with oxygen [3-101, 3-102]. NiTi was evaluated in Hank's solution (KCl 400 mg/l, KH_2PO_4 60 mg/l, Na_2HPO_4 48 mg/l, $NaHCO_3$ 350 mg/l, NaCl 8000 mg/l, $CaCl_2$ 140 mg/l, $MgCl\cdot6H_2O$ 100 mg/l, $MgSO_4\cdot7H_2O$ 148 mg/l, $CH_2OH(CHO)_4CHO$ 1000 mg/l) for cyclic polarization. For wear corrosion tests, a 50 g dead-weight ruby ball was applied at 10 rpm. It was found that (i) pitting corrosion resistance can be improved by ion implantation modification, (ii) wear corrosion resistance was improved through ion implantation and heat treatment, but (iii) nanopores (3 $\times$ 10^{17} ions/cm^{-2}) on the material surface are introduced at higher dose ion implantation, acting as the source of pitting corrosion and impairing pitting corrosion resistance [3-93].

The corrosion performance of NiTi SME alloy in human body simulating fluids (BSF) was evaluated in comparison with other implant materials (Ti-6Al-4V, Co-Cr, stainless steel) by Rondelli [3-102]. As for the passivity current (i.e., current density for passivation onset) in potentiostatic conditions, it was noted that the value is about three times higher for NiTi than for Ti-6Al-4V and stainless steel. Regarding the localized corrosion, while potentiondynamic scans indicated that, for the NiTi alloy good resistance to pitting attack which is similar to Ti-6Al-4V, the scratch tests by the abruptly damaging the surface passive film (i.e., potentiostatic scratch test and modified ASTM F746) pointed out that the characteristic of the passive film formed on NiTi alloy are not as good as those on Ti-6Al-4V, but are comparable or inferior to those on austenitic stainless steels (like 304L or 316 type stainless steels) [3-102]. Therefore, it is suggested that the self-healing (i.e., repassivation) capability of NiTi was inferior to that of Ti-6Al-4V.

Moreover, there are studies on corrosion behaviors of heavily alloyed Ti materials. Three Ti-based alloys (50.7Ni-49.4Ti, 51.4Ni-48.4Ti-0.19Cr, and 45.1Ni-49.6Ti-0.29Cr-4.97Cu) were electrochemically evaluated in 0.9% NaCl and 1% lactic acid solutions. It was concluded that small amounts of Cr and Cu changed the superelastic characteristics, but did not change the corrosion resistance of the

NiTi alloy [3-103]. Using the ASTM F746 "scratch" test, in 40°C 0.9% NaCl solution [3-104], 50Ni-50Ti, 44Ni-51Ti-5Cu, and 88Ti-6.5Mo-3.5Zr-2Sn [3-105] were examined through electrochemical tests. It was concluded that (i) both NiTi and NiTi-Cu wires exhibit low corrosion potential (50−150 mV vs. SCE), indicating an inferiority to that of TiMo alloy (which agreed with results reported in [3-102]) and (ii) TiMo proved to be immune to localized corrosion attacks up to 800 mV [3-105].

Different electrochemical studies were carried out for Zr, Ti-50a/oZr (or Ti-65.6w/oZr) and Zr-2.5w/oNb in solutions simulating physiologic media − Ringer's solution and PBS (phosphate-buffered saline) solution. The results from rest-potential measurements showed that (i) the three materials are spontaneously passivated in both solutions and (ii) the Ti-50Zr alloy has the greatest tendency for spontaneous oxide formation, which could be a mixture of ZrO_2 and TiO_2. Some corrosion parameters (such as the pitting and repassivation potentials) were obtained via cyclic voltammetry in both solutions, revealing that the Ti-50Zr has the best corrosion protection while Zr has the worst. On the other hand, it was reported that the pre-anodization (up to 8 V vs. SCE) of the alloys in a 0.15 mol/l Na_2SO_4 solution led to a significant improvement in their protection against pitting corrosion when exposed to the Ringer's solution [3-106].

In the last decade, the AC EIS technique has been utilized more frequently in corrosion studies. The *in vitro* and *in vivo* electrochemical behaviors of CpTi were characterized using a specialized osseous implant in conjunction with EIS measurement techniques. Capacitance and polarization resistance measurements by EIS methods suggested that a biofilm formed on the surface of CpTi in *in vivo* and *in vitro* situations is passive and stable [3-107]. The corrosion behaviors of CpTi, Ti-6Al-4V, and NiTi were studied in a buffered saline solution using anodic polarization and EIS techniques [3-108]. Pitting potential as low as +250 mV (vs. SCE) was recorded for Ti-45Ni. The initiated pits continued to propagate at potentials as low as −150 mV (vs. SCE). It was possible to increase the pitting potential of Ti-45Ni to values greater than +800 mV using a H_2O_2 surface treatment procedure; however, due to unstable passivation, this surface-modification process had no beneficial effect on the rate of pit repassivation. It was reported that the surface oxide layer consists of a porous outer layer and an inner barrier layer. The nature of this porous layer was found to depend on the nature of the electrolyte material and the presence of phosphate anions in the saline-buffered solution. Much higher porous layer resistances were recorded for CpTi and also for Ti-6Al-4V in the absence of the phosphate anions [3-108].

As mentioned briefly above, the addition of protein in corrosion media would alter the corrosion behavior remarkably. The ability of Ti-6Al-4V, Ti-13Nb-13Cr, and Ti-6Al-7Nb to repassivate was investigated in PBS, bovine albumin solutions in PBS, and 10% foetal calf serum in PBS at different pH values and at different albumin concentrations by Khan *et al.* [3-109]. It was concluded that (i) an increase in pH had a greater effect on the corrosion behavior of Ti-6Al-4V and Ti-6Al-7Nb than Ti-13Nb-13Cr in PBS, (ii) the addition of protein to the PBS reduced the influence of pH on the corrosion behavior of all the alloys, and (iii) the proteins in the environment appear to interact with the repassivation process at the surface of these alloys and influence the resulting surface properties [3-109].

Furthermore, the effect of proteins on corrosion rates of 316L stainless steel, CpTi, and Ti-6Al-4V was studied in the static and fretting modes. It was reported that (i) proteins had an insignificant effect on the static corrosion rate of Ti-6Al-4V samples, but caused a 31% increase in $1/R_p$ (a reverse of polarization resistance, and hence a reverse of corrosion resistance) of the CpTi samples, and (ii) for both CpTi and Ti-6Al-4V specimens, the presence of proteins had an insignificant effect on corrosion rates in the fretting mode [3-44].

Williams *et al.* prepared the protein solutions using crystallized bovine albumin and 95% clottable bovine fibrinogen [3-67, 3-68]. These were made up as 0.1% solutions in a buffer composed of 0.1 M sodium chloride and 0.01 M sodium phosphate at pH 7.4. Pure metallic elements (Al, Co, Cu, Cr, Mo, Ni, and Ti) were examined in the prepared solution. It was concluded that (i) serum albumin and a small amount of fibrinogen can have a significant influence on the corrosion behavior of metals, (ii) with some metals, such as Co and Cu, the corrosion rate increased significantly, (iii) with Cr and Ni the rates are moderately raised, but with Mo, the corrosion is inhibited, and (iv) there is little effect on Al and Ti [3-67, 3-68], which are normally covered with the passive oxide films.

Biotribology (friction, wear, and lubricant actions in a biological environment) is becoming an important concern; in particular, the toxicity of wear debris (likely as toxicity of corrosion products) should be considered.

Metallic implants inserted into the body will eventually undergo some degradation by a variety of mechanisms including abrasion and corrosion. The biological effect of the material released into the tissue is a subject of much concern and investigation. It is apparent that many patients have sensitivity (and some of them develop a hyper-sensitivity) to the metals in the implants and that the presence of metal(s) in a sensitive animal or human will elicit inflammatory responses and sometimes the formation of foreign body giant cells. This may have an adverse effect on the performance of the implant with pain, swelling, and tissue necrosis at the site, and, in some cases, loosening of the implant. Metal sensitivity reaction is not to the small metal ions but to a complex of metal ions and host tissue. The nature of the bonding of the metal ion to tissue or cells, the distribution of the metal ions or metal complexes in the body, and the biological responses to these complexes are of concern. According to Khan *et al.* [3-110], the *in vitro* and *in vivo* studies were undertaken on the binding of metal ions as metal salts or as corrosion products, and it is evident that metals bind mostly to albumin. The ability of these different metals to bind to red cells and white cells varies, with Cr^{+6} binding most strongly to cells. The chrome from corrosion products binds strongly to cells, and thus appears to be a Cr^{6+}. The interaction of serum and cells with chromium on the surface of the implant may markedly effect its corrosion and the biological activity of the complexes formed. Some of the tissues and proteins may be altered by changes in the tissue pH during inflammatory responses or infection [3-110].

Wear debris has been implicated in the pathogenesis of loosening and the osteolysis of total joint replacements by stimulating a foreign body and chronic inflammatory reaction capable of bone resorption. Effects of wear particles on bone have not been well documented. It was reported that 316 stainless steel exhibited wear toxicity by detecting Ni, Mn, and Cr ions dissolved out of stainless steel; while no Ti ion was detected to be dissolved from the wear debris [3-111]. A presence of the wear debris depends strongly on the wear resistance of materials. Khan *et al.* [3-112] performed the sliding-wear tests in both non-corrosive and corrosive environments. A simple pin-on-disc type of wear apparatus was designed and built to simulate the co-joint action of corrosion and sliding-wear. It was found that (i) the mixed phase alpha−beta alloys (Ti-6Al-4V and Ti-6Al-7Nb) possessed the best combination of both corrosion and wear resistance, and (ii) CpTi and the near-beta (Ti-13Nb-13Zr) and beta (Ti-15Mo) alloys displayed the best corrosion resistance properties [3-112].

Goodman *et al.* [3-113] performed a histomorphological and semi-quantitative morphometric analysis of the reaction of bone to different concentrations of phagocytosable particles of high-density polyethylene (averaged particle size: 4.7 ± 2.1 μm) and Ti-6Al-4V alloy implanted in a rabbit tibia. Suspensions of 10^6-10^9 particles per milliliter were mixed in saline, sterilized, and introduced into the proximal tibia of 30 mature female rabbits for 16 weeks. It was found that (i) phagocytosable particles of Ti-6Al-4V and high-density polyethylene, in concentrations of 10^6-10^9 particles per ml, induced a histocytic reaction without extensive fibrosis, necrosis, or granuloma formation, and (ii) this reaction occurred without disturbing the normal repair processes of bone formation and resorption to the surgical insult [3-113].

Lastly, there are studies on corrosion−fatigue interactions. To clarify the influence of NaCl aqueous solution in corrosion−fatigue crack initiation of mill-annealed Ti-6Al-4V, Ebara *et al.* [3-114] conducted rotating−bending fatigue tests of notched specimens with stress concentration factors in the range of 1.5−3.5, in aerated and deaearted 3% NaCl solutions at 80°C. Ultrasonic fatigue tests up to 10^{10} cycles were also conducted in aerated 3% NaCl solution at room temperature. It was reported that the fatigue limit in air and corrosion fatigue strength at 10^8 cycles decreased with increases in the stress concentration factor. The reduction in fatigue strength of the notched specimen ($K_t = 3.5$) by the environment at 5×10^7 cycles was at most 20%. No differences between aerated and deaearted conditions were observed. It was therefore concluded that (i) corrosion−fatigue crack initiation of the notched Ti-6Al-4V was predominantly influenced by the stress concentration factor and not by chlorine ions or dissolved oxygen content in the NaCl aqueous solution, and (ii) fatigue crack initiation of Ti-6Al-4V was not significantly influenced by the NaCl solution at room temperatrure even in long fatigue life up to 10^{10} cycles [3-114].

Removable partial dentures (made of CpTi or Ti-6Al-4V) are affected by fatigue because of the cyclic mechanism of masticatory system and frequent insertion and removal. Ti materials have been used in manufacturing of denture frameworks; however, preventive agents with fluorides are considered to attack Ti materials, as discussed previously. Tests were done at 70% load of the 0.2% offset yield strength in room temperature synthetic saliva and fluoride-added synthetic saliva. It was found that (i) Ti-6Al-4V achieved 21,269 cycles against 19,175 cycles for CpTi, (ii) there were no significant differences between either metal in corrosion−fatigue life for dry specimens, but (iii) when the solutions were present, the fatigue life was significantly reduced, probably because of the production of corrosion pits caused by superficial reactions [3-115], providing a localized stress-concentrated site and, therefore, a potential crack initiation site.

3.4. METAL ION RELEASE AND DISSOLUTION

From the biological point of view, titanium is a strange element. It is widely distributed in the earth's crust, and yet exists in only minimal amounts in animal and plant tissues. There is no evidence to suggest that it is an essential trace element, but on the other hand, it is extremely well tolerated by tissues and is essentially non-toxic. There is no clear evidence of titanium metabolism. Therefore, it may be described as a physiologically indifferent metal. With this characteristic, coupled with the excellent corrosion resistance, it is known that titanium materials exhibit excellent biocompatibility [3-116]. Knowledge about the release of various elements from metallic biomaterials in the oral cavity with respect to levels, kinetics, and chemical state is essential in the evaluation of possible local or systemic effects pertinent to individuals having such biomaterials as dental prothesis, dental implant, and orthopedic devices and implants as well. Certain types of metallic biomaterials are suspected of being associated with allergies or may influence the number of T-lymphocytes, possibly affecting the immune system [3-117]. When metal ions are released, some stay in the local implant site, while others are transported in the blood. Blood-borne elements may remain in the blood as cell complexes, accumulate in organs, or be excreted by sweat or urine. Metal ions are known to be toxic, sensitizing, and oncogenic [3-118].

During the corrosion process, metallic ion(s) can be released from the substrate surface area, causing the dissolution into corrosive media or surrounding tissue (if the case is for implant). Hence, ion release is directly related to the stability of surface oxide (passive film) formed on the titanium substrate, and capability for repassivation when such a stable passive film is disrupted. On the polarization process, there should be two occasions when ions can be released (in other words; instability of passivation); active dissolution prior to the onset of passivation occurs at relatively low potential, and transpassivity (or breakdown of stable passive film) at relatively high potential. For example, the transpassivation occurs at or higher than 1.5 V (vs. SCE), depending on the electrolyte and type of titanium materials used. Such transpassivation is normally accompanied with the formation of pits.

Even though the metals used in implants are quite corrosion-resistant, there is still some interchange of metal ions into the tissues or tissue fluids [3-119]. The amout of metal ions released is related to the corrosion resistance of the metal, the environmental conditions (i.e., pH, chloride ion concentration, temperature, etc.), mechanical factors (i.e., pre-existing cracks, surface abrasion, and film adhesion), electrochemical effects (i.e., applied potential, galvanic effects, pitting, or crevices), and the dense cell concentrations around implants. When an implant protrudes through the tissue, the electrochemical effects may be even more severe. Implants which undergo fretting wear provide the possibility of releasing more ions into solutions than those that do not, due to the stress-assisted dissolution. The response of the tissues and the entire body to metal ions has been studied and continues to be an active area of biomaterials research [3-119]. Certain metal ions have been shown to be more harmful than others. The response of local tissues to metal ions is usually short-term, while accumulation of metal ions in more remote tissues such as the kidney, liver, or blood may require a longer time. Systemic effects characterized by sensitization may also occur due to the body's interaction with the metal ions. Titanium, which has a very strong, adherent, semi-amorphous surface oxide film, has been found by neutron activation analysis in some of the tissues surrounding titanium implants [3-120]. The clinical, radiologic, and laboratory records of the patients were carefully reviewed and it was found that titanium does not seem to have a particularly harmful effect on local tissues. The systemic effects and sensitization response attributed to titanium seem to be minimal as well [3-120].

The biological significance of the metal ion release is believed to be strongly related to the cytocompatibility and hemocompatibility. It is also generally believed that corrosion resistance is responsible for the biocompatibility. Tissue levels of chromium and titanium are often very high, indicating that the elements may form stable complexes that remain in local tissues [3-118]. Woodman *et al.* [3-121] reported a significant increase in serum aluminum but no increase in titanium and vanadium from porous-coated Ti-6Al-4V implants in baboons. It was also reported that (i) high levels of titanium were found in lung and liver tissue, as well as regional lymph nodes, (ii) a substantial and progressive increase in aluminum content was found in lung nodes of baboons with porous Ti-6Al-4V implants, and (iii) titanium levels in the urine of baboons increased with porous titanium implants, but no significant increases were found in the levels of vanadium or aluminum [3-121].

An important factor related to the biocompatibility of metallic implants is the metal ion release rate from their surfaces. *In vitro* laboratory tests are often used to estimate the corrosion rates of implant materials. For these to be useful in predicting *in vivo* ion release, however, they should simulate *in vivo* conditions as closely as possible. Bundy *et al.* pointed out that a variable often not accounted for in such testing is the influence of applied stresses [3-122]. The effect of static stresses on the corrosion behavior of 316L stainless steel, Ti-6Al-4V, and Co-Cr-Mo alloy was investigated. The stress ratio (σ_{max}/σ_y) was selected in a range from 1.2 to 1.4, hence all sample surfaces were subjected to a plastic deformation. Although loading over the yield strength causes stress-enhanced ion release, it was mentioned that the stress-induced ion release may also be caused by elastic loading. The basic mechanisms for this phenomenon appear to be passive-film disruption followed by slow repassivation kinetics. Occurrence of pitting corrosion indicates that the surface oxide film was broken, and does not protect the underneath substrate. If effects similar to those observed apply to *in vivo* conditions, it was suspected that tests on unstressed alloys *in vitro* could grossly underestimate ion release rates of stressed implant devices *in vivo* [3-122]. It was also mentioned that ions are released but not necessarily in proportion to the alloy composition. Ni and Co bind to serum albumin, where as Cr is released with a valence of 6^+ and binds to red blood cells. Tissue levels of Cr and Ti are often very high indicating that these elements may form stable complex that remain in local tissues. The results of chemical analysis of urine from animals exposed to metal ions from injection of salts *in vivo* corrosion demonstrates a rapid excretion of Ni and Co; excretion of Cr is slow and represents a small percent of the total to which the animal is exposed [3-122]. The repassivation capability *in vitro* should not be the same as that found in the *in vivo* situation, since there are many evidence indicating that the oxide film formed on the placed Ti implant increases its thickness $7-20$ times, due to more oxygen availability *in vivo*, as seen in Chapter 4.

The metallic ion release from dental alloys has been transformed into one of the major problems in the health of dental patients who have metallic materials fitted in their mouths [3-123]. It is well known that metals are toxic in sufficient concentration and that they can cause inflammations [3-124], allergies, genetic mutations, or cancer [3-125]. The metallic elements that are released from oral implants have been detected in the tongue [3-124], the saliva [3-126], and in the gums adjacent to these alloys [3-127, 3-128]. However, it is not a case of a local problem but a general one due to their diffusion throughout the whole organism. The ions are conducted and can be excreted in part or entirely, or they may be accumulated selectively at the level of certain tissues [3-116, 3-121]. The harmful effects of certain metals and metallic composite materials were made clear in the epidemiological studies carried out by the nickel, chromium, cobalt, and copper industries and confirm that the action of metals in the form of ions, particles, and soluble or insoluble salts act at the cellular membrane level [3-129]. Metals from orthopedic implants are released into surrounding tissue by various mechanisms, including corrosion, wear, and mechanically accelerated electrochemical processes, such as stress implant failure, osteolysis, cutaneous allergic reactions, and remote site accumulation. Okazaki *et al.* [3-129] investigated the metal ion release from 316L stainless steel, Co-Cr-Mo, CpTi (grade 2), Ti-6Al-4V, Ti-6Al-7Nb, Ti-15Zr-4Nb-4Ta at 37°C in PBS solution, 0.9% NaCl aqueous solution, 1.2% L-cysteine solution, 1% lactic acid solution, and 0.01% HCl solution. It was concluded that (i) the quantity of Co released from the Co-Cr-Mo casting alloys was relatively small in all the solutions, (ii) the quantities of Ti released into PBS, calf serum, 0.9% NaCl and artificial saliva were much lower than those released into 1.2% L-cysteine, 1% lactic acid, and 0.01% HCl, (iii) the quantity of Al released from the Ti alloys gradually decreased with increasing pH, and a small amount of V was released into calf serum, PBS, artificial saliva, 1% lactic acid, and 0.01% HCl, and (iv) Ti-15Zr-4Nb-4Ta alloy, with its low metal release, is considered advantageous for long-term implants [3-129].

Ti-based implant materials show very high resistance to pitting corrosion in physiological solutions because of their state of passivity [3-130, 3-131]. The effect of temperature on the nucleation of corrosion pits on CpTi in Ringer's solution was investigated. Breakdown of the passivity (or transpassivation) of CpTi by nucleation of corrosion pits occurs in Ringer's solution at quite modest electrode potentials. It was reported that (i) the frequency of breakdown is very low at ambient temperature, but increases significantly with an increase in temperature (e.g., $20 \rightarrow 37 \rightarrow 50°C$), and (ii) the very slow overall release rate estimated from the data is consistent with previously measured release rates [3-130]. The vanadium-free Ti-15Zr-4Nb-4Ta alloy, containing 0.2Pd and Ti-6Al-4V ELI, were implanted in rat tibiae for 6−48 weeks. It was found that (i) the number of corrosion pits observed at the Ti-6-Al-4V ELI alloy surface tended to be slightly more than that of the Ti-15Zr-4Nb-4Ta alloy implant, and (ii) the concentrations of metal elements in the bone tissue containing the new bone tended to increase slightly more than in bones without the implants [3-132].

The formation of pits is related to the equilibrium potentials of (passive) film formation given as a function of the activities of the components of the film substance. The solubility of the product, with respect to the ions in the electrolyte, depends on the electrode potential, since oxidation states of the metal in the film and in the electrolyte are often different. Heusler [3-133] discussed the kinetics of uniform film formation and dissolution with respect to (1) equilibrium of all components across both interfaces, (2) partial equilibrium of one component at the outer film/electrolyte interface, and (3) irreversible ion transfer reactions at the outer interface. Examples are oxide films on iron (i.e., F_2O_3 and/or F_3O_4), titanium (TiO_2), and aluminum (Al_2O_3). It was mentioned that the processes during the incubation time (which should include chlorine ion adhesion and absorption) of pitting corrosion corresponded to non-uniform dissolution and formation of the passivating film [3-133].

Hence, it is important to form stable passive films to control the metal ion release. Chang and Lee [3-134] prepared as-received Ti-6Al-4V and modified Ti-6Al-4V with 300 μm thick containing Cu and Ni by vacuum brazing at 970°C for 2 h, and tested them in Hank's solution (NaCl 8 g, $CaCl_2$ 0.14 g, KCl 0.4 g, $NaHCO_3$ 0.35 g, glucose 1 g, $MgCl_2 \cdot 6H_2O$ 0.1 g, $Na_2HPO_4 \cdot 2H_2O$ 0.06 g, KH_2PO_4 0.06 g, and $MgSO_4 \cdot 7H_2O$ 0.06 g) with 8 mM ethylene diamine tetra-acetic acid (EDTA) at 37°C for immersion times of 1, 4, and 16 days. It was noted that the variation of chemical compositions of the alloy and nano-surface characteristics (chemistries of nano-surface oxide, amphoteric OH group adsorbed on oxides, and oxide thickness) was effected by surface modification and three passivation methods (34% HNO_3 passivation, 400°C heating in air, and aged in 100°C water). It was concluded that (i) passivation influences the surface oxide thickness and the early stage of ion dissolution rate of the alloy, (ii) the rate-limiting step can be best explained by metal-ion transport through the oxide film, rather than hydrolysis of the film, and (iii) variations of the chemistries of Ti alloy alter the electromotive force (EMF) potential of the metal, thereby affecting the corrosion and ion dissolution rate [3-134].

There is other evidence strongly supporting the importance of surface oxide to govern the ion release kinetics. During implantation, Ti releases corrosion products (which can be TiO, TiOH, or TiO_2) into the surrounding tissues and fluids even though it is covered by a thermodynamically stable oxide film. An increase in oxide thickness as well as the incorporation of elements from the extracellular fluids (P, Ca, and S) into the oxide has been observed as a function of implantation time. Moreover, changes in oxide stoichiometry, compositions, and thickness have been associated with the release of Ti corrosion products. Properties of the oxide, such as stoichiometry, defect density, crystal structure and orientation [3-135], surface defects, and impurities were suggested as factors determining biological performance [3-136].

From the above, it can be clearly stated that passivation stability can control the ion release kinetics under some factors such as chlorine ions, causing a premature passivation breakdown, as well as the formation of pits. Stable passivation can also be interfered by mechanical or physical actions [3-137, 3-138]. Okazaki [3-137] studied the effects of frictional forces on anodic polarization properties of several metallic biomaterials in eagle's medium and 1% lactic acid solutions. The results on Ti-6Al-4V demonstrated that, under static condition (in other words, no fractional force applied), I_{PASS} (passivation onset current density) was 4×10^{-6} A/cm^2, E_{PASS} (passivation potential) was about -400 mV (vs. SCE) and the established passivation was stable up to E_{BK} (breakdown potential of passivation) of 2000 mV (vs. SCE). However, when a kinetic frictional force of 20 N was applied, E_{PASS} (-700 mV) was about the same as the static E_{PASS}, and I_{PASS} showed a wide fluctuation between 2 and 7×10^{-5} A/cm^2 up to E_{BK} of approximately 2100 mV (vs. SCE). Oshida *et al.* [3-138] had investigated the effects of ultrasonic vibration on passivation stability. CpTi (grade 2) plates were polarized in an artificial saliva solution with and without ultrasonically vibrating the electrolyte. It was found that, under no vibration conditions, stable passivation was obtained from E_{PASS} of 150 mV (vs. SCE) up to E_{BK} of 1100 mV (vs. SCE). On the other hand, under vibration conditions, CpTi did not exhibit a stable passivity all the way up to 1500 mV (vs. SCE). Hence, the mechanical vibration hinders the stable establishment of passivation of the CpTi surface in 25°C artificial saliva solution [3-138].

For different Ti material groups there are several reports on ion release data. Levels of corrosion products released from dental alloys in natural or synthetic saliva, i.e., from amalgams, cobalt, gold, nickel, iron, or titanium-based alloys have been surveyed [3-117]. The corrosion behavior in the oral cavity is complex and involves several parameters, including, for example, composition and treatment of the alloy, pH, and oxygen variations in the oral cavity, presence of proteins, masticatory function, etc. Corrosion products and particles released from dental alloys in the oral cavity are eliminated from or absorbed by the human body. Thus, specific metals that may be accumulated within human tissues are subsequent to reabsorption in the gastrointestinal tract following the swallowing of saliva, after membrane diffusion in the Okla mucosa, or after migration to pulp through dentinal tubuli [3-118]. The amounts of implanted titanium that could be released into an electrolyte like artificial saliva seem, however, to be far from the intake of the element from food and drink. It is known that aluminum and titanium released from titanium alloy implants in baboons can accumulate in the lung tissue [3-139−3-141]. Titanium levels up to about 1000 ppm (dry weight) have been measured in osseous tissue surrounding various titanium implant materials after 6 months exposure in the trabecular bone of German shepherd dogs [3-142].

In studies done by Strietzel *et al.* [3-143], cast CpTi (grade 1) samples were subjected to immersion corrosion tests for 4 weeks in NaCl, NaF, NaSCN (sodim thiocyanate), lactic acid, acetic acid, oxalic acid, and tartaric acid. Atomic absorption was used to analyze the solutions weekly. It was noticed that (i) Ti reveals ion releases $[(0.01-0.1)\ \mu g/(cm^2 \times d)]$ in the magnitude of gold alloys, (ii) there is little influence of grinding and casting systems in comparison with organic acids or pH value, (iii) the ion release greatly increases [up to 500 $\mu g/(cm^2 \times d)$] in the presence of fluoride; and low pH values accelerate this effect even more, (iv) nevertheless, it is recommended that it is best to avoid the presence of fluoride or to reduce contact time, and (v) in prophylactic fluoridation of teeth, a varnish should be used [3-143].

In vitro metal leaching examinations were performed for Ti from CpTi, Ti, Al, and V from Ti-6Al-4V alloy, and Ti and Pd from CpTi (grade 7), for seven days. It was found that (i) in general, the low leaching of Ti, Al, and V metal ions and the absence of Pd from the different surfaces indicated excellent protection against corrosion by the surface oxide layer of Ti_2O_3, but (ii) leaching of Ti, Al, and V from metal surfaces over seven days in an isotonic medium indicated the possibility that leaching of metal ion may have detrimental effects on general health, suggesting the need for long-term studies on accumulation and toxicity of metal ions from various metal surfaces [3-144].

Although osteolysis adjacent to metal hip and knee prostheses is believed to result from release of wear debris into the periprosthetic region, it is suspected that metal ion released from the implant surface may also play some contributing role [3-122, 3-136, 3-145−3-147]. This suspicion arose in part from various retrieval studies of human total hip arthroplasties, which have shown that metal levels in the tissue surrounding metal implants are fairly high. Dorr *et al.* [3-148] have measured metal levels in the fibrous membrane encapsulating implants at up to 21 ppm Ti, 10.5 ppm Al, and 1 ppm V around Ti-6Al-4V and up to 2 ppm Co, 12.5 ppm Cr, and 1.5 ppm Mo around Co-Cr-Mo. Henning *et al.* [3-149] have found metal levels in the tissue surrounding loose Co-Cr-Mo femoral shafts to average 3.5 ppm Cr and 0.9 ppm Co. It was also reported that (i) metal ions released from the implant structure are suspected of playing some contributing role in loosening of hip and knee prostheses, and (ii) sublethal doses of the ionic constituents of Ti-6Al-4V alloy suppressed the osteoblastic phenotype and deposition of a mineralized matrix [3-150]. Thompson and Puleo [3-145] harvested bone marrow stromal cells from juvenile rats and exposed to time-staggered doses of a solution of ions representing Ti-6Al-4V alloy. Cells were cultured for 4 weeks and assayed for total protein, alkaline phosphatase, intra or extracellualr osteocalcin, and calcium. It was found that (i) the Ti-6Al-4V solutions were found to produce little difference from control solutions for total protein or alkaline phosphatase levels, but strongly inhibited osteocalcin synthesis, and (ii) calcium levels were reduced when ions were added before a critical point of osteoblastic differentiation (between 2 and 3 weeks after seeding). Based on these findings, it was concluded that ions associated with Ti-6Al-4V alloy inhibited the normal differentiation of bone marrow stromal cells to mature osteoblasts *in vitro*, suggesting that ions released from implants *in vivo* may contribute to implant failure by impairing normal bone deposition [3-145].

As for NiTi materials, it was reported that nickel ions were dissolved out [3-151, 3-152]. Kimura *et al.* [3-153] coated NiTi implants with oxide film to suppress dissolution of Ni, and estimated the corrosion resistance in 1% NaCl solution by means of anodic polarization measurement. It was found that (i) by coating, dissolution at low potential was suppressed and the dissolute current density decreased, (ii) a further decrease in current density, which results from stabilization of the passive state on the surface, was observed by the repeated polarization, and (iii) the oxide film showed close adhesion with the matrix, and did not form cracks or peel off by plastic deformation associated with the SME [3-153].

Huang *et al.* [3-154] evaluated four commercially available orthodontic superelastic NiTi wires in modified Fusayama artificial saliva — NaCl (400 mg/l), KCl (400 mg/l), $CaCl_2 \cdot 2H_2O$ (795 mg/l), $NaH_2PO_4 \cdot H_2O$ (690 mg/l), KSCN (300 mg/l), $Na_2S \cdot 9H_2O$ (5 mg/l) and urea (1000 mg/l). The value of pH was adjusted to be 2.5, 3.75, 5.0, and 6.25 by either lactic acid or sodium hydroxide. Tests were conducted at 37°C for a duration ranging from 1 to 28 days. It was concluded that (i) the released amount of metal ions increased with immersion period, (ii) the amount of metal ions released in pH >3.75 solution was much less than that in pH 2.5 solution, (iii) pre-existing surface defects in NiTi wire might be the preferred sites for corrosion, while NiTi wire, with a rougher surface, did not exhibit a higher ion release [3-154], and (iv) the average amount of Ni ions released per day from the tested NiTi wires was well below the critical concentration necessary to introduce allergy (600−2500 μg) [3-155] and under daily dietary intake level (300−500 μg) [3-156].

As demonstrated in the reviews above, the stability of oxide film formed on the surface of Ti appears to be crucial in governing the ion release equilibrium, as well as kinetics. One can find several proposed methods and technologies to enhance the oxide stability, thereby improving the ion release resistance. Ti-6Al-4V, CpTi, and TiN-coated Ti-6Al-4V by electron-beam PVD at 300−350°C were tested in 0.17 M NaCl plus 0.0027 M EDTA (ethyl-diamine tetraacetic acid, disodium salts) at 37°C. It was demonstrated that a substantial reduction in the release of metal ions may be achieved by aging the surface oxide in boiling distilled water or by thermal oxidation (400°C × 45 min in air) [3-135].

Healy *et al.* [3-136] prepared a titanium thin film (1 μm thick) by the deposition method onto silicon substrates by radio frequency sputter coating. Films were immersed into 23 ml of 8.0 mM ethyl-EDTA in simulated interstitial electrolyte solution and placed in an incubator maintained at 37°C, 10% O_2, 5% CO_2, and 97±3% humidity. It was concluded that (i) the dissolution of titanium in the presently simulated physiological environment occurred in two phases and the transition between phases appeared after 200−400 h of immersion, (ii) coincident with this transition was the onset of non-elemental P incorporation into the oxide, (iii) the dissolution rate decreased as the oxide thickness increased for immersion times less than 50 h, and (iv) the dissolution reaction depended on chemical reactions, including (1) the hydrolysis of the oxide and the exchange reactions between the Ti-OH surface sites and (2) the adsorbed P-containing species [3-136].

Ti-6Al-4V offers an excellent combination of mechanical properties for load-bearing applications such as hip prostheses, but there is concern about the slow accumulation of potentially harmful metal ions (such as aluminum and vanadium) in the soft tissue surrounding the prosthesis [3-157]. Furthermore, there is evidence that tissue reactions adjacent to Ti-6Al-4V alloy are less natural than those in the vicinity of CpTi [3-158]. Metal ion release is considered to occur via chemical dissolution of the surface oxide film, and it was shown that the dissolution rate is influenced by passivation treatments and surface coating such as TiN [3-135, 3-159]. These effects may be associated with structural changes of the surface oxide. The aging of the oxide on the surface of CpTi, Ti-6Al-4V, and TiN-coated Ti-6Al-4V was investigated. Johansson *et al.* [3-158] studied the influence of the surface oxide on the dissolution of the substrate material in saline solution. It was demonstrated that (i) a substantial reduction in the release of metal ions may be achieved by aging the surface oxide, which was treated in boiling distilled water or by thermal oxidation, and (ii) a reduction in the dissolution of metal ions from titanium-based implant materials in saline solution (0.17 M NaCl plus 0.0027 ml EDTA, at 37°C) may be achieved by aging the surface oxide in boiling distilled water or by thermal oxidation. It was further speculated that the change in the dissolution behavior is associated with the transformation of the surface oxide of TiO_2 from the anatase form (tetragonal, a=0.378 Å, c=0.951 Å) to the more compact rutile structure (tetragonal, a=0.459 Å, c=0.296 Å) [3-159], since the c-axis ratio between anatase TiO_2 and rutile TiO_2 is 3.21, while a-axis ratio is only 1.21, so that the rutile structure is a more closed packed crystalline structure.

There is other study on chemical and thermal surface modifications in order to control the metal ion release, performed by Spriano *et al.* [3-160]. The surfaces of Ti-6Al-7Nb were first sand-blasted with corundum, followed by chemical treatments for passivation in 35 wt.% nitric acid for 30 min, and in 5 M NaOH at 60°C for 24 h. Thermal treatments were also applied as an in-air oxidation at 600°C for 1 h in 5×10^{-1} bar. Studies on morphology, crystallographic structure, porosity and wettability, interaction with SBF (simulated body fluid), cytotoxicity, protein adsorption tests, *in vitro* fibroblast, and osteblast-like cell culture were performed. It was concluded that (i) the surfaces treated by the blasted-passivated-in-air and *in vacuo* oxidation can be described as constituted by a micro-textured oxide presenting high wettability and low metal ion release as well as moderate bone-like apatite induction ability, (ii) at the same time, a good chemical interaction of the surface with the culture medium, with a higher total protein binding than on the untreated one, was observed, suggesting that the multi-step thermo-chemical treatment can be used as an alternative to traditional coatings in order to prepare low metal release protheses with enhanced osteointegrability [3-160].

3.5. GALVANIC CORROSION

There are two types of galvanic cell: the first type is the micro-cells (or local cells) and the second type is the macro-cells. As for micro-cells, within a single piece of metal or alloy, there exist different regions of varying composition and hence different electrode potentials. This is true for different phases in heterogeneous alloys [3-161]. The positive way of applying this phenomenon is by etching the polished surface of metal to reveal the microstructures for metallographic observations. As severely deformed portion within one piece of a material can serve as an anodic site [3-162]. For example, a portion of a fully annealed nail was bent, and the whole nail immersed into a NaCl aqueous solution. After several hours, it was easily noticed that the localized plastic-deformed portion gets rusted. For macro-cells, which take place at the contact point of two dissimilar metals or alloys with different electrode potentials, the so-called galvanic corrosion is probably the best known of all the corrosion types [3-161]. Titanium is the metal choice for implant material due to excellent tissue compatibility; however, superstructures of such Ti implants are usually made of different alloys. This polymetallism leads to detectable (in macro-level) galvanic corrosion.

The intensity of the galvanism phenomenon is due to a number of factors, such as the electrode potential, extent of the polarization, the surface area ratio between anodic site and cathodic site, the distance between the electrodes, the surface state of the electrodes, conductivity of the electrolyte, and passage of a galvanic current on the electrolytic diffusion, stirring, aeration or deareation, temperature, pH, and compositions of the electrolytic milieu, coupling manner, and an accompanying crevice corrosion. When two solid bodies are in contact, it is fair to speculate that there could be marginal gaps between them, because they are not subject to any bonding method like diffusion bonding or fusion welding. Therefore, when the galvanic corrosion behavior is studied in a dissimilar material couple, the crevice corrosion is another important localized corrosion measurement accompanied with galvanic corrosion [3-162−3-168].

Among these numerous influencing factors, the surface area ratio between anodic surface and cathodic surface appears to be the most important factor to be considered. If the ratio of surface A_C/A_A (cathode area/anode area) is large, the increased corrosion caused by coupling can be considerable. Conductivity of the electrolyte and geometry of the system enter the problem because only that part of the cathode area is effective for which resistance between anode and cathode is not a controlling factor [3-168].

Al-Ali *et al.* [3-165] prepared CpTi (grade 2) samples coupled with type IV Au alloy, Au-Ag-Pt alloy and Ag-Au-Pd alloy, and effects of two coupling mode (laser welding and mechanical adhering method − in other words, without and with crevice situation between materials) and investigated their corrosion behaviors when electrochemically polarized in 37°C Ringer's solution. It was found that (i) significant differences in corrosion behavior were observed between the two different coupling methods, and crevice associated with adhesive joining made the galvanic couples more corrosive, (ii) by the laser welding method, the values of coupled I_{CORR} for all three couples were somewhat in the middle of I_{CORR} values of uncoupled materials, while by the mechanical adhering method, the coupled I_{CORR} was higher than those for uncoupled CpTi as well as uncoupled precious alloys, (iii) a simplified mixed theory [3-166] was not applicable to estimate the mixed potential and current, (iv) the type IV gold side of the Ti/Au couple for both methods was discolored to some extent, and (v) overall, the Ti/Ag-Au-Pd couple exhibited the best corrosion resistance for both joining methods [3-165]. As for the effects of surface area ratio between anodic and cathodic metals [3-165, 3-169], it was found that (i) I_{CORR} showed a bowl-shaped curve, (ii) I_{CORR} increased toward both ends (in other words, the uncoupled materials' sides), but (iii) I_{CORR} exhibited the lowest value when exposed surface areas of anodic and cathodic metals became equal.

There have been several reported failure cases of restorative devices due to a lack of understanding of galvanic or coupled corrosion properties of restorative materials [3-170, 3-171]. As the success of implants leads to their increasing use in restorative dentistry, attention should be devoted to the galvanic combination of restorative materials with titanium [3-172]. At present, most of the literature shows that galvanic corrosion could occur when titanium is coupled to less noble alloys rather than more noble alloys; that the corrosion current density of the titanium was lowered by the lower current density of noble metal. The *in vitro* investigations of these materials have demonstrated susceptibility to galvanic corrosion phenomena. Thus, the integrity of restorative dentistry devices and intraoral prostheses may be comprised because of galvanic corrosion [3-173]. The depolarization effect when the materials were coupled was overcome by bubbling nitrogen through the artificial saliva solution. The nitrogen purge also reduced fluctuation in the oxygen concentration of the electrolyte [3-174, 3-175]. It was hypothesized that the deaeration could also create a lack of oxygen for repassivation if a material's passive layer breaks down due to a galvanic coupling. This results in a reduced probability of the anodic component repassivating if its surface layer is disrupted. Crevices in such devices and wound sites have been documented to have, at the worst case, a pH of 3.0 or less [3-175−3-179]. When the two materials were coupled, the corrosion potential of both these materials converged to a common coupled corrosion potential [3-166]. This common potential was between the corrosion potentials of the materials involved in the couple. This is true for the case when at least one component in the couple is polarized [3-180]. The corrosion potential of the more noble material drops to the lower common corrosion potential value. This drop in potential of the noble component can serve as a driving force for the corrosion potential of the more active component to the higher coupled corrosion potential. Ti-based implant materials show very high resistance to pitting corrosion in physiological solutions because of their state of passivity [3-130, 3-131].

The electrochemical behavior of seven alloy superstructures (including high and low Au-based alloys, Ag-based alloys, Co-Cr alloys) with CpTi and Ti-6Al-4V implants was tested in Fusayama−Meyer deaerated saliva and Carter−Brugirard non-deaerated saliva. It was found that very low corrosion rates were obtained, ranging from 10^{-6} to 10^{-8} A/cm^2. Other possible types of corrosion are considered in addition to a galvanic corrosion, for example, pitting corrosion and crevice corrosion. These types of corrosion are specific to each alloy and are to be taken into account in clinical choices. Moreover, the biological characteristics of each individual represent a variable that cannot be reproduced easily *in vitro* (composition and acidity of saliva, dental hygiene, eating habits, administration of medicine). For this reason, it was remarked that the most favorable superstructure/implant couple is the one which is capable of resisting the most extreme conditions that could possibly be encountered in the mouth [3-181].

3.6. MICROBIOLOGY-INDUCED CORROSION

Microbiology-induced corrosion (MIC) is defined as the deterioration and degradation of metallic structures and devices by corrosion processes that occur directly or indirectly as a result of the activity of living microorganisms, which either produce aggressive metabolites, or are able to participate directly in the electrochemical reactions occurring on the metal surface. Corrosion has long been thought to occur by one of the three means: oxidation, dissolution, or electrochemical interaction. To this list, the newly discovered microbiological corrosion must be added [3-182, 3-183]. Oxidation is somewhat beneficial in the case of stainless steel (mainly the Cr element) and titanium because they create a non-porous layer of passive oxide film that protects the substrate surface from further oxidation; Cr_2O_3 or $(Fe,Ni)O.(Fe,Cr)_2O_3$ for stainless steel and TiO_2 for titanium materials. However, dissolution and electrochemical reactions are responsible for most of the detrimental corrosion. MIC has received increased attention by corrosion scientists and engineers in recent years. MIC is due to the presence of microorganisms on a metal surface, which leads to changes in the rates, and sometimes also the types of the electrochemical reactions which are involved in the corrosion processes. It is, therefore, not surprising that many attempts have been made to use electrochemical techniques (corrosion potential, the redox potential, the polarization resistance, the electrochemical impedance, electrochemical noise, and polarization curves including pitting) to study the details of MIC and determine its mechanism. Applications range from studies of the corrosion of steel pipes in the presence of sulfate-reducing bacteria to investigating the formation of biofilms and calcareous deposits on stainless steel in seawater, to the destruction of concrete pipes in sewers by microorganisms producing very low pH solutions, to dental prostheses exposed to various types of bacteria intraorally [3-184, 3-185].

Study on MIC is a typical interdisciplinary subject that requires at least some understanding in the fields of chemistry, electrochemistry, metallurgy, microbiology, and biochemistry. In many cases, microbial corrosion is closely associated with biofouling phenomena, which are caused by the activity of organisms that produce deposits of gelatinous slime or biogenically induced corrosion debris in aqueous systems. Familiar examples of this problem are the growth of algae in cooling towers and barnacles, mussels, and seaweed on marine structures. These growths either produce aggressive metabolites or create microhabitats suitable for the proliferation of other bacterial species, e.g., anaerobic conditions favoring the well-known sulfate-reducing bacteria. In addition, the presence of growths or deposits on a metal surface encourage the formation of differential aeration or concentration cells between the deposit and the surrounding environment, which might stimulate existing corrosion processes [3-182]. It is probable, however, that microbiological corrosion rarely occurs as an isolated phenomenon, but is coupled with some type of electrcochemical corrosion. For example, corrosion induced by sulfate-reducing bacteria usually is complicated by the chemical action of sulfides. Corrosion by an aerobic organism, by definition, always occurs in the presence of oxygen. To further confound the problem, microbiocidal chemicals also may have some conventional properties as corrosion inhibitors (e.g., film-forming). The materials on which a biofilm is developed have higher, or more noble, E_{CORR} values than those obtained in the absence of biofilms [3-186].

Microbes seem to accelerate the corrosion process. They locate susceptible areas, fix anodic sites, and produce or accumulate chemical species that promote corrosion. The microscopic heterogeneity of many materials — whether created intentionally or as an artifact — is quite clear on the scale of microbes, and is an important and overlooked factor in microbiologically influenced corrosion. Weld regions are particularly attractive to microbes in many of the systems [3-187], since the weld line normally exhibits slight differences in chemical compositions as well as microstructures, resulting in a possibility for galvanic corrosion occurrence. Microbes are extremely tenacious. They can exist over a wide range of temperatures and chemical conditions, they are prolific, and they can exist in large colonies. They form synergistic communities with other microbes or higher life forms and accomplish remarkably complex chemical reactions in consort. Microbes can metabolize metals directly, and they require many of the chemical species produced by the physical chemistry of corrosion processes for their metabolisms. Microbes can also act to depolarize anodic or cathodic reactions, and, in addition to catalyzing existing corrosion mechanisms, many microbes produce organic or mineral acids as metabolic products [3-187]. Accordingly, it is difficult to specify just what the important variables are when corrosion is influenced by microorganisms. In addition to the variables most important for the type of corrosion under consideration, variables such as dissolved oxygen, pH, temperature, and nutrients, which affect the life cycle of the bacteria, become important for both corrosion testing and corrosion mechanism. The critical situation involved in MIC is limited to the interfacial thin layer (usually with thickness in the $10-500$ μm range) between the biofilm and metallic surface, since the chemistry of the electrolyte solution at the metal surface is everchanging. Therefore, the bulk electrolyte properties may have little relevance to the corrosion as influenced by organisms within the film. The organism right at the metal surface influences corrosion activity, and those organisms multiply so rapidly on the surface that a low density of organisms in the bulk quickly becomes irrelevant [3-188].

Bacteria must adhere, spread, and proliferate on metallic surfaces in order to cause corrosive reactions. The process of bacterial adhesion is mediated by both an initial physiochemical interaction phase (i.e., electrostatic forces and hydrophobicity) and specific mechanisms (i.e., adhesion–receptor interactions, which allows bacteria to bind selectively to the surface), followed by a subsequent molecular and cellular interaction phase, which is normally followed by a plaque formation in dentistry [3-189]. The molecular and cellular interactions are complicated processes affected by many factors, including the characteristics of the bacteria themselves, the target material surface, and environmental factors such as the presence of serum proteins or bacterial substances. Microbial adhesion to surfaces is a common phenomenon. Many bacteria demonstrate a preference for the surface regions of solids, teeth, plant fibers, roots, etc., where nutrients may be concentrated by the adsorption of dissolved organic materials. Since microorganisms act as colloidal particles, their interaction with solid surfaces can be anticipated, based on colloidal theory. However, microorganisms are more complex than typical colloids as they are capable of independent locomotion, growth in different shapes, and production of extracellular polymeric materials that aid in anchoring the microbe to the surface [3-189]. Adhesion is a situation where bacteria adhere firmly to a surface by complete physiochemical interactions between them, including an initial phase of reversible physical contact and a time-dependent phase of irreversible chemical and cellular adherence [3-190]. Bacteria and other microorganisms have a natural tendency to adhere to surfaces as a survival mechanism. This can occur in many environments including the living host, industrial systems, and natural waters. The general outcome of bacterial colonization of surfaces is the formation of an adherent layer (biofilm) composed of bacteria embedded in an organic matrix [3-191]. The starvation and growth phase influence bacterial cell surface hydrophobicity. Both the number and kind of microorganisms that colonized metal surfaces depend on the type of metal and the presence of an imposed electrical potential. No significant differences in attachment and growth of a pure culture were observed when metal surfaces were dipped in an exogenous energy source. The chemical composition of naturally occurring adsorbed organic films on metal surfaces was shown to be independent of surface composition and polarization [3-192].

There are several experimental evidences on bacterial attachment on Ti materials. Despite the wide use of dental implants, the understanding of the mechanisms of bacterial attachment to implant surfaces and of the factors that affect such attachment is limited. The attachment of oral bacteria, including *Streptococcus sanguis*, *Actinomyces viscosus*, and *Porphyromonas gingivalis*, to titanium discs with different surface morphology (smooth, grooved, or rough) was examined by SEM. The greatest bacterial attachment was observed on the rough BSA-coated Ti surfaces (bovine serum albumin: coated by 0.25 ml of 0.15% glycine − 0.01% BSA at 4°C overnight). The smooth surfaces promoted poor attachment for *S. sanguis* and *A. viscocus*. However, *P. gingivalis* attached equally well to both the smooth and grooved coated Ti surfaces. Cell-surface finbriae (which may play a role in adhesion) of both *A. viscosus* and *P. gingivalis* were observed to be associated with Ti surfaces. It was concluded that Ti surface characteristics appeared to influence oral bacteria attachment *in vitro* [3-193].

Steinberg *et al.* [3-194] investigated the adhesion of radioactively labeled *A. viscosus*, *Actinobacillus actinomicetemcomitans*, and *P. gingivalis* to CpTi, Ti-6Al-4V coated with albumin or human saliva. It was reported that (i) all tested bacteria displayed greater attachment to Ti-6Al-4V than CpTi, (ii) *P. gingivalis* exhibited less adhesion to CpTi and Ti-6Al-4V than did the other bacterial stains, (iii) adhesion of *A. viscosus* and *A. actinomicetemcomitans* was greatly reduced when CpTi and Ti-6Al-4V were coated with albumin or saliva, and (iv) *P. gingivalis* demonstrated a lesser reduction in adhesion to albumin or saliva-coated surfaces. It was therefore suggested that oral bacteria have different adhesion affinities for CpTi and Ti-6Al-4V and that both albumin and human saliva reduce bacterial adhesion [3-194].

The adherence of *Streptococcus mutans* to six implant materials (polycrystal alumina, single crystal alumina, Ti-6Al-4V, Co-Cr-Mo alloy, hydroxyapatite HA, heat-curing PMMA resin) was studied *in vitro*. The change of free energy, which corresponds to the adherence process, was also evaluated for hydrophobic interaction. It was found that (i) adherence of *S. mutans* to polycrystal alumina was lower than adherence to other materials, and (ii) the adherence of *S. mutans* to the test materials was highly correlated with the change of free energy, which suggests that hydrophobic interaction plays an important role in the adherence of *S. mutans* to the implant materials [3-195].

The relatively low toxicity and impressive biocompatibility of Ti in human tissues has favored its use as an implant material for the replacement of missing teeth [3-196]. The design of oral implants requires communication with the oral cavity through the mucosal or gingival tissues. Exposure in the mouth presents a unique surface that can interact with host indigenous bacteria, leading to plaque formation. Early colonization of the tooth surface starts by adhesion of salivary bacteria to the pellicle [3-197, 3-198]. Plaque development on exposed surfaces of teeth and restoration materials begins with an acquired pellicle. The acquired pellicle consists primarily of glycoproteins, mucins, and enzymes present in saliva [3-199]. Electrostatic-type interaction between charged ions is one of the main routes in the formation of acquired pellicle on teeth and other materials. Thus, the greater quantities of salivary proteins adsorbing to enamel may be explained by the greater distribution of phosphate groups and calcium ions on enamel surfaces in relation to the Ti oxide layer. This results in a reduced affinity between Ti and charged salivary proteins [3-200, 3-201]. After the adsorption of the proteins, bacteria begin to adhere to the acquired pellicle [3-202]. In the oral cavity, different adsorbed salivary proteins may dictate which oral bacteria first adhere and how the plaque will develop. Since Ti and enamel differ in their engagement of salivary components, the plaque developing on a Ti surface and its potential pathogenicity may also differ [3-198]. Based on the above background, Leonhardt *et al.* [3-198] placed CpTi, HA (hydroxyapatite), and amalgam samples in Co-Cr splints and kept intraorally in each individual for 10 min, 1, 3, 6, 24, and 72 h. It was found that (i) qualitative or quantitative differences in early plaque formation were seen on the surfaces, and (ii) the different kinds of materials did not seem to have any antibacterial properties *in vivo*, and it seems as if the adherent bacterial plaque had a similar composition, regardless of material [3-198]. Sordyl *et al.* [3-203] collected bacterial samples from the superior aspect of implants after bar and abutment were removed. They found that the plaque specimens contained a predominantly gram-positive microflora, although gram-negative species were present in low numbers [3-203]. Balkin *et al.* [3-204] examined the peri-implant microboita of human mandubular subperosteal dental implants. The study established that gram-positive facultative *Streptococcus* and *Actinomyces* spp. are characteristic of subperiosteal implants with long-term clinical stability. Al Saleh *et al.* [3-205] studied the amount of plaque formation that occurs on CpTi, Ti-6Al-4V, gold, and enamel by using an artificial mouth system. *S. mutans* and *S. sanguis* were used as the test bacteria. The metals and control teeth were exposed to *S. mutans* for 2 days and *S. sanguis* for 3 days. It was reported that (i) a mean *S. mutans* coverage was 44% on the enamel, 98.5% on the gold, 91.8% on the CpTi, and 91.2% on the Ti-6Al-4V alloy, (ii) there was a statistically significant difference between enamel and the three metals, (iii) the mean of *S. sanguis* coverage was as follows: 48.3% on enamel, 50.4% on gold, 13% on CpTi, and 27.3% on Ti-6Al-4V alloy, and (iv) CpTi harbored less *S. sanguis* than enamel and gold, whereas enamel harbored less *S. mutans* than gold, CpTi and Ti-6Al-4V alloy [3-205].

While there is considerable information about bacteria adhesion on enamel, little is known about the mechanism of bacterial interactions with implant materials in the oral cavity [3-194]. It was found that in edentulous mouths, the presence of gram-positive cocci around implants was significantly greater than that of other bacteria [3-206]. In partially edentulous, healthy or diseased mouths, implants and teeth harbored the same type of bacteria [3-207, 3-208]. No significant differences were found between the subgingival flora of teeth and titanium implants in an *in vivo* study [3-209] of a periodontally healthy mouth. Microbial plaque accumulation surrounding dental implants may develop into peri-implantitis or peri-impantoclasia, which is defined as inflammation or infection around an implant, with accompanying bone loss [3-210]. Peri-implant disease can be as destructive as periodontal disease, and adequate diagnosis and timely treatment are essential. The microflora around successful implants is similar to healthy sulci, while that associated with failing implants is similar to periodontally diseased sites. Implant microflora is similar to the tooth microflora in the partially edentulous mouth. The microflora of implants in partially edentulous mouths differs from that in edentulous mouths. This seems to indicate a bacterial reservoir around the teeth and the possibility of reinfection of the implant sulcus by periodontal pathogens [3-211]. It is, therefore, important to maintain plaque-free surfaces on both supra- and subgingival portions of dental implants to prevent pre-implantitis. There are at least two methods of inhibiting the formation of microbial plaque. The first method is to inhibit the initial adhesion of oral bacteria. The second method is to inhibit the colonization of oral bacteria, which involves surface antibacterial activity. Yoshinari *et al.* [3-212] studied the antibacterial effect of surface modifications to Ti on *P. gingivalis* and *A. actinomycetemcomitans*. Surface modifications with the dry process included (1) ion implantation (Ca^+, N^+, F^+), (2) oxidation (anodic oxidation, titania spraying), (3) ion plating (TiN, alumina), and (4) ion beam mixing (Ag, Sn, Zn, Pt) with Ar^+ on CpTi plates. It was found that (i) F^+-implanted specimens significantly inhibited the growth of both *P. gingivalis* and *A. actinomycetemcomitans* than the untreated Ti, (ii) the other surface-modified specimens did not exhibit effective antibacterial activity against both bacteria, and (iii) no release of fluoride ions were detected from F^+-implanted specimens under dissolution tests. It was thus indicated that surface modifications by means of a dry process are useful in providing antibacterial activity of oral bacteria to Ti implants exposed to the oral cavity [3-212].

Both anaerobic and aerobic bacteria can participate in the process of microbiology-related corrosion. Sulfate-reducing bacteria are most prevalent under anaerobic conditions. Lekholm *et al.* found that the microbial species around stable versus failing implants are similar to the patterns observed around healthy versus periodontally involved teeth [3-213]. Quirynen *et al.* found no significant differences in the bacterial morphotypes between implants and natural teeth in partially edentulous patients (with implants and teeth in the same jaw). The percentages of coccoid cells, motile rods, spirochetes, and other bacteria were 65.8, 2.3, 2.1 and 29.8% for implants and 55.6, 4.9, 3.6, and 34.9% for teeth, respectively. The results suggested that teeth may serve as a reservoir for the bacterial colonization of Ti implants in the same mouth [3-207].

There exist several *in vitro* MIC-related corrosion tests. In an *in vitro* study using *Staphylococcus epidermidis*, a simple-producing strain and its isogenic slime-negative *mutant*, it was found that (i) both strains adhere to CpTi discs with significantly higher colony counts for the slime-producing strain, (ii) the colony count was dependent on temperature, time, and stain, and (iii) slime production is important for adherence and subsequent accumulation of *S. epidermidis* onto CpTi discs *in vitro* [3-214]. Drake *et al.* [3-215] prepared CpTi surfaces by pre-treating (SiC polished, sand-blasted, 30% NHO_3 passivated), followed by exposing them to UV sterilization, steam autoclaving at 121°C, ethylene gas treatment at 55°C, and plasma cleaning in argon gas. These surface-modified CpTi were investigated in terms of wettability, roughness, mode of sterilization, and the ability of the oral bacterium *S. sanguis* to colonize. It was concluded that (i) Ti samples that exhibited rough or hydrophobic (low wettability) surfaces, along with all autoclaved surfaces, were preferentially colonized, (ii) Ti surfaces that had been repeatedly autoclaved were colonized with the levels of bacteria 3 to 4 orders of magnitude higher than other modes of sterilization, and (iii) this may have implications relative to the commonly used method of autoclaving titanium implants, which may ultimately enhance bacterial biofilm formation on these surfaces [3-215].

Bacterial reaction on teeth is different from that on titanium materials. Gibbons and Van Houte [3-197] noted that the interaction of host flora with teeth involves a highly selective process related to specific interactions between tooth-bound salivary pellicles and bacterial surface adhesions. The ion distribution and charge character of titanium should be different from enamel. Therefore, salivary pellicle formation at each of these surfaces may in turn be different. It was found that (i) using an *in vitro* adherence model significantly lowered numbers of *A. viscosus* bound to the salivary-treated Ti surface were found when compared to that of the similarly treated enamel, (ii) the binding of *S. sanguis* to Ti or enamel did not vary significantly, (iii) a comparison of the percentage of cells bound to the titanium surface revealed that *S. sanguis* cells attached in significantly higher numbers when compared to the *A. viscosus* cells, and (iv) formation of salivary protein pellicles on Ti may be quite different from that of enamel. It was further speculated that the changes seen in the pellicle formation are the result of an alteration in the rate of pellicle formation on Ti or the weakening of interactions of the salivary proteins with the surface, as a result of the lowered surface free energy [3-196], supporting observations by Fujioka-Hirai *et al.* [3-195].

Chang *et al.* [3-216] investigated electrochemical behaviors of CpTi, Ti-6Al-4V, NiTi, Co-Cr-Mo, 316L stainless steel, 17-4 PH stainless steel, and Ni-Cr in (1) sterilized Ringer's solution, (2) *S. mutans* mixed with sterilized Ringer's solution, (3) sterilized tryptic soy broth, and (4) byproduct of *S. mutans* mixed with sterilized tryptic soy broth. It was concluded that (i) among the four electrolytes, the byproduct mixed with sterilized trypic soy broth was the most corrosive media, leading to an increase in I_{CORR} and reduction in E_{CORR}, (ii) CpTi, Co-Cr-Mo, and stainless steel increased their I_{CORR} significantly, but Ti-6Al-4V and NiTi did not show remarkable increase in I_{CORR} [3-216].

Koh [3-164] studied effect of surface area ratios on bacteria galvanic corrosion of commercially pure titanium coupled with other dental alloys. CpTi was coupled with a more noble metal (type IV dental gold alloy) and a less noble metal (Ni-Cr alloy) with different surface area ratios (ranging from 4:0 to 0:4 — in other words, 4:0 uncouple indicates that entire surface area of a noble metal was masked, while 0:4 uncouple indicates that only CpTi surface was masked, and area ratios between 4:0 and 0:4 indicate that both surfaces were partially masked to produce different surface areas). Galvanic couples were then electrochemically tested in bacteria culture media and culture media containing bacteria byproducts. It was found that I_{CORR} (and hence, corrosion rate) profiles as a function of surface area ratio showed a straight-line trend (meaning surface area ratio independence) for Ti/Au couples, whereas the representative curve for the

Ti/Ni-Cr alloy was bowl shaped. The latter curve indicates further higher corrosion rates at both ends (larger area of either CpTi or Ni-Cr), meaning that lowest corrosion rate was exhibited at the bottom of bowl (at equal surface area ratio) [3-164], supporting results obtained by Al-Ali *et al.* [3-165] and Garcia and Oshida [3-169].

3.7. REACTION WITH HYDROGEN

It is generally agreed that the solubility of hydrogen in μ-phase titanium and titanium alloys is quite low, namely on the order of $20-2000$ ppm at room temperature [3-217]. However, severe problem can arise when Ti-based alloys pick up large amounts of hydrogen, especially at elevated temperatures [3-218–3-222]. The strong stabilizing effect of hydrogen on the μ-phase field results in a decrease of the β (HCP crystalline structure) $\rightarrow$ β (BCC crystalline structure) transformation temperature from 885°C to a eutectoid temperature of 300°C [3-218]. Hydrogen absorption from the surrounding environment leads to degradation of the mechanical properties of the material [3-223]. This phenomenon has been referred to as hydrogen embrittlement over the past years. Hydrogen absorption often becomes a problem for high-strength steels even in air [3-224], and it also occurs for Ti in methanol solutions contaning hydrochloric acid [3-225–3-229]. Hydrogen damage of Ti and Ti-based alloys is manifested as a loss of ductility (embrittlement) and/or reduction in the stress-intensity threshold for crack propagation [3-217]. CpTi is very resistant to hydrogen embrittlement, but it becomes susceptible to hydrogen embrittlement under an existence of notch, at low temperature, high strain rates, or large grain sizes [3-230]. Eventually, all of these thermal and mechanical situations make the material more brittle in nature. For α and near α Ti materials, it is believed that such hydrogen embrittlement occurs due to precipitation of brittle hydride phases [3-231]. For $\alpha+\beta$ phase alloys, when a significant amount of μ phase is present, hydrogen can be preferentially transported with the μ lattice, and will react with μ phase at α/β boundaries. Under this condition, degradation will generally become severe [3-232]. For β alloys, hydrogen embrittlement was observed to occur in the Ti-Mo-Nb-Al [3-233], Ti-V-Cr-Al-Sn [3-234], and Ti-V-Fe-Al alloys [3-235], due to μ-hydride phase formation, which is brittle at low temperatures [3-217].

It was confirmed that a shear crack initiated at the root of the thread and propagated into the inner section of the screw in the dental implant systems. Gas chromatography revealed that the retrieved screw and absorbed a higher amount of hydrogen than the as-used CpTi sample. It was then suggested that CpTi in a biological environment absorbs hydrogen, and that this may be the reason for delayed fracture of Ti implants [3-224]. Such absorbed hydrogen might have two different sources. The first source of hydrogen is the internal hydrogen that is already present in the material, either in solution or as hydrides. The second source is external hydrogen, which is produced through the interaction between titanium and hydrogen or hydrogen producing environments (including cathodic reaction during the corrosion reaction) during service. It is generally assumed that hydrogen from the environment is taken up by the titanium, thereby becoming internal hydrogen. Thus, the two effects are obviously very similar. As an alloying element in titanium, hydrogen provides very little increase in strength but can seriously impair both ductility and toughness [3-236−3-238].

There are some processes by which hydrogen could develop, and possibly be absorbed, under passive conditions: direct absorption of hydrogen produced by water hydrolysis and absorption of atomic hydrogen produced by the corrosion processes to produce an oxide. Under anodic conditions, when passive corrosion prevails, the corrosion potential for passive titanium must rise at a value at which water reduction can cause titanium to oxidize, $Ti + 2H_2O \rightarrow TiO_2 + 2H_2$. Since Ti hydrides are thermodynamically stable at these potentials, the passive film can only be considered as a transport barrier, and not an absolute barrier. The rate of hydrogen absorption will be controlled by the rate of corrosion reaction, which dictates the rate of production of absorbable hydrogen. Since TiO_2 is extremely insoluble, the corrosion reaction will be effectively limited to an oxide film growth reaction [3-239]. Titanium corrosion processes are often accompanied with hydrogen production, introducing the possibility of the absorption of hydrogen into the material. Crevice corrosion, once initiated, is supported by both the reduction of oxygen on passive surfaces external to the crevice and the reduction of protons ($Ti + 4H^+ \rightarrow Ti^{4+} + 2H_2$) inside the crevice, leading to the absorption of atomic hydrogen in sufficient quantities to produce extensive hydride formation. For passive non-creviced or inert crevice conditions, corrosion could be sustained by reaction with water under neutral conditions ($Ti + 2H_2O \rightarrow TiO_2 + 2H_2$), and should proceed at an extremely slow rate. In the first step to possible failure by hydrogen-induced cracking, the hydrogen generated must pass through the TiO_2 film before absorption into the underlying titanium alloy. For absorption to proceed, redox transformation ($Ti^{4+} \rightarrow Ti^{3+}$) in the oxide film is necessary. This requires significant cathodic polarization of the metal, generally achievable by galvanic coupling to active materials (such as carbon steel), or application of cathodic protection potential [3-240].

Ogawa *et al.* [3-223] conducted immersion tests on a beta Ti alloy (77.8Ti-11.3Mo-6.6Zr-4.3Sn) under sustained tensile load (0−900 MPa) for various periods at room temperature, in 0.2 and 2.0% APF (2% $NaF + 1.7\%$ H_3PO_4). Hydrogen absorption behavior of in acid fluoride solution has been analyzed by hydrogen thermal desorption. It was reported that (i) in the case of an immersion time of 60 h, the amount of absorbed hydrogen exceeded 10,000 ppm, while the amount of hydrogen absorbed in the 0.2% APF solution was several times smaller than that in 2% solution, (ii) for immersion in 0.2% APF, hydrogen absorption saturated after 48 h, and (iii) during the later stage of immersion, the amount of absorbed hydrogen markedly increased under higher applied stress, although the applied stress did not enhance hydrogen absorption during the early stage of immersion. These results of hydrogen absorption behavior are consistent with the delayed fracture characteristics of the beta Ti alloy [3-223].

Lane *et al.* [3-241] tested various Ti alloy cantilever beam specimens, and reported that seawater embrittlement behavior is related to aluminum content, aging in the range of $480-700$°C, and the presence of isomorphous beta stabilizers (Mo, V, Co). Test results indicate that (i) seawater embrittlement is an environment-dependent brittleness triggered by an aqueous corrodent, (ii) seawater sensitive Ti alloys include CpTi, Ti-7Al-2Nb-1Ta, Ti-7Al-3Nb, Ti-6Al-2.5Sn, and Ti-6Al-3Nb-2Sn, whereas sea water insensitive Ti alloys include Ti-7Al-2.5Mo, Ti-6Al-2Sn-1Mo-1V, Ti-6.5Al-5Zr-1V, Ti-6Al-4V, and Ti-6Al-2Sn-1Mo-3V, and (iii) in order to reduce embrittlement, it is recommended to lower Al content, and to add elements (Mo or V) that suppress the formation of coherent Ti_3Al [3-241].

Superelastic NiTi wire is widely used in orthodontic clinics, but delayed fracture in the oral cavity has been observed. Because hydrogen embrittlement is known to cause damage to Ti alloy systems, Yokoyama *et al.* [3-242] treated orthodontic wires cathodically in 0.9% NaCl solution for charging hydrogen, followed by conducting tensile tests and fractographic observations. It was reported that (i) the strength of the Co-Cr alloy and stainless steel used in orthodontic mechanotreatment was not affected by the hydrogen charging; however, NiTi wire showed significant decreases in its strength, and (ii) the fractured surface of the alloy with severe hydrogen charging exhibited dimple patterns similar to those in the alloys from patients. In view of the galvanic current in the mouth, the fracture of the NiTi alloy might be attributed to the degradation of the mechanical properties due to hydrogen absorption [3-242]. Hydrogen-induced embrittlement of NiTi was also found when it was tested in acidic solutions, and fluoride-containing solutions such as APF [3-243, 3-244]. However, it was suggested that the work-hardening NiTi alloy was less sensitive to hydrogen embrittlement compared with NiTi superelastic alloy [3-226].

Titanium reaction with hydrogen is not necessarily a useless phenomenon. The formation of titanium hydrides can provide an engineering advantage. It is well known that titanium is one of the most promising materials for hydrogen storage due to a stable formation of non-stoichiometric titanium hydrides (δ, ε, or μ type) over the composition range from $TiH_{1.53}$ to $TiH_{1.99}$ [3-225].

Finally, the unique reaction of titanium with hydrogen can be used favorably for the grain refining, which has a sequence of hydrogen absorption $\rightarrow$ heat treatment and working $\rightarrow$ de-hydrogen treatment. For example, Ti-3Al-2.5Sn containing 0.5

mass% of hydrogen was heated at 780°C and solid-solution treated, followed by water-quenched to have martensitic phase. The alloy was further subjected to hot-rolling in the $\alpha + \beta$ dual structure, followed by heating at 600°C in vacuum to de-hydrogen treatment [3-229]. Such treated Ti-3Al-2.5Sn shows uniaxial microstructrue with the grain size in a range from 0.5 to 1 μm, which is suitable for the superelastic forming.

REFERENCES

[3-1] Brockhurst PJ. Dental materials: new territories for materials science. In: Metals forum. Aust Inst Met 1980;3:200−210.

[3-2] Takemoto S, Hatori M, Yoshinari M, Kawada E, Oda Y. Corrosion behavior and surface characterization of titanium in solution containing fluoride and albumin. Biomaterials 2005;26:829−837.

[3-3] Lausmaa J, Kasemo B, Hansson S. Accelerated oxide growth on titanium implants during autoclaving caused by fluorine contamination. Biomaterials 1985;6:23−27.

[3-4] Wilhelmsen W, Grande AP. The influence of hydrofluoric acid and fluoride ion on the corrosion and passive behavior of titanium. Electrochim Acta 1987;32:1469−1472.

[3-5] Oda Y, Kawada E, Yoshinari M, Hsegawa K, Okabe T. The influence of fluoride concentration on the corrosion of titanium and titanium alloys. Jpn J Dent Mater 1996;15:317−322.

[3-6] Nakagawa M, Matsuya S, Udoh K. Effect of fluoride and dissolved oxygen concentration on the corrosion behavior of pure titanium and titanium alloys. Dent Mater J 2002;21:83−92.

[3-7] Harzer W, Schroter A, Gedrange T, Muschter F. Sensitivity of titanium brackets to the corrosive influence of fluoride-containing toothpaste and tea. Angle Orthod 2001;71:318−323.

[3-8] Adell R, Lekholm U, Rockler B, Brånemark PI. A 15-year study of osseo-integrated implants in the treatment of the edentulous jaw. Int J Oral Surg 1981;10:387−416.

[3-9] Bard JA. Titanium. Vol. V. New York: Marcel Dekker. 1976. pp. 305−331.

[3-10] Barralis J, Meder G. Precis de Metallurgie, Elaboration Structurale, Properties et Normalisation. Paris: Nathan. 1991. pp. 35−49.

[3-11] Watanabe I, Watanabe E. Surface changes induced by fluoride prophylactic agents on titanium-based orthodontic wires. Am J Orthod Dentofacial Orthop 2003;123:653−656.

[3-12] Rosentritt M, Lang R, Plein T, Behr M, Handel G. Discoloration of restorative materials after bleaching application. Quintessence Int 2005;36: 33−39.

[3-13] Goldstein GR, Kiremidjian-Schuhmacher L. Bleaching: is it safe and effective? J Prosthet Dent 1993;69:325−328.

[3-14] Attin T. The security and use of carbamide peroxide gels in bleaching therapy. Dtsch Zahnärztl Z 1998;53:11−16.

[3-15] Haywood VB, Heymann HO. Nightguard vital bleaching: how safe is it? Quintessence Int 1991;22:515−523.

[3-16] Dahl JE, Pallesen U. Tooth bleaching − a critical review of the biological aspects. Crit Rev Oral Biol Med 2003;14:292−304.

[3-17] Oshida Y, Sellers CB, Mirza K, Farzin-Nia F. Corrosion of dental materials by dental treatment agents. Mater Sci Eng 2005;C25:343−348.

[3-18] Albrektsson T. The response of bone to titanium implants. CRC Crit Rev Biocompatibility 1985;1:53−84.

[3-19] Sellers CB, Oshida Y. Effects of fluoride treatment agents on dental metallic materials. AADR/IADR meeting, Honolulu HI. 2004 (Abstract No. 0066).

[3-20] Boere G. Influence of fluoride on titanium in an acidic environment measured by polarization resistance technique. J Appl Biomater 1995;6: 283−288.

[3-21] Johansson BI, Bergman B. Corrosion of titanium and amalgam couples: effect of fluoride, area size, surface preparation and fabrication procedures. Dent Mater 1995;11:41−46.

[3-22] Nakagawa M, Matsuya S, Shiraishi T, Ohta M. Effect of fluoride concentration and pH on corrosion behavior of titanium for dental use. J Dent Res 1999;78:1568−1572.

[3-23] Pröbster L, Kin W, Hütterman H. Effect of fluoride prophylactic agents on titanium surfaces. Int J Oral Maxillofac Implants 1992;7:390−394.

[3-24] Reclaru L, Meyer JM. Effects of fluorides on titanium and other dental alloys in dentistry. Biomaterials 1998;19:85−92.

[3-25] Siirila HS, Kononen M. The effect of oral topical fluorides on the surface of commercially pure titanium. Int J Oral Maxillofac Implants 1991;6: 50−54.

[3-26] Thomas DE, Bomberger HB. The effect of chlorides and fluorides on titanium alloys in simulated scrubber environments. Mater Perform 1983;22: 29−36.

[3-27] Tounmelin-Chemla F, Rouelle F, Burdairon G. Corrosive properties of fluoride-containing odontological gels against titanium. J Dent 1996;24: 109−115.

[3-28] Schiff N, Grosgogeat B, Lissac M, Dalard F. Influence of fluoridated mouth-washes on corrosion resistance of orthodontics wires. Biomaterials 2004;25:4535−4542.

[3-29] Mirza K, Oshida Y. Corrosiveness of bleaching agent on dental metallic materials. AADR/IADR meeting, Honolulu, HI. 2004 (Abstract No. 0621).

[3-30] Oshida Y, Mirza K, Sellers CB, Panyayong W, Farzin-Nia F. Effects of dental treatment agents on discoloration/corrosion behavior of metallic dental materials. In: Medical device materials. Shrivastava S, editor. Materials Park, OH: ASM International. 2004. pp. 467−470.

[3-31] Yoon SJ, Kim CW, Lim BS. Electrochemical corrosion of dental titanium in fluoride ion. J Dent Res 1997;76:81 (Abstract No. 538).

[3-32] Nakagawa M, Matono Y, Matsuya S, Udoh K, Ishikawa K. The effect of Pt and Pd alloying additions on the corrosion behavior of titanium in fluoride-containing environments. Biomaterials 2005;26:2239−2246.

[3-33] Kaneko K, Yokoyama K, Moriyama K, Asaoka K, Sakai J, Nahumo M. Delayed fracture of beta titanium orthodontic wire in fluoride aqueous solutions. Biomaterials 2003;24:2113−2120.

[3-34] Huang H-H. Effects of fluoride concentration and elastic tensile strain on the corrosion resistance of commercially pure titanium. Biomaterials 2002;23:59−63.

[3-35] Pan J, Liao H, Leygraf C, Thierry D, Li J. Variations of oxide films on titanium induced by osteoblast-like cell culture and the influence of an H_2O_2 pretreatment. J Biomed Mater Res 1998;40:244−256.

[3-36] Mustafa K, Pan J, Wroblewski J, Leygraf C, Arvidson K. Electrochemical impedance spectroscopy and x-ray photoelectron spectroscopy analysis of titanium surfaces cultured with osteoblast-like cells derived from human mandibular bone. J Biomed Mater Res 2001;59:655−664.

[3-37] Matasa CG, Attachment corrosion and its testing. J Clin Orthod 1995;29:16−23.

[3-38] Barton K. Protection against atmospheric corrosion. New York: Wiley. 1973. p. 202.

[3-39] Nilner K, Holland RI. Electrochemical potentials of amalgam restorations *in vivo*. Scand J Dent Res 1985;93:357−359.

[3-40] Corso PP, German RM, Simmons HD. Corrosion evaluation of gold-based dental alloys. J Dent Res 1985;64:854−859.

[3-41] Reclaru L, Meyer JM. Zonal coulometric analysis of the corrosion resistance of dental alloys. J Dent Res 1995;23:301−311.

[3-42] Ewers GJ, Thornber MR. The effect of a simulated environment on dental alloys. J Dent Res 1983;62:330 (Abstract No. 330).

[3-43] Pourbaix M. Atlas of electrochemical equilibria in aqueous solutions. London: Pergamon Press, 1966.

[3-44] Williams RL, Brown SA, Merritt K. Electrochemical studies on the influence of proteins on the corrosion of implant alloys. Biomaterials 1988;9:181−186.

[3-45] Brown SA, Simpson JP. Crevice and fretting corrosion of stainless steel plates and screws. J Biomed Mater Res 1981;15:867−878.

[3-46] Devine TM, Wulff J. Cast vs. wrought cobalt−chromium surgical implant alloys. J Biomed Mater Res 1975;9:151−167.

[3-47] Lucus LC, Buchanan RA, Lemons JE, Griffin CD. Susceptibility of surgical cobalt-based alloys to pitting corrosion. J Biomed Mater Res 1982;16: 799−810.

[3-48] Man HC, Gabe DR. A study of pitting potentials for some austenitic stainless steels using a potentiodynamic technique. Corr Sci 1981;21: 713−721.

[3-49] Oshida Y, Sachdeva R, Mitazaki S. Microanalytical characterization and surface modification of NiTi orthodontic arch wires. J Biomed Mater Eng 1992;2:51−69.

[3-50] Sherwin MP, Taylor DE, Waterhouse RB. An electrochemical investigation of fretting corrosion in stainless steel. Corr Sci 1971;11:419−429.

[3-51] Bandy R, Cahoon JR. Effect of composition on the electrcochemical behavior of austenitic stainless steel in Ringer's solution. Corrosion 1977; 33: 204−208.

[3-52] Cahoon JR, Bandyopadhya R, Tennese L. The concept of protection potential applied to the corrosion of metallic orthopedic implants. J Biomed Mater Res 1975;9:259−264.

[3-53] Cahoon JR, Chaturved MC, Tennese L. Corrosion studies on metallic implant materials. J Assoc Ad Med Instrument 1973;7:131−135.

[3-54] Mueller HJ, Greener EH. Polarization studies of surgical materials in Ringer's solution. J Biomed Mater Res 1970;4:29−41.

[3-55] Sutow EJ, Pollack SR, Korostoff A. An *in vitro* investigation of the anodic polarization and capacitance behavior of 316L stainless steel. J Biomed Mater Res 1976;10:671−693.

[3-56] Syrett BC, Wing SS. Pitting resistance of new and conventional orthopedic implants − effect of metallurgical conditions. Corrosion 1978;34:138−145.

[3-57] Syrett BC, Wing SS. An electrochemical investigation of fretting corrosion of surgical implant material. Corrosion 1978;34:379−386.

[3-58] Hoar TP, Mears DC. Corrosion-resistant alloys in chloride solutions: materials for surgical implants. Proc Royal Soc 1966;294A:486−510.

[3-59] Al-Ali S. Galvanic corrosion with and without crevice corrosion in titanium and dental precious alloy couples. Indiana University Master Thesis. 2002.

[3-60] Dunigan-Miller J. Electrochemical corrosion behavior of commercially pure titanium materials fabricated by different methods in three electrolytes. Indiana University Master Thesis, 2004.

[3-61] Alkhateeb E, Virtanen S. Influence of surface self-modification in Ringer's solution on the passive behavior of titanium. J Biomed Mater Res 2005;75A: 934−940.

[3-62] Steinemann S. Corrosion of surgical implants − *in vivo* and *in vitro* tests. In: Evaluation of biomaterials. Winters GD, Leray JL, deGroot K, editors. Chishester: Wiley. 1980. pp. 1−34.

[3-63] Revie RW, Greene ND. Comparison of the *in vivo* and *in vitro* corrosion of 18-8 stainless steel and titanium. J Biomed Mater Res 1969;3: 465−470.

[3-64] Colman RF, Herrington J, Scales JT. Concentration of wear products in hair, blood and urine after total hip replacement. Br Med J 1975;1:527−529.

[3-65] Merritt K, Brown SA, Sharkey NA. Blood distribution of nickel, cobalt and chromium following intramuscular injection into hamsters. J Biomed Mater Res 1984;18:991−1004.

[3-66] Samitz MH, Katz SA. Nickel dermatitis hazards from prostheses, *in vivo* and *in vitro* solubilization studies. Br J Dermatol 1975;92:287−290.

[3-67] Williams DF, Askill IN. The adsorption and desorption of proteins on clean metal surfaces. Trans Soc Biomater 1983;6:119−121.

[3-68] Williams DF, Clark GCF. The accelerated corrosion of pure metals by proteins. Trans Soc Biomater 1981;4:17−20.

[3-69] Clark GCF, Williams DF. The effect of proteins on metallic corrosion. J Biomed Mater Res 1982;16:125−134.

[3-70] Speck KM, Fraker AC. Anodic polarization behavior of Ti-Ni and Ti-6Al-4V in simulated physiological solutions. J Dent Res 1980;59:1590−1595.

[3-71] Brown SA, Merritt K. Electrochemical corrosion in saline and serum. J Biomed Mater Res 1980;14:173−175.

[3-72] Poubaix M. Electrochemical corrosion of metallic biomaterials. Biomaterials 1984;5:122−134.

[3-73] Kasemo B, Lausmaa J. Surface science aspects on inorganic biomaterials. CRC Crit Rev Biocompatibility 1986;2:335−380.

[3-74] Koike M, Fujii H. The corrosion resistance of pure titanium in inorganic acids. Biomaterials 2001;22:2931−2936.

[3-75] Merritt K, Rodrigo JJ. Immune response to synthetic materials, Sensitization of patients receiving orthopedic implants. Clin Orthop 1996;326:71−79.

[3-76] Urban RM, Jacobs JJ, Tomlinson MJ, Gavrilovic J, Black J, Peoch M. Dissemination of wear particles to the liver, spleen, and abdominal lymph nodes of patients with hip or knee replacement. J Bone Joint Surg Am 2000;82:457−476.

[3-77] Lalor PA, Revell PA, Gray AB, Wright S, Railton GT, Freeman MAR. Sensitivity to titanium, a cause of implant failure. J Bone Joint Surg Br 1991;73−B:25−28.

[3-78] Abdallah HI, Balsara RK, O'Riordan AC. Pacemaker contact sensitivity: clinical recognition and management. Ann Thorac Surg 1994;57: 1017−1018.

[3-79] Yamauchi R, Morita A, Tsuji T. Pacemaker dermatitis from titanium. Contact Dermatitis 2000;42:52−53.

[3-80] Basketter DA, Whittle E, Monk B. Possible allergy to complex titanium salt. Contact Dermatitis 2000;42:310−311.

[3-81] Simonis A, Krüämer A, Netuschil L, Schlachta T, Geis-Gerstorfer J. In-vivo-Korrsionuntersuchungen am Zahüärztlichen Legierungen unter Berücksichtigung des Speichel-pH-Werts. Dtsch Zahnüärztl Z 1990;45: 485−490.

[3-82] Suga S. Illustrated Cariology. Tokyo: Ishiyaku Pub., Inc. 1990. pp. 120−127.

[3-83] Ravnholt G. Corrosion current and pH rise around titanium coupled to dental alloys. Scand J Dent Res 1988;96:466−472.

[3-84] Imfeld Th. Evaluation of the carcinogenicity of confectionery by intra-oral wire-telemetry. Scweiz Monatsschr Zahnheilkd 1977;87:437−464.

[3-85] Miura I, Ida K. Dental uses of titanium. Tokyo: Quintessence Book, Inc. 1988. pp. 28–41.

[3-86] Gluszek J, Jędrkowiak J, Markowski J, Masalski J. Galvanic couples of 316L steel with Ti and ion plated Ti and TiN coatings in Ringer's solutions. Biomaterials 1990;11:330−335.

[3-87] Hernandez FE. Corrosion behavior of commercially pure titanium with different grades and from different material suppliers. Indiana University Master Thesis, 2003.

[3-88] Kuphasuk C, Oshida Y, Andres CJ, Hivijitra ST, Barco MT, Brown DT. Electrochemical corrosion of titanium and titanium-based alloys. J Prosthet Dent 2001;85:195−202.

[3-89] Güleryüz H, Çimenoğlu H. Effect of thermal oxidation on corrosion and corrosion-wear behaviour of a Ti-6Al-4V alloy. Biomaterials 2004;25: 3325−3333.

[3-90] Starosvetsky D, Gotman I. Corrosion behavior of titanium nitride coated Ni-Ti shape memory surgical alloy. Biomaterials 2001;22:1853−1859.

[3-91] Huang H-H. Corrosion resistance of stressed Ni-Ti and stainless steel orthodontic wires in acid artificial saliva. J Biomed Mater Res 2003;66A: 829−839.

[3-92] Carroll WM, Kelly MJ. Corrosion behavior of nitinol wires in body fluid environments. J Biomed Mater Res 2003;67A:1123−1130.

[3-93] Huang H-H. Surface characterizations and corrosion resistance of nickel-titanium orthodontic archwires in artificial saliva of various degrees of acidity. J Biomed Mater Res 2005;74A:629−639.

[3-94] Tan L, Dodd RA, Crone WC. Corrosion and wear-corrosion behavior of NiTi modified by plasma source ion implantation. Biomaterials 2003;24:3931−3939.

[3-95] Stone P, Bennett RA, Bowker M. Reactive re-oxidation of reduced $TiO_2(110)$ surfaces demonstrated by high temperature STM movies. New J Phys 1999;1:1−12.

[3-96] Hanawa T, Ota M. Calcium phosphate naturally formed on titanium in electrolyte solution. Biomaterials 1991;12:767−776.

[3-97] Demri B, Hage-Ali M, Moritz M, Muster D. Surface characterization of C/Ti 6Al 4V coating treated with ion beam. Biomaterials 1997;18:305−310.

[3-98] Green SM, Grant DM, Wood JV, Johanson A, Johnson E, Sarholt-Kristensen L. Effect of N^+ implantation on the shape memory behavior and corrosion resistance of an equiatomic Ni-Ti alloy. J Mater Sci Lett 1993;12: 618−619.

[3-99] Chen J. Conrad JR, Dodd RA. Methane plasma source ion implantation (PSII) for improvement of tribiological and corrosion properties. J Mater Process Tech 1995;49:115−124.

[3-100] Riviere JP. Formation of hard coatings for tribological and corrosion protection by dynamic ion mixing. Surf Coat Tech 1998;108:276−283.

[3-101] Mandl S, Krause D, Thorwarth G, Sader R, Zeilhofer F, Horch HH, Rauschenbach B. Plasma immersion ion implantation treatment of medical implants. Surf Coat Tech 2001;142:1046−1050.

[3-102] Rondelli G. Corrosion resistance tests on Ni-Ti shape memory alloy. Biomaterials 1996;17:2003−2008.

[3-103] Iijima M, Endo K, Ohno H, Mizoguchi I. Effect of Cr and Cu addition on corrosion behavior of Ni-Ti alloys. Dent Mater J 1998;17:31−40.

[3-104] ASTM F 746-87 Standard Test method for pitting or crevice corrosion of metallic surgical implant materials. In: Annual book of ASTM standards. Vol. 13.01. Philadelphia, PA: American Society for Testing and Materials. 1996. pp. 80−85.

[3-105] Rondelli G, Vicentini B. Effect of copper on the localized corrosion resistance of Ni−Ti shape memory alloy. Biomaterials 2002;23:639−644.

[3-106] Oliveira NTC, Biaggio SR, Rocha-Filho RC, Bocchi N. Electrochemical studies on zirconium and its biocompatible alloys Ti-50Zr wt.% and Zr-2.5Nb wt.% in simulated physiologic media. J Biomed Mater Res 2005;74A: 397−407.

[3-107] Fox WC, Miller MA. Osseous implant for studies of biomaterials using an *in vivo* electrochemical transducer. J Biomed Mater Res 1993;27:763−773.

[3-108] Aziz-Kerrzo M, Conroy KG, Fenelon AM, Farrell AT, Breslin CB. Electrochemical studies on the stability and corrosion resistance of titanium-based implant materials. Biomaterials 2001;22:1531−1539.

[3-109] Khan MA, Williams RL, Williams DF. The corrosion behaviour of Ti-6Al-4V, Ti-6Al-7Nb and Ti-13Nb-13Zr in protein solutions. Biomaterials 1999;20:631−637.

[3-110] Khan MA, Williams RL, Williams DF. Conjunct corrosion and wear in titanium alloy. Biomaterials 1999;20:765−772.

[3-111] Merritt K, Brown SA. Biological effect of corrosion products from metals. ASTM STP 859. In: Corrosion and degradation of implant materials: second symposium. Fraker AC, Griffin CD, editors. American Society for Testing and Materials. 1985. pp. 195−207.

[3-112] Khan MA, Williams RL, Williams DF. *In-vitro* corrosion and wear of titanium alloys in the biological environment. Biomaterials 1996;17: 2117−2126.

[3-113] Goodman SB, Davidson JA, Song Y, Martial N, Fornasier VL. Histo-morphological reaction of bone to different concentrations of phagocytosable particles of high-density polyethylene and Ti-6Al-4V alloy *in vivo*. Biomaterials 1996;17:1943−1947.

[3-114] Ebara R, Yamada Y, Goto A. Corrosion-fatigue behavior of Ti-6Al-4V in a sodium chloride aqueous solution. In: Corrosion fatigue: mechanics, metallurgy, electrochemistry, and engineering. Crooker TW, Leis BN, editors. Philadephia: American Society for Testing and Materials, ASTM STP 801. 1983. pp. 135−146.

[3-115] Zavanelli RA, Pessanha GE, Ferreira I, de A. Rollo JMD. Corrosion-fatigue life of commercially pure titanium and Ti-6Al-4V alloys in different storage environments. J Prosthet Dent 2000;84:274−279.

[3-116] Williams DF. The properties of titanium and its uses in cardiovascular surgery. J Cardiovas Technol 1978;20:52−65.

[3-117] Brune D. Metal release from dental biomaterials. Biomaterials 1986;7:163−75.

[3-118] Brown SA, Merritt K, Farnsworth LJ, Crowe TD. Biological significance of metal ion release. In: Quantitative characterization and performance of

porous implants for hard tissue applications. Lemonds, JE, editor. ASTM STP 953. 1987. pp. 163−181.

[3-119] Van Orden AC, Fraker AC, Sung P. The influence of small variations in composition on the corrosion of cobalt-chromium alloys. Proc Soc Biomater 1982;5:108−112.

[3-120] Meachim G, Williams DF. Tissue changes adjacent to titanium implants. J Biomed Mater Res 1973;7:555−572.

[3-121] Woodman JL, Jacobs JJ, Galante JO, Urban RM. Metal ion release from titanium-based prosthetic segmental replacements of long bones in baboons: a long-term study. J Orthop Res 1984;1:421−430.

[3-122] Bundy KJ, Williams CJ, Luedemann RE. Stress-enhanced ion release − the effect of static loading. Biomaterials 1991;12:627−639.

[3-123] Cortada M, Giner L, Costa S, Gil FJ, Rodríguez D, Planell JA. Metallic ion release in artificial saliva to titanium oral implants coupled with different superstructures. J Biomed Mater Eng 1997;7:213−220.

[3-124] Goyer RA. Toxic effects of metals, In: Cassarett and Doull's toxicology. Klaassen CD, Amdur MO, Doull J, editors. New York: Mcmillan. 1986. pp. 582−635.

[3-125] Beach JD. Nickel carbonyl inhibition of RNA synthesis by a chromatin −RNA polymerase complex from hepatitic nuclei. Cancer Res 1970;30: 48−50.

[3-126] Stenburg T. Release of cobalt from cobalt-chromium alloy constructions in the oral cavity of man. Scand J Dent Res 1982;90:472−479.

[3-127] Rechmann P. LAMMS and ICP-MS detection of dental metallic compounds in non-discolored human gingival. Dent Res 1992;71:599−603.

[3-128] Hao SQ, Lemons JE. Histology of dog dental tissues with Cu-based crowns. Dent Res 1989;68:322−327.

[3-129] Okazaki Y, Gotoh E. Comparsion of metal release from various metallic biomaterials *in vitro*. Biomaterials 2005;26:11−21.

[3-130] Burstein GT, Liu C, Souto RM. The effect of temperature on the nucleation of corrosion pits on titanium in Ringer's physiological solution. Biomaterials 2005;26:245−256.

[3-131] Ong JL, Lucas LC, Raikar KN. Gregory JC. Electrochemical corrosion analyses and characterization of surface-modified titanium. Appl Surf Sci 1993;72:7−13.

[3-132] Okazaki Y, Nishimura E, Nakada H, Kobayashi K. Surface analysis of Ti-15Zr-4Nb-4Ta alloy after implantation in rat tibia. Biomaterials 2001;22: 599−607.

[3-133] Heusler KE. The influence of electrolyte composition on the formation and dissolution of passivating films. Corr Sci 1989;29:131−147.

[3-134] Chang E, Lee TM. Effect of surface chemistries and characteristics of Ti6Al4V on the Ca and P adsorption and ion dissolution in Hank's ethylene diamine tetra-acetic acid solution. Biomaterials 2002;23:2917−2925.

[3-135] Wisbey A, Gregson PJ, Peter LM, Tuke M. Effect of surface treatment on the dissolution of titanium-based implant materials. Biomaterials 1991;12: 470−473.

[3-136] Healy KE, Ducheyne P. The mechanisms of passive dissolution of titanium in a model physiological environment. J Biomed Mater Res 1992;26:319−338.

[3-137] Okazaki Y. Effect of friction on anodic polarization properties of metallic biomaterials. Biomaterials 2002;93:2071−2077.

[3-138] Oshida Y, Farzin-Nia F, Ito M, Panyaypong W. Passivation stability of titanium. In: Medical device materials II. Helmus M, Medlin D, editors. Materials Park, OH: ASM International. 2005. pp. 375−379.

[3-139] Solar RJ, Pollack SR, Korostoff E. Titanium release from implants: a proposed mechanism. In:Corrosion and degradation of implant materials. Kansas City, MO: Symp American Society for Testing and Materials. 1978. pp. 2−3.

[3-140] Woodman JR, Jacobs JJ, Galante JO, Urban RM. Titanium release from fiber metal compositions in baboons − a long term study. Trans Orthop Res Soc 1982;7:166−171.

[3-141] Woodman JR, Lacobs JJ, Urban RM, Galante JO. Vanadium and aluminum release from fiber metal composites in baboons − a long term study. Trans Orthop Res Soc 1983;8:238−243.

[3-142] Ducheyne P, Willems G, Martens M, Helsen L. *In vivo* metal-ion release from porous titanium-fiber material. J Biomed Mater Res 1984;18: 293−308.

[3-143] Strietzel R, Hösch A, Kalbfleisch H, Buch D. *In vitro* corrosion of titanium. Biomaterials 1998;19:1495−1499.

[3-144] Boths SJ, Coetzee WJC. Metal leaching from the Ti_2O_3 surfaces of different Ti-Metals. J Dent Res 1996;75(Special Issue):75 (Abstract No. 460).

[3-145] Thompson GJ, Puleo DA. Ti-6Al-4V ion solution inhibition of osteogenic cell phenotype as a function of differentiation timecourse *in vitro*. Biomaterials 1996;17:1949−1954.

[3-146] Betts F, Wright T, Salvati EA, Boskey A, Bansal M. Cobalt-alloy metal debris in periarticular tissues from total hip revision arthroplasties. Clin Orthop 1992;276:75−82.

[3-147] Margevicius KJ, Bauer T, McMahan JT, Brown SA, Merritt K. Isolation and characterization of debris in membranes around total hip prostheses. J Bone Joint Surg 1994;76-A:1664−1675.

[3-148] Dorr LD, Bloebaum R, Emmanual J, Meldrum R. Histologic, biochemical, and ion analysis of tissue and fluids retrieved during total hip arthroplasty. Clin Orthop 1990;261:82−95.

[3-149] Hennig FF, Raithel HJ, Schaller KH, Döhler JR. Nickel-chrome- and cobalt-concentrations in human tissue and body fluids of hip prosthesis patients. J Trace Elem Electrolytes Health Dis 1992;6:239−243.

[3-150] Thompson GJ, Puleo DA. Effects of sub-lethal metal ion concentrations on osteogenic cells derived from bone marrow stromal cells. J Appl Biomater 1995;6:249−258.

[3-151] Park HY, Shearer TR. *In vitro* release of nickel and chromium from simulated orthodontic appliances. Am J Orthod 1983;84:156−159.

[3-152] Kerosuo H, Moe G, Kleven E. *In vitro* release of nickel and chromium from different types of simulated orthodontic appliances. Angle Orthod 1995;65:111−116.

[3-153] Kimura H, Sohmura T. Improvement in corrosion resistance of Ti-Ni shape memory alloy by oxide film coating. Jpn J Dent Mater 1988;7: 106−110.

[3-154] Huang H-H, Chiu Y-H, Lee T-H, Wu S-C, Yang H-W, Su K-H, Hsu C-C. Ion release from NiTi orthodontic wires in artificial saliva with various acidities. Biomaterials 2003;24:3585−3592.

[3-155] Kaaber K, Veien NK, Tjell JC. Low nickel diet in the treatment of patients with chronic nickel dermatitis. Br J Dermatol 1978;98:197−201.

[3-156] Schroeder HA, Balassa JJ, Tipton IH. Abnormal trace metals in man-nickel. J Chronic Dis 1962;15:51−62.

[3-157] Black J. Systemic effects of biomaterials. Biomaterials 1984;5:11−18.

[3-158] Johansson C, Lausmaa J, Ask M, Hansson HA, Albrektsson T. Ultra-structural differences of the interface zone between bone and Ti-6Al-4V or commercially pure titanium. J Biomed Eng 1989;111:3−8.

[3-159] Wisbey A, Gregson PJ, Peter LM, Tuke M. Titanium release from TiN coated implant materials. International Conference on Changing Role of Engineering in Orthopedics (MEP) London, UK. 1989. pp. 9−14.

[3-160] Spriano S, Bosetti M, Bronzoni M, Vernè E, Maina G, Bergo V, Cannas M. Surface properties and cell response of low metal ion release Ti-6Al-7Nb alloy after multi-step chemical and thermal treatments. Biomaterials 2005;26;1219−1229.

[3-161] Wranglén G. An introduction to corroison and protection of metals. New York: Chapman and Hill. 1985. pp. 85−87.

[3-162] Van Black LH. Elements of materials science and engineering. Reading, MA: Addison-Wesley Pub. 1989. pp. 509−511.

[3-163] Garcia I. Galvanic corrosion of dental implant materials. Indiana University Master Degree Thesis, 2000.

[3-164] Koh Il-W. Effects of bacteria-induced corrosion on galvanic couples of commercially pure titanium with other dental alloys. Indiana University Mater Degree Thesis, 2003.

[3-165] Al-Ali S, Oshida Y, Andres CJ, Barco MT, Brown DT, Hovijitra S, Ito M, Nagasawa S, Yoshida T. Effects of coupling methods on galvanic corrosion behavior of commercially pure titanium with dental precious alloys. J Biomed Mater Eng 2005;15:307−316.

[3-166] Fontana MG, Greene ND. Corrosion engineering. New York: McGraw-Hill Book Co. 1967. pp. 20−25, and 330−338.

[3-167] Tomashov ND. Theory of corrosion and protection of metals. New York: MacMillan Co. 1966. pp. 212−218.

[3-168] Uhlig HH, Revie RW. Corrosion and corrosion control. New York: Wiley 1985. pp. 101−105.

[3-169] Garcia I, Oshida Y. Surface area ratio effects on galvanic corrosion of titanium couples. J Dent Res 2001;80(Abstract No. 798).

[3-170] American Society for Testing and Materials: Standard guide for development and used of galvanic series for predicting galvanic corrosion performance. Vol. 03.02. 1991. pp. 334−340.

[3-171] Johansson BI, Lucas LC, Lemons JE. Corrosion of copper, nickel, and gold dental casting alloys: an *in vitro* and *in vivo* study. J Biomed Mater Res 1989;23:349−361.

[3-172] Lemons JE. Dental implant retrieved analyses. In: Proceeding of the 1988 consensus development conference on dental implants. Risso AA, editor. J Dent Res 1988. pp. 748−756.

[3-173] Venugopalan R, Lucas LC. Evaluation of restorative and implant alloys galvanically coupled to titanium. Dent Mater 1998;14:165−172.

[3-174] Leinfelde KF, Lemons JE. Clinical restorative materials and techniques. Philadelphia, PA: Lea and Febiger. 1988. pp. 139−159.

[3-175] Lucas LC, Lemons JE. Biodegradation of restorative metallic systems. Adv Den Res 1992;6:32−37.

[3-176] Silver IA, Murrills RJ, Etherington DJ. Microelectrode studies on the acid microenvironment beneath adherent macrophages and osteoclasts. Exp Cell Res 1988;175:266−276.

[3-177] Bagambiosa FB, Joos U, Schilli W. Interaction of osteogenic cells with hydroxylapatite implant materials *in vitro* and *in vivo*. Int J Oral Maxillofac Implants 1990;5:217−226.

[3-178] Blair HC, Teitelbaum SL, Tan HL, Koziol CM, Schlesinger PH. Passive chloride permeability charge coupled to (H^+)-ATPase of avian osteoclast ruffled membrane. Am J Phys 1991;260:1315−1319.

[3-179] Bundy K, Laycock Y, Gomez-Novoa C, Bucalo V. Mechanisms of surface attack of molecular THRs: a study of galvanic contact and crevice conditions. In: Proceedings of the symposium on compatibility of biomedical implants. Vol. 94-15. Philadelphia, PA: Electrochemical Society. 1994. pp. 307−318.

[3-180] Uhlig HH. Corrosion control. New York: Wiley. 1965. pp. 92−96.

[3-181] Grosgogeat B, Reclaru L, Lissac M, Dalard F. Measurement and evaluation of galvanic corrosion between titanim/Ti6Al4V implants and dental alloys by electrochemical techniques and auger spectroscopy. Biomaterials 1999;20: 933−941.

[3-182] Tiller AK. Aspects of microbial corrosion. In: Corrosion processes. Parkins, RN, editor. London: Applied Science Pub. 1982. pp. 115−119.

[3-183] Matasa CG. Stainless steels and direct-bonding brackets, III: microbiological properties. Inf Orthod Kierferorthop 1993;25:269−271.

[3-184] Mansfeld F, Little BJ. A technical review of electrochemically techniques applied to microbiologically influenced corrosion. Corr Sci 1991;32:247−272.

[3-185] Dexter SC, Duquette DJ, Siebert OW, Videla HA. Use and limitations of electrochemical techniques for investigation microbiological corrosion. Corrosion 1991;47:308−318.

[3-186] Pope DH, Stoecker JG. Microbiologically influenced corrosion. Proceedings of Industries Corrosion NACE. 1986. p. 235.

[3-187] Walsh D, Pope D, Danford M, Huff T. The effect of microstructure on micro-biologically influenced corrosion. JOM 1993;45:22−30.

[3-188] Dexter SC. Microbiological effects. In: Corrosion tests and standards: application and interpretation. R. Baboian, editor. ASTM Manual Series MNL 20. 1995. pp. 419−429.

[3-189] Gibbons RJ, Cohen L, Hay DI. Strains of *Streptococcus mutans* and *Streptococcus sobrinus* attach to different pellicle receptors. Infect Immun 1986;52:555−561.

[3-190] An YH, Friedman RJ. Concise review of mechanisms of bacterial adhesion to biomaterial surfaces. J Biomed Mater Res 1998;43:338−348.

[3-191] Tsibouklis J, Stone M, Thorpe AA, Graham P, Peters V, Heerlien R, Smith JR, Green KL, Nevell TG. Preventing bacterial adhesion onto surfaces: the low-surface-energy approach. Biomaterials 1999;20: 1229−1235.

[3-192] Little BJ, Wagner P. Factors influencing the adhesion of microorganism to surfaces. J Adhesion 1986;20:187−210.

[3-193] Wu-Yuan CD, Eganhouse KJ, Keller JC, Walters KS. Oral bacterial attachment to titanium surfaces: a scanning electron microscopy study. J Oral Implant 1995;21:207−213.

[3-194] Steinberg D, Sela MN, Klinger A, Kohavi D. Adhesion of periodontal bacteria to titanium, and titanium alloy powders. Clin Oral Impl Res 1998;9:67−72.

[3-195] Fujioka-Hirai Y, Akagawa Y, Minagi S, Tsuru H, Miyake Y, Sugunaka H. Adherence of *Streptococcus mutans* to implant materials. J Biomed Mater Res 1987;21:913−920.

[3-196] Wolinsky LE, de Camargo PM, Erard JC, Newman MG. A study of *in vitro* attachment of *Streptococcus sanguis* and *Actinomyces viscosus* to saliva-treated titanium. Int J Oral Maxillofac Implants 1989;4:27−31.

[3-197] Gibbons RJ, Van Houte J. Bacterial adherence and the formation of dental plaques. In: Bacteriasl adherence. Beachey EH, editor. New York: Chapman and Hall Co. 1980. pp. 61−104.

[3-198] Leonhardt Å, Olsson J, Dahlén. Bacterial colonization on titanium, hydroxyl-apatite, and amalgam surfaces *in vivo*. J Dent Res 1995;74: 1607−1612.

[3-199] Kohavi D, Klinger A, Steinberg D, Sela MN. Adsorption of salivary proteins onto prosthetic titanium components. J Prosthet Dent 1995;74: 531−534.

[3-200] Rolla G. Formation of dental integuments − some basic chemical considerations. Swed Dent J 1977;1:241−251.

[3-201] Kasemo B. Biocompatibility of titanium implants: surface science aspects. J Prosthet Dent 1983;49:832−837.

[3-202] Kolenbrander PE, London J. Adhere today, here tomorrow: oral bacterial adherence. J Bacteriol 1993;175:3247−3252.

[3-203] Sordyl CM, Simons AM, Molinari JA. The microbial flora associated with stable endosseous implants. J Oral Implant 1995;21:19−22.

[3-204] Balkin BE, Robert TW, Feik D, Molzan AK, Ram TE. Microbiology of human mandibular subperiosteal dental implants in healthy and disease. J Dent Res 2000;79:168 (Abstract No. 195).

[3-205] Al Saleh KA, Simonsson T, Nilner K, Bondsson HE. Plaque formation on titanium, *in vitro* study. J Dent Res 1994;78:374 (Abstract No. 2174).

[3-206] Mombelli A, Buser D, Lang NP. Colonization of osseointegrated titanium implants in edentulous patients. Early results. Oral Microbio Immunol 1988;3:113−120.

[3-207] Quirynen M, Listgarten M. Distribution of bacterial morphotypes around natural teeth and titanium implant of modum Branemark. Clin Oral Impl Res 1990;1:8−13.

[3-208] Apse P, Ellen RP, Overall CM, Zarb GA. Microbiota and crevicular fluid collagenase activity in the osseointegrated dental implant sulcus: a comparison of sites in edentulous and partially edentulous patients. J Periodont Res 1989;24:96−105.

[3-209] Kohavi D, Greenberg R, Raviv E, Sela MN. Sub- and supragingival microbial flora around healthy osseointegrated implants in partially edentulous patients. Int J Oral Maxillofac Implants 1994;9:673−678.

[3-210] Haanaes HR. Implants and infections with special reference to oral bacteria. J Clin Periodontol 1990;17:516−524.

[3-211] Bauman GR, Mills M, Rapley JW, Hallmon WW. Palque-induced inflammation around implants. Int J Oral Maxillofac Implants 1992;7:330−337.

[3-212] Yoshinari M, Oda Y, Kato T, Okuda K. Infleunce of surface modifications to titanium on antibacterial activity *in vitro*. Biomaterials 2001;22: 2043−2048.

[3-213] Lekholm U, Ericsson I, Adell R, Slots J. The condition of the soft tissues at tooth and fixture abutment supporting fixed bridges, a microbiological and histological study. J Clin Periodont 1986;13:558−562.

[3-214] König DP, Perdreau-Remington F, Rütt J, Stoßberger P, Hilgers R-D, Plum G. Slime production of *Staphylococcus epidermidis*; increased bacterial adherence and accumulation onto pure titanium. Acta Orthop Scand 1998;69: 523−526.

[3-215] Drake DR, Paul J, Keller JC. Primary bacterial colonization of implant surfaces. Int J Oral Maxillofac Implants 1999;14:226−232.

[3-216] Chang J-C, Oshida Y, Gregory RL, Andres CJ, Barco TM, Brown DT. Electro-chemical study on microbiology-related corrosion of metallic dental materials. J Biomed Mater Eng 2003;13:281−295.

[3-217] Moodey NR, Costa JE. Microstructure/property relationship in titanium alloys and titanium aluminies. Kim YW, Boyer RR, editors. Warrendale, PA: TMS. 1991. pp. 587−604.

[3-218] Tal-Gutelmacher E, Eliezer D. The hydrogen embrittlement of titanium-based alloys. J Metals 2005;57:46−49.

[3-219] Williams DN, Jaffee RI. Relationships between impact and low-strain-rate hydrogen embrittlement of titanium alloys. J Less-Common Metals 1960:2:42−48.

[3-220] Hardie D, Quyang S. Effect of hydrogen and strain rate upon the ductility of mill-annealed Ti6Al4V. Corr Sci 1999;41:155−177.

[3-221] Solovioff G, Eliezer D, Desch PB, Schwarz RB. Hydrogen-induced cracking in an Al-Al$_3$Ti-Al$_4$C$_3$ alloy. Scripta Metall Mater 1995;33:1315−1320.

[3-222] Froes FH, Eliezer D, Nelson HG. Hydrogen effects in materials. Thompson AW, Moody NR, editors. Warrendale PA: TMS. 1994. pp. 719−733.

[3-223] Ogawa T, Yokoyama K, Asaoka K, Sakai J. Hydrogen absorption behavior of beta titanium alloy in acid fluoride solutions. Biomaterials 2004;25: 2419−2425.

[3-224] Hirth JP. Effects of hydrogen on the properties of iron and steel. Metall Trans A 1980;11A:861−890.

[3-225] Yokoyama K, Kaneko K, Ogwa T, Moriyama K, Asaoka K, Sakai J. Hydrogen embrittlement of work-hardened Ni-Ti alloy in fluoride solutions. Biomaterials 2005;26:101−108.

[3-226] Paton NE, Williams LC. Effect of hydrogen on titanium and its alloys. In: Hydrogen in metals. Columbus, OH: American Society for Metals. 1974. pp. 409−432.

[3-227] Ebtehaj K, Hardie D, Parkins RN. The stress corrosion and pre-exposure embrittlement of titanium in methanolic solutions of hydrochloric acid. Corr Sci 1985;25:415−429.

[3-228] Hollis AC, Scully JC. The stress corrosion cracking and hydrogen embrittlement of titanium in methanol−hydrochloric acid solutions. Corr Sci 1993;34:821−835.

[3-229] Nakahigashi J, Tsuru K, Yoshimura H. Application of hydrogen-treated titanium alloy to dental prosthesis. Experimental manufacture of crown by superplastic forming. J Jpn Soc Dental Mater Devices 2005;24: 291−295.

[3-230] Gerard DA, Koss DA. The combined effect of stress state and grain size on hydrogen embrittlement of titanium. Scripta Metall Mater 1985;19: 1521−1524.

[3-231] Shih DS, Robertson IM, Birnbaum HK. Hydrogen embrittlement of α titanium: *in situ* TEM studies. Acta Metall 1988;36:111−124.

[3-232] Nelson HG. Hydrogen embrittlement of α titanium: *in situ* TEM studies. Met Trans 1973;4:364−367.

[3-233] Young GA, Scully JR. Effects of hydrogen on the mechanical properties of a Ti-Mo-Nb-Al alloy. Scripta Metall Mater 1993;28:507−512.

[3-234] Kolman DG, Scully JR. Effects of the environment on the initiation of crack growth. West Conshohocken, PA: ASTM. 1997. pp. 61−73.

[3-235] Costa JE, Williams JC, Thompson AW. Metall Trans 1987;A23:1421−1430.

[3-236] Nakamura H, Koiwa M. Hydride precipitation in titanium. Acta Metall 1984;32:1799−1807.

[3-237] Wipf H, Kappesser B, Werner R. Hydrogen diffusion in titanium and zirconium hydrides. J Alloys Comp 2000;310:190−195.

[3-238] Tsai MM, Howe JM. Lengthening kinetics of (0110)γ-TiH precipitates in α-Ti in the temperature range of 25°C to 80°C. Metall Trans 1995;A26:2219−2226.

[3-239] Hua F, Mon K, Pasupathi P, Gordon G, Scoesmith D. Modeling the hydrogen-induced cracking of titanium alloys in nuclear waste repository environments. J Metals 2005;57:20−26.

[3-240] Murai T, Ishikawa M, Miwa C. The absorption of hydrogen into titanium under cathodic polarization. Corr Eng 1977;26:177−183.

[3-241] Lane IR, Cavallaro JL, Morton AGS. Sea-water embrittlement of titanium. In: ASTM STP 397, Stress-corrosion cracking of titanium. 1966. pp. 246−258.

[3-242] Yokoyama K, Hamada K, Moriyama K, Asaoka K. Degradation and fracture of Ni-Ti superelastic wire in an oral cavity. Biomaterials 2001;22: 2257−2262.

[3-243] Yokoyama K, Kaneko K, Moriyama K, Asaoka K, Sakai J, Nagumo M. Delayed fracture of Ni-Ti superelastic alloys in acidic and neutral fluoride solutions. J Biomed Mater Res 2004;69A:105−113.

[3-244] Yokoyama K, Kaneko K, Moriyama K, Asaoka K, Sakai J, Nagumo M. Hydrogen embrittlement of Ni-Ti superelastic alloy in fluoride solution. J Biomed Mater Res 2003;65A:182−187.

Chapter 4
Oxidation and Oxides

4.1.	Formation of Titanium Oxides	81
	4.1.1 Air-Formed Oxide	82
	4.1.2 Passivation	84
	4.1.3 Oxidation at Elevated Temperatures	85
4.2.	Influences on Biological Process	86
4.3.	Crystal Structures of Ti Oxides	87
4.4.	Characterization of Oxides	90
4.5.	Unique Applications of Titanium Oxide	91
4.6.	Oxide Growth, Stability, and Breakdown	91
4.7.	Reaction with Hydrogen Peroxide	94
References		97

Chapter 4
Oxidation and Oxides

4.1 FORMATION OF TITANIUM OXIDES

Titanium is a highly reactive metal and will react within microseconds to form an oxide layer when exposed to the atmosphere [4-1]. Although the standard electrode potential was reported in a range from -1.2 to -2.0 V for the Ti $\leftrightarrow$ Ti^{3+} electrode reaction [4-2, 4-3], due to strong chemical affinity to oxygen, it easily produces a compact oxide film, ensuring high corrosion resistance of the metal. This oxide, which is primarily TiO$_2$, forms readily because it has one of the highest heats of reaction known (ΔH = $-$915 kJ/mol) (for 298.16$-$2000$°$ K) [4-4–4-6]. It is also quite impenetrable by oxygen (since the atomic diameter of Ti is 0.29 nm, the primary protecting layer is only about 5–20 atoms thick) [4-7].

The formed oxide layer adheres strongly to the titanium substrate surface. The average single-bond strength of the TiO$_2$ to Ti substrate was reported to be about 300 kcal/mol, while it is 180 kcal/mol for Cr$_2$O$_3$/Cr, 320 kcal/mol for Al$_2$O$_3$/Al and 420 kcal/mol for both Ta$_2$O$_5$/Ta and Nb$_2$O$_5$/Nb [4-8]. Adhesion and adhesive strength of Ti oxide to substrates are controlled by oxidation temperature and thickness of the oxide layer as well as the significant influence of nitrogen on oxidation in air. In addition, adhesion is greater for oxidation in air than in pure oxygen [4-9], suggesting that the influence of nitrogen on the oxidation process is significant.

In addition to these oxidation conditions, it is well known that several parameters can modify the accommodation of the stresses developed during the oxidation of a metal and consequently play an important role in maintaining the protective properties of oxide layers. The results show that the adhesion of the oxide layers to the metal substrate decreases as the layer thickness increases. It is shown that the adhesion of the oxide layers decreases when the oxidation temperature increases, despite the increase in oxide plasticity [4-10]. Adhesive strength between Ti substrate and TiO$_2$ is also related to the thermal mismatching when the Ti/TiO$_2$ couple was subjected to the elevated temperatures. Although TiO$_2$ does not exhibit any phase transformation up to its melting point (i.e., 1885$°$C), its substrate Ti metal has an allotropic phase transformation at 885$°$C (below which it is HCP – hexagonal closed-packed crystalline structure – and above which it becomes BCC – body-centered cubic crystalline structure). Accordingly, when Ti substrate is exposed beyond this phase transformation temperature, the bond strength between Ti and TiO$_2$ will be weakened due to the significant differences in coefficients of thermal expansion, particularly during the cooling stage passing through the 885$°$C transus temperature.

The performance of titanium and its alloys in surgical implant applications can be evaluated with respect to their biocompatibility and capability to withstand the corrosive species involved in fluids within the human body [4-11]. This may be considered as an electrolyte in an electrochemical reaction. It is well documented that the excellent corrosion resistance of titanium materials is due to the formation of a dense, protective, and strongly adhered film – which is called a passive film. Such a surface situation is referred to a passivity or a passivation state. The exact composition and structure of the passive film covering titanium and its alloys is controversial. This is the case not only for the "natural" air oxide, but also for films formed during exposure to various solutions, as well as those formed anodically. The "natural" oxide film on titanium ranges in thickness from 2 to 7 nm, depending on such parameters as the composition of the metal and surrounding medium, the maximum temperature reached during the working of the metal, the surface finish, etc.

Passivity is a property of a metal commonly defined in two ways. One of these is based on a change in the electrochemical behavior of the metal, and the other on its corrosion behavior [4-12]. Passivity is an unusual phenomenon observed during the corrosion of certain metals, indicating a loss of chemical or electrochemical reactivity under certain environmental conditions [4-13]. Such passivity appears on certain the so-called passivating metals, many of which are "transition metals", characterized by an unfilled group of electrons in an inner electron shell. Passivating metals are, for example, Fe (covered with Fe_2O_3 and/or Fe_3O_4), Cr (with Cr_2O_3), Zr (with ZrO_2), and Ti (with TiO_2) [4-14]. This type of passivity is ascribed to an invisibly thin but dense and semiconducting oxide film on the metal substrate surface, displacing the electrode potential of the metal strongly in the positive (or more noble) direction (e.g., Ti passive film exhibits its stable passivity up to 1.5–2.0 V). Hence, comparing the electrode potential (-1.2 to -1.5 V) to this stable passivity potential, surface nobility was remarkably improved.

Using the electron diffraction techniques (TED: transmission electron diffraction method on stripped films, and RED: reflection electron diffraction method on bulk specimens with formed oxide film) and other surface microanalytical techniques, composition and structure of the oxide film that forms on titanium under various conditions were investigated. The findings are summarized as follows.

4.1.1 Air-formed oxide

When fresh titanium is exposed to the atmosphere by such cutting acts as lathing, milling, or sawing, an oxide layer begins to form within nanoseconds [4-7]. After only 1 s, a surface oxide (with 2–7 nm in thickness) will be formed. Air oxidation at room temperature produced titanium monoxide (TiO) with small quantities of titanium oxide, Ti_3O_5. Titanium has proved to be a highly successful material for implants inserted into human bone. Under certain conditions, titanium will establish and maintain a direct contact with the bone tissue in a process called osseointegration. The mechanisms underlying this behavior are not yet understood. However, it is clear that the properties of the implant surface are of vital importance for a successful osseointegration. The properties of titanium implant surfaces are determined by the thin (20–70 Å) oxide film that covers the metal. Thus, the biocompatibility of titanium implants is associated with the surface titanium oxide and not with the bulk titanium metal [4-1, 4-15–4-17]. The surface oxide is also formed during the implant preparation procedures. During the machining of the implants, pure titanium metal is exposed to air and is rapidly oxidized. In the following cleaning and autoclaving procedures, oxide film is then modified and grows. In order to obtain an oxide film that is reproducible with respect to the chemical composition, structure and thickness, it is important that the different steps of the implant preparation are performed under carefully controlled conditions. Even small changes in the preparation procedures might lead to considerable changes of the implant surface [4-18].

As mentioned previously, because the surface air-formed oxide is nearly impenetrable, once this thin passivating film has formed, oxygen is prevented from reaching the metal beneath and further oxide layer thickening is quickly halted because the oxide film is dense and semiconductive (not like an electron-conductive metal substrate). The compositional structure and exact thickness of the passivating oxide layer is dependent on many factors associated with its formation. These include such factors as type of machining, surface roughness, coolants used during machining, and sterilization procedures. It should be no surprise that reported biocompatibility of devices is prepared by different investigators, and therefore may vary to a significant degree [4-7].

It is well recognized that the excellent tissue–bone compatibility of titanium is mainly due to the properties of its stable surface oxide layer. Biocompatibility of implant materials relies on the chemical and electrochemical stability of this surface oxide layer, which interfaces with the soft and hard tissue and bone structure. Oxide layers formed on titanium change from lower to higher oxides as oxidation progresses and temperature increases. The following phases form in air: $Ti+O \rightarrow Ti(O) \rightarrow Ti_6O \rightarrow Ti_3O \rightarrow Ti_2O \rightarrow TiO \rightarrow Ti_2O_3 \rightarrow Ti_3O_5 \rightarrow TiO_2$. Among these oxides of different stoichiometry (e.g., TiO, Ti_2O_3, TiO_2), TiO_2 is the most common and stable thermodynamically. TiO_2 can have three different crystal structures – rutile, anatase, and brookite – but also can be amorphous. TiO_2 is very resistant against chemical attack, which makes titanium one of the most corrosion resistant metals.

Another physical property that is unique to TiO_2 is its high dielectric constant, which ranges from 14 to 110 (in 10^6 cycles; dielectric strength: 350 V/mil, volume resistivity: 10^{14}–10^{16} Ω/cm) depending on crystal structure [4-19]. This high dielectric constant would result in considerably stronger van der Waal's bonds on

TiO_2 than on other oxides. TiO_2, like many other transition metal oxides, is catalytically active for a number of inorganic and organic chemical reactions, which also may influence the interface chemistry [4-20].

4.1.2 Passivation

Passive films can be formed by either chemically or electrochemically (or anodic) treating titanium surfaces. Fraker and Ruff [4-21] found that more rigid oxidizing conditions produce higher oxides of titanium alloy (thin films) in saline water (3.5% NaCl), at a temperature range between 100° and 200°C, using a transmission electron microscopy. They found that oxides ranged from Ti_2O to TiO_2 (anatase), with the higher oxides corresponding to the higher temperatures. Concerning oxide films formed anodically on titanium, most investigators agree that the film consists of TiO_2. All three crystalline forms of TiO_2 (rutile, anatase, and brookite) have been reported, as will be discussed later. It has also been reported that the anodic film is either oxygen-excess or oxygen-deficient. The controversy is widened by reports that the film is hydrated or contains mixed oxides. In addition to variations in oxygen stoichiometry, the films may contain various amounts of elements other than titanium and oxygen, or incorporate such additional element(s) into TiO_2 structure. Thus, it is most probable that the anodic film is not necessarily stoichiometric TiO_2, and that the films studied by different authors had different compositions due to differences in the conditions of oxidation [4-21].

When titanium alloys (Ti-6Al-7Nb and Ti-6Al-4V) were chemically treated, surface oxide films appear to be more complicated. Sittig *et al.* [4-22] treated these alloys along with CpTi in nitric acid hydrogen fluoride and identified formed oxides. Three types of oxides (TiO_2, Ti_2O_3, and TiO) were identified on the CpTi substrate, and it was found that the dissolution rate depends on grain orientations. On the other hand, for Ti-6Al-7Nb and Ti-6Al-4V alloys, Al_2O_3, Nb_2O_5 or V-oxides such as V_2O_5, were formed in addition to TiO_2 oxide, and it was found that the selective α-phase dissolution and enrichment of the β-phase appears to occur [4-22].

Studies on the nature and properties of thin films formed by the electrolytic oxidation on titanium, which, in turn, determine the behavior of the electrode in the open circuit, indicate the duplex nature of these films [4-23]. In both the acid and alkaline media, the outer layer of the anodic oxide film was found to be more susceptible to dissolution than the inner layer, thereby indicating a more defective structure of the outer layer. The surface reactivity was found to be higher in oxygen-saturated vs. nitrogen-saturated solutions. Galvanostatic anodization of titanium has been studied in N_2-deaerated, 1 N solution of NaOH, Na_2SO_4, H_2SO_4, HCl, $HClO_4$, HNO_3, and H_2PO_4 at 25°C at current densities ranging from 4 to 50×10^6 A/cm^2. From formation rate values, the following arrangement of passivation was obtained: (from the highest rate to the lowest rate) $NaOH > Na_2SO_4 > HClO_4 > NHO_3 > HCl > H_3PO_4 > H_2SO_4$ [4-23].

Metikoš-Huković *et al.* [4-24] employed structurally sensitive *in situ* methods, such as photopolarization and impedance, to examine the passivation process and the properties of the protective oxide layers on titanium. The thickness of anodic oxidation and the non-stoichiometry of the surface oxide were correlated. It was mentioned that the composition of the anodic film on titanium changes with the relative potential from lower to higher oxidation stages according to the equation: $TiH_2 + TiO \rightarrow nTi_2O_3 \cdot nTiO_2 \rightarrow Ti_5O_9$ or $Ti_6O_{11} \rightarrow Ti_3O_5 \rightarrow TiO_2$. The characteristic behavior of titanium can easily be seen at higher anodic potential ($\sim +1.5$ V vs. SCE in 5 mol H_2SO_4) when the electrode is covered with a nearly stoichiometric TiO_2 layer. The semiconducting properties of TiO_2 were investigated using an anodic film stabilized at $+2$ V (vs. SCE: saturated calomel electrode with 0.2444 V vs. NHE), and it was found that TiO_2, like the lower titanium oxides, is an *n*-type (metal-excess) semiconductor under anodic polarization [4-24].

Repassivation capability [4-25] associated with titanium substrate can be considered as another reason for the high corrosion resistance and biocompatibility of titanium materials. Although titanium owes its high corrosion resistance to the presence of an extremely thin oxide, when the thin film is disrupted mechanically, underlying metal substrate will react quickly with the surroundings to reform the oxide film. The mechanical and electrochemical parameters associated with the fracture and repassivation of titanium oxide films were studied [4-26]. The scanning electrochemical microscope has been modified to evaluate the high-speed mechanical disruption of the oxide film of titanium and the associated repassivation process. A mechanically polished Ti-6Al-4V surface was immersed and potentiostatically held in phosphate-buffered saline. An 18-μm radius diamond stylus was used to apply 30–50 μm scratches with a controlled load and the resultant current transient was measured. It was found that (i) current transients were detected at the lowest loads applied (0.008 N), and (ii) topographic images showed that probe loads as low as 0.1 N caused permanent deformation in the surface of about 0.1 μm, and below 0.05 N no deformation could be seen [4-26]. Such capability of repassivation is sometimes called a self-healing ability.

4.1.3 Oxidation at elevated temperatures

Above 100°C, the surface is a lower oxide type, such as TiO. At elevated temperatures (above 200°C), complex oxides are formed, such as Al_2TiO_5 on Ti-6Al-4V and $NiTiO_3$ on TiNi, in addition to rutile-type TiO_2 [4-15, 4-16]. Oxidation in air at 875–1050°C produced a flaky scale of TiO, Ti_2O_3, and rutile (TiO_2). It was reported that the high oxide TiO_2 appears on the surface of titanium in the presence of the most rigid conditions of oxidation; in less rigid conditions a lower oxide ($3Ti_2O_3 \cdot 4TiO_2$) forms, while in still weaker oxidizing conditions, the formation of even lower oxides (e.g., TiO) are possible [4-9].

The characteristics of high-temperature oxidation of Ti are the main obstacle to strong Ti-ceramic bonding [4-27]. A thin oxide layer (~32 nm) with good adherence to the substrate was formed when CpTi was oxidized at 750°C in 0.1 atm, and a thick layer (~1 μm) with poor adherence was formed when oxidized at 1000°C [4-28]. In another study [4-29], the thickness of TiO_2 on the CpTi surface increased with an increase in oxidation temperatures. Ti surface hardness also increases substantially after oxidation at and above 900°C [4-28]. Therefore, the substrate is now in the BCC structure range. Lower Ti-ceramic bond strength was attributed to a thick TiO_2 zone on the metal surface when Ti was oxidized at higher temperatures [4-30]. Enhanced bond strength between titanium and ceramic was reported when porcelain was fired on cast Ti in a reduced argon atmosphere [4-31]. Modifying Ti surfaces to control high-temperature oxidation has been examined. Published studies show that Ti surface nitridation [4-32] or a thin Cr coating [4-33] is effective methods in limiting Ti oxidation at high temperatures. Improved Ti-ceramic bonding was found when a thin layer of Si_3N_4 was deposited on Ti surface [4-34].

4.2. INFLUENCES ON BIOLOGICAL PROCESS

Sundgren *et al.* [4-35] and McQueen *et al.* [4-36], using Auger electron spectroscopy (AES) to study the change in the composition of the titanium surface during implantation in human bone, observed that the oxide formed on titanium implants grows and takes up minerals during the implantation. The growth and uptake occur even though the adsorbed layer of protein is present on the oxide, indicating that mineral ions pass through the adsorbed protein. Liedberg *et al.* [4-37], using Fourier transform infraRed reflection absorption spectroscopy (FTIR-RAS), showed that phosphate ions are adsorbed by the titanium surface after the protein has been adsorbed. Using X-ray photoelectron spectroscopy (XPS), Hanawa [4-38] demonstrated that oxides on titanium and titanium alloy (Ti-6Al-4V) change into complex phosphates of titanium and calcium containing hydroxyl groups which bind water on immersion in artificial saliva (pH 5.2). All these studies [4-35–4-38] indicate that the surface oxide on titanium reacts with mineral ions, water, and other constituents of biofluids, and that these reactions in turn cause a remodeling of the surface.

It was shown that titanium is in almost direct contact to bone tissue, separated only by an extremely thin cell-free non-calcified tissue layer. Transmission electron microscopy revealed an interfacial hierarchy that consisted of a 20–40 nm thick proteoglycan layer within 4 nm of the titanium oxide, followed by collagen bundles as close as 100 nm and Ca deposits within 5 nm of the surface [4-39]. To reach the steady-state interface described, both the oxide on titanium and the adjacent tissue undergo various reactions. The physiochemical properties of titanium have been associated with the unique tissue response to the materials: these include the biochemistry of released corrosion products, kinetics of release and the oxide stoichiometry, crystal defect density, thickness and surface chemistry [4-40].

While the conditions to produce an optimal oxide formation need to be better understood, in general, the titanium passivating layer not only produces good corrosion resistance, but it seems also to allow physiological fluids, proteins, and hard and soft tissue to come very close and/or deposit on it directly. Reasons for this are still largely unknown, but it may have something to do with things such as the high dielectric constant for TiO_2 (50–170 vs. 4–10 for alumina and dental porcelain) which should result in considerably stronger van der Waal's bonds on TiO_2 than other oxides; TiO_2 may be catalytically active for a number of organic and inorganic chemical interactions influencing biological processes at the implant interface [4-20]. The TiO_2 oxide film may permit a compatible layer of biomolecule to attach [4-41, 4-42].

During implantation, titanium releases corrosion products into the surrounding tissue and fluids even though it is covered by a thermodynamically stable oxide film [4-43–4-45]. An increase in oxide thickness, as well as incorporation of elements from the extracellular fluid (P, Ca, and S) into the oxide, has been observed as a function of implantation time [4-35]. Moreover, changes in the oxide stoichiometry, composition, and thickness have been associated with the release of titanium corrosion products *in vitro* [4-46]. Properties of the oxide, such as stoichiometry, defect density, crystal structure and orientation, surface defects, and impurities were suggested as factors determining biological performance [4-17, 4-21, 4-47].

4.3. CRYSTAL STRUCTURES OF TI OXIDES

It should be clear that the proven high biocompatibility of titanium as an implant materiasl is connected with the properties of its surface oxide. In air or water Ti quickly forms an oxide thickness of 2–7 nm at room temperature. TiO_2 possesses three crystalline structures: anatase, rutile, and brookite. Anatase-type TiO_2 is a tetragonal crystalline system with $a_0 = 3.78$ Å and $c_0 = 9.50$ Å; the rutile-type TiO_2 is also a tetragonal structure, but the lattice constants are quite different from those of anatase-type (i.e., $a_0=4.58$ Å and $c_0=2.98$ Å). The third type is brookite-type, and has an orthorhombic crystalline structure with $a_0=9.17$ Å, $b_0=5.43$ Å, and $c_0=5.13$ Å [4-48]. Among these oxides, rutile is known to be the most stable phase [4-49].

Identification of crystalline structures of formed oxides can be achieved by several ways. TFXRD (thin-film X-ray diffraction) or XRD if the oxide film is reasonably thick enough, so that diffracted intensities can be obtained which are high enough to be identified and differentiated from each other and background noise. The alternative way will be diffraction using accelerated electron. If an appropriate sample holding device can be manufactured and installed inside the transmission electron microscopy, then the oxidized sample with surface oxide can be tilted almost parallel to the accelerated electron beam, resulting in a half ring of RED (reflection electron diffraction) [4-50–4-52], from which one can identify the crystalline structure(s) of the oxide(s). Another way for crystalline identification of formed oxides is based on stripping (or isolating) formed oxide from the substrate by chemically etching the backside (the interface side of formed oxide and substrate) to be isolated from the metallic substrate. Normally for stainless steel, 10% bromine in anhydrated methanol solution is used [4-52, 4-53], and for titanium materials, a mixed acid of HNO_3 and HF is used [4-54] or 10% bromine in ethyl acetate [4-49]. The stripped thin oxide film is carried on a copper mesh and TED can be performed [4-54, 4-55].

For the diffraction method, the Bragg's law [4-56, 4-57] is fully used to calculate the inter-distance between diffracted planes, d, in the equation: $n\lambda=2d \sin \theta$ (where n is the diffraction order and normally equal to 1, λ the wavelength of used diffraction beam, and θ a diffraction Bragg's angle). For the X-ray diffraction, the wavelength λ is ranged roughly from 1 to 2 Å (e.g., $\lambda_{Cu}=1.54$ Å, $\lambda_{Mo}=0.71$ Å, $\lambda_{Fe}=1.94$ Å), so that 2 θ reaches its limitation of 180° with $d=0.5$ Å. On the other hand, for electron diffraction, λ can be approximated by $\lambda=(150/V)^{1/2}$, where V is an accelerated voltage. If $V=50$ kV, $\lambda \approx 0.055$ Å, and if $V=100$ kV, $\lambda \approx 0.038$ Å. Accordingly, with such an accelerated electron, 2θ can be 5×10^{-2} radius $\approx 3°$ for a crystalline structure with $d=0.5$ Å. Therefore, 2θ is so small that $n\lambda=2d \sin \theta$ can be simplified as $\lambda=2d\theta$ (if $n=1$, as normal). Because θ is small, it can be substituted by r (which is a radius of obtained diffraction ring). Furthermore, if λ is kept constant (which is usually the case), we have $\lambda=2dr=$constant. For most studies (e.g., [4-53–4-55]), a thin fully annealed gold foil (Au) is used as a standard sample, with which $2dr=$constant can be further developed to $2d^{Au} r^{Au} =$ constant $= 2d^{sample} r^{sample}$. Once all diffracted d-spacings are determined, the crystalline structure can be identified by referring to the XRD Data File [4-58].

Information on titanium oxides of anatase-type, rutile-type, brookite-type, or a mixture of these, appears to be scattered. It is useful here to review and summarize these data to determine some common findings. The plasma-assisted chemical vapor deposition process appears to form *anatase*-type oxide [4-59, 4-60].

By dry oxidation, the oxides are identified as *rutile*-type oxide, oxidized at 300°C for 0.5 h [4-15, 4-16], 400°C×0.75 h [4-61], 500°C×3 h [4-62], 750°C [4-50], and 875–1050°C [4-9].

On the other hand, crystalline structures of oxides formed through wet oxidation are varied. *Only rutile* type of TiO_2 was identified by boiling in 5 wt.% H_2SO_4 or 10 wt.% HCl [4-55], anodizing with 0.5 M H_2SO_4 at 5–10 V [4-63], boiling in 10% H_2SO_4 [4-7], boiling in 10% HCl [4-64], and treating in 0.5–1 M HCl [4-65]. A mixture of rutile- and anatase-type oxide was found when CpTi was treated in $HF/HNO_3/H_2O$, followed by heating in 5 mol NaOH at 75°C [4-55], followed by air-oxidation at 600°C×1 h [4-55], or boiling in 0.1 wt.% H_2SO_4 [4-59]. On the other hand, mainly anatase, with a small fraction of rutile-type mixture, was identified by anodizing in 40% H_2SO_4 at 8 V [4-22], and boiling in 10% CrO_3 or 65% HNO_3×3 h [4-22]. When CpTi was treated in alkaline or mildly acid solution, formed oxides were identified to be *solely anatase*-type. It can be formed by treating CpTi in an electrolyte containing $Ca(H_2PO_4)_2$, $Ca(COCH_3)_2$ and Na(EDTA) (pH 14) [4-60], 1–10% H_2O_2, followed by in-air oxidation at 500°C×3 h [4-62], anodizing in HNO_3 solution [4-64], 0.1 M Na_2CO_3 or 0.01 M HCl [4-65], anodizing in 0.1 M H_2SO_4 at 12.5 mA/cm² [4-66], anodizing in 0.1 M H_2SO_4 at 5 V [4-67], or anodizing in 1 M H_2SO_4 at 155 V [4-68]. It was also reported that when CpTi was boiled in 0.2 wt% HCl for 24 h, a *mixture of anatase and brookite* was formed [4-69].

Summarizing the aforementioned information, as seen in Table 4.1, rutile-type oxides are favored when the CpTi surface is exposed to in-air oxidation at elevated temperature. Processes involved in plasma spraying or vapor deposition make CpTi surfaces covered with anatase-type oxide. In wet oxidation (by either simply boiling or electrochemically anodizing), if the electrolyte (or solution) is a strong acid, the rutile-type oxide is favorable, whereas if the treating solution is mild or alkaline, anatase becomes dominant, with a small fraction of rutile type.

Table 4.1. Summary of crystalline structures of titanium oxide.

Condition	Type of crystalline structure of formed titanium oxide		
Physical depositon			**Anatase**
Dry oxidation	**Rutile**		
Wet oxidation	**Rutile**	**R + A**	**Anatase**
Solution pH	Low pH (acid)	$\longrightarrow$	High pH (alkaline)

4.4. CHARACTERIZATION OF OXIDES

Thermally and electrochemically formed oxides on Ti-6Al-4V were investigated by XPS, scanning AES, and SIMS, including depth profiling. The oxide layers on the alloy are predominantly TiO_2, but have considerable concentration of the alloying elements included in the oxide. Al, but not V, is observed in the outermost atomic layers of the oxide. Both Al and V are present at relatively high maximum atomic concentration (Al/Ti ~0.17 and V/Ti ~0.07) inside anodic oxides. The V concentration varies laterally over the surface, reflecting the variation of the V concentration in the underlying metal due to its two phases [4-70].

The *in situ* Raman spectra of electrochemically anodized (in H_2SO_4 and HNO_3) formed oxide (rutile) films on CpTi electrodes were measured. It was found that (i) at the potential of 10 V, the spectra showed clear evidence for the formation of rutile, and (ii) anatase spectra were found after emersion and drying of the oxide film and were also observed when the oxide film was formed in HNO_3 [4-64]. The chemical compositions of the anodic oxoide films formed on CpTi (grade 1) during electropolishing (in methanol + *n*-butanol + perchloric acid) and anodic oxidation (in H_2SO_4, H_3PO_4 or acetic acid) were investigated by microanalytical means. It was found that the formed oxide is mainly TiO_2. However, chemical composition can be modified by anion adsorption and/or incorporation when H_2SO_4 or H_3PO_4 electrolytes are used. Modification of the anodic film composition also occurs during sterilization. Increased Ca and H levels are also observed after autoclaving. It was also found that the oxide thickness depends linearly on the anodizing potential in the range 5–80 V [4-71]. Electrochemical *in situ* ellipsometry with a focused laser beam (microellipsometry) is used to determine properties of oxide layers on single grains of Ti. The data differ from grain to grain and also a strong dependence on the sample rotation around the surface normal occurs due to the anisotropy of the Ti/TiO_2 system [4-72].

Dolata *et al.* [4-74] developed a new technique to evaluate the equivalent circuit elements of a semiconductor–electrolyte interface, based on impedance measurements over a wide frequency range [4-73]. The EIS tests were carried out over a wide frequency range (from 0.03 Hz to 10 kHz), analyzing the whole spectrum. It was found that the high frequency response was mainly determined by semiconductor properties of the anatase and rutile layers, whereas the low frequency one reflects particular morphology of the films [4-74].

For understanding the complex interfacial phenomena between Ti and a biological system, Hanawa [4-38] prepared CpTi, Ti-6Al-4V, NiTi, 316L stainless steel, Co-30Cr-5Ni, Ni-20Cr, Au-9Cu-6Ag, Ag-20Pd-15Cu-12Au which were polished and immersed in Hank's electrolyte (pH 7.4) at 37°C for 1 h, 1 day, and

30 days. Such sample surfaces were microanalyzed. It was found that (i) calcium phosphate layers are formed on the passive films of Ti and its alloys, stainless steel, Co-Cr, Ni-Cr alloy, in a neutral electrolyte solution. The calcium phosphate formed on Ti, more so than that formed on the other alloys, is similar to apatite. In all these materials, the formation of calcium phosphates is related to the bio-compatibility of the material, while the passive oxide films are related to the cor-rosion resistance [4-38].

4.5. UNIQUE APPLICATIONS OF TITANIUM OXIDE

As an interesting application of TiO_2 (titania), we can find TiO_2 as an oil paint pig-ment additive (known as titania white). Such white permanent pigment provides good covering power in paints to improve the durability of paints. Titania white pigment is also added to toothpaste to make it whiter, and plastics to improve the durability. Paints made with titania are also excellent reflectors of infrared radia-tion, and are therefore used extensively by astronomers and in exterior paints. It is also used in cement, in gemstones, and as a strengthening filler in paper [4-20]. Since a fine powder of TiO_2 crystals can scatter ultraviolet rays, fine TiO_2 powder is mixed into the ultraviolet prevention cosmetics (sun-screen agents). It is used as an ineffective ingredient in over-the-counter-top cold medicine. TiO_2 powder is also added to cosmetics to brighten and intensify the color of make-up, and pierc-ings and jewelry. Recently, it was found that TiO_2 possesses a photo-catalytic func-tion. A photo-catalyst is a material which absorbs light to bring it to higher energy level, and provides such energy to a reacting substance to make a chemical reac-tion occur. When the photo-catalyst absorbs the light, positive holes are created in the valence electron band to react with electrons. Such positive holes in TiO_2 react with water or dissolved oxygen to generate OH radicals, which decompose toxic substances. The OH radical is a stronger oxidant than chlorine or ozone for steril-ization or disinfection purposes [4-75].

4.6. OXIDE GROWTH, STABILITY, AND BREAKDOWN

The oxide layer forms spontaneously in air, and normally has a thickness of between 2 and 7 nm, as mentioned previously. During machining, the relatively high surface temperature can produce a much thicker oxide layer [4-76]. There is much evidence suggesting this layer grows steadily *in vivo*. The stoichiometry of the oxide is similar to titanium oxide (TiO_2) at the surface, and changes to a mix-ture of oxides at the metal/oxide interface [4-1, 4-77]. TiO_2 is an *n*-type (metal-excess or donor type) semiconductor, while TiO and Ti_2O_3 are listed as amphoteric conductors [4-78]. The surface is also known to have at least two types of hydroxyl groups attached to it [4-40]. TiO_2 is non-conducting, but electrons can tunnel through the layer. Thin oxide layers can allow the passage of electrons, leading to conformational changes and denaturing of proteins [4-79].

For implants located in cortical bone, the thickness of the interfacial oxide layer remains unaffected, while it increases by a factor of 3–4 on samples located in bone marrow [4-35]. In general, when foreign agents, such as implant-material surface particles, are exposed to host tissue, circulating neutrophils and/or monocytes are recruited from the intravascular compartment to the location of the exposure [4-80]. Following recognition, the particles are encapsulated – literally engulfed by the phagocyte – and lysosomal granules, along with the particles, from a complex unit, the phagolysosome. Simultaneously, in the cytplasmic vacuoles, enzyme releases occurs and, as a result, degrading of several components takes place. Upon recognition, neutrophils and monocytes experience a "respiratory burst" and during the period, almost 20-fold oxygen consumption by the cells is observed [4-81]. There is convincing evidence that the consequent increase in oxygen secretion by these cells is mainly the result of this initial respiratory activity [4-82]. In addition, polymorphonuclear cells have been found to secrete superoxide anion (O_2^-) and hydrogen peroxide (H_2O_2) upon activation induced by several stimuli, including immunoglobulins and opsonized bacteria, among others [4-83].   .

The interactions between solid surfaces and biological systems relatively unexplored [4-84]. The combination of electrochemical methods with surface analytical techniques offers an insight into the composition, thickness and structure of the surface oxide layer. All reports using X-ray XPS techniques [4-41, 4-85–4-91] confirmed that the oxide layer on top of Ti-6Al-4V was predominantly TiO_2, which contains a small amount of suboxides TiO and Ti_2O_3 closer to the metal/oxide interface. The presence of Al and V in the oxide layer was also noted. Sodhi *et al.* [4-90] reported that Al and V were present throughout the oxide layer; and Al in the passivated layer was greatly enhanced (26 wt.%), while V was reduced (1 wt.%). Ask *et al.* [4-70] observed Al but not V at the outermost surface, and within the oxide, both alloying elements were enriched. Sundarajan *et al.* [4-86] observed oxidized Al, but not oxidized V in the layer. On the other hand, Okazaki *et al.* [4-87] observed a small amount of oxidized V. Evidently the presence of V is relatively ambiguous. Moreover, Maeusli *et al.* [4-91] did not observe V at the outermost surface of oxide by XPS or AES, although they detected it by SIMS. In all reports, the oxidation state of Al was Al^{3+} (i.e., Al_2O_3), whereas that of V was reported between V^{3+} and V^{5+} (i.e., V_2O_3 and V_2O_5) [4-85]. Al and V might be present either as Al_2O_3 and V_2O_5, respectively, or as ions at interstitial or substitutional sites in the TiO_2 matrix [4-70].

O'Brien *et al.* [4-92] passivated nitinol (NiTi) wires and vascular stent components in 10% nitric acid solution, and evaluated electrochemical corrosion behaviors. It was reported that (i) potentiodynamic polarization tests demonstrated a significant increase in breakdown potential for passivated samples (600–700 V vs. SCE), compared to heat-treated surfaces (250–300 V vs. SCE), (ii) surface analysis indicated that the passivation reduces Ni and NiO content in the oxide and increases TiO_2 content, (iii) long-term immersion tests demonstrated that Ni release from the surface of the material decreases with time and the quantity of Ni released is lower for passivated samples, and (iv) the improved corrosion resistance is maintained after prolonged immersion in 37 V 0.9% NaCl solution [4-92].

Pure elements of Ti, Nb, V, Al (with 99.9% purity), Ti-6Al-4V and Ti-6Al-6Nb were tested in 37°C aerated simulated physiological Hank's solution at pH of 6.9. It was found that (i) the excellent passivating properties of the anodically formed oxide film on CpTi (grade 4), and high corrosion resistance of the Ti-6Al-6Nb alloy have been attributed to the stabilizing effect of Nb^{5+} cations on the passive film, by annihilation of stoichiometric defects (anion vacancies) caused by the presence of titanium suboxides, (ii) localized corrosion sensitivity of the Ti-6Al-4V alloy has been correlated to the dissolution of vanadium at the surface film/electrolyte interface coupled with generation of cation vacancies and their diffusion through the film as a part of the solid-state diffusion process, and (iii) the presence of a high concentration of chloride ions (0.15 g/l) in electrolyte further accelerates these processes [4-93].

Ohtsuka *et al.* [4-94] investigated cathodic reduction behavior of the anodic oxide film on Ti using *in situ* ellipsometry combined with electrochemistry. It was reported that (i) in acidic sulfate solution, the anodic oxide film reductively dissolved into the solution as Ti^{3+}, resulting in a reduction in thickness without any significant change in the optical properties of the remaining film, while (ii) in neutral phosphate solution, the anodic oxide film absorbs hydrogen in the hydrogen evolution potential region, resulting in a change in the optical properties without thickness reduction [4-94].

The equilibrium potentials of film formation are given as a function of the activities of the components of the film substance. Its solubility product with respect to the ions in the electrolyte depends on the electrode potential, since the states of oxidation of the metal in the film and in the electrolyte are often different. Heusler discussed [4-95] three cases for the kinetics of uniform film formation and dissolution, (1) equilibrium of all components across both interfaces, (2) partial equilibrium of one component at the outer film/electrolyte interface, and (3) irreversible ion transfer at the outer interface. It was found that (i) oxide films formed on Fe, Ti and Al indicate the specific adsorption of anions at passivating oxide films, (ii) the processes during the incubation time of pitting corrosion are shown to correspond to non-uniform dissolution and formation of the passivating film, and (iii) during the incubation time, electrochemical noise due to chloride ions at passive Fe can be measured [4-95]. Pit generation on Ti with anodic oxide film, which is formed potentiostatistically in 0.5 M H_2SO_4 solution, has been studied in 1 M NaBr solution. It was reported that (i) the distribution of pitting potential obeys the normal probability distribution and pitting potential exhibits a maximum around a formation potential of 6.0 V, and (ii) the pit generation process occurs by the parallel birth and death stochastic model, and depends on film formation potential, i.e., the film thickness and properties [4-96].

Sul *et al.* [4-97] studied electrochemical dissolution of anodic oxide films on Ti in H_2SO_4 by the use of an optical-electrolytic cell adapted for *in situ* ellipsometric and electrochemical measurements. Films were grown on electropolished Ti surfaces by anodic oxidation in 0.5 mol dm^{-3} H_2SO_4 for 30 s in the potential range from 0 to 100 V. Electrochemical dissolution of these films was performed in the potential range between −0.4 and −0.9 V vs. SCE. The potential at which electrochemical dissolution is fastest and homogeneous is determined. In addition it was shown that for potentials more cathodic than −0.8 V, TiH_2 is formed on the electrode surfaces if polarization lasts sufficiently long [4-97].

The surface properties of anodic oxides formed on CpTi screw implants as well as air-formed natural oxides on turned CpTi implants were investigated. Anodic oxides were prepared in CH_3COOH to the high-forming voltage of dielectric breakdown and spark formation. It was reported that (i) the oxide thicknesses were in the range of about 200–1000 nm, and (ii) the crystal structures of the Ti oxide was amorphous, anatase, and a mixture of anatase and rutile [4-98].

4.7. REACTION WITH HYDROGEN PEROXIDE

The insertion of an implant is inevitably associated with an inflammatory response due to the surgical trauma. Whether this reaction will subdue or persist may very well be dependent on the material selected, as well as the site of implantation and the loads put on it [4-99]. It is most probable that the interaction between material and inflammatory cells, either directly or via mediators of cell activation and migration such as complement factors [4-100], is of main importance for the biocompatibility of an implanted material. Further, experimental studies suggest that oxygen-free radicals generated in large amounts during the respiratory burst by inflammatory cells are important for the tissue damage and persistence of the inflammatory reaction [4-101]. Further, it has been observed that an oxide-like layer grows on titanium as well as stainless-steel implants *in vivo* during the implantation [4-35, 4-102]. Model experiments have indicated that hydrogen peroxide in buffered saline solution in the millimolar concentration range produces an oxide-like layer on titanium surfaces *in vitro* [4-103], and at higher H_2O_2 concentrations several oxidation intermediates are distinguished in the Ti-H_2O_2 system. It is well known that hydrogen peroxide is the one of the strongest oxidizing agents. These observations lead one to believe that *in vivo* the interaction of H_2O_2 and oxygen with the titanium would be the cause of the oxide growth observed [4-83]. It is well known that Ti^{4+} ions will not produce hydroxyl radicals from H_2O_2; however, in the presence of the superoxide radical (O_2^-), both H_2O_2 and OH· may be produced through the action of superoxide dismutase and a cyclic Fenton reaction [4-104–4-107]. Such superoxide radical (O_2^-) generated during the metablic activation is (in the body) enzymatically dismuted by hydrogen peroxide through the reaction: $O_2^- + O_2^- + 2H^+ \rightarrow H_2O_2 + O_2$. Superoxide or hydrogen peroxide alone is not potent enough in degrading adducts and, therefore, several possible mechanisms, starting from the reaction above have been suggested [4-107]. As a removal mechanism of bacteria inside the body, neutrophil activates such superoxidants of the cells to generate hydrochlorous acid, which possesses the strongest antibacterial agent among various chlorine compounds. The most interesting is the Fenton type of reaction: $M^{n+} + H_2O_2 \rightarrow M^{(n+1)+} + OH· + OH^-$, where the potent hydroxyl radical, OH·is formed. In the presence of O_2^- , the oxidized metal ion, $M^{(n+1)+}$ is reduced again (to M^{n+}), and a cyclic continuous production of OH· would occur [4-108].

Hence, hydroxyl radicals formed from hydrogen peroxide during the inflammatory response are potent agents for cellular deterioration. The behavior of implanted material in terms of its ability to sustain or stop free radical formation may be therefore very important. *In vitro* studies of Ti, which is known to be biocompatible and osseointegrates into human bone, were carried out. The production of free radicals from H_2O_2 at Ti and TiO_2 surfaces was measured by spin trapping techniques. It was found that there is no sustained hydroxyl radical production at a titanium (oxide) surface, due to the quenching of the Fenton reaction through both trapping and oxidation of superoxide radicals in a TiO·OH adduct. Janzén *et al.* [4-108] concluded that the degree of surface-induced hydroxyl radical formation from H_2O_2 through the Fenton reaction may be of importance for the behavior of implanted materials.

Tengvall *et al.* [4-109] investigated the role of Ti and TiO_2, as well as other metals, in the inflammatory response through the Fenton reaction. The TiO·OH matrix formed traps the superoxide radical so that no or very small amounts of free hydroxyl radicals are produced. Ellipsometry and spin trapping with spectrophotometry and electron spin resonance were used to study the interaction between Ti and H_2O_2. Spectrophotometry results indicated that Ti, Zr, Au, and Al are low free OH-radical producers. It was suggested that a hydrated TiO·OH matrix, after the inflammatory reaction, responded to good ion exchange properties, and extracellular components may interact with the Ti^{4+}-H_2O_2 compound before matrix formation. The TiO·OH matrix is formed when the H_2O_2 coordinated to the Ti^{4+}-H_2O_2 complex is decomposed to water and oxygen [4-109].

It has been reported that (i) Ti implants inserted in the human body for many years have shown high oxidation rates at the areas relatively adjacent to the Ti implant, and (ii) the implants have shown a dark pigmentation due to the interaction with H_2O_2 [4-110]. It is well known that H_2O_2 is a strong oxidant that will increase the redox potential of the system. Gas evolution was observed on the Ti surface in the presence of H_2O_2 in the PBS (phosphate-buffered saline: 8.77NaCl g/l, 3.58 $Na_2HPO_4 \cdot 12H_2O$, 1.36KH_2PO_4, pH 7.2–7.4) solution. This is due to the decomposition of H_2O_2 into water and molecular oxygen according to: $2H_2O_2 \rightarrow H_2O + O_2$, which is catalyzed by the presence of TiO_2 on the surface due to the combined effect of the electron-donating and electron-accepting properties of TiO_2 occurring simultaneously [4-109], and the formation of some Ti-complex, which can lead to a higher electrode potential [4-111]. Pan *et al.* [4-112] performed electrochemical measurements, XPS, and scanning tunneling microscopy analyses to study the effect of hydrogen peroxide on the passivity of CpTi (grade 2) in a phosphate-buffered saline (PBS) solution. The results indicate that the passive film formed in the PBS solution – with and without addition of H_2O_2 – may be described with a two-layer structure model. The inner layer has a structure close to TiO_2, whereas the outer layer consists of hydroxylated compounds. The introduction of H_2O_2 in the PBS solution broadens the hydroxylate-rich region, probably due to the formation of a Ti^{4+}-H_2O_2 complex. Furthermore, the presence of H_2O_2 results in enhanced dissolution of Ti and a rougher surface on a microscopic scale. It was also observed that H_2O_2 addition furthermore seems to facilitate the incorporation of phosphate ions into the thicker porous layer [4-112].

The influence of H_2O_2 generated by an inflammatory response also has been suggested as an explanation for the high rate of Ti oxidation/corrosion *in vivo* [4-101, 4-109, 4-113].

The incorporation of mineral ions into the oxide film is believed to be important for the osseointegration behavior, which can lead to direct bone–titanium contact and strong bonding that in turn can contribute to the load-bearing capacity of titanium implants [4-114].

To analyze titanium's response to representative surgical wound environments, Bearinger *et al.* [4-115] examined CpTi and Ti-6Al-4V, which were exposed to phosphate-buffered saline (PBS) with 30 mM with hydrogen peroxide (H_2O_2) addition. The study was characterized by simultaneous electrochemical atomic force microscopy and step-polarization impedance spectroscopy. It was found that (i) surfaces were covered with protective oxide domes that indicated topography changes with potential and time of immersion, and (ii) less oxide dome coarsening was noted on surfaces treated with PBS containing H_2O_2 than on surfaces exposed to pure PBS. In both types of solutions, oxide (early) resistances of CpTi samples were higher than Ti-6Al-4V oxide resistances, but CpTi oxide resistance was lower in the hydrogen peroxide solution compared to pure PBS. Differences in electrical properties between CpTi and Ti-6Al-4V surfaces suggest that CpTi, but not Ti-6Al-4V, has catalytic activity on H_2O_2, and that the catalytic activity of CpTi oxide affects its ability to grow TiO_2 [4-115].

REFERENCES

[4-1] Kasemo B. Biocompatibility of titanium implants: surface science aspects. J Pros Dent 1983;49:832–837.

[4-2] Tomashov ND. Theory of corrosion and protection of metals: the science of corrosion. New York: The MacMillan Co. 1966. p. 144.

[4-3] Weast RC. CRC handbook of chemistry and physics. The 47th edition. Cleveland: The Chemical Rubber Co. 1966. p. D-54.

[4-4] Ashby MF, Jones DRH. Engineering materials. An Introduction to their properties and applications. New York: Pergamon Press. 1980. pp. 194–200.

[4-5] Weast RC. CRC Handbook of chemistry and physics. The 47th edition. Cleveland: The Chemical Rubber Co. 1966. p. D-27.

[4-6] West JM. Basic corrosion and oxidation, 2nd ed. London: Wiley. 1986. pp. 27–30.

[4-7] Lautenschlager EP, Monaghan P. Titanium and titanium alloys as dental materials. Int Dent J 1993;43:245–253.

[4-8] Douglass DL. Oxidation of metals and alloys. Columbus OH: American Society for Metals. 1971. p. 46.

[4-9] Coddet C, Chaze AM, Beranger G. Measurements of the adhesion of thermal oxide films: application to the oxidation of titanium. J Mater Sci 1987;22:2969–2974.

[4-10] Douglass DL. Oxidation of metals and alloys. Columbus, OH: American Society for Metals. 1971. pp. 142–143.

[4-11] Solar RJ. Corrosion resistance of titanium surgical implant alloys: a review. In: Corrosion and degradation of implant materials, ASTM STP 684. Syrett BD, editor. Philadelphia, PA: American Society for Testing and Materials. 1979. pp. 259–273.

[4-12] Uhlig HH. The corrosion handbook. New York: Wiley. 1966. p. 20.

[4-13] Fontana MG, Greene ND. Corrosion engineering. New York: McGraw-Hill Book. 1967. pp. 319–321.

[4-14] Wranglén G. An introduction to corrosion and protection of metals. London: Chapman & Hall. 1985. pp. 65–66.

[4-15] Oshida Y, Sachdeva R, Miyazaki S. Changes in contact angles as a function of time on some pre-oxidized biomaterials. J Mater Sci: Mater Med 1992;3:306–312.

[4-16] Oshida Y, Sachdeva R, Miyazaki S, Daly J. Effects of shot peening on surfaces contact angles of biomaterials. J Mater Sci: Mater Med 1993;4: 443–447.

[4-17] Albrektsson T, Brånemark PI, Hansson HA, Kasemo B, Larsson K, Lundstroem I, McQueen DH, Skalak R. The interface zone of inorganic implants in vivo: titanium implants in bone. Ann Biomed Eng 1983;11:1–27.

[4-18] Lausmaa J, Kasemo B, Hansson S. Accelerated oxide growth on titanium implants during autoclaving caused by fluorine contamination. Biomaterials 1985;6:23–27.

[4-19] Weast RC. CRC handbook of chemistry and physics. The 47th edition. Cleveland, OH: The Chemical Rubber Co. 1966. p. E-58.

[4-20] http://en.wikipedia.org/wiki/Titanium.

[4-21] Fraker AC, Ruff AW. Corrosion of titanium alloys in physiological solutions. In: Titanium science and technology. Vol. 4 New York: Plenum Press. 1973. pp. 2447–2457.

[4-22] Sittig C, Textor M, Spencer ND, Wieland M, Vallotton P-H. Surface characterization of implant materials c.p. Ti, Ti-6Al-7Nb and Ti-6Al-4V with different pretreatments. J Mater Sci: Mater Med 1999;10:35–46.

[4-23] Mazhar AA, Heakal FEl-T, Gad-Allah AG. Anodic behavior of titanium in aqueous media. Corrosion 1988;44:705–710.

[4-24] Metikoš-Huković M, Ceraj-Cerić M. Anodic oxidation of titanium: mechanism of non-stoichiomeric oxide formation. Surf Technol 1985;24:273–283.

[4-25] Hanawa T, Asami K, Asaoka K. Repassivation of titanium and surface oxide film regenerated in stimulated bioliquid. J Biomed Mater Res 1998;40:530–538.

[4-26] Gilbert JL, Buckley CA, Lautenschlager EP. Electrochemical reaction to mechanical disruption of titanium oxide films. J Dent Res. Special Issue 1995;74:92 (Abstract No. 645).

[4-27] Hruska AR, Borelli P. Quality criteria for pure titanium casting, laboratory soldering, intraoral welding, and a device to aid in making uncontaminated castings. J Prosthet Dent 1999;66:561–565.

[4-28] Adachi M, Mackert JR Jr, Parry EE, Fairhurst CW. Oxide adherence and porcelain bonding to titanium and Ti-6Al-4V alloy. J Dent Res 1990;69:1230–1235.

[4-29] Kimura H, Horng CJ, Okazaki M, Takahashi J. Oxidation effects on porcelain-titanium interface reactions and bond strength. Dent Mater J 1990;9:91–99.

[4-30] Hautaniemi JA, Herø H. Porcelain bonding on Ti: its dependence on surface roughness, firing and vacuum level. Surf Interface Anal 1993;20:421–436.

[4-31] Atsü S, Berksun B. Bond strength of three porcelain to two forms of titanium using two firing atmospheres. J Prosthet Dent 2000;84:567–574.

[4-32] Oshida Y, Hashem A. Titanium–porcelain system. Part I: oxidation kinetics of nitrided pure titanium, simulated to porcelain firing process. J Biomed Mater Eng 1993;3:185–198.

[4-33] Wang RR, Fung KK. Oxidation behavior of surface-modified titanium for titanium–ceramic restorations. J Prosthet Dent 1997;77:423–434.

[4-34] Wang RR, Welsch G, Monteiro O. Silicon nitride coating on titanium to enable titanium–ceramic bonding. J Biomed Mater Res 1999;46: 262–270.

[4-35] Sundgren J-E, Bodö P, Lundström I. Auger electron spectroscopic studies of the interface between human tissue and implants of titanium and stainless steel. J Colloid Interf Sci 1986;110:9–20.

[4-36] McQueen D, Sundgren JE, Ivarsson B, Lundstrom I, af Ekenstam CB, Svensson A, Brånemark PI, Albrektsson T. Auger electron spectroscopic studies of titanium implants. In: Lee AJC, editor. Clinical applications of biomaterials. New York: Wiley. 1982. pp. 179–185.

[4-37] Liedberg B, Ivarsson B, Lundstrom I. Fourier transform infrared reflection absorption spectroscopy (FTIR-RAS) of fibrinogen adsorbed on metal and metal oxide surfaces. J Biochem Biophys Method 1984;9:233–243.

[4-38] Hanawa T. Titanium and its oxide film: a substrate for formation of apatite. In: The bone-biomaterial interface. Davies JE, editor. Toronto: University of Toronto Press. 1991. pp. 49–61.

[4-39] Healy KE, Ducheyne P. Hydration and preferential molecular adsorption on titanium in vitro. Biomaterials 1992;13:553–561.

[4-40] Healy KE, Ducheyne P. The mechanisms of passive dissolution of titanium in a model physiological environment. J Biomed Mater Res 1992;26:319–338.

[4-41] Brånemark PI. Introduction to osseointegration. In: Tissue-integrated protheses. Brånemark PI, editor. Chicago: Quintessence Publishers. 1985. pp. 63–70.

[4-42] Kasemo B, Lausmaa J. Metal selection and surface characteristics. In: Tissue-integrated protheses. Brånemark PI, editor. Chicago: Quintessence Publishers. 1985. pp. 108–115.

[4-43] Meachim G, Williams DF. Changes in non-osseous tissue adjacent to titanium implants. J Biomed Mater Res 1973;7:555–572.

[4-44] Ferguson AB, Akahoshi Y, Laing PG, Hodge ES. Characteristics of trace ions released from embedded metal implants in the rabbit. J Bone Joint Surg 1962;44-A:323–336.

[4-45] Ducheyne P, Williams G, Martens M, Helsen J. *In vivo* metal-ion release from porous titanium-silver material. J Biomed Mater Res 1984;18:293–308.

[4-46] Ducheyne P, Healy KE. Surface spectroscopy of calcium phosphate ceramic and titanium implant materials. In: Surface characterization of biomaterials. Ratner BD, editor. Amsterdam: Elsevier Science. 1988. pp. 175–192.

[4-47] Albreksson T, Hansson HA. An ultrastructural characterization of the interface between bone and sputtered titanium or stainless steel surfaces. Biomaterials 1986;7:201–205.

[4-48] Wyckoff RWG. The structure of crystals. New York: Chemical Catalog Co. 1931. pp. 134, 239.

[4-49] Tang H, Prasad K, Sanjinès R, Schmid PE, Lévy F. Electrical and optical properties of TiO_2 anatase thin films. J Appl Phys 1994;75:2042–2047.

[4-50] Douglass DL, van Landuyt J. The structure and morphology of oxide films during the initial stages of titanium oxidation. Acta Met 1966;14: 491–503.

[4-51] Nakayama T, Oshida Y. Effects of surface working on the structure of opxide films by wet oxidation in austentic stainless steels. Trans Jpn Inst Metals 1971;12:214–217.

[4-52] Mahla EM, Nielsen NA. A study of films isolated from passive stainless steels. Trans Elctrochem Soc 1948;81:913–916.

[4-53] Nakayama T, Oshida Y. Identification of the initial oxide films on 18-8 stainless steel in high temperature water. Corrosion 1968;24:336–337.

[4-54] Kuphasuk C, Oshida Y, Andres CJ, Hovijitra ST, Barco MT, Brown DT. Electro-chemical corrosion of titaniuim and titanium-based alloys. J Prosthet Dent 2001;85:195–202.

[4-55] Lim YJ, Oshida Y, Andres CJ, Barco MT. Surface chacterization of variously treated titanium materials. Int J Oral Maxillofac Implants 2002;16:333–342.

[4-56] Cullity BD. Elements of X-ray diffraction. Reading, MA: Addison-Wesley. 1978. pp. 86–89.

[4-57] Schwartz LH, Cohen JB. Diffraction from materials. Heidelberg: Springer, 1987. pp. 46–49.

[4-58] American Society for Testing and Materials. X-ray powder data file. Philadelphia: American Society for Testing and Materials. 1960.

[4-59] Głuszek J, Masalski J, Furman P, Nitsch K. Structural and electrochemical examinations of PACVD TiO_2 films in Ringer solution. Biomaterials 1997;18:789–794.

[4-60] Liu X, Ding C. Plasma sprayed wollastonite/TiO_2 composite coatings on titanium alloys. Biomaterials 2002;23:4065–4077.

[4-61] Browne M, Gregson PJ. Surface modification of titanium alloy implants. Biomaterials 1994;15:894–898.

[4-62] Takemoto S, Yamamoto T, Tsuru K, Hayakawa S, Osaka A, Takashima S. Platelet adhesion on titanium oxide gels: effect of surface oxidation. Biomaterials 2004;25:3845–3892.

[4-63] Allen KW, Alsalim HS. Titanium and alloy surfaces for adhesive bonding. J. Adhesion 1974;6:229–237.

[4-64] Felske A, Plieth WJ. Raman spectrcoscopy of titanium dioxide layers. Electrochim Acta 1989;34:75–77.

[4-65] Matthews A. The crystallization of anatase and rutile from amorphous titanium dioxide under hydrothermal conditions. Am Miner 1976;61:419–424.

[4-66] Yahalom J, Zahavi J. Electrolytic breakdown crystallization of anodic oxide films on Al, Ta and Ti. Electrochem Acta 1970;15:1429–1435.

[4-67] Ohtsuka T. Structure of anodic oxide films on titanium. Surface Sci 1998;12:799–804.

[4-68] Liang B, Fujibayashi S, Neo M, Tamura J, Kim H-M, Uchida M, Kikubo T, Nakamura T. Histological and mechanical investigation of the bone-bonding ability of anodically oxidized titanium in rabbits. Biomaterials 2003;24: 4959–4966.

[4-69] Koizumi T, Nakayama T. Structure of oxide films formed on Ti in boiling dilute H_2SO_4 and HCl. Corr Sci 1968;8:195–199.

[4-70] Ask M, Lausmaa J, Kasemo B. Preparation and surface spectroscopic characterization of oxide films on Ti6Al4V. Appl Surf Sci 1988/1989;35: 283–301.

[4-71] Lausmaa J, Kasemo B, Mattsson H, Odelius H. Multiple-technique surface characterization of oxide films on electropolished and anodically oxidized titanium. Appl Surf Sci 1990;45:189–200.

[4-72] Michaelis A, Schultze JW. Effect of anisotropy on microellipsometry in the Ti/TiO_2 system. Thin Solid Films 1993;233:86–90.

[4-73] Tomkiewicz M. Relaxation spectrum analysis of semiconductor-electrolyte interface-TiO_2 J Electrochem Soc 1979;126:2220–2225.

[4-74] Dolata M, Kedzierzawski P, Augustynski J. Comparative impedance spectroscopy study of rutile and anatase TiO_2 film electrodes. Electrochim Acta 1996;41:1287–1293.

[4-75] http://www.highbeam.com/doc./1G1:102604482/Photocatalyst.

[4-76] Sutherland DS, Forshaw PD, Allen GC, Brown IT, Williams KR. Surface analsyis of titanium implants. Biomaterials 1993;14:893–899.

[4-77] Lausmaa GJ, Kasemo B. Surface spectroscopic characterization of titanium implant materials. Appl Surf Sci 1990;44:133–146.

[4-78] Kubaschewski O, Hopkins BE. Oxidation of metals and alloys. London: Butterworths. 1962. p. 24.

[4-79] Tummler H, Thull R. Model of the metal/tissue connection of implants made of titanium or tantalum. Biological and biochemical performance of biomaterials. Netherlands: Elsevier. 1986. pp. 403–408.

[4-80] Eliades T. Passive film growth on titanium alloys: physiochemical and biologic considerations. Int J Oral Maxillofac Implant 1997;12: 621–627.

[4-81] Drath DB, Karnovsky ML, Superoxide production by phagocytic leukocytes. J Exp Med 1975;144:257–261.

[4-82] Root BK, Metcalf JA. H_2O_2 release from human granulocytes during phagocytosis: Relationship to superoxide anion formation and cellular catabolism of H_2O_2. Studies with normal and cytochalasin B-treated cells. J Clin Invest 1977;60:1266–1279.

[4-83] DeChatelet LR. Oxide bactericidal mechanisms of PMN. J Infect Dis 1975;131:295–303.

[4-84] Milošev I, Metikoš-Huković M, Strehblow H-H. Passive film on orthopaedic TiAlV alloy formed in physiological solution investigated by X-ray photoelectron spectroscopy. Biomaterials 2000;21:2103–2113.

[4-85] Kovacs, P. Electrochemical techniques for studying the corrosion behaviour of metallic implant materials. Corrosion '92, NACE Conference Proceedings 1992;214:1–214.

[4-86] Sundararajan T, Kamachi MU, Mair KGM, Rajeswari S, Subbiyan M. Surface characterization of electrochemically for passive film on nitrogen implanted Ti6Al4V alloy. Mater Trans JIM 1998;39:756–761.

[4-87] Okazaki Y, Tateishi T, Ito Y, Corrosion resistance of implant alloys in pseudo physiological solution and role of alloying elements in passive films. Mater Trans JIM 1998;38:78–84.

[4-88] Hernardez de Gatica NL, Jones GL, Gardella JA Jr. Surface characterization of titanium alloys sterilized for biomedical applications. Appl Surf Sci 1993;68:107–121.

[4-89] Pham MT, Zyganow I, Matz W. Corrosion behaviour and microstructure of titanium implanted with α and β stabilizing elements. Thin Solid Films 1997; 310:251–259.

[4-90] Sodhi RNS, Weninger A, Davis JE, Sreenivas K. X-ray photoelectron spectroscopic comparison of sputtered Ti, Ti6Al4V, and passivated bulk metals for use in cell culture techniques. J Vac Sci Technol A 1991;A9: 1329–1333.

[4-91] Maeusli P-A, Bloch PR, Geret V. Steinmann SG. In: Biological and biomechanical performance of biomaterials. Christel P, Meunier A, editors. Amsderdam: Elsevier. 1986. p. 57.

[4-92] O'Brien B, Carroll WM, Kelly MJ. Passivation of nitinol wire for vascular implants – a demonstration of the benefits. Biomaterials 2002;23: 1739–1748.

[4-93] Metikoš-Huković M, Kwokal A, Piljac J. The influence of niobium and vanadium on passivity of titanium-based implants in physiological solution. Biomaterials 2003;24:3765–3775.

[4-94] Ohtsuka T, Masuda M, Sato N. Cathodic reduction of anodic oxide films formed on titanium. J Electrochem Soc 1987;134:2406–2410.

[4-95] Heusler KE. The influence of electrolyte composition on the formation and dissolution of passivating films. Corr Sci 1989;29:131–147.

[4-96] Arsov Lj D. Dissolution electrochimique des films anodiques du titane dans l'acide sulfurique. Electrochim Acta 1985;12:1645–1657.

[4-97] Sul Y-T, Johansson CB, Petronis S, Krozer A, Jeong Y, Wennerberg A, Albrektsson T. Charactristics of the surface oxides on turned and electrochemically oxidized pure titanium implants up to dielectric breakdown: the oxide thickness, micropore configuration, surface roughness, crystal structure and chemical composition. Biomaterials 2002;23: 491–501.

[4-98] Shibata T, Zhu Y-C. The effect of film formation potential on the stochastic processes of pit generation on anodized titanium. Corr Sci 1994;36:153–163.

[4-99] Tengvall P, Elwing H, Sjöqvist L, Lundström I. Interaction between hydrogen peroxide and titanium: a possible role in the biocompatibility of titanium. Biomaterials 1989;10:118–120.

[4-100] Jacob HS, Craddock PR, Hammerschmidt DE, Moldow CF. Induced granulocyte aggregation – unsuspected mechanism of disease. New Engl J Med 1980;302:789–794.

[4-101] Del Masestro RF, An approach to free radicals in medicine and biology. Acta Physiol Scand (Suppl) 1980;492:153–168.

[4-102] Sundgren J-E, Bodö P, Lundström I, Berggren A, Hellem S. Auger electron spectroscopic studies of stainless steel implants. J Biomed Mater Res 1985;19:663–671.

[4-103] Tengvall P, Bjursten LM, Elwing H, Lunström I, Free radicals oxidation of metal implants and biocompatibility. 2nd International conference of biointercations '87. London. 1987.

[4-104] Jaeger CD, Bard AJ. Spin-trapping and electron spin resonance detection of radical intermediates in the photodecompositions of water at TiO_2 particulate systems. J Phys Chem 1979;83:3146–3151.

[4-105] Barb WG, Baxendale JH, George P, Hargrave KR. Recations of ferrous and ferric ions with hydrogen peroxide. Trans Farady Soc 1951;47:462–500.

[4-106] Bielski BHJ. Fast kinetic studies of dioxygen-derived species and their metal-complexes. Phil Trans R Soc Lond B 1985;311:473–482.

[4-107] Armstrong D. Free radicals in molecular biology, aging and disease. New York: Raven Press. 1984.

[4-108] Janzén EG, Wang YY, Shetty RW. Spin-trapping with α-(4-pyridyl-1-oxide)-N-tert-nutylnitrone in aqueous solutions: a unique electron spin resonance spectrum for the hydroxyl radical adduct. J Am Chem Soc 1978;100: 2923–2925.

[4-109] Tengvall P, Lundström I, Sjöqvist L, Elwing H, Bjursten LM. Titanium-hydrogen peroxide interaction: model studies of the influence of the inflammatory response on titanium implants. Biomaterials 1989;10:166–175.

[4-110] Pan J, Thierry D, Leygraf C. Electrochemical and XPS studies of titanium for biomaterial applications with respect to the effect of hydrogen peroxide. J Biomed Mater Res 1994;28:113–122.

[4-111] Bianchi G, Malaguzzi S. Cathodic reduction of oxygen and hydrogen peroxide on titanium. Proceedings of 1st international congress of metallic corrosion London, April 10–15. 1961. pp. 78–83.

[4-112] Pan J, Thierry D, Leygraf C. Hydrogen peroxide toward enhanced oxide growth on titanium in PBS solution: blue coloration and clinical relevance. J Biomed Mater Res 1996;30:393–402.

[4-113] Pan J, Liao H, Leygraf C, Thierry D, Li J. Variation of oxide films on titanium induced by osteoblast-like cell culture and the influence of an H_2O_2 pretreatment. J Biomed Mater Res 1998;40:244–256.

[4-114] Albrektsson T. The response of bone to titanium implants. CRC Crit Rev Bicompat 1984;1:53–84.

[4-115] Bearinger JP, Orme CA, Gilbert JL. Effect of hydrogen peroxide on titanium surfaces: in situ imaging and step-polarization impedance spectroscopy of commercially pure titanium and titanium, 6-aluminum, 4-vanadium. J Biomed Mater Res 2003;67A:702–712.

Chapter 5
Mechanical and Tribological Behaviors

5.1.	Fatigue	107
5.2.	Fracture and Fracture Toughness	111
5.3.	Biotribological Actions	113
	5.3.1 General	113
	5.3.2 Friction	114
	5.3.3 Wear and Wear Debris Toxicity	116
References		120

Chapter 5
Mechanical and Tribological Behaviors

Biomechanics includes all facets of musculoskeletal, cardiovascular, respiratory, implant–tissue interface, mechanics of prostheses, and mechanisms of chewing. Given such a wide range of coverage, the area of interest becomes much narrower when considering titanium applications only. Simultaneously, however, biomechanics is a multidisciplinary field, requiring various basic knowledge including statics and dynamics, elasticity and plasticity, fatigue, fracture mechanics and toughness, and tribological actions. Materials in bioengineering, in a broad sense, range widely from very small units, such as a virus or bacteria, to very large and complex, such as a healthy human body. They also cover the various products manufactured by using the methodologies in connection with biological disciplines.

Based on the above argument, in this chapter, fatigue, fracture, and biotribological actions, such as friction and wear as well as wear debris toxicity, will be discussed.

5.1. FATIGUE

Fatigue is the loss of strength and other important properties as a result of stressing over a quite long period of time. Fatigue is very important, as it is the single largest cause of failures in metals, estimated as about 90% of metallic failures, polymers and ceramics are also susceptible to this type of failure. Fatigue is a very general phenomenon in most materials and even living organisms. The 'Fatigue' is originally from Latin 'Fatigare'. The term itself is derived from fatigue meaning 'to tire'. The root 'fati' refers literally to an opening or yawning phenomenon vividly descriptive of fatigue in people, and especially, in view of ubiquity of crazing and cavitation, in materials as well [5-1].

Fatigue has often been observed in Ti and Ti alloys, and is considered as an important cause to the failure of the materials. For example, fatigue failure has been reported in dental implants [5-2], removable partial denture (RPD) framework [5-3], clasps [5-4, 5-5], plates, pedicle screws and cables [5-6–5-9]. The cyclic loading applied to orthopedic implants during body motion results in an alternating localized plastic deformation of microscopically small zones of stress concentration produced by notches or microstructural inhomogeneities [5-9]. Mechanical environments related to fatigue may be divided into two distinct groups: strain-controlled and stress-controlled. Although practically most fatigue environments in human body are a combination of the two, the highly compliant nature of biological materials tends to place them toward the direction of strain-controlled fatigue [5-10]. Examples for strain-controlled application include pacemaker leads, which require a conductive metal that can survive very high numbers of flexing motions without breaking, and clasps of RPD, which require adequate retention for insertion and removal of the device [5-11]. Lin *et al.* [5-12] employed the tension-to-tension stress mode for fatigue tests on titanium materials (CpTi, Ti-13Nb-13Zr, Ti-6Al-4V, and Ti-7.5Mo) in air under R (stress ratio = $\sigma_{min}/\sigma_{max}$) of 0.1, with frequency of 10 Hz. It was found that (i) Ti-6Al-4V and CpTi have higher stress-controlled fatigue[1] resistance, but lower strain-controlled fatigue (see footnote 1) resistance than Ti-7.5Mo and Ti-13Nb-13Zr, (ii) Ti-7.5Mo demonstrates the best strain-controlled fatigue performance, and (iii) the fatigue cracks almost always initiate from casting-induced surface/subsurface pores [5-12].

Clasps undergo permanent deformation to cause fatigue fracture under repeated flexures during denture insertion and removal, and masticatory actions [5-13–5-16]. The fatigue life of cast clasps made of CpTi was reported to be shorter than that of Co-Cr and gold alloy clasps [5-11]. However, the fatigue fracture and permanent deformation of cast claps made of Ti-based alloy has not been sufficiently assessed in relation to stress distribution, and little is known about how these clasps would function in long-term clinical use. Permanent deformation and fatigue fracture are caused by the stress created in the clasp [5-17, 5-18]. The stress distribution may depend on the elastic modulus of the alloy, dimensions and curvature of the clasp [5-19, 5-20], and amount and direction of deflection in relation to the abutment undercut [5-21]. Mahmoud *et al.* [5-22] investigated the gold alloy and Ti-6Al-7Nb, which were subjected to cyclic deflection of pre-set values of 0.25, 0.50, and 0.75 mm for 10^6 cycles. It was concluded that the gold alloy (70Au-4.4Pt-13.5Ag-8.8Cu-2Pd-0.1Ir) clasps exhibited significantly longer fatigue lives, while the Ti-6Al-7Nb clasps showed significantly greater resistance to permanent deformation under cyclic deflection [5-22].

---

[1] Stress-controlled fatigue test is conducted in such a way that repeated load is controlled between pre-determined upper and lower stress levels. If the tested material is fully annealed to start with and a work-hardening type, the material's strength increases by repeated stressing. Hence, the severity of fatigue process is getting milder. On the other hand, if the starting material's condition is a work-softening type, its original strength decreases by cycles, resulting in shortening fatigue life. On the contrary, by the strain-controlled fatigue testing, each cycle is controlled between pre-set upper and lower strain. If the fully annealed and work-hardening material is tested under the strain-controlled fatigue mode, because the material strength increases by cyclic loading, the applied stress needs to increase to maintain the pre-set level of upper strain. On the other, by the work-softening type materials, applied stress level decreases.

The fatigue resistances of experimentally prepared NiTi (50.8%Ni-49.2%Ti) cast clasps as well as CpTi, Co-Cr alloy, and Au-Ag-Pd-Cu alloy clasps were evaluated in a simulated clinical situation [5-4]. The change in force required to remove the NiTi clasps was recorded under a repeated placement–removal test on steel model abutment teeth. The tips of the clasps were located in the 0.25 and 0.5 mm undercut area of the abutments. It was found that (i) no significant change in the retentive force was found in NiTi clasps in the 1010 repeated cycles, but (ii) the other three types of clasps revealed a significant decrease in the force required for removal during the test, suggesting that the cast NiTi clasp may be suitable in removable prosthodontic constructions because of its significantly less permanent deformation during service [5-4].

The fatigue failure of hip joint prostheses will be expected to assume more importance in second generation implants aimed at younger, more active patients. Furthermore, new designs and material combinations including coatings (e.g., hydroxyapatite) may introduce fatigue problems that as of yet have not been considered. Styles *et al.* [5-23] conducted corrosion-fatigue tests on a model four point bendbar test-piece (milled-annealed Ti-6Al-4V) in a physiological solution (i.e., Ringer's solution) at 37°C. It was found that the introduction of superimposed block overload (50 cycles) to signify stair ascent/descent of fast walking and single overloads to signify fit/stand movements or stumbling were found to reduce fatigue life by >50% [5-23].

CpTi (grade 2) Nd:Yag laser weldment was subjected to fatigue cycling at a stress ratio of R = −1 under the frequency of 10 Hz. The tests were conducted in synthetic saliva (pH 7) and fluoride synthetic saliva (pH 7) with NaF of 1000 ppm or 2.223 g/1000 ml, compared to in-air tests. It was concluded that (i) the laser–welding procedure significantly reduced the fatigue life of CpTi and Ti-6Al-4V alloy specimens, and (ii) synthetic saliva and fluoride saliva adversely affected fatigue life compared with the control in-air tests [5-24].

Surface roughness of cast CpTi and Ti-6Al-4V was evaluated when they were submitted to conventional or electrolytic polishing, correlating the results with corrosion-fatigue (10 Hz, R=1, indicating that the test was conducted in tension–tension mode) at strength testing performed in artificial fluoride saliva. Specimens were also tested in air at room temperature to evaluate the effectiveness of the corrosion-fatigue test model. It was found that (i) electrolytic polishing (5% HF–35% HNO_3 – 60% H_2O) provided lower surface roughness (0.24 μm) than conventional polishing using titanium polishing paste (0.32 μm), (ii) regardless of the polishing conditions, surface roughness of Ti-6Al-4V (0.25 μm) was lower than that of CpTi (0.31 μm), and the fluoridated environment did not influence fatigue performance, and (iii) there was no correlation between fatigue performance and surface roughness. Based on these findings, it was concluded that surface roughness of Ti-6Al-4V was lower than CpTi. For cast Ti frameworks, the electrolytic polishing regimen was found to be more effective than the manufacturer's polishing instructions with abrasives and rotary instruments. After polishing, it was found that differences in surface roughness values did not affect corrosion-fatigue performance [5-25].

It is widely understood that the mechanical properties of titanium alloys are very sensitive to the characteristics of the microstructure. The type of phases, grain size and shape, morphology, and distribution of the fine microstructure ($\alpha+\beta$ phases) determine the properties and therefore the application of titanium alloys [5-26]. Microstructural modification by thermomechanical treatment is widely used to optimize the properties of high-strength titanium alloys for specific applications. However, because the surface of a mechanically loaded titanium part often experiences different conditions than the bulk, it makes sense in many applications to modify only the surface microstructure. By combining thermal treatments with mechanical working, stability of the resulting tailored surface is enhanced. Such stability is especially significant for resistance to fatigue crack nucleation and growth. Akahori *et al.* [5-27] found that, for Ti-6Al-7Nb alloy, by increasing the volume fraction of the primary α-phase from 0 to 70%, the fatigue limit is raised. It was also reported that changing the cooling rate after solution treatment from air-cooling to water-quenching improved the fatigue limit Ti-6Al-7Nb alloy significantly [5-27], due to increased surface layer hardness.

Koahn *et al.* [5-28] used acoustic emission (AE) phenomenon to record AE events and event intensities (e.g., event amplitude, counts, duration, and energy counts) and analyzed these parameters during fatigue loading of uncoated and porous-coated Ti-6Al-4V alloy. AE provides the ability to spatially and temporally locate multiple fatigue cracks, in real time. Fatigue of porous-coated Ti-6Al-4V is governed by a sequential: fracture process of transverse fracture in the porous coating $\rightarrow$ sphere/sphere and sphere/substrate debonding $\rightarrow$ substrate fatigue crack initiation $\rightarrow$ slow and rapid substrate fatigue crack propagation. Because of the porosity of the coating, the different stages of fracture within the coating may occur in a discontinuous fashion. Therefore, the AE events generated are intermittent and the onset of each mode of fracture in the porous coating can be detected by increases in AE events rate. Changes in AE events rate also correspond to changes in crack extension rate, and may therefore be used to predict failure. It was mentioned that (i) the magnitude of the AE event intensities increased with increasing fracture and microvoid coalescence generated the highest AE event amplitudes (100 dB), whereas plastic flow and friction generated the lowest AE event amplitudes (55–65 dB), and (ii) fractures in the porous coating can be characterized by AE event amplitude of less than 80 dB [5-28].

There are several ways to improve fatigue strength, including modifying surface roughness and hardness, surface residual stress magnitude, and altering materials design, as well as surface coating. Mechanical surface treatment such as shot peening, polishing, or surface rolling can improve the endurance limit in titanium-based materials, but the changes they induce can have contradictory influences on fatigue strength. For example, surface roughness determined whether fatigue strength is primarily crack-nucleation controlled (smooth) or crack-propagation controlled (rough). For smooth surfaces, a work-hardened surface layer can delay crack nucleation because of its increase in strength. But on rough surfaces, a work-hardened surface layer cannot retard crack propagation because of its reduced ductility. However, near surface residual compressive stresses significantly retard crack growth whether the surface is smooth or rough. It was also reported that by grain refining of Ti-8Al alloy, the fatigue strength is improved to 325 MPa, compared to a coarse grain of 275 MPa at 10^6 cycles [5-29, 5-30].

A new surface-coating method by which CaP invert glass is used to improve the bioactivity of titanium alloys has been developed recently [5-31]. In this method, the powder of CaP invert glass (CaO-P_2O_5-TiO_2-Na_2O) is coated on the surface of titanium alloy samples and heated between 800 and 850°C. With this treatment, a calcium phosphate layer mainly containing β-$Ca_3(PO_4)_2$ phase can be coated easily on titanium alloy samples. The effect of such coating was evaluated on the fatigue properties of Ti-29Nb-13Ta-4.6Zr (a new metastable β alloy for biomedical applications). It was reported that the fatigue endurance limit of the coated alloy was found to be about 15% higher than that of uncoated alloy as a result of the formation of a hard ($\alpha+\beta$) layer and a small amount of the ω phase during the coating process, (ii) the coating exhibits excellent adhesion to the substrate during the tensile and fatigue tests, and (iii) subsequent aging at 400°C for 3 days greatly improves the fatigue resistance of the coated alloy due to isothermal ω phase precipitation [5-31].

5.2. FRACTURE AND FRACTURE TOUGHNESS

The ideal biomaterial for implant applications, especially for joint replacements, is expected to exhibit excellent properties such as no adverse tissue reactions, excellent corrosion resistance in the body fluid medium, high mechanical strength and fatigue resistance, low modulus, low density, and good wear resistance [5-32]. As for searching V-free Ti-based alloys, Ti-6Al-7Nb and Ti-5Al-2.5Fe were developed for orthopedic applications. These two alloys exhibited mechanical properties similar to Ti-6Al-4V and were primarily developed in response to concerns of potential cytotoxicity and adverse tissue reactions caused by V [5-33, 5-34]. Further studies have shown the release of both V and Al ions from Ti-6Al-4V alloy might cause long-term health problems, such as peripheral neuropathy, osteomalacia, and Alzheimer diseases [5-35, 5-36].

The fracture behavior of the two cast Ti materials was investigated by AE analysis during the fracture toughness testing. Specimens for the test were cast by the lost wax method using a specially designed Ti casting machine of pressure-different method for dental use. A fatigue crack was inserted from the machine notch tip into the body of a specimen in the range of 0.45–0.55 a/W. AE signals released during the fracture toughness tests were detected by two sensors attached to both ends of the specimen. It was found that (i) from the early stage of the fracture toughness tests, AE signals started to be released in all types of specimens, and (ii) a reaction layer with the investment materials of about 50–100 μm was thought to be the result of the AE release from an early stage of the fracture toughness test [5-37].

The mechanical properties of titanium alloys strongly depend on their phase composition and structure, which result from casting and/or thermomechanical processing, and can be significantly influenced by the following heat treatment. Treatments that include heating into the beta-phase, which would provide full transformation of the as-received microstructure, are not commonly used because of a fast grain growth in the beta-phase. This drawback leads to significant deterioration of the mechanical properties. Moreover, large beta-grains are typically associated with the formation of coarse-grained structures by both martensitic $\beta \rightarrow \alpha$ or diffusional $\beta \rightarrow (\alpha+\beta)$ transformations on cooling. It was shown that a significant improvement in the mechanical properties of high-strength titanium alloys can be reached by using a special heat treatment based on rapid heating into the beta-region. Such a treatment allows the production of a high-temperature beta-phase with a fine-grained microstructure and optimal chemical inhomogeneity, which strongly determines the mechanism and kinetics of phase transformations on cooling and further annealing or aging. A wide range of final microstructures with an enhanced combination of mechanical properties can be obtained [5-38].

Banerjee *et al.* [5-32] evaluated Ti-34Nb-9Zr-8Ta and Ti-13Mo-7Zr-3Fe for their heat-treatment capabilities. In the homogenized condition, both alloys exhibited a microstructure consisting primarily of a beta matrix with grain boundary alpha precipitates and a low-volume fraction of intra-granular alpha precipitates. On aging the homogenized alloys at 600°C for 4 h, both alloys exhibited the precipitation of refined scale secondary alpha precipitates homogeneously in the beta matrix. It was reported that, while the hardness of the Ti-Mo-Zr-Fe alloy marginally increased, the hardness of Ti-Nb-Zr-Ta alloy decreased substantially as a result of the aging treatment [5-32].

Ti single crystal planes of different atomic density have been reported to show different oxidation characteristics, as discussed in Chapter 4. The differences in oxide characteristics have further been demonstrated to lead to differences in osteoblast attachment. Mante *et al.* [5-39] investigated the preferred crystallographic planes of titanium for osteoblast attachment in order to optimize the surfaces of single-crystal and polycrystalline titanium implants for anchoring various prostheses. Nanoindentation techniques were used to determine mechanical properties of two crystallographic planes of Ti of different atomic density. It was reported that (i) modulus of elasticity (MOE) of 128 GPa was obtained for polycrystalline Ti and 123 and 124 GPa for the basal plane and pyramidal planes, respectively, (ii) the variation of MOE with crystal orientation was not greater than the statistical variation in the data, (iii) surface hardness values were 2.1 GPa for polycrystal and 1.6 and 1.9 GPa for basal and pyramidal planes, respectively, and (iv) hardness depth profile (0–2000 nm) showed a sharp increase at shallow depths and may reflect changes caused by oxidations of the Ti surfaces [5-39].

It has been found that every machining operation produces a distinctive residual stress in the surface layer. When any metallic material is subjected to wire-drawing, cutting, pressing or rolling, there will be a residual strain, which is normally relieved by an annealing process. If such residual strain is not removed, the grains become distorted. Distorted microstructures decrease the plastic formability under repeated stressing, causing a premature crack. When titanium wire is formed to manufacture dental implants by the wire-drawing process, the work strain can remain in the entire microstructures. Besides, if cutting is further performed, relatively large amounts of residual strain would remain. It is reported that heating at 600°C for 1 h is an optimum stress relief heat treatment for CpTi (grade 2) without any remarkable grain coarsening [5-40]. Residual stress is not always adverse, rather it should be beneficial if it is compressive residual stress, so that cracks tend to closure and applied stress can increase. Rolling (creating surface compressive residual stress) alone increases the notched fatigue limit from the low value of 400 to 1100 MPa [5-29, 5-30].

5.3. BIOTRIBOLOGICAL ACTIONS

5.3.1 General

Biotribology is a compound word of biology and tribology. In 1966, a term "tribology"[2] was introduced, with an origin of Greek $\tau\rho\iota\beta o\sigma$ (to rub), defined as the science and technology dealing with the friction, wear, and lubrication.

Tribology is the science dealing with the interaction of surfaces in tangential motion. Hence, tribology includes the nature of surfaces from both a chemical and physical point of view, including topography, the interaction of surfaces under load and the changes in the interaction when tangential motion is introduced.

---

[2] The prefix "tri" normally refers to "three" in those words, such as "tripot", "trigonometry", "trilingual" "trifocal", etc. At the same time we just know that tribology is a science about three phenomena (friction, wear, and lubrication), so that it is easily misinterpreted that the tribology is a science dealing with the forementioned three phenomena, But this is not true. The Greek $\tau\rho\iota\beta o\sigma$ does not have a prefix meaning 'three'. This is just a matter of coincidence.

Macroscopically, the interactions are manifested in the phenomena of friction and wear. Modification of the interaction through the interposition of liquid, gaseous, or solid films is known as the lubrication process. Hence, from a macroscopic point of view, tribology includes lubrication, friction, and wear [5-41]. There are four major clinical reasons to remove the implants: (1) fracture, (2) infection, (3) wear, and (4) loosening. Among these, the removal due to the infection generally occurs in relatively early stages after implantation, while the other three incidents typically increase gradually by years, because they are somewhat related to biotribological reactions. Human joints are one example of natural joints, and show low wear and exceedingly low friction through efficient lubrication. Disease or accident can impair the function at a joint, and this can lead to the necessity for joint replacement [5-41].

Special tribological element problems arise in many mechanical devices used in various areas, for example, computer devices and medical implants (prostheses, assists and artificial organs). Medical implants, such as prostheses, should demonstrate very good tribological properties, as well as biological inertness. Their longevity depends significantly on the wear of the rubbing elements and must be extended as much as possible. The developments of total replacement hip, knee, ankle, elbow, shoulder, and hand joints (and the appropriate surgical techniques) has been the major success of orthopedic surgery, and would not have been possible without extensive in *in vitro* and *in vivo* studies of the tribological problems, especially wear, associated with such artificial joints.

5.3.2 *Friction*

Zwicker *et al.* [5-42] investigated the friction behavior of a hip prosthesis head fabricated from Ti-5Al-2.5Fe alloy in contact with an ultrahigh molecular weight polyethylene cup. It was shown that (i) the frictional behavior of a head coated with an oxide layer about 1–3 μm thick produced by induction heating of the surface was equal to that of a head fabricated from alumina ceramic, (ii) the oxide layer was still present after 10^7 cycles under maximum load of 4000 and 3200 N in 0.9% NaCl solution, and (iii) examination of the microstructure of the oxide layer after testing showed that it was unchanged by friction test under overload service conditions. Therefore, the production of such oxide layers on Ti and Ti alloy implants that are locally stressed by friction during service is recommended [5-42].

Another important incidence involving friction in dentistry is between the orthodontic bracket and archwires. While orthodontic mechanotherapy is active, sliding of bonded brackets along archwires occurs in all orthodontic cases. When these sliding movements occur, frictional forces between the archwire and the bracket are produced. Different combinations of materials have varying coefficients of friction, leading to different amounts of frictional force created within the bracket/archwire system. This force either retracts or protracts between brackets.

It is thought that static, rather than kinetic, frictional forces may be more relevant in orthodontic tooth movement. This is due to the bracket/archwire system having stopping and starting movements as the teeth are moved. *In vitro* testing of orthodontic archwires and brackets can be performed in either a wet or dry environment. Wet testing is usually done by submerging the archwire/bracket system in saliva, salivary substitute, or glycerin solution. Stannard *et al.* [5-43] measured kinetic coefficients of friction for stainless steel, beta Ti, NiTi, and Co-Cr archwires on a smooth stainless steel or Teflon surface. A universal material testing instrument was used to pull rectangular archwire (0.17×0.025 in) through pneumatically controlled binding surfaces. Coefficients of friction were determined under dry and wet (artificial saliva) conditions. It was found that (i) frictional force values (and thus coefficients of friction) were found to increase with increasing normal force for all materials, (ii) beta-Ti and stainless steel were sliding against stainless steel, and stainless-steel wire on Teflon consistently exhibited the lowest dry friction values, (iii) artificial saliva increased friction for stainless steel, beta-Ti and NiTi wires sliding against stainless steel, but (iv) artificial saliva did not increase friction of Co-Cr, stainless steel sliding against stainless steel, or stainless-steel wire on Teflon compared to the dry condition, and (v) stainless steel and beta-Ti wires sliding against stainless steel, and stainless-steel wire on Teflon, showed the lowest friction values for the wet conditions [5-43].

Stainless steel, Co-Cr, NiTi, and beta-Ti wires were tested in narrow single (0.050 in), medium (0.130 in), and wide twin (0.180 in) stainless-steel brackets in both 0.018 and 0.022 in slots. It was reported that (i) beta-Ti and NiTi wires generated greater amounts of frictional forces than stainless steel or Co-Cr wires did for most wire size, and (ii) increase in wire size generally resulted in increased bracket-wire friction [5-44]. However, it was reported that wires in ceramic brackets generated significantly stronger frictional forces than did wires in stainless steel-brackets [5-45].

Kusy *et al.* [5-46] evaluated coefficients of friction in the dry and wet (saliva) states of stainless steel, Co-Cr, NiTi, and beta-Ti wires against either stainless steel or polycrystalline alumina brackets. It was found that (i) in the dry state and regardless of slot size, the mean kinetic coefficients of friction were smallest for all the stainless-steel combinations (0.14) and largest for the beta-Ti wire combination (0.46), (ii) the coefficients of polycrystalline alumina combinations were generally greater than the corresponding combinations that included stainless steel-brackets, (iii) in the wet state, the kinetic coefficients of the all stainless-steel combinations increased to 0.05 over the dry state, but (iv) all beta-Ti wire combinations in the wet state decreased to 50% of the values in the dry state [5-46].

Surface roughness and static frictional force resistance of orthodontic NiTi archwires along with beta-Ti alloy wire, stainless-steel and Co-Cr alloy wires were measured. It was found that (i) the Co-Cr alloy and the NiTi alloy wires exhibited the lowest frictional resistance, (ii) the stainless-steel alloy and the beta-Ti wires showed the highest frictional resistance, and (iii) no significant correlation was found between arithmetic average roughness and frictional force values [5-47]. Oshida *et al.* [5-48] conducted friction tests to compare the wet static frictional forces of low friction "colors" (surfaces of wires are colored by anodizing process) TMA (Ti-Mo alloy) archwires with archwires of other materials (stainless steel, NiTi, and uncoated TMA), and to test the effects of repetitive sliding. The results showed that uncoated TMA wires produce the highest wet static frictional forces in all cases. In most cases, NiTi wires produced the next highest force levels followed by the colored TMA wires, and then stainless steel. Repetition seemed to have little to no effect on all of the archwires with the exception of NiTi and uncoated TMA. NiTi wires showed a decrease in force values as the wire was subjected to more repetitions. Uncoated TMA, to the contrary, showed that more friction developed between runs and a trend toward increasing friction as the wire had repeated use [5-48]. It is then recommended that further studies are required to correlate types of oxide(s) formed on Ti surfaces to frictional behaviors.

5.3.3 *Wear and wear debris toxicity*

Unlike dental implants, most orthopedic implants will be subjected to biotribological actions. Metal ions released from the implant surface are suspected of playing some contributing role in the loosening of hip and knee prostheses, which are substantially subjected to biotribological environments. Thompson and Puleo demonstrated that sublethal doses of the ionic constituents of Ti-6Al-4V alloy suppressed expression of the osteoblastic phenotype and deposition of a mineralized matrix [5-49]. Bone marrow stromal cells (which were cultured for 4 weeks) were harvested from juvenile rats and exposed to time-staggered doses of a solution of ion representing the Ti-6Al-4V alloy. It was reported that (i) Ti-6Al-4V solutions were found to produce little difference from control solutions for total protein or alkaline phosphatase levels, but strongly inhibited osteocalcin synthesis, and (ii) calcium levels were reduced when ions were added before a critical point of osteoblastic differentiation (between 2 and 3 weeks after seeding). On the basis of these findings, it was indicated that ions associated with the Ti-6Al-4V alloy inhibited the normal differentiation of bone marrow stromal cells to mature osteoblasts *in vitro*, suggesting that ions released from implants *in vivo* may contribute to implant failure by impairing normal bone deposition [5-49].

Although devices made of Ti-6Al-4V have been remarkably successful, primarily in orthopedic and dental applications, clinical reports have implicated the biological response to release metal from this class of metals as a cause of failure. Bianco *et al.* [5-50] hypothesized that in the absence of wear, the amount of titanium released is small and will preferentially accumulate in local tissues. One important implication is that measurable quantities of titanium in serum and urine that have been observed in clinical studies result from mechanically induced or -assisted release phenomena. In order to test this hypothesis, Bianco *et al.* [5-50] determined titanium levels in various tissues and fluids of animals both with and without titanium implants. Titanium fiber felts were implanted into the tibia of rabbits. At various time points, serum and urine samples were collected from these rabbits as well as from two groups of control rabbits. The samples were analyzed for titanium concentration using electrochemical absorption spectrophotometry. It was reported that titanium levels in serum and urine do not increase in comparison to control up to 1 year after implantation [5-50].

In total joint replacement, a polished metal surface generally articulates against an ultrahigh molecular weight polyethylene (UHMWPE) counter-bearing surface. Metals used include 316L stainless steel, Co-Cr-Mo alloy, and Ti-6Al-4V alloy (particularly with a hardened N^+ ion-implanted surface to form TiN layer). For long-term performance it is desired to minimize friction and UHMWPE wear, and it is also desirable to minimize metal ion release, which results from constant removal and reformation of passive surface oxides and oxyhydroxides during articulation. It was reported that (i) the inert ceramic bearing surfaces eliminate the oxidative wear of UHMWPE and Ti-6Al-4V, and also resistant to potential three-body wear from bone cement debris or potential stray porous metal coating material, but (ii) ceramic materials (e.g., alumina and zirconia) are available only for total hip replacement. For total knee replacement, although it is difficult and expensive to manufacture a monolithic ceramic knee surface, surface coating such as plasma-sprayed alumina and zirconia, TiN and amorphous diamond-like coatings through PVD/CVD methods, or nitrogen ion implantation and oxygen or nitrogen diffusion hardening can be applicable [5-51]. The wear of balls (of femoral stem implant) of nitrogen ion-implanted CpTi and Ti-6Al-4V alloy bearings against cups (of acetabular component) of UHMWPE was investigated using a simulator to approximate the conditions of an artificial hip joint. It was found that (i) differences in polymer wear rates related to variations in the surface finish of the balls, but not to ion implantation, were observed, (ii) the wear resistance of the femoral heads was significantly improved by the ion implantation, (iii) after one million cycles in the simulator (which is equivalent to approximately 1 year's use of a hip prosthesis in a patient), significant parts of the ion-implanted layer had worn off on parts of the specimen, and (iv) this metallic wear was more extensive with the CpTi samples than with those of Ti-6Al-4V [5-52].

Ti-6Al-7Nb discs were subjected to pin-on-disc wear tests against a high-fusing porcelain which is normally used as denture. The wear test was interrupted at 10,000 cycles, 50,000, 100,000, 150,000 and 200,000 cycles for X-ray diffraction examination. Two planes (010) and (011) were chosen for investigating the progressive wear damage characterization. It was found that (i) the plane (010) showed more sensitive than the plane (011) to the step-wise wear damage, (ii) the half-value width of the plane (010) showed continuously line-broadening, indicating either or both of non-uniform micro strain increasing and particle size decreasing, (iii) the plane (010) shifted toward higher 2θ angle (or lower d-spacing), indicating that crystalline grain was distorted during the wear action [5-53].

Metallic particles are clearly cytotoxic *in vitro*, whether being produced from titanium-based [5-54 – 5-57] or Co-Cr alloys [5-54, 5-58, 5-59]. This cytotoxicity is probably secondary to intracellular dissolution at low pH levels, since it can be affected by blocking H^+ release [5-60]. Discussions of biological response to wear debris have tended to emphasize the number of particles produced by a device in a period of time or, occasionally, the total weight or rate of production (by weight) of the debris. The *in vitro* investigations of macrophage response to small particles suggest that the total surface area of the debris may be an important parameter [5-61]. Wear debris particle size, produced by adhesion, tends to vary inversely with material rigidity (i.e., MOE), thus metallic and ceramic materials can be expected to produce smaller particles than the polyethylene debris released by conventional bearings [5-62].

A recent study comparing the induction of macrophage apoptotic cell death by alumina, zirconia, and polyethylene particulates found the response to be concentration and size dependent, and particle composition (alumina, zirconia, or polyethylene) had no effect [5-63]. However, in a rabbit model, Kubo *et al.* found marked histiocytic response around particles of UHMPE (11 μm), stainless steel (3.9 μm), and Co-Cr (3.9 μm). A significant reduction in the histiocytic response was found surrounding particles of alumina ceramics (3.9 μm) and titanium (3.9 μm) [5-64]. Catels *et al.* [5-65] found that larger particles (>2 μm) induced greater cell death than smaller particles, and inflammatory mediator (TNF-a release) was significantly higher with polyethylene particles when compared to alumina and zirconia. Osteolysis has been reported in association with ceramic wear debris. Yoon *et al.* [5-66] performed total hip arthroplastics with insertion of a ceramic femoral head and acetabular component in 96 patients to determine the radiographic prevalence of osteolysis. It was reported that (i) 10 patients required revision with gross and microscopic evidence of surface wear, and (ii) histology and SEM analyses of the periprosthetic membrane demonstrated "abundant" ceramic wear particles with a mean particle size of 0.71 μm (in a range from 0.13 to 7.2 μm), supporting the concept that ceramic wear particles can stimulate a foreign-body response and periprosthetic osteolysis.

The fresh surface will be revealed due to the formation of fraction/wear products, and there may be a chemical-reaction taking place between the freshly revealed surface of implants, and the surrounding environments, and the selective dissolution of the alloy constituents into the surrounding tissues. Besides, it is generally believed that the solid powder such as wear/friction products particles have allergic reaction to the living tissue, so that the biological effects of wear products (debris) should be considered separately from the biocompatibility of the implants. Particulate wear debris is detected in histocytes/macrophages of granulomatous tissues adjacent to loose joint prostheses. Such cell–particle interactions have been simulated *in vitro* by challenging macrophages with particles doses according to weight percent, volume percent, and number of particles. Macrophages stimulated by wear particles are expected to synthesize numerous factors affecting events in the bone–implant interface. Wear debris from orthopedic joint implants initiates a cascade of complex cellular events that can result in aseptic loosening of the prosthesis [5-67]. Macrophages are cells, which are mainly involved in phagocytosis and signaling and maintaining inflammation that leads to cell damage in soft tissues and bone resorption [5-68, 5-69]. The presence of large particles provides the major stimulus for cell recruitment and granuloma formation, whereas the small particles are likely to be the main stimulus for the activation of cells which release pro-inflammatory products. Apoptosis can be morphologically recognized by a number of features, such as loss of specialized membrane structures, blebbing, condensation of the cytoplasm, condensation of nuclear chromatin, and splitting of the cell into a cluster of membrane-bound bodies [5-68, 5-69]. Recent advances in the understanding of the mechanisms of osteoclastogenesis and osteoclast activation at the cellular and molecular levels have indicated that bone marrow-derived macrophages may play a dual role in osteolysis associated with the total joint replacement: (1) the major cell in host defense responds to UHMWPE particles via the production of cytokines, and (2) the precursors for the osteoclasts responsible for the ensuing bone resorption [5-70].

To avoid such wear debris toxicity, the wear resistance needs to be improved. Güleryüz *et al.* [5-71] performed a comparative investigation of thermal oxidation treatment for Ti-6Al-4V to determine the optimum oxidation conditions for evaluation of corrosion-wear properties. Characterization of modified surface layers was made by means of microscopic examinations, hardness measurements and X-ray diffraction analysis. Optimum oxidation condition was determined according to the results of accelerated corrosion tests made in 5 M HCl solution It was reported that (i) Ti-6Al-4V alloy exhibited excellent resistance to corrosion after oxidation at 600°C for 60 h, and (ii) this oxidation condition achieved 25 times higher wear resistance than the untreated alloy during a reciprocating wear test conducted in a 0.9% NaCl solution [5-71].

REFERENCES

[5-1] Suresh S. Fatigue of materials. Cambridge: Cambridge University Press, 1991.

[5-2] Piattellei A, Piattelli M, Scarano A, Montesani L. Light and scanning electron microscopic report of four fractured implants. Int J Oral Maxillofac Implants 1998;13:561–564.

[5-3] Guichet DL, Caputo AA, Choi H, Sorensen JA. Passivity of fit and marginal opening in screw- or cement-retained implant fixed partial denture designs. Int J Oral Maxillofac Implants 2000;15:239–246.

[5-4] Kotake M, Wakabayashi N, Ai M, Yoneyama T, Hamanaka H. Fatigue resistance of titanium-nickel alloy cast clasps. Int J Prosthodont 1997;10:547–552.

[5-5] Rodrugues RCS, Ribeiro RF, da G.C.de Mattos M, Bezzon OL. Comparative study of circumferential clasp retention force for titanium and cobalt-chromium removable partial dentures. J Prosthet Dent 2002;88: 290–296.

[5-6] Lewis AJ. Fatigue of removable partial denture casting during service. J Prosthet Dent 1978;39:147–149.

[5-7] Rohlmann A, Bergmann G. Graichen F, Mayer HM. Placing a bone graft more posteriorly may reduce the risk of pedicle screw breakage: analysis of an unexpected case of pedicle screw breakage. J Biomech 1998;31:763–767.

[5-8] Garcia R, Gorin S. Failure of posterior titanium atlantoaxial cable fixation: case studies. Spine J 2003;3:166–170.

[5-9] Long M, Rack HJ. Titanium alloys in total joint replacement: a materials science perspective. Biomaterials 1998;19:1621–1639.

[5-10] Duerig TW, Pelton AR, Stöckel D. The use of superelasticity in medicine. Metall 1996;50:569–574.

[5-11] Vallittu PK, Kokkonen M. Deflection fatigue of cobalt-chromium, titanium and gold aloys cast denture clasp. J Prosthet Dent 1995;74:412–419.

[5-12] Lin C-W, Ju C-P, Lin J-H C. A comparison of the fatigue behavior of cast Ti-7.5Mo with c.p. titanium, Ti-6Al-4V and Ti-13Nb-13Zr alloys. Biomaterials 2005;26:2899–2907.

[5-13] Keltjens HM, Mulder J, Kayser AF, Creugers NH. Fit of direct retainers in removable partial dentures after 8 years of use. J Oral Rehabil 1997;24: 138–142.

[5-14] Saito M, Notani K, Miura Y, Kawasaki T. Complications and failures in removable partial dentures: a clinical evaluation. J Oral Rehabil 2002;29:627–633.

[5-15] Hogmann E, Behr M, Handel G. Frequency and costs of technical failures of clasp- and double crown-retained removable partial dentures. Clin Oral Investig 2002;6:104–108.

[5-16] Carr AB, McGivney GP, Brown DT. McCracken's removable partial prosthodontics. 11th edition. St. Louis: Elsevier. 2005. pp. 79–115.

[5-17] Vallittu PK. Fatigue resistance and stress of wrought-steel wire clasps. J Prosthet Dent 1996;6:186–192.

[5-18] Gapido CG, Kobayashi H, Mitakawa O, Kohno S. Fatigue resistance of cast occlusal rests using Co-Cr and Ag-Pd-Cu-Au alloys. J Prosthet Dent 2003;90:261–269.

[5-19] Yuasa Y, Sato Y, Ohkawa S, Nagasawa T, Tsuru H. Finite element analysis of the relationship between clasp dimensions and flexibility. J Dent Res 1990;69:1664–1668.

[5-20] Sato Y, Yuasa Y, Akagawa Y, Ohkawa S. An investigation of preferable taper and thickness ratios for cast circumferential clasp arms using finite element analysis. Int J Prosthodont 1995;8:392–397.

[5-21] Sato Y, Abe Y, Yuasa Y, Akagawa Y. Effect of friction coefficient on Akers clasp retention. J Prosthet Dent 1997;78:22–27.

[5-22] Mahmoud A, Wakabayashi N, Takahashi H, Ohyama T. Deflection fatigue of Ti-6Al- 7Nb, Co-Cr, and gold alloy cast clasps. J Prosthet Dent 2005; 93:183–188.

[5-23] Styles CM, Evans SL, Gregson PJ. Development of fatigue lifetime predictive test methods for hip implants: Part I. Test methodology. Biomaterials 1998;19:1057–1065.

[5-24] Zavanelli RA, Guilherme AS, Pessanha-Henriques GE, Antônio De Arruda Nóbilo M, Mesquita MF. Corrosion-fatigue of laser-repaired commercially pure titanium and Ti-6Al-4V alloy under different test environments. J Oral Rehabil 2004;31:1029–1034.

[5-25] Guilherme AS, Henriques GEP, Zavanelli RA, Mesquita, MF. Surface roughness and fatigue performance of commercially pure titanium and Ti-6Al-4V alloy after different polishing protocols. J Prosthet Dent 2005;93:378–385.

[5-26] Malinov S, Sha W. Modeling thermodynamics, kinetics, and phase transformation morphology while heat treating titanium alloys. J Metals 2005; 67:42–45.

[5-27] Akahori T, Niinomi M, Fukunaga K-I, Inagaki I. Biomedical [alpha/beta] Titanium alloys. TMS Metall Mater Trans A: 2000;31A:1949–1958.

[5-28] Koahn DH, Ducheyne PH, Awerbuch J. Acoustic emission during fatigue of porous-coated Ti-6Al-4V implant alloy. J Biomed Mater Res 1992;26: 19–38.

[5-29] Wagner L, Gregory JK. Improve the fatigue life of titanium alloys. Part I. Adv Mater Processes 1994;146:35–36.

[5-30] Wagner L, Gregory JK. Improve the fatigue life of titanium alloys. Part II. Adv Mater Processes 1994;146:50–53.

[5-31] Li SJ, Niinomi M, Akahori T, Kasuga T, Yang R, Hao YL. Fatigue characteristics of bioactive glass-ceramic-coated Ti-29Nb-13Ta-4.6Zr for biomedical application. Biomaterials 2004;25:3369–3378.

[5-32] Banerjee R, Nag S, Stechschulte J, Fraser HL. Strengthening mechanisms in Ti-Nb-Zr-Ta and Ti-Mo-Zr-Fe orthopaedic alloys. Biomaterials 2004; 25:3414–3419.

[5-33] Steinmann SG. Corrosion of surgical implants – *in vivo* and *in vitro* tests. In: Evaluation of biomaterials. Winter GD, Leray JL, de Groot K, editors. New York: Wiley, 1980.

[5-34] Steinmann SG. Corrosion of titanium, titanium alloy for surgical implants. Titanium 84': Science and technology Vol. 2, Munich: Desche Ges Meatllkd EV. 1985. pp. 1373–1379.

[5-35] Rao S, Uchida T, Tateishi T, Okazaki T, Asao Y. Effect of Ti, Al and V ions on the relative growth rate of fibroblasts (L929) and osteblasts (MC3T3-E1) cells. J Biomed Mater Eng 1996;6:79–86.

[5-36] Walker PR, LeBlanc J, Sikorska M. Effects of aluminum and other cations on the structure of brain and liver chromatin. Biochemistry 1989;28: 3911–3915.

[5-37] Kim K-H, Choi M-Y, Kishi T. Fracture analysis of cast pure Ti and Ti-6Al-4V alloy for dental use. J Biomed Mater Eng 1997;7:271–276.

[5-38] Ivasishin OM, Markovsky PE. Enhancing the mechanical properties of titanium alloys with rapid heat treatment. J Metals 1996;48:48–61.

[5-39] Mante FK, Baran GR, Lucas B. Nanoindentation studies of titanium single crystals. Biomaterials 1999;20:1051–1055.

[5-40] Tanaka S, Takahashi Y, Tajima S, Shiratori N, Ito M. Relationship between heat treatment and properties for titanium implant material. J Jpn Soc Oral Implantol 2004;17:202–208.

[5-41] Dumbleton JH. Tribology Series 3: Tribology of natural and artificial joints. New York: Elsevier, 1981.

[5-42] Zwicker U, Breme J. Investigations of the friction behaviour of oxidized Ti-5%Al-2.5%Fe surface layers on implant material. J Less-Common Metals 1994;100:371–375.

[5-43] Stannard JG, Gau JM, Hanna MA. Comparative friction of orthodontic wires under dry and wet conditions. Am J Orthod 1986;89:485–491.

[5-44] Kapila S, Angolkar PV, Duncanson MG, Nanda RS. Evaluation of friction between edgewise stainless steel brackets and orthodontic wires of four alloys. Am J Orthod Dentofacial Orthop 1990;98:117–126.

[5-45] Angolkar PV, Kapila S, Duncanson MG, Nanda RS. Evaluation of friction between ceramic brackets and orthodontic wires of four alloys. Am J Orthod Dentofacial Orthop 1990;98:499–506.

[5-46] Kusy RP, Whitley JQ, Prewitt MJ. Comparison of the frictional coefficients for selected archwire-bracket slot combinations in the dry and wet state. Angle Orthod 1991;61:293–301.

[5-47] Prososki RR, Bagby MD, Erickson LC. Static frictional force and surface roughness of nickel-titanium arch wires. Am J Orthod Dentofacial Orthop 1991;100:341–348.

[5-48] Oshida Y, Kotona TR, Farzin-Nia F, Rosenthall MR. Comparison of frictional forces between three grades of low friction "colors" TMA. In: Medical device materials. Shrivastava S, editor. Materials Park: ASM International, 2004. pp. 455–458.

[5-49] Thompson GJ, Puleo DA. Ti-6Al-4V ion solution inhibition of osteogenic cell phenotype as a function of differentiation timecourse *in vitro*. Biomaterials 1996;17:1949–1954.

[5-50] Bianco DP, Ducheyne P, Cuckler JM. Titanium serum and urine levels in rabbits with a titanium implant in the absence of wear. Biomaterials 1996;17:1937–1942.

[5-51] Davidson JA, Mishra AK, Surface modification issues for orthopaedic implant bearingsurfaces. In: Surface modification technologies. Sudarshan TS, editor. Cambridge: The University Press. 1992. pp. 1–14.

[5-52] Lausmaa J, Roestlund T, McKellop H. Wear of ion-implanted pure titanium and Ti-6Al-4V alloy against UHMWPE. In. Surface modification technologies. Sudarshan, TS, editor. The Institute of Materials. Cambridge: The University Press. 1992. pp. 139–154.

[5-53] Bartolovic DJ. *In Vitro* wear characterization of Ti-6Al-7Nb for dental prostheses against high-fusing Porcelain. Indiana University Master Thesis, 2005.

[5-54] Haynes DR, Boyle SJ, Rogers SD, Susan D, Howie DW, Barrie V-R. Variation in cytokines induced by particles from different prosthetic materials. Clin Orthop 998;352:223–230.

[5-55] Nakashima Y, Sun DH, Trindale MC, Trindale MCD, Chun LE, Song Y, Goodman SB, Schuman DJ, Maloney WJ, Smith RL. Induction of macrophase C-C chemokine expression by titanium alloy and bone cement particles. J Bone Joint Surg Br 1999;81:155–162.

[5-56] Shanbhag AS, Jacobs JJ, Black J. Human monocyte response to particulate biomaterials generated *in vivo* and *in vitro*. J Orthop Res 1995;13: 792–801.

[5-57] Shanbhag AS, Jacobs JJ, Black J, Galante JO, Glant TT. Macrophage/particle interactions: effect of size, composition and surface area. J Biomed Mater Res 1994;28:81–90.

[5-58] Howie DW, Rogers SD, McGee MA, Haynes DR. Biologic effects of cobalt chrome in cell and animal models. Clin Orthop 1996;329S:S217-S232.

[5-59] Maloney WJ, Smith RL, Castro F, Schurman DJ. Fibroblast response to metallic debris *in vitro*. Enzyme induction cell proliferation, and toxicity. J Bone Joint Surg Am 1993;75:835–844.

[5-60] Haynes DR, Rogers SD, Howie DW, Pearcy MJ, Barrie V-R. Drug inhibition of the macrophage response to metal wear particles *in vitro*. Clin Orthop 1996;323:316–326.

[5-61] Shanbhag AS, Jacobs JJ, Black J, Galante JO, Glant TT. Effects of particles on fibroblast proliferation and bone resorption *in vitro*. Clin Orthop 1997; 342:205–217.

[5-62] Black J. Biological performance of materials: fundamental of biocompatibility. New York: Marcel Dekker, 1992.

[5-63] Catelas I, Petit A, Zukor DJ, Marchand R, Yahia L'Hocine, Huk OL. Induction of macrophage apoptosis by ceramic and polyethylene particles *in vitro*. Biomaterials 1999;20:625–630.

[5-64] Kubo T. Sawada K, Hirakawa K, Shimizu C, Takamatsu T, Hirasawa Y. Histiocyte reaction in rabbit femurs to UHMWPE, metal, and ceramic particles in different sizes. J Biomed Mater Res 1999;45:363–369.

[5-65] Catelas I, Huk OL, Petit A, Zukor DJ, Marchand R, Yahia L'Hocine. Flow cytometric analysis of macrophage response to ceramic and polyethylene particles: effects of size, concentration, and composition. J Biomed Mater Res 1998;41:600–607.

[5-66] Yoon TR, Rowe SM, Jung ST, Seon KJ, Maloney WJ. Osteolysis in association with a total hip arthroplasty with ceramic bearing surfaces. J Bone Joint Surg Am 1998;80:1459–1468.

[5-67] Stea S, Visentin M, Granchi D, Cenni E, Ciapetti G, Sudanese A, Toni A. Apoptosis in peri-implant tissue. Biomaterials 2000;21:1393–1398.

[5-68] Chiba J, Rubash H, Kim KJ, Iwaki Y. The characterization of cytokines in the interface tissue obtained from failed cementless total hip arthroplasty with and without femoral osteolysis. Clin Orthop Rel Res 1994;300: 304–312.

[5-69] Perry M, Murtuza FY, Ponsford FM, Elson CJ, Atkins RM, Learmonth D. Properties of tissue from and around cemented joint implants with erosive and/or linear osteolysis. Arthroplasty 1997;12:670–676.

[5-70] Ingham E, Fisher J. The role of macrophages in osteolysis of total hip joint replacement. Biomaterials 2005;26:1271–1286.

[5-71] Güleryüz H, Çimenoğlu H. Effect of thermal oxidation on corrosion and corrosion–wear behaviour of a Ti-6Al-4V alloy. Biomaterials 2004;25: 3325–3333.

Chapter 6
Biological Reaction

6.1.	Toxicity	127
6.2.	Cytocompatibility (Toxicity to Cells)	130
6.3.	Allergic Reaction	133
6.4.	Metabolism	136
6.5.	Biocompatibility	138
6.6.	Biomechanical Compatibility	143
References		148

Chapter 6
Biological Reaction

Titanium materials are exposed to biological environments, and their behaviors in such environments need to be known. This chapter will discuss biological reactions in general.

6.1. TOXICITY

Toxicity can be measured by the effects on the target (organism, organ, or tissue). Because individuals typically have different levels of response to the same dose of a toxin, a population-level measure of toxicity is often used which relates the probability of an outcome for a given individual in a population. One such measure is the LD50 (LD stands for lethal dose), which is a concentration measure for a toxin at which 50% of the members of an exposed population die from exposure [6-1]. When such data does not exist, estimates are made by comparison to known similar toxic substances, or to similar exposures in similar organisms. Then the safety factors must be built in to protect against the uncertainties of such comparisons, in order to improve protection against these unknowns. Toxicity can refer to the effect on a whole organism (such as a human, a bacterium, or a plant), or to a substructure (such as the liver). In the science of toxicology, the subject of such study is the effect of an external substance or condition and its deleterious effects on living things: organisms, organ systems, individual organs, tissues, cells, and subcellular units. Toxicity of a substance can be affected by many different factors, such as the route of administration, the time of exposure, the number of exposures, the physical form of the toxin (solid, liquid, gaseous), the genetic makeup of an individual, an individual's overall health, and many others.

There are generally three types of toxic entities: chemical, biological, and physical.

(1) *Chemicals* include both inorganic substances, such as lead, hydrofluoric acid, and chlorine gas, as well as organic compounds such as ethyl alcohol, most medications, and poisons from living things.

Chemicals (which can be toxic) should include corrosion product such as oxides, chlorides, or sulfides as well as dissolved metallic ions through anodic reaction. The toxicity is related to primary biodegradation products (simple and complex cations and anions), particularly those of higher atomic weight metals. Factors to be considered include (1) the amount dissolved by biodegradation per time unit, (2) the amount of material removed by metabolic activity in the same time unit, and (3) quantities of solid particles and ions deposited in the tissue and any associated transfers to the systemic system. Hence, the extent of toxicity is indirectly related to corrosion rate [6-2–6-5]. An *in vivo* corrosion experiment can always be followed up by a tissue examination. This type of tissue reaction is observed for both corrosion-resistant materials (Mo-containing stainless steel and Co-Cr-Mo alloys) and for the strongly corroding metals Fe, Mo, and Al. It must be assumed that the unwanted corrosion product itself is the determining factor. Fe, Mo, and Al, as non-toxic metals, do not trigger an extreme tissue reaction despite high corrosion rates (about 300 times that of stainless steel). On the other hand, the very low corrosion rates of stainless steel and Co-Cr alloys, comparable to that of Ti, Nb, and Zr, release sufficient quantities of the highly toxic elements nickel and cobalt, to trigger a noticeable tissue reaction. It should be mentioned at this point that no limiting concentration is commonly admitted for an allergic reaction. Gold and silver are also to be found in this middle group, as these noble metals can also corrode in living tissue. As demonstrated above, a low corrosion rate alone is not sufficient to guarantee compatibility, nor is the elemental dose the only determining factor. It would appear that the intrinsic toxicity of the element and its ability to bind to macromolecules are equally important. Chemical data for Ti, Zr, Nb, and Ta show that organic complexes of this type are unlikely, and it is also known that organo-metallic compounds of these elements, as far as they exist, are unstable [6-6].

There are certain types of elements which are essential for living organisms and types of elements are different between kinds of living substances. For plants, there are macronutrients inlcuding N, P, S, Ca, Mg, and Fe, and micronutrients such as Mn, Zn, Cu, Mo, B, and Cl [6-7]. On the other hand, essential trace elements for living organisms (particularly, the higher animals) include Fe, F, Si, Zn, Sr, Pb, Mn, Cu, Sn, Se, I, Mo, Ni, B, Cr, As, Co, and V. Among these elements, the following nine elements are considered as the most important bioelements, since they are able to exist as a constituting element for metalloproteins, metalloenzymes, or even vitamins: Fe, Zn, Mn, Cu, Se, Mo, Ni, Co, and V [6-8, 6-9].

Despite reports associating tissue necrosis with implant failure, the degree to which processes, such as metal toxicity, negatively impact implant performance is unknown. Hallab *et al.* [6-10] evaluated representative human peri-implant cells (i.e., osteoblasts, fibroblasts, and lymphocytes) when challenged by Al^{3+}, Co^{2+}, Cr^{3+}, Fe^{3+}, Mo^{5+}, Ni^{2+}, and V^{3+} chloride solutions (and Na^{2+} as a control) over a wide range of concentrations (0.01–10.0 mM). It was reported that (i) differential effects were found to be less a function of the cell type than of the composition and concentration of metal challenge, (ii) no preferential immuno-suppression was demonstrated, and (iii) soluble Co released from Co-Cr alloy and V released from

Ti-6Al-4V alloy could be implicated as the most likely to mediate cell toxicity in the peri-prosthetic milieu [6-10].

Maurer *et al.*[6-11] using diluted concentration of vanadate and titanium tetrachloride as salts, investigated the cellular uptake of titanium and vanadium. It was found that (i) titanium was not observed to be toxic to the cells, but (ii) vanadium was toxic at levels greater than 10 μg/ml. The percentage of cellular association of titanium was shown to be about 10 times that of vanadium. The percentage of cellular association of either element was greater from fretting corrosion than from the addition of salts. It was also mentioned that (i) the presence of vanadium did not affect the cellular uptake of titanium, but (ii) the presence of titanium decreased the cell association of vanadium [6-11].

Corrosion resistance is one of the most important properties determining the biocompatibility of metallic biomaterials. The high corrosion resistance of metallic materials is attributed to a passive film on their surfaces. However, the passive film on these metals may be continuously damaged by wear when these materials are used as bone plates and screws, artificial joints, or dental restorations. The wear on metallic biomaterials can generate debris and accelerate metal ion release. Powders of Co-Cr-Mo alloy were produced by grinding larger particles for 8 h in water, serum, or joint fluid, and were administered in low doses (0.05–0.5 mg/ml) for 1–6 days to human dermal fibroblasts *in vitro*. Particles were ground in a biological fluid in order to simulate conditions in an artificial hip joint. It was found that (i) such particles adhered to, or were phagocytosed, by the cells far less than those ground in water, and (ii) the toxicity of the alloy was linked with a failure of test cells to grow as quickly as the controls – particles ground in water were the most toxic [6-12].

(2) *Biological* toxicity can be more complicated to measure. In a host with an intact immune system the inherent toxicity of the organism is balanced by the host's ability to fight back; the effective toxicity is then a combination of both parts of the relationship. A similar situation is also present with other types of toxic agents. In particular, toxicity of cancer-causing agents is problematic, since for many such substances it is not certain if there is a minimal effective dose or whether the risk is just too small to see. Here, too, the possibility exists that a single cell transformed into a cancer cell is all it takes to develop the full effect. Yang *et al.* [6-13] demonstrated an increase in levels of elemental Ti in the blood and lung of rats with a Ti alloy implant. Rats were implanted with Ti alloy disks for 4 weeks. The levels of elemental Ti in the blood and lung were especially increased compared with other tissues. The Ti alloy implant enhanced lung injury related to endotoxin from Gram-negative bacteria (lipopolysaccharide, LPS), which was characterized by lung edema and other histological changes such as recruitment of neutrophils, interstitial edema, and alveolar hemorrhage in the lung. In the presence of endotoxin, an

increase of nitrite production was shown in the plasma and bronchoalveolar lavage fluid of rats implanted with a Ti alloy. Moreover, the Ti alloy implant further enhanced the induction of inducible nitric oxide (NO) synthase (iNOS) protein expression induced by LPS in the lung. These endotoxin-related responses in the presence or absence of the Ti alloy implant could be inhibited by aminoguanidine (an iNOS inhibitor). It was first indicated that circulating Ti released from Ti alloy implants had an ability to affect pulmonary iNOS protein expression, and enhance the pathogenesis of acute lung injury during endotoxemia [6-13].

(3) *Physically* toxic entities include things not usually thought of as such by the lay person: direct blows, concussion, sound and vibration, heat and cold, non-ionizing electromagnetic radiation such as infrared and visible light, ionizing non-particulate radiation, such as X-rays and gamma rays, and particulate radiation, such as alpha, beta, and cosmic rays [6-14].

Besides these, particulate macrophages should be included here; even species per se are not toxic, but because of size, they will exhibit physical toxicity. Particulate wear-debris is detected in histiocyte/macrophages of granulomatous tissues adjacent to loose joint prostheses. Such cell-particle interactions have been simulated *in vitro* by challenging macrophages with particles dosed according to weight percent, volume percent, and number of particles. Each of these dosage methods has inherent shortcomings due to varying size and density of challenging particles of different compositions. In studies done by Shanbhag *et al.* [6-15], P388D1 macrophages with titania (TiO_2) and polystyrene particles (<2 μm), with dosage based on the ratio of the cells, were challenged. The effect of size and composition on the bone reabsorbing activity, and fibroblast proliferation was studied. It was reported that (i) macrophage response to particulate debris appears to be dependent on particle size, composition, and dose as given by surface area ratio, and (ii) macrophages stimulated by wear particles are expected to synthesize numerous factors affecting events in the bone–implant interface [6-15].

6.2. CYTOCOMPATIBILITY (TOXICITY TO CELLS)

Over the years, many metal and polymer implants have been developed for internal fracture fixation. However, there are some problems associated with their application, such as implant loosening or infection. Cytotoxicity of soluble and particulate Ti, V, and Ni by biochemical functional analysis and by microscopic morphology with micro-area elemental analysis *in vitro* using human neutrophils as probes and *in vivo* in animals was compared. It was concluded that (i) the biochemical analysis of LDH (lactate dehydrogenase), superoxide anion, inflammatory mediator TNF-α, and SEM showed that Ni in solution destroys the cell membranes of neutrophils, whereas Ti and V in solution stimulate neutrophils and increase the quantity of released superoxide anions, (ii) fine Ti particles (1–3 μm), which are smaller than neutrophils (~5 μm), were phagocytized by the cells, and the results were similar *in vivo*, and (iii) these results showed that the cytotoxic effect of Ti particles is size dependent, and that they must be smaller than that of cells [6-16].

Kumazawa *et al.* [6-17] studied the morphology and adhesion of both fibroblasts and osteoblasts to manufactured commercially pure, medical implant-quality anodized titanium surfaces, and five modified titanium surfaces, and examined their relative cytocompatibility to assess which modified surface could potentially be used *in vivo*. It was reported that (i) small variations were observed both qualitatively and quantitatively in the spreading and adhesion of fibroblasts and osteoblasts to the studied surfaces, but (ii) noticeable fibroblast and osteoblast spreading and adhesion were found on the anodized surfaces, and (iii) coating medical implant-quality anodized titanium surfaces could probably improve soft-tissue adhesion and/or osseointegration of bone *in vivo* [6-17].

The cytotoxicity and mutagenicity of metal salts were systematically evaluated by Yamamoto *et al.* [6-18, 6-19]. Metal ions released from the wear debris of Ti alloys [6-20], and the cytotoxicity of ceramic particles [6-21] were investigated, and it was confirmed that active oxygen species generated by macrophages phagocytosing high density polyethylene particles induced metal ion release from Ti [6-22].

Manceur *et al.* [6-23] reported on the biocompatibility of NiTi single crystals, since certain orientations of single crystals present several advantages over the polycrystalline form in terms of maximal strain, fatigue resistance, and temperature range of superelasticity. The *in vitro* biocompatibility of 50.8 wt% Ni-Ti single crystals in the orientation <001> were conducted after four different heat treatments in a helium atmosphere followed by mechanical polishing. The study was performed on the material extracts after immersion of the specimens in cell culture medium (DMEM) for 7 days at 37°C. Cytotoxicity studies were performed on L-929 mouse fibroblasts using the MTT assay. J-774 macrophages were used to assess the potential inflammatory effect of the extracts by IL1-β and TNF-α dosage (sandwich ELISA method). It was found that (i) exposure of L-929 to material extracts did not affect cell viability, and (ii) IL1-β and TNF-α secretion was not stimulated after incubation with NiTi extracts compared to the negative controls, concluding that TiNi single crystals do not trigger a cytotoxic reaction [6-23].

Titanium alloys have been proven to behave poorly in friction due to detection of wear particles in tissues and organs associated with titanium implants. Bordji *et al.* [6-24] investigated three surface treatments in order to improve the wear resistance and the hardness of Ti-6Al-4V and Ti-5Al-2.5Fe: (1) glow discharge nitrogen implantation (10–17 atoms/cm^2), (2) plasma nitriding by plasma diffusion treatment (PDT), and (3) deposition of TiN layer by plasma-assisted chemical vapor deposition (PACVD) additionally to PDT. A cell culture model using human cells was chosen to study the effect of such treatments on the cytocompatibility of these materials. The results showed that (i) Ti-5Al-2.5Fe alloy was as cytocompatible as the Ti-6Al-4V alloy, and the same surface treatment led to identical biological consequences on both alloys, (ii) nitrogen implantation did not at all modify the cellular behavior observed on untreated samples, (iii) after the two nitriding treatments, cell proliferation and viability appeared to be significantly reduced, and the scanning electron microscopy study revealed somewhat irregular surface states, but (iv) osteoblast phenotype expression and protein synthesis capacity were not affected. Hence, it was concluded that plasma diffusion nitriding treatment and TiN deposition by chemical vapor process may be an interesting alternative to the physical vapor deposition technique [6-24]. Wever *et al.* [6-25] carried out a dilution minimal essential medium extract cytotoxicity test, a guinea-pig sensitization test and two genotoxicity (Salmonella reverse mutation and the chromosomal aberration) tests on NiTi. It was found that the NiTi alloy showed no cytotoxic, allergic, or genotoxic activity similar to the clinical reference control material AISI 316 stainless steel.

By the prolonged use of NiTi material in a human body, deterioration of the corrosion resistance of the materials becomes a critical issue because of the increasing possibility of deleterious ions released from the substrate to living tissues. It was proven that a certain surface modifications, such as nitrogen, acetylene, and oxygen plasma immersion ion implantation (PIII or PI^3), can improve the corrosion resistance and mechanical properties of the materials. With this modification, it was reported that (i) the release of nickel is drastically reduced as compared with the untreated control, and (ii) the *in vitro* tests show that the plasma-treated surfaces are well tolerated by osteoblasts [6-26, 6-27].

A number of ternary NiTi-X alloys have been introduced with the aim to improve the fatigue properties and to decrease the thermal hysteresis range; well-known ternary alloying elements are Cu and Fe [6-11, 6-13, 6-28]. Particularly, Cu has been shown to dissolve in the B2 (austenitic) phase in concentrations up to 30 at%. However, NiTi-Cu solid solutions containing more than 10 at% are characterized by poor formability, so that alloys of technical interest usually contain Cu in the range from 5 to 10 at%. NiTi (57.6 w/o Ni, 42.4 w/o Ti) and NiTiCu (50.7 w/o Ni, 42.4 w/o Ti, 6.9 w/o Cu) were evaluated by the Tada method [6-28]. It was concluded that (i) a significant difference in the *in vivo* biocompatibility between the binary and ternary alloy was shown and (ii) the presence of the binary alloy lead to few morphological changes and a slight reduction of dehydrogenase activity of epithelial cell cultures, whereas the effect of the ternary alloy appears to be more pronounced, obviously due to the releasing of copper ions [6-29].

Newly developed β-type titanium alloys, Ti-29Nb-13Ta-4.6Zr, CpTi, and Ti-6Al-4V were evaluated in terms of cytotoxicity [6-30]. Each plate specimen was put on zirconia balls in a vessel, and in an autoclave. The vessel with 10 ml Eagle's culture solution at a temperature of 310° K was rotated with a speed of 240 rpm, and extraction periods were 7 and 14 days. As-extracted solution and filtrated extract solution using a 0.22 μm membrane filter were prepared. In the As-extracted solution and filtrated extracted solution, the survival rate of L929 cells derived from mice was evaluated using the NR [6-31] and MTT [6-32] methods. The exposure period of L929 cells to extract the solutions was 2 days. The MTT method is a method for evaluating cell respiration by the mitochondria, while the NR method is a method for evaluating the ratio of sound cells of which cell membranes have no lesion by the amount of Neutral Red (NR) pigment. The survival rate of L929 cells indicates the cyto-toxicity level of the extracts, and it was also reported that the new alloy is equivalent to that of CpTi and Ti-6Al-4V [6-30].

Spriano *et al.* [6-33] treated Ti-6Al-7Nb as prosthetic implant material by various methods: surfaces were (1) mechanically polished with SiC paper, (2) passivated in 35 wt% HNO_3 for 30 min, (3) treated in alkali 5 M NaOH at 60°C for 24 h, (4) heated at 600°C for 1 h in air, and (5) heated at 600°C for 1 h in vacuum. These treated surfaces were characterized in order to evaluate their surface properties and cell interaction, in order to assess if the treatments were effective in obtaining a surface presenting at the same time bone-like apatite induction ability, and low metal ion release. Wettability was examined by surface contact angle measurements. The untreated surface showed 55°, which was the highest value of surface contact angle, indicating that the untreated surface is not at an active stage. On SiC-polished surface, it was 50°, then it follows very low values of contact angle; the in-air oxidized surface exhibits almost zero degree of contact angle, indicating very high wettability. It was also reported that such active surface was covered with a film of a composition close to standard Ca/P ratio bone-like apatite after the surfaces were immersed in SBF (simulated body fluid) for 30 days [6-33].

6.3. ALLERGIC REACTION

Allergy is the opposite of AIDS (acquired immune deficiency syndrome). Allergy symptoms are the result of too much immunity. The immune system produces antibodies to fight infections. If one has AIDS, he has too little immunity to defend himself against getting sick. AIDS is too little immunity, and allergy is too much. Metal sensitivity is thought to be a very important factor in the overall biocompatibility of implants. Although titanium has not been found to show sensitivity, other materials such as nickel from stainless steel, other high nickel alloys, and cobalt from cobalt-based alloys have shown to be sensitizers in skin dermatitis and might, therefore, show an allergic response upon implantation [6-34]. Allergic contact dermatitis to metals is a common skin disease in many countries of the world. Ni allergy is most frequent, and the prevalence is reported to be 10% in females and 1% in men [6-35]. The incidence of Ni allergy has increased especially in the young female generation, where the cause is probably due to the increased habit of ear piercing in females in recent years [6-36, 6-37]. Cr is another well-known metal allergen. Especially in men, occupational chromium dermatitis occurs among cement workers, Cr platers, and workers dealing with leather tanning [6-38].

Ti is considered as a non-sensitizing or less sensitizing metal, and the application of Ti materials is increasing rapidly in a variety of products such as spectacle' rims, wrist watches, and dental/medical materials. However, there are some reports of contact dermatitis due to Ti [60-39, 6-40]. Titanium allergy is barely recognized in mainstream medicine – yet it was reported that as many as 1 in 10 people can be affected by it. For those afflicted with titanium allergy, the symptoms can be multiple and bewildering. These can range from simple skin rashes to muscle pain and fatigue. From foodstuff to medicine, titanium is now an everyday metal. Several brands of candy, such as Skittles and M&M, have titanium dioxide in the coating – often described by its E-number: E171. Some brands of toothpaste and cold medicine contain titanium particles, as discussed in Chapter 4. Like all metals, titanium releases particles through normal corrosion. These metals become ions in the body and then bind to body proteins. For those who react, the body will try to attack this structure. This starts a chain reaction which can lead to many symptoms including chronic fatigue syndrome (CFS) or, in the most severe cases, multiple sclerosis (MS). The MELISA® test is the only scientifically proven test which can diagnose titanium allergy and measure its severity. Those who test positive should avoid exposure or remove the titanium from their body to stop the internal reaction. This can be simple, like changing a brand of toothpaste. Or it may be more complex, such as replacing titanium implants [6-41].

Identification of the causative chemical of contact dermatitis, comparison of the sensitization potency of chemicals and the finding of cross-reactivity among these chemicals are important to develop a safe material for the metal-sensitized patients. For this purpose, reproducible animal models are useful. Cr sensitization is easily induced in animals by using potassium dichromate, which is recognized as a strong sensitizer [6-42–6-46]. Although the sensitization rate to Ni in humans is the highest among various allergens, the sensitization potential of Ni salts was reported to be weak to mild in the animal studies [6-42, 6-43, 6-47–6-50]. On the other hand, there is little knowledge about the sensitization potential of zirconium and titanium salts by animal assays using these salts. Ikarashi *et al.* [6-51] evaluated contact sensitization capacity of four metal salts, nickel sulfate ($NiSO_4$), potassium dichromate ($K_2Cr_2O_7$), titanium chloride ($TiCl_4$), and zirconium chloride ($ZrCl_4$), using guinea pigs and mice. In the guinea-pig sensitization tests, an injection concentration to 1% for all chemicals was set up, and the challenge concentration was changed. Guinea pigs were sensitized with $NiSO_4$, $K_2Cr_2O_7$, and $TiCl_4$. It was found that (i) among the test metal salts, $K_2Cr_2O_7$ showed the highest sensitization rate and strongest skin reactions, (ii) $ZrCl_4$ did not cause any sensitization responses under our experimental conditions, (iii) minimal challenge concentration to cause a skin response was $<0.25\%$ for $K_2Cr_2O_7$, 0.5% for $NiSO_4$, 2% for $TiCl_4$, and (iv) in the sensitive mouse lymph node assay, $TiCl_4$ as well as $ZrCl_4$ caused mild lymph node responses, but was classified as a non-sensitizer. Based on these findings, it was concluded that the order of sensitization potential was $K_2Cr_2O_7>NiSO_4>TiCl_4>ZrCl_4$ [6-51].

Titanium brackets are used in orthodontic patients with an allergy to nickel and other specific substances. Harzer *et al.* [6-52] studied the corrosive properties of fluoride-containing toothpastes with different pH values on titanium brackets. Molar bands were placed on 18 orthodontic patients. In these same patients, titanium brackets were bonded on the left quadrants and stainless-steel brackets on the right quadrants of the upper and lower arches. Fifteen patients used Gel Kam containing soluble tin fluoride (pH 3.2), whereas three used fluoride-free toothpaste. The brackets were removed for evaluation by light microscopy and scanning microscopy 5.5–7.0 months and 7.5–17 months after bonding. It was observed that (i) the matte gray color of titanium brackets dominating over the silver gleam of the steel brackets was noticed, (ii) the plaque accumulation on titanium brackets was high, and (iii) pitting corrosion and crevices were observed in only 3 of the 165 brackets tested. It was concluded that the *in vivo* study confirms the results of *in vitro* studies, but the changes are so minor that titanium brackets can safely be used for up to 18 months [6-52].

Although no cytotoxic effect caused by NiTi alloy has been reported [6-53], information concerning the biological side effects of Ni is available in literature [6-54–6-57]. Ni is capable of eliciting toxic and allergic responses [6-54]. Ni can produce more allergic reactions than all other metal elements [6-55]. The *in vivo* study, NiTi alloys show cytotoxic reactions [6-56]. Cases also show the conversion of Ni-non-sensitive subjects into Ni-sensitive subjects following the use of NiTi wires [6-57].

It is well documented that some metals are known to produce inflammatory responses (e.g., Cr, Co, Cu, Ni, Pd, Ti, Zr, and their based alloys) *in vitro* [6-58–6-60] and allergic reactions (e.g., Cr, Co, Ni, Ti, and NiTi) *in vivo* [6-61, 6-62]. Several studies highlight that elements can diffuse from appliances and accumulate in tissues [6-63, 6-64]. Ti-6Al-4V is the most frequently employed as an orthopedic implant material as well as device. However, it was reported that Al is an element involved in severe neurological, e.g., Alzheimer's disease [6-65] and metabolic-bone disease, e.g., osteomalacia [6-66]. Al accumulated in bone is only a temporal soft-tissue deprivation [6-67]. Bone remodeling will produce further Al release, even after the appliances no longer leach out Al, perpetrating the dangerous side affect of this element [6-65, 6-66]. Zaffe *et al.* [6-67] studied CpTi plates, Ti-6Al-4V screws, and surrounding tissues to evaluate the release and accumulation of ions. Optical microscopy, SEM, and X-ray microanalysis were carried out to evaluate the plates and screws removed from patients presenting inflammation and biopsies. It was reported that (i) ions released from metallic appliances or leaching from granules toward soft tissues was observed, (ii) an accumulation of Al but not Ti was found in soft tissues, and (iii) a peculiar accumulation of Al in the dense lamella of newly formed bone was recorded, indicating that biological perturbations may be related to Al release from the tested biomaterials [6-67].

6.4. METABOLISM

Most attempts to determine the concentration of titanium in animals and plant tissues have met with little success because of the very low levels involved [6-68]. In man, the highest levels of titanium appear to be associated with the spleen and adrenals, at around 10 μg/100 g with much lower amounts in the liver (3 μg) and kidneys (1.5 μg). It was demonstrated that there is increased titanium content in the lungs with increasing age in humans, but there was little evidence of accumulation in other tissues [6-69]. Very little exists concerning the metabolism of titanium and indeed there is some confusion over this subject. Generally, it appears there is little absorption of titanium in the gastrointestinal system, although the little that is absorbed appears to be stored in the heart, lungs, spleen, and kidneys of experimental animals [6-70].

It has been shown that titanium oxide (TiO_2) is transported in blood by phagocytic monocytes and deposited in organs such as liver, spleen, and lung 6 months after intraperitoneal injection [6-71]. Furthermore, it is well known that exposure to metal traces alters the cellular redox status. Rats were intraperitoneally injected with 1.60 g/100 g body weight of TiO_2 in saline solution. Organs (liver, spleen, lung) were processed for histological evaluation. Reactive oxygen species in alveolar macrophages obtained by bronchoalveolar lavage were evaluated using the nitroblue tetrazolium test and quantitative evaluation by digital image analysis. It was found that (i) the histological analysis of organs revealed the presence of titanium in the parenchyma of these organs with no associated tissue damage, and (ii) although in lung alveolar macrophages TiO_2 induced a significant rise in reactive oxygen species generation, it failed to cause tissue alteration [6-72].

Schlegel *et al.* [6-73] evaluated the Ti level within the muco-periosteal flaps covering submerged Ti implants. It was assumed that implant-covering flap was never absolutely immobile, and the existing small movement may have led to an eraser-like effect which impregnated the tissue continuously with titanium. Since Ti transfer to the muco-periosteal flap during insertion is only a minor possibility due to the standardized surgical technique, the eraser-like effect may be the only remaining explanation. The investigated material included 38 biopsies taken after 2.4–18 months after implant insertion. Any influence of the implantation trauma itself was excluded due to the evident time delay between implantation and taking the biopsy. A comparison between Ti impregnation of the investigated biopsies did not demonstrate any remarkable influence of the surface differences. This can be explained by the fact that only the top diameter and not the implant surface as a whole was the contact area with the excised tissue. Ti in the biopsies was analyzed in terms of its effect histologically and regarding the Ti quantity by spectrophotometry. It was reported that (i) even the highest Ti contamination was observed, it was considered as a negative effect on the muco-periosteal cover flaps, and (ii) the results highlighted the biological acceptance of Ti [6-73].

The increasing use of commercially pure titanium (CpTi) and the Ti-6Al-4V alloy for material applications in dental and orthopedic implants warrant the study of the storage and elimination of these elements. To understand the *in vivo* storage and elimination of metals, one must first understand the environment and then the metal. The *in vivo* environment is essentially a saline (0.9% NaCl) solution with numerous proteins and enzymes [6-74]. The pH of an extracellular fluid environment normally ranges from 7.00 to 7.70 [6-75]. The implantation of a pure metal into a physiological environment may cause it to ionize or form a metal compound. Hamsters were injected with titanium, aluminum, and vanadium salts either intraperitoneally or intramuscularly to study the transport, storage, and elimination of these metals. Blood samples were taken at 4 or 24 h, and urine samples were at 24, 48, and 72 h. The hamsters were then injected weekly for 5 weeks after the initial injection. Blood and portions of the kidneys, liver, lung, and spleen were taken at sacrifice. All samples were analyzed for titanium, aluminum, and vanadium concentrations using graphic furnace atomic adsorption spectroscopy. It was reported that (i) although titanium was not found to be excreted in the urine which might be related to its insolubility in the physiological environment, it was found in low levels in the blood, and was elevated over control in the kidney, liver, and spleen, (ii) aluminum detection showed wide standard deviations and high levels in controls; however, aluminum was found to be excreted in the urine and to be transported by the blood in the experimental animals, and a small amount accumulated in the liver and spleen, and (iii) vanadium was excreted in high levels in the urine, being related to its solubility in physiological conditions [6-75].

6.5. BIOCOMPATIBILITY

The oral environment with saliva and bacteria is corrosive for dental restorative materials. For chemical and biological evaluations, appropriate biological tests and standards are developed. Such tests and standards, which have been developed in the past 10–15 years, serve as the basis for recommending any dental restorative material. Until a few years ago, almost all national and international dental standards and testing programs focused entirely on physical and chemical properties. Today, however, dental materials standards require biological testing as well. The science of dental materials now encompasses a knowledge and appreciation of certain biological considerations associated with the selection and use of materials designed for use in the oral cavity. According to existing standards, all dental materials should pass primary tests (screening to indicate cellular response), secondary tests (evaluating tissue responses), and usage tests in animals before being evaluated clinically in humans. Testing programs for dental materials are based on specifications or standards established by national or international standards organizations, such as the American National Standards Institute (ANSI) and International Organization for Standardization (ISO). The oldest and largest of these programs has been operated continuously by the ADA (American Dental Association) since the late 1920s. Evaluation of dental products for safety and efficacy has historically been the scope of both the ADA and the FDA (Food and Drug Administration). In accordance with these regulations (US Medical Device Amendments, 1976), all dental materials are reviewed for safety and effectiveness, and classified by the FDA as Class I, II, or III. Class I materials are those considered to be of low risk in causing adverse reactions. Materials in Class II must satisfy the requirements outlined in the current ANSI/ADA specifications. The most extensive testing is required for Class III materials, which includes full safety and efficacy assessments prior to marketing. The FDA regulates the components of dental amalgam into Class II (special controls), whereas the USP (United States Pharmacopoeia) grade mercury into Class I (general controls) [6-76].

Biological compatibility (or in short, biocompatibility) is the ability of a material to perform with an appropriate host response in a specific application. There are three definitions of biocompatibility; (1) the ability of a material to perform with an appropriate host response in a specific application, (2) the quality of not having toxic or injurious effects on biological systems, and (3) comparison of the tissue response produced through the close association of the implanted candidate material to its implant site within the host animal to that tissue response recognized and established as suitable with control materials. All these definitions deal with materials and not with devices. This is a drawback since many medical devices are comprised of many materials. Much of the pre-clinical testing of the materials is not conducted on the devices, but rather the material itself. At some stage, however, the testing will need to include the device, since the shape and geometry of the device will also affect the biocompatibility [6-77]. The biocompatibility of a long-term implantable medical device refers to the ability of the device to perform its intended function, with the desired degree of incorporation in the host, without eliciting any undesirable local or systemic effects in that host. On the other hand, the biocompatibility of short-term implantable medical device that is intentionally placed within the cardiovascular system for transient diagnostic or therapeutic purposes refers to the ability of the device to carry out its intended function within flowing blood, with minimal interaction between device and blood that adversely affect device performance, and without inducing uncontrolled activation of cellular or plasma protein cascades. Furthermore, the biocompatibility of tissue-engineering products (such as a scaffold or matrix) must meet more engineering requirements and refers to the ability to perform as a substrate that will support the appropriate cellular activity, including the facilitation of molecular and mechanical signaling systems, in order to optimize tissue regeneration, without eliciting any undesirable effects in those cells, or inducing any undesirable local or systemic responses in the eventual host.

There are researches on Ti materials focusing on biocompatibility evaluation. Compared with the presently used ($\alpha + \beta$) Ti alloys for biomedical applications, β-Ti alloys have many advantageous mechanical properties, an improved wear resistance, a high elasticity and an excellent cold and hot formability, thus promoting their increased application as materials for orthopedic joint replacement. Not all elements with β-stabilizing properties in Ti alloys are suitable for biomaterial applications – corrosion and wear processes cause a release of these alloying elements (Nb, Ta, Ti, Zr, and Al) to the surrounding tissue. Eisenbarth *et al.* [6-78] evaluated the biocompatibility of alloying elements for β-Ti and near β-Ti alloys in order to estimate their suitability for biomaterial components, with CpTi (grade 2) and the implant steel (316L stainless steel) as reference materials. The investigation included the corrosion properties of the elements, proliferation, mitochondrial activity, cell morphology, and the size of MC3T3-E1 cells and GM7373 cells. It was reported that the biocompatibility range of the investigated metals is (in decreasing biocompatibility): niobium (as a β-phase stabilizing element) $\rightarrow$ tantalum (as a β-phase stabilizing element) $\rightarrow$ titanium $\rightarrow$ zirconium (as an α-phase stabilizing element) $\rightarrow$ aluminium (as an α-phase stabilizing element) $\rightarrow$ 316L (Mo containing 18Cr-8Ni stainless steel with low carbon content) $\rightarrow$ molybdenum [6-78].

The biocompatibility of metallic implant surfaces is governed in large part by the interfacial kinetics associated with metal release and protein binding. The kinetics of metal release from, and protein binding to, Co-Cr-Mo and Ti implants (CpTi and Ti-6Al-4V) in human serum were investigated. The studies were carried out by (1) measuring the temporal release of Cr and Ti into serum from Co-Cr-Mo and Ti implants, (2) examining the composition of human serum proteins adsorbed onto the surfaces of Co-Cr-Mo and Ti implants, and (3) identifying the serum proteins associated with the binding of soluble Cr and Ti degradation products. It was reported that (i) Cr was released from Co-Cr-Mo implant at an order of magnitude higher than Ti was released from Ti implants, (ii) serum became saturated with soluble Cr and Ti at levels as high as 3250 ng/ml Ti from CpTi, 3750 ng/ml Ti from Ti-6Al-4V, and 35 400 ng/ml Cr from Co-Cr-Mo degradation, and (iii) both Cr and Ti released from Co-Cr-Mo and Ti materials exhibited a bimodal binding pattern to both low molecular weight serum protein(s) (<32 kDa), and to higher molecular weight protein(s) in the 180–250 kDa range. Based on these findings, it was concluded that identification of metal alloy-dependent biofilm compositions and dissolution products provides the basis for understanding the bioavailability and bioreactivity of these implant alloys and their degradation products [6-79].

Krupa *et al.* [6-80] investigated the corrosion resistance and biocompatibility of titanium after surface modification by the ion implantation of calcium, phosphorus, or calcium + phosphorus. Calcium and phosphorus ions were implanted in a dose rate of 10^{17} ions/cm^2 at the ion beam energy of 25 keV. The corrosion resistance was examined by electrochemical methods in a simulated body fluid (SBF) at a temperature of 37°C. The biocompatibility was evaluated *in vitro*. The results of the electrochemical examinations (Stern's method) indicated that (i) the calcium, phosphorus, and calcium + phosphorus implantation into the surface of titanium increases its corrosion resistance in stationary conditions after short- and long-term exposures in SBF, (ii) potentiodynamic tests showed that the calcium-implanted samples undergo pitting corrosion during anodic polarization, and (iii) the breakdown potentials measured were high (2.5–3V) [6-80], which is unfairly too high when compared to the reasonable range of intraoral potential (0–50 mV) (see Chapter 3).

For enhancing the biocompatibility of Ti materials, there are several methods proposed. Li *et al.* [6-81] coated a thin hydroxyapatite (HA) layer on a microarc oxidized titanium substrate by means of the sol–gel method. The HA sol was aged fully to obtain a stable and phase-pure HA, and the sol concentration was varied to alter the coating thickness. Through the sol–gel HA coating, the Ca and P concentrations in the coating layer increased significantly. It was reported that (i) the microarc oxidation (anodizing) enhanced the biocompatibility of the Ti, and the bioactivity was improved further by the sol–gel HA coating on the anodized Ti, and (ii) the proliferation and alkaline phosphatase activity of the osteoblast-like cells on the microarc-oxidized Ti/HA sol–gel-treated Ti were significantly higher than those on the microarc-oxidized Ti without the HA sol–gel treatment [6-81].

Schwarz *et al.* [6-82] evaluated the influence of plaque biofilm removal on the mitochondrial activity of human SaOs-2 osteoblasts grown on titanium surfaces. Volunteers wore acrylic splints with structured titanium disks for 72 h to build up plaque biofilms. Specimens were randomly instrumented using (1) ultrasonic system + chlorhexidine, or (2) plastic curettes + CHX. Specimens were incubated with SaOs-2 cells for 6 days. Residual plaque biofilm/clean implant surface areas (%), mitochondrial cell activity (counts/second), and cell morphology were assessed. It was concluded that (i) mitochondrial cell seemed to be impaired by the presence of residual plaque biofilm areas, and (ii) its removal alone might not be the crucial step in the re-establishment of the biocompatibility of titanium surfaces [6-82].

Nitinol (or NiTi)-based shape memory alloys were introduced to medicine in the late 1970s. Despite unique physical, chemical, and mechanical properties associated with NiTi, the wide medical application has been hindered for a long time because of the lack of knowledge on the nature of the biocompatibility of NiTi alloys. A review of biocompatibility with an emphasis on the most recent studies, combined with the results of X-ray surface investigations, allows us to draw conclusions on the origin of the good biological response observed *in vivo*. The tendency of nitinol surfaces to be covered with TiO_2 with only a minor amount of nickel under normal conditions is considered to be responsible for these positive results. A certain type of toxicity, usually observed *in vitro* studies, may result from the much higher *in vitro* Ni concentrations which are probably not possible to achieve *in vivo*. The essentiality of Ni as a trace element may also contribute to the nitinol (NiTi) biocompatibility within the human body tissues [6-83].

The unique properties of NiTi have provided the enabling technology for many applications in the medical and dental industries, including implants within the bloodstream. Two particularly successful NiTi medical devices are the Simon Nitinol Filter and the Mitek bone suture anchor. The Simon Nitinol Filter is an umbrella-shaped device deployed via the shape memory effect to entrap blood clots in the vena cava. Mitek bone suture anchors have revolutionized the field of orthopedic surgery by providing a secure, stable attachment for tendons, ligaments, and other soft tissue to bone. Based on more than 20-years successful clinical performance, the excellent biocompatibility, very high corrosion resistance, and excellent cytocompatibility of NiTi has been proven. The nickel in NiTi is chemically joined to the titanium by a strong intermetallic bond, so the risk of reaction, even in patients with nickel-sensitivity, is extremely low [6-84–6-86]. There are several studies revealing a lower biocompatibility of NiTi under certain conditions [6-56, 6-87]. The surface preparation techniques have been shown to have a significant effect on the biocompatibility of the alloy [6-87]. Ion release measurements on NiTi alloy have shown that the initial rate of Ni ion release is high, but falls rapidly within 2 days, which is similar to that of Ni released from 316L stainless steel [6-53, 6-88], although 316L stainless steel contains only 8 wt%. Armitage *et al.* [6-89] prepared NiTi which was mechanically polished, followed by buff-polishing with the diamond paste, in order to study the influences of surface modifications on the biocompatibility. For another modification, surfaces were shot-peened with 160 μm spherical glass media, and electropolished in 30% HNO_3 in CH_3OH. It was reported that (i) cytotoxicity and cytocompatibility studies with both fibroblast and endothelial cells showed no differences in the biocompatibility of any of the NiTi surfaces, (ii) the cytotoxicity and cytocompatibility of all surfaces were favorable compared to the untreated controls, (iii) the hemolysis caused by a range of NiTi surfaces was no different from that caused by polished 316L stainless steel or polished titanium surfaces (iv) heat treatment (by in-air oxidation at 400C for 30 min or 600°C for 30 min) of NiTi was found to significantly reduce thrombogenicity to the level of the control, (v) the XPS results showed a significant decrease in the concentration of surface nickel with heat treatment and changes in the surface nickel itself from a metallic to an oxide state, and (vi) surface contact angle of both pre-oxidized surfaces showed much lower (36.3 and 13.5 for 400 and 600°C oxidation, respectively) than polished surfaces (64.6 for buff-polished and 42.5°C for electro-polished) [6-89], indicating that pre-oxidation makes the Ti surface more active.

Disks consisting of macroporous NiTi alloy are used as implants in clinical surgery for fixation of spinal dysfunctions. Prymak *et al.* [6-90] preformed the biocompatibility studies by co-incubation of porous NiTi samples with isolated peripheral blood leukocyte fractions (polymorphonuclear neutrophil granulocytes, PMN; peripheral blood mononuclear leukocytes, PBMC) in comparison with control cultures without NiTi samples. The cell adherence to the NiTi surface was analyzed by fluorescence microscopy and scanning electron microscopy. The activation of adherent leukocytes was analyzed by measurement of the released cytokines using enzyme-linked immunosorbent assay. It was found that (i) the cytokine response of PMN (analyzed by the release of IL-1ra and IL-8) was not significantly different between cell cultures with or without NiTi, (ii) there was a significant increase in the release of IL-1ra, IL-6, and IL-8 from PBMC in the presence of NiTi samples, but (iii) in contrast, the release of TNF-α by PBMC was not significantly elevated in the presence of NiTi, and IL-2 was released from PBMC only in the range of the lower detection limit in all cell cultures [6-90].

Filip *et al.* [6-91] characterized the surface and the bulk structure of NiTi implants by SEM, TEM, XPS, and AES (scanning Auger electron microprobe analysis). NiTi implants were compared with otherwise identically prepared non-implanted specimens, sputter-cleaned, and reoxidized samples. Non-implanted and implanted samples had essentially the same surface topography and microstructure. Ti, O, and C were the dominant elements detected on the surface. Trace amounts (~1 at%) of Ni and Ca, N, Si, B, and S were also detected. Ti was present as TiO_2 on the surface, while nickel was present in metallic form. A significant difference in Ni peak intensity was observed when retrieved or non-implanted control samples (very low nickel content) were compared with sputter-cleaned and reoxidized samples (well-detected Ni). It is mentioned that (i) the method of passivation is crucial for Ni loosening, and (ii) no major changes occurred in the NiTi sample bulk structure or in the surface oxide during the implantation periods investigated [6-91].

Early passive or active motion programs after flexor tendon repair have been shown to be extremely beneficial [6-92, 6-93]. However, the application of early active motion increases the stress on the repair, leading to a significant number of repair ruptures [6-94]. It seems to be that two factors are involved in producing a strong mechanical repair, the technique used and the suture material. Stainless steel has better tensile strength and good knot-holding security, but it is difficult to use and is weakened by kinking [6-95–6-98]. NiTi SME alloy has been introduced as a possible new suture material for flexor tendon surgery [6-99–6-101]. NiTi sutures were implanted for biocompatibility testing into the right medial gastrone tendon in 15 rabbits for 2, 6, and 12 weeks. Additional sutures were implanted in subcutaneous tissue for measuring strength in order to determine the effect of implantation on strength properties of NiTi suture material. Braided polyester sutures of approximately the same diameter were used as control. It was found that (i) encapsulating membrane formation around the sutures was minimal in the case of both materials, (ii) the breaking load of NiTi was significantly greater compared to braided polyester, and (iii) implantation did not affect the strength properties of either material [6-102].

6.6. BIOMECHANICAL COMPATIBILITY

Biomaterials (especially, implantable materials) must function under biomechanical environments. If there is no compatibility between placed foreign biomaterials and the receiving host tissue, the intended biofunctionality cannot be accomplished. This is one of three important compatibilities required for successful and biofunctional implants; it is needless to say that the other two compatibilities are biological compatibility (which has been discussed already) and morphological compatibility (which will be discussed in Chapter 7). Figure 6-1 shows a linear

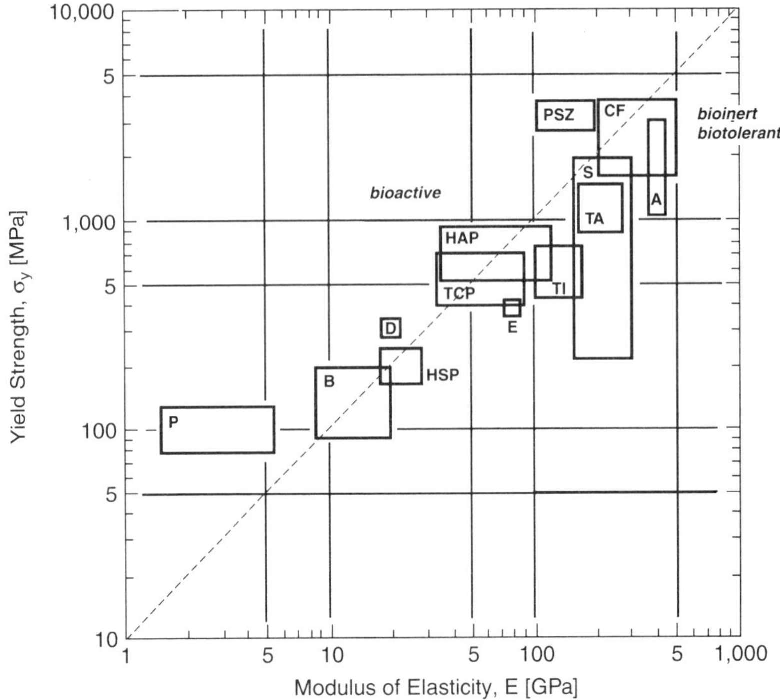

Figure 6-1. Linear relationship between strength and rigidity of various biomaterials in log–log plot. P: polymeric materials, B: bone, HSP: high strength polymers (Kelvar, Kapton, PEEK, etc.), D: dentin, TCP: tricalcium phosphate, HAP: hydroxyapatite, E: enamel, TI: commercially pure titanium (all unalloyed grades), TA: titanium alloys (e.g., Ti-6Al-4V), S: steels (e.g., 304-series stainless steel), A: alumina, PSZ: partially stabilized zirconia, CF: carbon fiber.

relationship between strength (in terms of yield strength) and rigidity (in terms of modulus of elasticity) among various types of biomaterials.

From the above figure, several interesting and important things can be pointed out. (1) There is a linear relationship between the strength and rigidity of biomaterials by factor of 10, regardless of the type of material. (2) HAP and TCP are positioned between TI and TA (as foreign implants material) and bone (as a receiving host substance). As is well documented, applying HAP or TCP on Ti materials is aimed to promote the osseointegration of titanium biomaterials. As far as the biological compatibility is concerned, there is no problem to have a HAP or TCP layer between the TI/TA substrate and bone tissue. But, when the mechanical behavior of the interface between these materials comes into consideration, there could be a large quantity of interfacial stress built up due to the remarkable differences in modulus of elasticity between the two materials. If such interfacial (discrete) stress exceeds the bonding strength (due to the osseointegration) between the implant material's surface and bone tissue, there may not be a successful biointegration. Fortunately, having a HAP or TCP coated on the TI or TA will tremendous reduce the differences in modulus of elasticity between these materials, because HA or TCP can be found between bone and TA or TI substrate in the above figure, serving as a sort of mechanical buffering function. This assists and promotes a mechanical biocompatibility, in addition to the original purpose of achieving the biological compatibility.

One important variable in the role of bone remodeling is the difference in modulus of elasticity between bone and its replacement [6-103]. Normally the larger the difference, the more rapidly bone changes take place. Ti is particularly good in this regard because of its lower modulus of elasticity [6-104]. However, this is not true when the above mechanical environment is concerned. Medical grade titanium samples were examined using X-ray photoelectron spectroscopy before and after immersion in various proteins. Additionally, an implant removed from a patient following clinical failure was examined using scanning ion and electron microscopy. The surface of the as-received sample was found to be mainly TiO_2, with contaminations of H_2O/OH^-, and also calcium and nitrogen, which remained after autoclaving. The failed clinical sample was found to be partially fibrously encapsulated, with evidence of calcification. Small amounts of TiOOH were detected at the fibrous periphery [6-104], supporting the theory of Tengvall *et al.* [6-105, 6-106].

The surface layer of implants should be biomechanically compatible with the mechanical properties of vital soft/hard tissues, as mentioned before. For accomplishing this specific aim, we have already discussed that coating of the material (with HAP or TCP) is very effective. Another technique was introduced by Asaoka *et al.* [6-107] who prepared titanium powder with a granule diameter of 420–500 μm and fabricated porous titanium specimens made from this powder. The mechanical properties of these specimens were examined. It was reported that (i)   the compressive strength and low cyclic compressive fatigue strength were 182 and 40 MPa, respectively, (ii) the compressive strength of the porous-titanium-coated dental implants was 230 MPa, but fatigue strength was not improved. The possibility of using porous titanium on dental implant can be considered with respect to biomechanical compatibility and strength, (iii) materials with an elastic modulus of 70–200 GPa have most frequently been used for dental implants, but porous titanium has a lower elastic modulus than this value, however (iv) a thick core with an ingrowth of tissue into void channels raises the elastic modulus of the material [6-107].

Once a biomechanical compatibility is established, the placed implant needs to bond and incorporate firmly to the surrounding tissue structure. There are four basic methods to fix implants to bone. The *first* method is based on modifying the implant's surface form and texture to promote fixation by mechanical interlocking on the macro or micro scale. For example, plasma-sprayed coatings of metal or other materials (including ceramics) on their surface promote macrointerlock and microinterlock. The *second* basic method of bone–implant fixation involves the use to some type of "bioactive" biomaterials for the implant, either as a surface coating or as the bulk material, to promote stronger bone–implant attachment through the physical and/or chemical bonding between implant surface and tissue constituents. The *third* fixation concept is "osseointegration", which indicates a direct structural and functional connection between vital bone and the surface of a load-carrying implant through the macrointerlocking as well as micro-interlocking. In the last few years, the main advancements in implantology were achieved by creating optimal surface qualities of dental implants. New osteo-inductive and osteo-conductive surfaces have been developed [6-108–6-113]. The reason for working on an optimal surface was to achieve a better result of osseointegration and therefore a better long-term stability of dental implants. To achieve a full osseointegration and good peri-implant gingival was the main task of scientific work. Good peri-implant gingival has a protective function for the bone–implant interface [6-49, 6-55]. The *fourth* fixation concept is "fibro-osseo integration" ("pseudo-periodontal ligament"). This is a connection of implant to tissue by means of a region of non-mineralized connective tissue which forms adjacent to the implant and attaches it to surrounding bone [6-114].

While it is known that dental implants can work due to the osseointegrtaion, implants can also fail. In designing a successful dental implant, the main objective is to ensure that the implant can support biting forces and deliver them safely to interfacial tissues over the long term. Biomechanics are central in this design problem. Key topics include: (1) the nature of the biting forces on the implants (2) how the biting forces are transferred to the interfacial tissues, and (3) how the interfacial tissues react, biologically, to stress transfer conditions. For biting forces on dental implants, the basic problem is to determine the *in vivo* loading components on implants in various prosthetic situations, e.g., for implants acting as single tooth replacements or as multiple supports of loaded bridgework. Significant progress has been made as several theoretical models have been presented for determining the partitioning of forces among dental implants supporting bridgework. Interfacial stress transfer and interfacial biology represent more difficult problems. While many engineering studies have shown that variables such as implant shape, elastic modulus, extent of bonding between implant, and bone can affect the stress transfer conditions, the unresolved question is whether there is any biological significance to such differences. Recent research suggests that the research for a more detailed hypothesis regarding the relationship between interface mechanics and biology should take into account basic bone physiology, e.g., wound healing after implantations plus basic processes of bone modeling and remodeling [6-115]. The successful clinical results achieved with osseointegrated dental implants underscore the fact that such implants easily withstand considerable masticatory loads. In fact, one study showed that bite forces in patients with these implants were comparable to those in patients with natural dentitions. Studies of the interface between titanium implants and bone show that a close contact between a titanium implant develops during the healing period. It is important to analyze macroscopic stress distribution and load transfer mechanisms where close apposition of bone to titanium implants is provided at the microscopic level. The close apposition of bone to the titanium implant is the essential feature that allows a transmission of stress from the implant to the bone without any appreciable relative motion or abrasion. The absence of any intermediate fibrotic layer allows stress to be transmitted without any progressive change in the bond or contact between the bone and implant. The osseointegrated implant provides a direct contact with bone, and therefore will transmit any stress waves or shocks applied to the fixtures. For this reason, it is advisable to use a shock-absorbing material such as acrylic resin in the form of acrylic resin artificial teeth in the fixed partial denture. This arrangement allows for the development of a stiff and strong substructure with adequate shock protection on its outer surface [6-114].

Since dental implants must withstand relatively large forces and moments in function, a better understanding of *in vivo* bone response to loading would aid implant design. Brunski [6-116] pointed out that the following topics are essential in this problem: (1) theoretical models and experimental data are available for understanding implant loading as an aid to case planning, (2) at least for several months after surgery, bone healing in gaps between implant and bone, as well as in pre-existing damaged bone, will determine interface structure and properties, (3) an interfacial cement line exists between the implant surface and bone for titanium and hydroxyapatite, (4) excessive interfacial micromotion early after implantation interferes with local bone healing and predisposes the implants to a fibrous tissue interface instead of osseointegration, and (5) large strains can damage bone. For implants that have healed *in situ* for several months before being loaded, data support the hypothesis that interfacial overload occurs if the strains are excessive in interfacial bone [6-116].

Beside the surface qualities, there are other important factors for achieving perfect peri-implant bone and gingival conditions. In most cases eccentric forces, overloading of implants, and stress forces are the main reason for peri-implant bone loss in the crestal region, and therefore bad peri-implant results. So it is important to avoid large and eccentric force to prevent peri-implant bone damage. In some cases, it is not possible to avoid small tension forces caused by technical production of superstructures. This tension can only be minimized by biokinetic abutments positioned in the implant neck region. These "intramobile elements" were used in the last years as interpolated silicon shock absorbing elements [6-115, 6-117–6-119]. The silicon abutments were in direct contact to the peri-implant gingival, although this material is not suited as a bioinert implant material in this region. Therefore, it is necessary to develop a new implant system with good surface qualities in the intraosseous and trans-gingival region that includes a central shock-absorbing element without contact to the peri-implant region. It was reported that good functional properties are essential in dental implantology. Biokinetics elements are imitating dental resilience. The mobile implant (SIS Inc., Klagenfurt, Australia) is a self-cutting conical screw implant with an integrated biokinetic element. The shock absorber is a central part of the implant and a titanium ring closes the shock-absorbing unit within the implant. The resilience of the implant was tested by axial and horizontal loading in a special testing unit. Furthermore, a survival test of the elastic titanium ring in the most exposed cervical part of the implant was performed. After the region was examined by electron microscopy after 122 million movements in the axial and horizontal direction, it was reported that (i) a progressive shock absorption was registered during horizontal and axial movements, (ii) the maximum movements were 0.06 mm in the axial and 0.16 mm in horizontal direction, and (iii) no signs of material destruction were observed in the electron microscopic analysis [6-120].

REFERENCES

[6-1] Venugopal B, Luckey TD. Metal toxicity in mammals. New York, NY: Plenum Press. 1978.

[6-2] Plenk H, Zitter H. Material considerations. In: Endosseous implants: scientific and clinical aspects. Watzek G, editor. Chicago: Quintessence. 1996.

[6-3] Lemons JE. Dental implant retrieval analyses. J Dent Educ 1988;52: 748–756.

[6-4] Lemons JE, Lucas LC, Johansson B. Intraoral corrosion resulting from coupling dental implants and restorative metallic systems. Implant Dent 1992;1:107–112.

[6-5] Zitter H, Plenk H. The electrochemical behavior of metallic implant materials as an indicator of their biocompatibility. J Biomed Mater Res 1987;21: 881–896.

[6-6] Schroeder A, Sutter F, Krekeler G. Oral implantology. New York: Geor Thieme Verlag Stuttgart. 1991. pp. 45–48.

[6-7] www.soils.wisc/edu.

[6-8] Sakurai H. Why are metals important to human body? Tokyo: Kodansha. 2000. pp. 59–65.

[6-9] Fuwa K. Living organisms and heavy metals. Tokyo: Kodansha-Scientific. 1980. pp. 33–56.

[6-10] Hallab NJ, Anderson S, Caicedo M, Brasher A, Mikecz K, Jacobs JJ. Effects of soluble metals on human peri-implant cells. J Biomed Mater Res 2005;74A:124–140.

[6-11] Maurer AM, Merritt K, Brown SA. Cellular uptake of titanium and vanadium from addition of salts or fretting corrosion *in vitro*. J Biomed Mater Res 1994;28:241–246.

[6-12] Evans EJ, Benjamin M. The effect of grinding conditions on the toxicity of cobalt-chrome-molybdenum particles *in vitro*. Biomaterials 1987;8:377–384.

[6-13] Yang RS, Tsai KS, Liu SH. Titanium implants enhance pulmonary nitric oxide production and lung injury in rats exposed to endotoxin. J Biomed Mater Res 2004;69A:561–566.

[6-14] http://en.wikipedia.org/wiki/Toxicity.

[6-15] Shanbhag A, Jacobs JJ, Black J, Galante JO, Glant TT. Macrophage/particle ineractions: effects of seize, composition and surface area. J Biomed Mater Res 1994;28:81–90.

[6-16] Kumazawa R, Watari F, Takashi N, Tanimura Y, Uo M, Totsuka Y. Effects of Ti ions and particles on neutrophil function and morphology. Biomaterials 2002;23:3757–3764.

[6-17] Harris LG, Patterson LM, Bacon C, ap Gwynn I, Richards RG. Assessment of the cytocompatibility of different coated titanium surfaces to fibroblasts and osteoblasts. J Biomed Mater Res 2005;73A:12–20.

[6-18] Yamamoto A, Honma R, Sumita M. Cytotoxicity evaluation of 43 metal salts using murine fibroblasts and osteoblastic cells. J Biomed Mater Res 1998;39:331–340.

[6-19] Yamamoto A, Kobayashi T, Sumita M, Yamamoto A, Hanawa T. Mutagenicity evaluation of forty-one metal salts by the *umu* tests. J Biomed Mater Res 2002;59:176–183.

[6-20] Okazaki Y, Gotoh E. Effect of metal released from Ti alloy wear powder on cell viability. Mater Trans JIM 2000;41:1247–1255.

[6-21] Yamamoto A, Honma R, Sumita M, Hanawa T. Cytotoxicity evaluation of ceramic particles of different sizes and shapes. J Biomed Mater Res 2004;68A:244–256.

[6-22] Mu Y, Kohyama Y, Hanawa T. Metal ion releases from titanium with active oxygen species generated by rat macrophages *in vitro*. J Biomed Mater Res 2000;49:238–243.

[6-23] Manceur A, Chellat F, Merhi Y, Chumlyakov Y, Yahia L'H. *In vitro* cytotoxicity evaluation of a 50.8% NiTi single crystal. J Biomed Mater Res 2003;67A:641–646.

[6-24] Bordji K, Jouzeau JY, Mainard D, Payan E, Netter P, Rie KT, Stucky T, Hage-Ali M. Cytocompatibility of Ti-6Al-4V and Ti-5Al-2.5Fe alloys according to three surface treatments, using human fibroblasts and osteoblasts. Biomaterials 1996;17:929–940.

[6-25] Wever DJ, Veldhuizen AG, Sanders MM, Schakenraad JM. Cytotoxic, allergic and genotoxic activity of a nickel–titnium alloy. Biomaterials 1997;18:1115–1120.

[6-26] Yeung KWK, Poon JWY, Liu XY, Ho JPY, Chung CY, Chu PK, Lu WW, Chan D, Cheung KMC. Investigation of nickel suppression and cytocompatibility of surface-treated nickel–titanium shape memory alloys by using plasma immersion ion implantation. J Biomed Mater Res 2005;72A: 238–245.

[6-27] Yeung KWK, Poon RWY, Liu XY, Ho JPY, Chung CY, Chu PK, Lu WW, Chan D, Cheung KMC. Corrosion resistance, surface mechanical properties, and cyto-compatibility of plasma immersion ion implantation-treated nickel–titanium shape memory alloys. J Biomed Mater Res 2005;75A:256–267.

[6-28] Tada H, Shibo O, Kuroshima K, Koyama M, Tsukamoto K. An improved colorimetric assay for interleukin-2. Immunol Methods 1986;93:157–165.

[6-29] Es-Souni M, Es-Souni M, Brandies HF. On the transformation behavior, mechanical properties and biocompatibility of tow NiTi-based shape memory alloys: NiTi$_{42}$ and NiTi$_{42}$Cu$_7$. Biomaterials 2001;22:2153–2161.

[6-30] Niinomi M. Fatigue performance and cyto-toxicity of low rigidity titanium alloy, Ti-29Nb-13Ta-4.6Zr. Biomaterials 2003;24:2673–2683.

[6-31] Borenfreund E, Purener JA. Toxicity determined *in vitro* by morphological alternations and neutral red absorption. Toxicol Lett 1985;24:119–124.

[6-32] Mosman T. Rapid colorimetric assay for cellular growth and survival. Application of proliferation and cytotoxicity assay. J Immunol Methods 1983;65:55–63.

[6-33] Spriano S, Bosetti M, Bronzoni M, Vernè E, Maina G, Bergo V, Cannas M. Surface properties and cell response of low metal ion release Ti-6Al-7Nb alloy after multi-step chemical and thermal treatments. Biomaterials 2005;26: 1219–1229.

[6-34] Wood JFL. Patinet sensitivity to alloy used in partial denture. Br Dent J 1974;136: 423–424.

[6-35] Menne T. Christophersen J. Green A. Epidemiology of nickel dermatitis. In: Nickel and the skin: immunology and toxicology. Maibach HI, Menne T editor. Boca Raton, FL: CRC Press. 1989. pp. 109–116.

[6-36] Romaguera C, Grimalt F, Vilaplana J. Contact dermatitis from nickel: an investigation of its source. Contact Dermatitis 1988;19:52–57.

[6-37] Meijer C, Bredberg M, Fischer T, Widstrom L. Ear piercing and nickel and cobalt sensitization, in 520 young Swedish men doing compulsory military service. Contact Dermatitis 1995;32:147–179.

[6-38] Gammelgaard B, Fullerton A, Avnstorp C, Menne T. Permeation of chromium salts through human skin *in vitro*. Contact Dermatitis 1992;27: 302–310.

[6-39] Peters MA, Schroeter AL, van Hale HM, Broadbent JC. Pacemaker contact sensitivity. Contact Dermatitis 1984;11:214–222.

[6-40] Layor PA, Revell PA, Gray AB, Wright S, Railton GT, Freeman MA. Sensitivity to titanim. A case of implant failure? J Bone Joint Surg 1991;73: 25–28.

[6-41] http://www.melisa.org.

[6-42] Magnusson B, Kligman AM. The identification of contact allergens by animal assay. The Guinea pig maximization test. J Invest Dermatol 1969;52:268–276.

[6-43] Goodwin BJG, Crecel RWR, Johnson AW. A comparison of three guinea-pig sensitization procedures for the detection of 19 reported human contact sensitizers. Contact Dermatitis 1981;7:248–258.

[6-44] Mor S, Ben-Efraim S, Leibovici J, Ben-David A. Successful contact sensitization to chromate in mice. Int Arch Allergy Appl Immunol 1988;85: 452–457.

[6-45] Vreeburg KJJ, de Groot K, van Hoogstraten IMW, von Blomberg BME, Scheper RJ. Successful induction of allergic contat dermatitis to mercury and chromium in mice. Int Arch Allergy Appl Immunol 1991;96: 179–183.

[6-46] Basketter DA, Sholes EW. Comparison of the local lymph node assay with the guinea-pig maximization test for the detection of a range of contact allergens. Food Chem Toxicol 1992;30:65–69.

[6-47] Samits MH, Katz SA, Sheiner DM, Lewis JE. Attempts to induce sensitization in guinea pigs with nickel complexes. Acta Derm-Venereol 1975;55: 478–480.

[6-48] Moller H. Attempts to induce contact allergy to nickel in the mouse. Contact Dermatitis 1984;10:65–68.

[6-49] Rohold AE, Nielsen GD, Andersen KE. Nickel-sulphate-induced dermatitis in the guinea pig maximization test: a dose–response study. Contact Dermatitis 1991;24:35–39.

[6-50] Zissu D, Cavelier C, De Ceaurriz J. Experimental sensitization of guinea-pigs to nickel and patch testing with metal samples. Food Chem Toxicol 1987;25:83–85.

[6-51] Ikarashi Y, Momma J, Tsuchiya T, Nakamura A. Evaluation of skin sensitization potential of nickel, chromium, titanium and zirconium salts using guinea-pigs and mice. Biomaterials 1996;17:2103–2108.

[6-52] Harzer W, Schröter A, Gedrange T, Muschter F. Sensitivity of titanium brackets to the corrosive influence of fluoride-containing toothpaste and tea. Angle Orthod 2001;71:318–323.

[6-53] Ryhanen J. Niemi E, Serlo W, Niemela E, Sandvik P, Pernu H, Salo T. Biocompatibility of nickel titanium shape memory metal and its corrosion behavior in human cell cultures. J Biomed Mater Res 1997;35: 451–457.

[6-54] McKay GC, Macnair R, MacDonald C, Grant MH. Interactions of orthopaedic metals with an immortalized rat osteoblast cell line. Biomaterials 1996;17:1339–1344.

[6-55] Kerosuo H, Kullaa A. Kerosuo E, Kanerva L, Hensten-Ptterson A. Nickel allergy in adolescents in relation to orthodontic treatment and piercing of ears. Am J Orthod Dentofacial Orthop 1996;109:148–154.

[6-56] Berger-Gorbet M, Broxup B, Rivard C, Yahia L'H. Biocompatibility testing of NiTi screw using immuno histochemistry on sections containing metallic implants. J Biomed Mater Res 1996;32:243–248.

[6-57] Bass JK, Fine H, Cisnero GJ. Nickel hypersensitivity in the prosthodontics patient. Am J Orthod Dentofacial Orthop 1993;103:280–285.

[6-58] Rogers SD, Howie DW, Graves SE, Pearcy MJ, Haynes DR. *In vitro* human monocyte response to wear particles of titanium alloy containing vanadium or niobium. J Bone Joint Surg 1997;79B:311–315.

[6-59] Horowitz SM, Luchetti WT, Gonzales JB, Ritchie CK. The effects of cobalt chromium upon macrophages. J Biomed Mater Res 1998;41:468–473.

[6-60] Schmalz G, Schuster U, Schweikl H. Influence of metals in IL-6 release *in vitro*. Biomaterials 1998;19:1689–1694.

[6-61] Burden DJ, Eedy DJ. Orthodontic headgear related to allergic contact dermatitis; a case report. Br Dent J 1991;170:447–448.

[6-62] Veien NK, Borchorst E, Hattel T, Laurberg G. Stomatitis or systemically induced contact dermatitis from metal wire in orthodontic materials. Contact Dermatitis 1994;30:210–213.

[6-63] Torgerson S, Gjerdet NR, Erichsen ES, Bang G. Metal particle and tissue changes adjacent to miniplates. A retrieval study. Acta Odontol Scand 1995;53: 65–71.

[6-64] Alpert B, Seligson D. Removal of asymptomatic bone plates used for orthognathic surgery and facial fractures. J Oral Maxillofac Surg 1996;54: 618–621.

[6-65] Mjoberg B, Hellquist E, Mallmin H, Lindh U. Aluminum, Alzheimer's disease and bone fragility. Acta Orthop Scand 1997;68:511–514.

[6-66] Boyce BF, Byars J, McWilliams S, Mocan MZ, Elder HY, Boyle IT, Junor BJ. Histo-logical and electron microprobe studies of mineralization in aluminum-related osteomalacia. J Clin Pathol 1992;45:502–550.

[6-67] Zaffe D, Bertoldi C, Consolo U. Accumulation of aluminium in lamellar one after implantation of titanium plates, Ti-6Al-4V screws, hydroxyapatite granules. Biomaterials 2004;25:3837–3844.

[6-68] Williams DF. Biological effects of titanium. In: Systemic aspects of biocompatibility. Willimas DF, editor. Boca Raton: CRC Press. 1981. pp. 169–178.

[6-69] Schroeder HA, Bolassa JJ, Hipton IH. Abnormal trace metals in man: Titanium. J Chronic Diseases 1963;16:55–72.

[6-70] Williams DF. The properties of titanium and its uses in cardiovascular surgery. J Cardiovasc Technol 1978;20:52–65.

[6-71] Olmedo DG, Fernández, Guglielmotti MB, Cabrini RL. Macrophages related to dental implant failure. Implant Dent 2003;12:75–80.

[6-72] Olmedo DG, Tasat DR, Guglielmotti MB, Cabrini RL. Effect of titanium dioxide on the oxidative metabolism of alveolar macrophages: an experimental study in rats. J Biomed Mater Res 2005;73A:142–149.

[6-73] Schlegel KA, Eppeneder S, Wiltfang J. Soft tissue findings above submerged titanium implants – a histological and spectroscopic study. Biomaterials 2002;23:2939–2944.

[6-74] Merritt K, Margevicius RW, Brown SA. Storage and elimination of titanium, aluminum, and vanadium salts, *in vivo*. J Biomed Mater Res 1992;26:1503–1515.

[6-75] Ganong WF. Regulation of extracellular fluid composition and volume. In: Review of medical physiology. 12th ed. Los Altos, CA: Lange Medical Publication. 1985. p. 104.

[6-76] Anusavice KJ. Phillips' science of dental materials. Saunders: An Imprint of Elsevier. 2003. pp. 185–196.

[6-77] Williams DF. Revisiting the definition of biocompatibility. Med Device Technol 2003;14:10–13.

[6-78] Eisenbarth E, Velten D, Müller M, Thull R, Breme J. Biocompatibility of β-stabilizing elements of titanium alloys. Biomaterials 2004;25:5705–5713.

[6-79] Hallab NJ, Skipor A, Jacobs JJ. Interfacial kinetics of titanium- and cobalt-based implant alloys in human serum: metal release and biofilm formation. J Biomed Mater Res 2003;65A:311–318.

[6-80] Krupa D, Baszkiewicz J, Kozubowski JA, Lewandowska-Szumieł M, Barcz A, Sobczak JW, Biliński A, Rajchel A. Effect of calcium and phosphorous ion implantation on the corrosion resistance and biocompatibility of titanium. J Biomed Mater Eng 2004;14:525–536.

[6-81] Li L-H, Kim H-W, Lee S-H, Kong Y-M, Kim K-E. Biocompatibility of titanium implants modified by microarc oxidation and hydroxyapatite coating. J Biomed Mater Res 2005;73A:48–54.

[6-82] Schwarz F, Papanicolau P, Rothamel D, Beck B, Herten M, Becker J. Influence of plaque biofilm removal on reestablishment of the biocompatibility of contaminated titanium surfaces. J Biomed Mater Res 2006;77A:437–444.

[6-83] Shabalovskaya SA. On the nature of the biocompatibility and on medical applications of NiTi shape memory and superelastic alloys. J Biomed Mater Eng 1996;6:267–289.

[6-84] Ryhanen J. Biocompatibility evaluation of nickel–titanium shape memory metal alloy. http://herkules.oulu.fi/isbn 9514252217.

[6-85] Oshida Y, Miyazaki S. Corrosion and biocompatibility of shape memory alloys. Corr Eng 1991;40:1009–1025.

[6-86] Castleman LS, Motzkin SM. Biocompatibility of nitinol. In: Biocompatibility of clinical implant materials. Williams DF, editor. Cleveland, OH: CRC Press. 1981, pp. 129–154.

[6-87] Assad M, Lomardi S, Bernèche S, DesRosiers EA, Yahia L'H, Rivard CH. Cytotoxicity testing of the nickel titanium shape memory alloy. Ann Chir 1994; 48:731–736.

[6-88] Wever DJ, Veldhuizen AG, deVries J, Busscher HJ, Uges DRA, van Horn JR. Electro-chemical and surface characterization of a nickel titanium alloy. Biomaterials 1998;19:761–766.

[6-89] Armitage DA, Parker TL, Grant DM. Biocompatibility and hemocompatibility of surface-modified NiTi alloys J Biomed Mater Res 2003;66A: 129–137.

[6-90] Prymak O, Bogdanski D, Köller M, Esenwein SA, Muhr G, Beckmann F, Donath T, Assad M, Epple M. Morphological characterization and *in vitro* biocompatibility of a porous nickel–titanium alloy. Biomaterials 2005;26: 5801–5807.

[6-91] Filip P, Lausmaa J, Musialek J, Mazanec K. Structure and surface of NiTi human implants. Biomaterials 2001;22:2131–2138.

[6-92] Elliot D, Moiemen NS, Flemming AF, Harris SB, Foster AJ. The rupture rate of acute flexor tendon repairs mobilized by the controlled actions motion regimen. J Hand Surg 1994;19B:607–612.

[6-93] Gelberman RH, Amifl D, Gonsalves M, Woo S, Akeson WH. The influence of protected passive mobilization on the healing of flexor tendons: a biochemical and minoangio-graphic study. Hand 1981;13:120–128.

[6-94] Small JO, Brennen MD, Colville J. Early active mobilization following flexor tendon repair in zone 2. J Hand Surg 1989;14B:383–391.

[6-95] Trail IA, Powell ES, Noble J. An evaluation of suture materials used in tendon surgery. J Hand Surg 1989;14B:422–427.

[6-96] Nystrom B, Holmlund D. Separation of sutured tendon ends when different suture techniques and different suture materials are used. An experimental study in rabbits. Scand J Plast Reconstr Surg 1983;17:19–23.

[6-97] Silfrerskiod KI, Anderssen CH. Two new methods tendon repair: an *in vitro* evaluation of tensile strength and gaps formation. J Hand Surg 1993; 18A:58–65.

[6-98] Aoki A, Manske PR, Pruitt DL, Larson BJ. Tendon repair using flexor tendon splints: an experimental study. J Hand Surg 1994;19A:984–990.

[6-99] Moneim MS, Firoozbakhsh K, Mustapha AA, Larsen K, Shahinpoor M. Flexor tendon repair using shape memory alloy suture: a biomechanical evaluation. Clin Orthop 2002;402:251–259.

[6-100] Miura F, Mogi M, Ohura Y, Hamanaka H. The super-elastic property of the Japanese NiTi alloy wire for use in orthodontics. Am J Orthod Dentofacial Orthop 1986;90:1–10.

[6-101] Duerig TW, Pelton AP, Stockel D. The utility of superelasticity in medicine. J Biomed Mater Eng 1996;6:255–266.

[6-102] Kujala S, Pajala A, Kallioinen M, Pramila A, Tuukkanen J, Ryhänen J. Biocompatibility and strength properties of nitinol shape memory alloy suture in rabbit tendon. Biomaterials 2004;25:353–358.

[6-103] Williams DF. The biological applications of titanium and titanium alloys. In: Materials science and implant orthopaedic surgery. Kossowsky R, Klemchuk PP, editors. Dordrecht, Netherlands: Martinus Nijhoff. 1986. pp. 107–116.

[6-104] Sutherland DS, Forshaw PD, Allen GC, Brown IT, Williams KR. Surface analysis of titanium implants. Biomaterials 1993;14:893–899.

[6-105] Tengvall P, Lundstrom I, Sjoqvist L, Elwing H. Titanium-hydrogen peroxide interactions: model studies of the influence of the inflammatory response on titanium implants. Biomaterials 1989;10:166–175.

[6-106] Tengvall P, Lundstrom I. Physio-chemical considerations of titanium as a biomaterial. Clin Mater 1992;9:115–134.

[6-107] Asaoka K, Kuwayama N, Okuno O, Miura I. Mechanical properties and bio-mechanical compatibility of porous titanium for dental implants. J Biomed Mater Res 1985;19:699–713.

[6-108] Cole BJ, Bostrom MP, Pritchard TL, Summer DR, Tomin E, Lane JM, Weiland AJ. Use of bone morphologenetic protein 2 on ectopic porous coated implants in the rat. Clin Orthop 21997;345:219–228.

[6-109] Denissen H, van Beek E, van den Bos T, de Blieck J, Klein C, van den Hooff A. Degradable biphosphonate-alkaline phosphatase-complexed hydroxyapatite implants *in vitro*. J Bone Miner Res 1997;12:290–297.

[6-110] Piztelli A, Scarano A, Di Alberti L, Piatelli M. Histological and histochemical analyses of acid and alkaline phosphatase around hydroxyapatite-coated implants: a time course study in rabbit. Biomaterials 1997;18: 1191–1194.

[6-111] Batzer R, Liu Y, Cochran DL, Szmuckler-Moncler S, Dean DD, Buyan BD, Schwartz Z. Prostaglandine mediate the effects of titanium surface roughness on MG63 osteoblast-like cells and alter cell responsiveness to a alpha, $25\text{-}(OH)_2D_3$. J Biomed Mater Res 1998;41:489–496.

[6-112] Overgaard S, Lind M, Glerup H, Bunger C, Soballe K. Porous-coated versus grit-blasted surface texture of hydroxyapatite-coated implants during controlled micromotion: mechanical and histomorphological results. J Arthroplasty 1998;13:449–458.

[6-113] Wie H, Herø H, Solheim T. Hot isostatic pressing-processed hydroxyapatite-coated titanium implants: light microscopic and scanning electron microscopic investigation. Int J Oral Maxillofac Implants 1998;13: 837–844.

[6-114] Skalak R. Biomechanical considerations in osseointegrated prostheses. J. Prosthetic Dent 1983;49:843–848.

[6-115] Kanth L. The cast (intramobile) screw implant. DDZ 1971;25:465–467.

[6-116] Brunski JB. *In vivo* bone response to biomechanical loading at the bone/dental–implant interface. Ad Dent Res 1999;13:99–119.

[6-117] Koch WL. 2-phasic endosseous implantation of intramobile cylindrical implants I. Quintessenz 1976;27:23–31.

[6-118] Kanth L. Titanium plasma coated intramobile conical implants. ZWR 1982;91:42–46.

[6-119] Kirsch A. The two-phase implantation method using IMZ intramobile cylinder implants. J Oral Implantol 1983;11:197–202.

[6-120] Gaggl A, Schultes G. Biomechnical properties in titanium implants with integrated maintenance free shock absorbing elements. Biomaterials 2001;22: 3061–3066

Chapter 7
Implant-Related Biological Reactions

7.1.	Bone Healing	162
7.2.	Hemocompatibility	164
7.3.	Cell Adhesion, Adsorption, Spreading, and Proliferation	166
7.4.	Roughness and Cellular Response to Biomaterials	179
7.5.	Cell Growth	182
7.6.	Tissue Reaction and Bone Ingrowth	185
7.7.	Osseointegration and Bone/Implant Interface	193
7.8.	Some Adverse Factors for Loosening Implants	199
References		200

Chapter 7
Implant-Related Biological Reactions

Dental implants are an ideal option for people in good general oral health who have lost a tooth (or teeth) due to periodontal disease, an injury, or some other reason. Dental implants (as an artificial tooth root and usually made from titanium materials) are biocompatible metal anchors surgically positioned in the jawbone underneath the gums to support an artificial crown where natural teeth are missing. There are main three different types of dental implants: (i) root form implants, (ii) plate form implants, and (iii) subperiosteal implants. Using the root form implants (the closest in shape and size to the natural tooth root), the healing period usually varies from as few as 3–6 months or more. During this time osseointegration occurs. The bone grows in and around the implant creating a strong structural support. Plate (or blade) form implants are usually used when the bone is so narrow that it may not be suitable for the root form implant, and the area is not suitable for bone grafting. Like root form implants, there is usually a healing period for osseointegration, although some plate form implants are designed for immediate restoration (or immediate loading). With very advanced jawbone resorption there may not be enough bone width or height for the root form or plate form implant, and the subperiosteal implant may be prescribed. The subperiosteal implant is custom made and designed to sit on top of the bone, but under the gums.

Generally, any edentulous (toothless) area having enough bone support can be an indication for restoring a missing tooth with an implant. A decision has to be made on case-by-case bases whether it is a good idea based on the patient's requirements and expectations. Even though there are practically no absolute contraindications for placement of implants, there are few factors that can put it in a high-risk situation. The followings are normally believed as contraindicative factors: (1) endocrine disorders, such as uncontrolled diabetes mellitus, pituitary and adrenal insufficiency, and hypothyroidism can cause considerable healing problems, (2) uncontrolled granulomatous diseases, such as tuberculosis and sarcoidosis may also lead to a poor healing response to surgical procedures, (3) patients with cardiovascular diseases, taking blood-thinning drugs and patients with uncontrolled hematological disorders, such as generalized anemia, hemophilia (Factor VIII deficiency), Factors IX, X, and XII deficiencies and any other acquired coagulation disorders are contraindicated to surgical procedures due to poor hemorrhage control, (4) patients with bone diseases, such as histiocytosis X, Paget's disease and fibrous dysplasia may not be good candidates for implants, because there is a higher chance for the implant to fail due to poor osseointegration,

(5) cigarette smoking, and (6) patients receiving radio and chemotherapy should not have implants within the 6 months period of therapy [7-1].

For most patients, the placement of dental implants involves two surgical procedures. *First* (surgical stage), implants are placed within the jawbone. For the first three to 6 months following surgery, the implants integrate with the jawbone – osseointegration (due to successful growing of osteoblasts on and into the rough surfaces of the implant, forming a structural and functional connection between the hard tissue and placed implant). After some months the implant is uncovered and a healing abutment and temporary crown is placed onto the implant. This encourages the gum to grow in the right scalloped shape to approximate a natural tooth's gums and allows assessment of the final aesthetics of the restored tooth. After the implant has integrated/bonded to the jawbone, the *second* (prosthetic) phase begins. The placed implants are uncovered and small posts are attached which will act as anchors for the artificial teeth. These posts protrude through the gums. When the artificial teeth (or permanent crowns) are placed, these posts will not be seen. The entire procedure usually takes 4–8 months.

There are four major factors influencing the success rates of placed implants. They include (1) correct indication and favorable anatomic conditions (bone and mucosa), (2) good operative technique, (3) patient cooperation (oral hygiene), and (4) adequate superstructure. Failure of a dental implant is usually related to failure to osseointegrate correctly. A dental implant is considered to be a failure if it is lost, mobile, or shows peri-implant bone loss of greater than 1 mm in the first year after implanting and greater than 0.2 mm a year after that. Dental implants are not susceptible to dental caries but they can develop a periodontitis condition called peri-implantitis where correct oral hygiene routines have not been followed. Unfortunately, it is true that many times a patient who was evaluated as a good implant patient candidate has a bad history of dental hygiene practice. Failure in dental implants is found to be increased in the anterior mandible where bone quality is not as dense. More rarely, an implant may fail because of poor positioning at the time of surgery, or may be overloaded initially causing failure to integrate. Implant surgery is believed to exhibit its success rate of more than 90% of the time. Occasionally, an implant fails to bond with the surrounding bone. This is usually discovered at the second-stage surgery when the implant is uncovered and the surgeon finds it is loose. In this case, the "failed" implant has to be removed, and another implant can be placed either immediately or later. Potential reasons for implants failing to integrate with surrounding bone include: surgical trauma, infection around the implant, smoking (this appears to decrease blood flow to the healing gums and bone, which could interfere with the bonding process), lack of healthy bone (if there is not enough bone for the implant to remain stable, the implant may move around within the bone and bonding will not occur), titanium allergies (these are extremely rare) [7-2, 7-3]. Problems also can develop years after implants are placed. For example, just like natural teeth and gums, the gums around implants can become infected by bacteria, leading to a form of periodontal disease called peri-implantitis. If that situation is left untreated, this condition can cause bone loss, which could cause the implant to become loose and have to be removed. Generally, this situation can be treated using procedures that are very similar to those used to treat periodontal disease affecting natural teeth. Another type of complication that can happen over time is the implant-supported restoration (crown, bridge, or denture) can break, or the implant itself can fracture. This usually happens if the occlusion is not aligned properly. If the occlusion is off, too much force might be placed on the restoration or implant. Broken restorations often can be repaired, but a fractured implant has to be removed. A broken implant or an implant that fails because of an infection can be replaced with a new implant [7-1].

On the other hand, orthopedic implants are used to replace damaged or troubled joints. Each implant procedure involves removal of the damaged joint and an artificial prosthesis replacement, including hip, knee, finger, elbow, shoulder, ankle, and trauma products. Orthopedic implants are mainly constructed of titanium alloys for strength and lined with plastic to act as artificial cartilage. Some are cemented into place and others (i.e., cementless) are pressed to fit and allow the patient's bone to grow into the implant for strength. The advantages of orthopedic implants should include an increased mobility, reduced pain, and a higher quality of life. The disadvantages can include a strict post-surgery recovery plan, infection, and possible malfunction. Each implant is designed to withstand the movement and stress associated with each individual joint and to provide increased mobility and decreased pain. The primary need for orthopedic implants is the result of osteoarthritis, also called degenerative joint disease. When cartilage is worn down, painful bone-to-bone contact occurs, or cartilage might break down as a result of excess body weight and or the lack of joint movement. Historically for the domain of the elderly, orthopedic implants are now increasingly used to replace arthritic and/or damaged joints. For hip implants, the form characteristics of the femoral cup and head are critical to its load-bearing capability. Typically, however, the main cause of failure in hip joints is due to wear, and in particular the particles that are generated, resulting in wear debris toxicity. For example, the most common cause of femoral bone loss is due to osteolysis. Although the total cause is not known, it has been attributed to a variety of factors including foreign body reaction to particulate debris, in particular to polymeric debris. As such, the avoidance of wear in hip joints is critical [7-4, 7-5].

Today, titanium and some of its alloys are considered to be among the most biocompatible materials. The fact that titanium is being used preferentially in many of the more recent applications in maxillofacial, oral, neuro-surgery and cardiovascular surgery, as well as gaining increasing preference in orthopedics, indicates superiority. Moreover, commercially they have been successfully used for orthopedic and dental implants. Several studies show that when titanium is inserted into human bone it forms an intimate contact with the bone in a process called osseointegration [7-6–7-8], due to good biocompatibility which is greatly relied on formation of stable passive oxide film in a biological system including chemical and mechanical environments. However, the presence of the passive film is only part of the answer when we consider the complex interfacial phenomena to be found between titanium and a biological system. In this chapter, several important points related to successful implantology will be reviewed.

7.1. BONE HEALING

For both dental and orthopedic implant surgeries, bone healing is the most common, and major, step immediately after implant placements. There is union bone fracture and nonunion bone fracture. For these implant surgeries, it is important to understand the nonunion bone healing mechanism. There are principle steps of peri-implant bone healing. The first and most important healing phase, osteoconduction, relies on the recruitment and migration of osteogenic cells to the implant surface, through the residue of the peri-implant blood clot. This formation of hematoma is similar to that seen elsewhere in the body. This is a clot of blood originating from the lacerated blood vessels in the bone and the periosteum. Among the most important aspects of osteo-conduction are the knock-on effects generated at the implant surface, by the initiation of platelet activation, which result in directed osteogenic cell migration. Necrotic bone is then resorbed by osteoclasts, and the hematoma by macrophages. As the hematoma is resorbed, granulation tissue develops from the periosteum and endosteum. When this has completed, pluripotent cells migrate into the granulation tissue. These cells become chondrocytes, and later osteocytes, producing cartilage and bone, respectively. The second healing phase (bone formation) results in a mineralized interfacial matrix equivalent to that seen in the cement line in natural bone tissue. These two healing phases, osteoconduction and *de novo* bone formation, result in contact osteogenesis and, given an appropriate implant surface, bone bonding. The structure surrounding the fracture site is now slightly harder; this is a provisional callus. The area can be called a proper callus as time goes on, and more and more woven bone is made by the osteoblasts. This woven bone is initially remodeled into lamellar bone. With time, the bone is remodeled over the next few months, and the callus becomes smaller, as the trabeculae are formed along lines of stress [7-9].

A material implanted in bone is always inserted into coagulating blood. Protein and cell interactions during this initial implantation time will govern later healing. Many studies have focused on the tissue surrounding implants. Eriksson *et al.* [7-10] developed a method for evaluation of healing around implants in bone by studying cells adhering to the implant surface. Hydrophilic (in other words, active) titanium disks were inserted into rat tibiae. Samples were retrieved after 1, 2, 4, and 8 days of implantation, and were analyzed by fluorescence microscopy techniques and scanning electron microscopy. Both proliferating and apoptotic cells were found on the surface. Generally, cells closest to the implant surface were nonviable, whereas cells in the fibrin network a distance from the surface were viable. Bone morphogenetic protein-2 (BMP-2) is an osteogenic substance. An increase in BMP-2-positive cells was seen during the implantation period, and a population of large BMP-2-positive cells appeared on the surface after 4 days of implantation. Hence, it was evaluated that this unique method was a suitable tool for rapid evaluation of the initial healing around implant material [7-10].

Osborn *et al.* [7-11] classified bio-response to endosseous implants into the following three groups: (1) biotolerant type, characterized by distanceosteogenesis, (2) bioinert type, characterized by contactosteogenesis, and (3) bioreactive type, characterized by physico-chemical linkage between the implant and the surrounding bone. Rutile-type oxide, which is formed on titanium as a titanium dioxide, is described as a stable crystalline form similar to ceramics in its bio-reactive behavior [7-12]. However, it is well recognized that Ti is known as a biotolerant biomaterial (which exhibits "distance osteogenesis", so that although such biomaterials are not rejected from the living tissue, they do not connect to the bone directly because they are covered with connective tissue), and when inserted to the living hard tissue, bone regeneration can be achieved either by bone conductive mechanism (using hydroxyapatite, tricalcium phosphate, decalcified frozen dried bone, or autologous-bone) or bone inductive mechanisms (using bone morphogenetic protein on scaffold or platelet-rich plasma on scaffold) [7-13].

Titanium nitride (TiN) has been used in many fields as a coating for surgical instruments, with the purpose of creating materials more resistant to wear and corrosion, and also reducing adhesion. Titanium implants (180 implants with 2×2 mm^2) were divided into the following three groups: Group 1 ($n=60$): 30 machined and 30 machined TiN-coated, Group 2 ($n=60$): 30 sandblasted and 30 sandblasted TiN-coated, and Group 3 ($n=60$): 30 titanium plasma sprayed, 30 titanium plasma sprayed and coated with TiN. The animals were killed after 5, 10, 20, 30, or 60 days. It was reported that (i) the healing around the TiN-coated implants was similar to that observed around the uncoated surfaces, and (ii) the TiN coating demonstrated a good biocompatibility, did not have unwanted effects on the peri-implant bone formation [7-14].

7.2. HEMOCOMPATIBILITY

Blood-contacting materials for long-term implantation in the body, such as artificial hearts, artificial valves, and stents, are required to have good blood compatibility. The correlations between Ti oxide layers on oxidized Ti substrates and blood platelet adhesion were examined. Untreated CpTi and three differently treated CpTi (treated in 1–10% hydrogen peroxide for 2–6 h, heated at 300, 400, and 500°C for 3 h, and treated in 3% H_2O_2 and heated) were tested. It was found that (i) titanium oxide layers formed on Ti by heating displayed less blood platelet adhesion than thin oxide layers on untreated NiTi, (ii) the largest number of adhesive platelets was found on the H_2O_2-oxidized surface, (iii) adhesive platelets could hardly be observed on Ti substrate treated with H_2O_2 and subsequently heated above 300°C, and (iv) the number of blood platelets adhering to these substrates is indicated as follows: H_2O_2-treated CpTi > untreated CpTi > in-air oxidized CpTi > H_2O_2-treated and subsequently oxidized CpTi [7-15]. Hemocompatibility is a key property of biomaterials that come in contact with blood. Surface modification has shown great potential for improving the hemocompatibility of biomedical materials and devices [7-15]. The hemocompatibility of the titanium oxide films was characterized in terms of clotting, time, blood platelet adhesion and activation, thrombin time, and thrombinogen time. Huang *et al.* [7-16] studied the behavior of protein adsorption by irradiation labeling with [125]I. Ti-O thin films were prepared by plasma immersion ion implantation (PI^3) and deposition, and by sputtering. Systematic evaluation of hemocompatibility, including *in vitro* clotting time, thrombin time, prethrombin time, blood platelet adhesion and *in vivo* implantation into a dog's ventral aorta or right auricle from 17 to 90 days, proved that Ti-O films have excellent hemocompatibility. It is, hence, suggested that the significantly lower interface tension among Ti-O films, blood and plasma proteins, and the semiconducting nature of Ti-O films, give them their improved hemocompatibility [7-16].

Already widely used in the orthopedic field as biomaterials, Ti-6Al-4V could be selected as the construction material for blood-contacting devices, such as the left ventricular assist device [7-17]. Several authors have discussed the biocompatibility and hemocompatibility of pure titanium [7-18–7-21]. Protein adsorption studies on titanium oxide (TiO_2) have been reported [7-22, 7-23]. Ti-6Al-4V alloy behavior toward bovine serum albumin (BSA) and bovine gamma-globulin has been studied using an enzyme-linked immunosorbent assay, and has demonstrated a greater protein binding for Ti-6Al-4V alloy than for stainless steel [7-24]. These results indicate that the Ti-6Al-4V alloy is well tolerated by blood [7-17]. It was also suggested that the Ti-6Al-4V alloy is well tolerated by the blood and, therefore, is a promising biocompatible material for construction of biomedical devices, either in the orthopedic field or in the cardiovascular one [7-17].

Muramatsu *et al.* [7-25] evaluated the fibrin deposition and blood platelet adhesion onto alkali- and heat-treated CpTi, alkali- and water-treated CpTi, and alkali- and heat-treated CpTi formed with apatite in simulated body fluid. They were evaluated by exposure to anticoagulated blood or washed platelet suspension under static conditions and subsequent observation with scanning electron microscopy. It was reported that (i) thrombus formation on alkali-treated CpTi surfaces, which were exposed to heparinized whole blood for 1 h, was significantly less than that on untreated control CpTi, on which pronounced depositions of fibrin-erythrocytes and lymphocytes were observed, (ii) no thrombus was observed on the apatite-like formed CpTi surface, and (iii) the apatite-like formed CpTi surface exhibits stronger antithrombogenic characteristics than CpTi and other materials examined in heparinized blood [7-25].

The role of apoptosis/cell death in the inflammatory response at the implanted materials is unexplored. Two surfaces with different cytotoxic potential and *in vivo* outcomes, titanium (Ti) and copper (Cu) were incubated *in vitro* with human monocytes and studied using a method to discriminate apoptotic and necrotic cells. It was found that Ti surfaces displayed enhanced monocyte survival and production of IL-10 and TNF-α. Cu adherent cells exhibited apoptotic signs as early as 1 h after incubation, but (ii) in contrast to Ti, after 48 h the predominance of apoptotic cells switched to apoptotic/necrotic cells on Cu surfaces, and (iii) material properties rapidly influence the onset of human monocyte apoptosis and progression to late apoptosis/necrosis, concluding that early detection of apoptosis and cell death may be important for the understanding of the biological response to implanted materials [7-26].

Ion implantation onto NiTi has been used to decrease the surface nickel concentration of this alloy and produce a titanium oxide layer [7-16]. Nothing is known yet about the blood compatibility of this surface and the suitability for implants in the blood vessels, like vascular stents. Maitz *et al.* [7-27] treated superelastic NiTi by oxygen or helium plasma-immersion ion implantation to deplete surface Ni, leading to the formation of a nickel-poor titanium-oxide surface with a nanoporous structure. Fibrinogen adsorption and conformation changes, blood platelet adhesion, and contact activation of the blood clotting cascade have been checked as *in vitro* parameters of blood compatibility; metabolic activity and release of cytokines IL-6 and IL-8 from cultured endothelial cells on these surfaces give information about the reaction of the blood vessel wall. It was found that (i) the oxygen-ion-implanted NiTi surface adsorbed less fibrinogen on its surface and activated the contact system less than the untreated nitinol surface, but (ii) conformation changes of fibrinogen were higher on the oxygen-implanted nitinol, (iii) no difference between initial and oxygen-implanted NiTi was found for the platelet adherence, endothelial cell activity, or cytokine release, and (iv) oxygen-ion implantation is seen as a useful method to decrease the nickel concentration in the surface of nitinol for cardiovascular applications [7-27].

The Ti-6Al-4V titanium alloy is widely employed as an implant material. Untreated Ti-6Al-4V and glow-discharge treated Ti-6Al-4V samples were tested on human peripheral blood mononuclear cells. It was found that (i) apoptosis, undetectable after 24 h contact of peripheral blood mononuclear cells with the two sample types, is induced after 48 h by treated samples, and, after 48 h, but in the presence of 1.5 μg/ml PHA, by both sample types, and (ii) although plasma-treated titanium alloy shows a better biocompatibility in comparison with the untreated one, attention must be paid to the careful control of the first signs of inflammation [7-28].

7.3. CELL ADHESION, ADSORPTION, SPREADING, AND PROLIFERATION

Ti implant systems have been exploited as bone-anchored implants for many years. Although titanium is well tolerated by the body, the reason for the successful use of this material is not clear. Beneficial effects due to the oxide that spontaneously forms on Ti have been implicated [7-29, 7-30, see Chapter 4 for details]. The success is not due to Ti being passively incorporated into the body, since the body recognizes Ti as foreign and Ti elicits an inflammatory response. When a biological system encounters an implant, reactions are induced at the implant–tissue interface. Such interfacial reactions include that (1) the body wants to keep the implant isolated, (2) a protective layer of macrophages, monocytes, and giant cells is formed, and (3) a wall of collagen fibers (capsule) is established, and the type of material may influence the fibrous capsule thickness. Surface and interface properties of interest are chemical composition, contamination and cleanliness, micro architecture and structure, etc. [7-6]. Fibroblasts lay down dense fibrous tissue (collagen, elastin, reticular) that is gradually replaced by less dense, normal connective tissue, but the body cannot build normal tissue all the way up to an implant.

Host hard/soft-tissue response to biomaterials should be considered first as an inflammatory reaction, as mentioned before. Any biomaterial implanted into the body will be perceived as a threat. The host defenses will attempt to eliminate it. This will not happen if the biomaterial is inert and cannot be degraded. Eventually, inert biomaterials will be integrated into the tissue or will become walled off. Biodegradable biomaterials are slowly resorbed over time and are eventually eliminated. Some biomaterials have chemical reactivity and will continue to stimulate over long periods of time. The aforementioned inflammatory processes should include four main stages: initial events (redness, swelling, pain), cellular invasion (white cells invade tissues), tissue remodeling, repair (being orchestrated by macrophages; occurs differently in different tissues; bone may completely remodel; usually complete within 3–4 weeks), and resolution (extrusion, resorption, integration, and encapsulation). Furthermore, cellular invasion has neutrophil action (main function is phagocytes; die at tissue site; release further inflammatory medicators; prostaglandins, leukotrienes) and macrophages (phagocytes; removal of dead cells; numbers and activity depend on pressure of particulate material). Moreover, resolution is an attempt to return to the original condition: extrusion and resorption of implant material for re-establishment of homeostasis, and integration and encapsulation with a layer of fibrous tissue to establish a steady state [7-31, 7-32].

The response, however, resembles normal wound healing regarding cell recruitment and persistence [7-33, 7-34]. During the surgical procedure of implantation, the biomaterial most likely will encounter blood. Almost instantly following contact with blood the implant surface will be covered with plasma proteins that become adsorbed to the surface [7-35, 7-36]. Ti in contact with serum or plasma is known to adsorb high molecular weight kininogen, Factor XII, fibrinogen, IgG, prekallikrein, and Clq [7-37, 7-38]. It was reported that, after 5 s, platelets were found on Ti surfaces exposed to whole blood, while it took about 10 min before polymorphonuclear granulocytes adhered [7-39]. Protein–platelet [7-40], leukocyte–platelet [7-41], and protein–leukocyte [7-42] are possible interactions at the surface. Furthermore, release of inflammatory mediators at the surface may influence both recruitment and activation of cells [7-40]. The polymorphonuclear granulocytes are the first leukocytes to be recruited to a Ti surface exposed to blood [7-39]. Adhesion and activation are two processes polymorphonuclear granulocytes may go through upon material contact. Adhesion involves receptor occupation with appropriate surface ligands [7-42, 7-43], and when polymorphonuclear granulocytes are challenged with different stimuli, they activate their NADPH (nicotinamide adenine dinucleotide phosphate) oxidase (which are a group of plasma membranes – associated enzymes formed in a variety of cells of mesodermal origin). By using oxygen, the NADPH oxidase produces reactive oxygen species that are used in the cell's defense against microorganisms [7-40, 7-44].

Close contact between bone and implant has been observed with titanium implants [7-45–7-47], but has also been reported with alumina oxide (Al_2O_3) [7-48] and zirconium implants [7-49]. Most authors report a noncellular layer of proteoglycans between bone and titanium [7-46]. There are indications that implant materials, including a thick proteoglycan layer formation, have a lower biocompatibility than implants producing a thin proteoglycan layer [7-50, 7-51]. Most of the articles on implant/bone interactions are presented and discussed based on interfacial reaction between metal surface and bone. However, this is not right, since the immediate contacting species of the implant surface is not metal, but its oxide. If the implant material is one of titanium materials, titanium oxide should be considered as an extreme surface, contacting the living hard/soft tissue. Titanium reacts immediately with oxygen when exposed to air, forming a very thin surface oxide film. This film, which increases during prolonged exposure to oxygen, consists primarily of titanium dioxide (TiO_2), but Ti_2O_3 and other oxides are also present [7-52, 7-53]. Titanium dioxide has physical/chemical characteristics that differ from metallic Ti, characteristics which are more closely related to ceramics than to metals. As titanium dioxide particles have been used in chromatography, it is well established that titanium oxide surfaces bind cations, particularly polyvalent cations [7-54]. Titanium dioxide surfaces have a net negative charge at the pH values encountered in animal tissues (around 4.0). This binding of cations is based on electrostatic interactions between titanium-linked O^- on the implant surface and cations. The oxide is highly polarized and attracts water and water-soluble molecules in general [7-51].

The immediate reactions on the implant surface after exposure in the tissues may influence later events, and are conceivably important for the clinical success of the implantation. An understanding of these immediate biological reactions may be significant. The chemical properties of the titanium oxide surface suggest that calcium ions may be attached to the oxide-covered surface by electrostatic interaction with O^-, as just discussed. Calcium deposits have been observed in direct contact with the titanium oxide [7-50]. According to the same model, calcium-binding macromolecules may adsorb selectively to the implant surface *in vivo* as the next sequence of events. Calcium-binding molecules are often acidic with surface-exposed carboxyl, phosphate, or sulfate groups. Proteoglycans, which contain carboxyl and sulfate groups, might bind to the TiO_2 surface by this mechanism [7-46, 7-50, 7-51]. Hydroxyapatite, the major mineral component of bone, also exhibits a surface dominated by negatively charged oxygen (P-bound) that can attract cations and subsequently anion calcium-binding macromolecules [7-55, 7-56]. Ellingen [7-22] treated CpTi surfaces with (1) 100 m mol/l $CaCl_2$ for 5 min, water-washed and air-dried, and (2) 100 m mol/l $CaCl_2$ for 5 min, followed by treating for 24 h with 0.2 mol/l EDTA, water-washed and air-dried, with untreated CpTi as a control. The reaction of human serum proteins with TiO_2 was examined and compared with the reaction with hydroxyapatite (HA). It was found that (i) the oxide covered Ti surfaces reacted with calcium when exposed to $CaCl_2$ and calcium was identified to a depth of 17 nm into the oxide layer, (ii) surface-adsorbed serum proteins were dissolved by EDTA and surfaces of TiO_2 and HA appeared to take up the same selectively from human serum, and (iii) albumin, pre-albumin, and IgG were identified by immunoelectrophoresis, suggesting that calcium binding may be one mechanism by which proteins adsorb to TiO_2, and it is well established that this is the case with HA [7-22].

During the first week after implantation of Ti in soft tissues, a fluid space, which contains proteins [7-57] and scattered inflammatory cells [7-58], separates the implant surface from the tissue. The majority of cells in the fluid space are not in direct contact with the surface during the first week of the healing [7-59]. However, during later time intervals, macrophages are adherent to the surface of Ti [7-58, 7-60] and other biomaterials [7-61]. One factor which may influence the regenerating tissue around the implant is the activity of the macrophages and the release of mediators from macrophages adherent on the implant surface. It is possible that these cells act as "messenger cells", transmitting information about the implant surface structure and composition to the surrounding tissue [7-62, 7-63].

Healy *et al.* [7-64] employed surface sensitive spectroscopies, Auger electron and X-ray photoelectron (XPS) techniques to characterize changes in Ti oxide composition, oxide stoichiometry, and adsorbed surface species as a function of exposure to human serum in a balanced electrolyte (serum/SIE), and 8.0 mM ethylenediaminetetraacetic acid in a balanced electrolyte (EDTA/SIE) at 37 V. It was reported that (i) before immersion, the oxide was near ideal TiO_2, covered by two types of hydroxyl groups: acidic OH(s) with oxygen doubly coordinated to Ti, and basic Ti-OH groups singly coordinated, and (ii) after extended exposure to both solutions, up to 5000 h (*ca.* 200 days), the surface concentration of OH groups increased and nonelemental P appeared. The adsorbed phosphate species were presumed to be either $Ti\text{-}H_2PO_4$ or $Ti\text{-}HPO^{4-}$. Hence, it was suggested that Ti metal is not in direct contact with the biological milieu, but rather there is a gradual transition from the bulk metal, near-stoichiometric oxide, Ca- and P- substituted hydrated oxide, adsorbed lipoproteins and glycolipids, proteoglycans, collagen filaments and bundles to cells [7-64]. This finding is very important to develop the gradual functioning implant surface structure, as will be described in Chapter 12.

The protein adsorption behavior of thin films of calcium phosphate bioceramic and Ti was investigated by Zeng *et al.* [7-65]. The thin films were produced with an ion beam sputter deposition technique using targets of hydroxyapatite, fluorapatite, and Ti. Fourier transform infrared spectroscopy (FTIR) with attenuated total internal reflectance was used to evaluate protein adsorption on these surfaces. It was shown that (i) surface composition and structure influenced the kinetics of protein adsorption and the structure of adsorbed protein, (ii) calcium phosphate adsorbed a greater amount of protein than the Ti surface, and caused more alteration of the structure of adsorbed BSA than did the Ti surface, and (iii) the differences in protein adsorption behavior could result in very different initial cellular behavior on calcium phosphate bioceramic and Ti implant surfaces [7-65].

MacDonald *et al.* [7-66] surface-modified titanium dioxide particles and investigated their effects on protein adsorption, using a model cell binding protein, human plasma fibronectin (HPF), which is an important matrix glycoprotein that plays a major role in cell and protein attachment. Titanium dioxide surfaces were modified by heating Titanium dioxide powder at 800°C for 1 h or treating with an oxidizing agent: peroxide ammonium hydroxide followed by peroxide in hydrochloric acid. Oxidized and control samples were further treated with butanol:water (by 9:1 fraction) for 30 min. It was found that (i) maximum HPF adsorption on modified or unmodified titanium dioxide occurred within 30 min, with greater protein adsorption occurring on butanol-treated samples, (ii) desorption was minimal with all modifications, and (iii) by modifying the physico-chemical properties of titanium dioxide surfaces, it may be possible to alter protein adsorption, and hence optimize cell attachment [7-66].

The effect of surface roughness (R_a: 0.320, 0.490, and 0.874 μm) of Ti-6Al-4V on the short- and long-term response of human bone marrow cells *in vitro* and on protein adsorption was investigated. Cell attachment, cell proliferation, and differentiation (alkaline phosphatase specific activity) were determined. The protein adsorption of BSA and fibronectin, from single protein solutions on rough and smooth Ti-6Al-4V surfaces was examined with XPS and radiolabeling. It was reported that (i) cell attachment and proliferation were surface roughness sensitive, and increased as the roughness of Ti-6Al-4V increased, (ii) human albumin was adsorbed preferentially onto the smooth substratum, and (iii) the rough substratum bound a higher amount of total protein (from culture medium supplied with 15% serum) and fibronectin (10-fold) than did the smooth one [7-67].

Eriksson *et al.* [7-29] treated CpTi surfaces (1) to have a rigid TiO_2 formation by boiling CpTi in $H_2O - H_2O_2 - NH_4OH$ (5:1:1) for 5 min, followed by heating at 700°C, or (2) to have a rough (facetted) Ti surface by etching in 10% HF to grow thick oxide film. CpTi surfaces were also treated in concentrated HNO_3 to grow a novel thin oxide, which resulted in a rough surface, too. It was observed that all surfaces consisted of TiO_2 covered by a carbonaceous layer. The surfaces were incubated with capillary blood for time periods of between 8 min and 32 h. It was reported that (i) leukocyte adhesion to the surfaces increased during the first hours of blood-material contact and then decreased, (ii) polymorphonuclear granulocytes were the dominating leukocytes on all surfaces, followed by monocytes, (iii) cells adhering to rough surfaces were higher than cells on smooth surfaces, (iv) maximum respiratory burst response occurred earlier on the smooth than on the rough surfaces, and (v) topography had a greater impact than oxide thickness on most cellular reactions investigated, where the latter often had a dampening effect on the responses [7-29].

Events leading to integration of an implant into bone, and hence determining the long-time performance of the device, take place largely at the interface formed between the tissue and the implant [7-68]. The development of this interface is complex and is influenced by numerous factors, including surface chemistry and surface topography of the foreign material [7-30, 7-46, 7-69–7-71]. For example, Oshida *et al.* [7-72] treated NiTi by acid-pickling in $HF:HNO_3:H_2O$ (1:1:5 by volume) at room temperature for 30 s to control the surface topology and selectively dissolve Ni, resulting in a Ti-enriched NiTi surface layer. Zinger *et al.* [7-73] investigated the role of surface roughness on the interaction of cells with titanium model surfaces of well-defined topography using human bone-derived cells (MG63 cells) to understand the early phase of interactions through a kinetic morphological analysis of adhesion, spreading, and proliferation of the cells. Scanning electron microscope (SEM) and double immunofluorescent labeling of vinculin and actin revealed that the cells responded to nanoscale roughness with a higher cell thickness and a delayed apparition of the focal contacts. A singular behavior was observed on nanoporous oxide surfaces, where the cells were more spread and displayed longer and more numerous filopods. On electrochemically microstructured surfaces, the MG63 cells were able to go inside, adhere, and proliferate in cavities of 30 or 100 μm in diameter, whereas they did not recognize the 10 μm diameter cavities. Cells adopted a three-dimensional shape when attaching inside the 30 μm diameter cavities. It was therefore concluded that nanotopography on surfaces with 30 μm diameter cavities had little effect on cell morphology compared to flat surfaces with the same nanostructure, but cell proliferation exhibited a marked synergistic effect of microscale and nanoscale topography [7-73].

Rat bone marrow stromal cells were cultured on either a smooth surface (R_a: 0.14 μm) or a rough surface (5.8 μm) of Ti-6Al-4V disks for 24 or 48 h. Cells on the smooth surface showed typical fibroblastic morphology, whereas cells on the rough surface were in clusters with a more epithelial appearance. RNA was extracted from the cells at both time points, and gene expression was analyzed by using a rat gene microarray. At 24 and 48 h, a similar number of genes were both up- and downregulated at least twofold on the rough surface compared to the smooth surface. It was noticed that (i) roughness did not appear to be a specific stimulator of osteogenesis because genes of both the bone and cartilage lineage were upregulated on the tough surface, and (ii) surface roughness alters the expression of a large number of genes in marrow stromal cells, which are related to multiple pathways of mesenchymal cell differentiation [7-74]. A highly porous Ti-6Al-4V was fabricated by a polymeric sponge replication method [7-75]. A polymeric sponge, impregnated with Ti-6Al-4V slurry prepared from Ti-6Al-4V powders and binders, was subjected to drying and pyrolyzing to remove the polymeric sponge and binders. The porous Ti-6Al-4V made by this method had a three-dimensional trabecular porous structure with interconnected pores mainly ranging from 400 to 700 μm and a total porosity of about 90%. The compressive strength was 10.3 $\pm$ 3.3 MPa and the elastic constant 0.8 $\pm$ 0.3 GPa. It was also reported that MC3T3-E1 cells attached and spread well in the inner surface of pores [7-75].

The adsorption of BSA on to Ti powder has been studied as a function of protein concentration and pH, and in the presence of calcium and phosphate ions. It was reported that (i) the maximum adsorption at pH 6.8 is 1.13 ± 0.21 mg/m^2, (ii) for the pH dependence of adsorption, the amount adsorbed increases with decreasing pH (at pH 5.15 the maximum adsorption was 1.31 ± 0.2 mg/m^2), indicating that hydration effects are important, and (iii) adsorption increases and decreases in the presence of calcium (at pH 6.8 the maximum adsorption was 1.73 ± 0.23 mg/m^2) and phosphate (where at pH 6.8 the maximum adsorption was 0.64 ± 0.14 mg/m^2) ions, indicating that electrostatic effects are important [7-76].

The *in vitro* study was conducted to investigate the adsorption of albumin and fibronectin on titanium surfaces, and the effect of pre-adsorbed BSA and bovine fibronectin on osteoblast attachment. Maximum adsorption of BSA and fibronectin on Ti surfaces was observed after the incubation for 180 min. In the presence of pre-adsorbed proteins, osteoblast attachment on Ti surfaces was observed to be enhanced compared to control Ti surfaces. However, cell attachment was affected by the types of protein adsorbed. Pre-adsorbed albumin was observed to have no significant effect on the amount of osteoblast cells attached. In comparison to the control Ti surface and those pre-adsorbed with albumin, the Ti surfaces pre-adsorbed with fibronectin for 15 min were observed to significantly increase osteoblast cell attachment, whereas the Ti surfaces pre-adsorbed with fibronectin for 180 min did not affect cell attachment. In addition, cell morphology of the attached cells on protein pre-adsorbed Ti surfaces was not affected by the type of protein used. Based on these findings, it was concluded that (i) the concentration of fibronectin adsorbed on Ti surfaces was higher compared to albumin, and (ii) the concentration of fibronectin on Ti surfaces plays a role in governing cell attachment [7-68].

Webster *et al.* [7-77] claimed that $CaTiO_3$ is a strong candidate to form at the interface between hydroxyapatite (HA) and titanium implants during many coating procedures. The ability of bone-forming cells (osteoblasts) to adhere on titanium coated with HA, resulting in the formation of $CaTiO_3$, was investigated. For formation of $CaTiO_3$, titanium was coated on HA disks and annealed either under air or an $N_2 + H_2$ environment. Results from cytocompatibility tests revealed increased osteoblast adhesion on materials that contained $CaTiO_3$ compared to both pure HA and uncoated titanium. The greatest osteoblast adhesion was observed on titanium-coated HA annealed under air conditions. Because adhesion is a crucial prerequisite to the subsequent functions of osteoblasts (such as the deposition of calcium containing minerals), it was suggested that orthopedic coatings that form $CaTiO_3$ could increase osseointegration with juxtaposed bone needed to enhance implant efficacy and longevity [7-77].

Extensive studies on the microstructure, chemical composition, and wettability of thermally and chemically modified Ti-6Al-4V alloy disks were done and correlated with the degree of radiolabeled fibronectin-alloy surface adsorption and subsequent adhesion of osteoblast-like cells [7-78]. Heating, either in pure oxygen or atmosphere (atm), resulted in an enrichment of Al and V within the surface oxide formed on Ti-6Al-4V [7-79]. As for the surface contact angle (or wettability), it was reported that (i) preheating (oxygen/atm) or hydrogen peroxide treatment brought a thicker oxide layer with more hydrophilicity when compared with passivated controls, but (ii) post-treatment with butanol resulted in less hydrophilic surfaces than those by heating or hydrogen peroxide treatment. It was therefore indicated that (i) an increase in the absolute content of Al and/or V (heat), and/or in the Al/V ratio (with little change in hydrophilicity; heat+butanol) is correlated with an increase in the fibronectin-promoted adhesion of an osteoblast-like cell line, and (ii) the thermal treatment-induced enhancement of cell adhesion in the presence of the protein is due to its increased biological activity, rather than a mass effect alone, which appear to be associated with changes in chemical composition of the metallic surface [7-78].

Nanophase materials are unique materials that simulate dimensions of constituent components of bone since they possess particle or grain sizes of less than 100 nm [7-76], and their interactions with osteoblasts, compared to conventional metals, were *in vitro* studied to synthesize, characterize, and evaluate osteoblast adhesion on nanophase metals (specifically, Ti, Ti-6Al-4V, and Co-Cr-Mo alloys), which are widely used in orthopedic applications. It was found that (i) the osteoblast adhesion on nanophase was more than that on conventional metals, and (ii) nanometal surfaces possessed similar chemistry, and only altered in degree of nanometer surface roughness when compared to their respective conventional counterparts [7-80].

Improving the biological performance of engineered implants interfacing tissues is a critical issue in implantology and its related science and engineering. Micromotion at the soft-tissue-implant interface has been shown to sustain an inflammatory response. To eliminate micromotion, it is desirable to promote cellular and extracellular matrix adhesion to the implant surface. Jain and von Recum [7-81] modified titanium surfaces topographically or chemically to effect cellular adhesion and to influence cellular interactions and function, and conducted an *in vitro* study to compare the independent effects of surface chemistry and topography on fibroblast-test specimen proximity. Titanium was sputter-coated in stepwise, increasing thicknesses (20–350 nm) onto a series of either smooth or microtextured polyethylene terephthalate (PET), resulting in a stepwise change from a PET surface to a Ti surface – as gradually functioning material. It was found that (i) fibroblast proximity to the coverslip surface increases as the Ti thickness increases on either smooth or textured test specimens, and (ii) fibroblasts were firmly attached to the ridge tops on the coated textured test specimens.

Therefore, fibroblast apposition is strongly enhanced by microtextured surfaces and Ti rather than smooth surfaces and polyethylene terephthalate [7-81].

Osteoblast adhesion on the implant material surface is essential for the success of any implant in which osseointegration is required. Surface properties of implant material have a critical role in the cell adhesion progress. Zhang *et al.* [7-82] employed the micro-arc oxidizing and hydrothermally synthesizing methods to modify the TiO_2 layer on the titanium surface, on which the mouse osteoblastic cell line (MC3T3-E1) was seeded to evaluate their effect on cell behavior. The surface structure of micro-arc oxidized samples exhibited micropores with a diameter of 1–3 μm, whereas the micro-arc oxidation/hydrothermal synthesized samples showed additional multiple crystalline microparticles on the microporous surface. Both treated surfaces possessed higher energy than that of untreated titanium. It was concluded that the micro-arc oxidation and micro-arc oxidation/hydrothermal synthesization methods change the surface energy of the TiO_2 layer on the titanium surface, indicating that this may have an influence on the initial cell attachment [7-82].

Surface characteristics play a vital role in determining the biocompatibility of materials used as bone implants. It is well documented that calcium ion implantation of titanium enhances osseointegration and bone formation *in vivo*. In order to measure the precise effects of ion implantation of titanium on bone cells *in vitro*, alveolar bone cells were seeded on the surface of polished titanium disks implanted with calcium, potassium, and argon ions. Using radioisotopically tagged bone cells, it was found that (i) although the calcium ion implanted surface reduced cell adhesion, it nevertheless significantly enhanced cell spreading and subsequent cell growth, but (ii) in contrast, few differences in bone cell behavior were observed between the potassium- and argon-implanted titanium and the control nonimplanted titanium disks. Based on these findings, it was suggested that the calcium-implanted surface may significantly affect the biocompatibility of titanium implants by enhancing bone cell growth; it was therefore concluded that surface modification by ion implantation could thus prove to be a valuable tool for improving the clinical efficacy of titanium for bone repair and regeneration *in vivo* [7-83]. Bioactivation of the titanium surface by sodium plasma immersion ion implantation and deposition (PI^3&D) was compared to that of the untreated, Na beam-line implanted and NaOH-treated titanium samples [7-84]. It was found that (i) from a morphological point of view, cell adherence on the NaOH-treated titanium is the best, (ii) on the other hand, the cell activity and protein production were higher on the nonbioactive surfaces, and (iii) the active surfaces support an osteogenic differentiation of the bone marrow cells at the expense of lower proliferation, due to the high alkaline phosphatase activity per cell [7-84]. Muhonen *et al.* [7-85] investigated the responses of mature osteoclasts cultured on austenite and martensite phases of NiTi shape memory implant material, using the sensitivity of osteoclasts to the underlying substrate and actin ring formation as an indicator of the adequacy of the implant surface. It was reported that osteoclasts tolerated well the austenite phase of NiTi, but the chemically identical martensitic NiTi was not as well tolerated by osteoclasts (e.g., indicated by diminished actin ring formation), concluding that certain physical properties specific to the martensitic NiTi have an adverse effect on osteoclast survival in this NiTi phase [7-85].

Living bone cells are responsive to mechanical loading. Consequently, numerous *in vitro* models have been developed to examine the application of loading to cells. However, not all systems are suitable for the fibrous and porous three-dimensional materials, which are preferable for tissue repair purposes, or for the production of tissue engineering scaffolds. For three-dimensional applications, mechanical loading of cells with either fluid flow systems or hydrodynamic pressure systems has to be considered. The response of osteoblast-like cells to hydrodynamic compression, while growing in a three-dimensional titanium fiber mesh scaffolding material, was evaluated by Walboomers *et al.* [7-86]. Bone marrow cells were obtained from the femora of young (12-day-old) or old (1-year-old) rats, and precultured in the presence of dexamethasone and β-glycerophosphate to achieve an osteoblast-like phenotype. Subsequently, cells were seeded onto the titanium mesh scaffolds, and subjected to hydrodynamic pressure, alternating from 0.3 to 5.0 MPa at 1 Hz, at 15-min intervals for a total of 60 min per day for up to 3 days. After pressurization, cell viability was checked. Afterward, DNA levels, alkaline phosphatase (ALP) activity, and extracellular calcium content were measured. Finally, all specimens were observed with scanning electron microscopy. Cell viability studies indicated that (i) the applied pressure was not harmful to the cells, and (ii) the cells were able to detect the compression forces [7-86].

Improved methods to increase surface hardness of metallic biomedical implants are being developed in an effort to minimize the formation of wear debris particles that cause local pain and inflammation. However, for many implant surface treatments, there is a risk of film delamination due to the mismatch of mechanical properties between the hard surface and the softer underlying metal. Hence, Clem *et al.* [7-87] developed a new surface modification of titanium alloy (Ti-6Al-4V), using microwave plasma chemical vapor deposition to induce titanium nitride formation by nitrogen diffusion. The result is a gradual transition from a titanium nitride surface to the bulk titanium alloy, without a sharp interface that could otherwise lead to delamination. The vitronectin adsorption, as well as the adhesion and spreading of human mesenchymal stem cells to plasma-nitrided titanium were evaluated, and they were equivalent to that of Ti-6Al-4V, while hardness is improved 3–4-fold. These *in vitro* results suggested that the plasma nitriding technique has the potential to reduce wear, and the resulting debris particle release, of biomedical implants without compromising osseointegration, thus minimizing the possibility of the implant loosening over time [7-87].

As we have seen above, the biological events occurring at the bone–implant interface are influenced by the topography, chemistry, and wettability of the implant surface. Advincula *et al.* [7-88] introduced other method to achieve the aforementioned specific aim. The surface properties of titanium alloy prepared by either surface sol–gel processing (SSP), or by passivation with nitric acid, were investigated. The bioreactivity of the substrates was assessed by evaluating MC3T3-E1 osteoblastic cell adhesion, as well as by *in vitro* formation of a mineralized matrix. It was found that (i) surface analysis of sol–gel-derived oxide on Ti-6Al-4V substrates showed a predominantly titanium dioxide (TiO_2) composition with abundant hydroxyl groups, (ii) the surface was highly wettable, rougher, and more porous compared to that of the passivated substrate, and (iii) significantly more cells adhered to the sol–gel-coated surface as compared with passivated surfaces after 1 and 24 h following cell seeding, and a markedly greater number of mineralized nodules were observed on sol–gel coatings [7-88].

Studies of blood plasma protein interactions with Ti oxides is one possible way to understand blood interactions and tissue integration better, since blood is the first medium encountered after the surgical procedure, and thus governs the subsequent cell attachment and wound healing. Adsorption of albumin and fibrinogen from human blood plasma onto CpTi surfaces with varying oxide properties was studied with an enzyme-linked immunosorbent assay (ELISA). The intrinsic activation of blood coagulation (contact activation) was studied *in vitro* using a kallinkrein-sensitive substrate. It was reported that (i) low fibrinogen and high albumin adsorption was observed for all titanium samples, except for the radio frequency plasma-treated and water-incubated samples, which adsorbed significantly lower amounts of both, and (ii) smooth samples with a surface roughness with less than 1 nm showed some correlation between surface wettability and adsorption of fibrinogen and albumin, whereas rough surfaces (>5 nm) did not [7-89]. The role of implant surface topography in cell attachment has been reviewed extensively [7-6, 7-90–7-92]. Rough or textured porous surfaces promote cell attachment [7-93, 7-94]. Surface roughness and composition have been considered the most important parameter for altering cell activity. There are observations that cell responses, such as adhesion, morphologym, and collagen synthesis differed on the two grades of CpTi. Between 4 and 24 h, the rate of cell attachment to CpTi (grade 1) differed significantly, compared to cell attachment to CpTi (grade 4). At 24 h, the percent of collagen synthesized was significantly more on grade 1 than on grade 4. Alkaline phosphatase activity was similar on all substrates. The calcium content was significantly higher on grade 1 than on 4 as well as on glass at 24 h and at 4 weeks. Based on these observations, it was concluded that commonly used CpTi induced differential morphologic and phenotypic changes in human osteoblast-like cells depending on the material grade [7-95].

Khakbaznejad *et al.* [7-96] investigated on osteogenic cells from newborn rat calvariae which were cultured on titanium surfaces, on which cell orientation could be manipulated, to measure the orientation angles of cells. Substrata included smooth surfaces and smooth regions (gaps) flanked by grooves of 47-μm pitch and 3, 10, or 30 μm depth to create micro-machined surfaces. There were several interesting findings reported. Grooves proved effective in orienting cells, but their orienting ability decreased above the ridge level. Cells on the smooth surface showed no preferred orientation. Cells in the gaps became oriented as a result of cell–cell interactions with the cells on the flanking grooves. Cells in the grooves produced oriented collagen fibers, but in the gaps, fibers could be parallel, perpendicular, or diagonal to the grooves. Collagen fibers on the smooth surfaces formed arrays of parallel fibers in a crisscross pattern. In long-term cultures, bone-like nodules were formed, but mostly above the ridge level. Based on these findings, it was reported that grooved surfaces can influence cell orientation both in cell populations above the cells in contact with the grooves and in cell populations adjacent to the grooves [7-96].

To investigate the roles of composition and characteristics of titanium surface oxides in cellular behavior of osteoblasts, the surface oxides of titanium were modified in composition and topography by anodic oxidation in two kinds of electrolytes; 0.2 M H_3PO_4, and 0.03 M calcium glycerophosphate + 0.15 M calcium acetate. It was found that (i) calcium and phosphorous ions are incorporated into the anodized surfaces in the form of phosphate and calcium phosphate, (ii) the geometry of the micro-pores in the anodized surfaces varied with diameters up to 0.5 μm in 0.2 M H_3PO_4, and to 2 μm in 0.03 M Ca-GP and 0.15 M CA, depending on voltages and electrolyte, and (iii) contact angles of all the anodic oxides were in the range of 60–90°, indicating a relatively hydrophobicity. As for cell culture studies, the absence of cytotoxicity and an increase of osteoblast adhesion and proliferation by the anodic oxides were found, and cells on the surfaces with micro-pores showed an irregular, polygonal growth and more lamellipodia [7-97].

Saldaña *et al.* [7-98] evaluated the biocompatibility of the oxidized (700°C×1 h and 500°C×1 h) surfaces of Ti-6Al-4V by assessing cell adhesion, proliferation, and differentiation of primary cultures of human osteoblastic cells. Compared with polished alloy, both thermal treatments increased osteoblast adhesion measured as cell attachment, β1 integrin, and FAK-Y397 expression, as well as cytoskeletal reorganization. It was found that (i) when compared with treatment at 500°C, thermal oxidation at 700°C enhanced cell adhesion, (ii) treatment at 700°C transiently impaired cell proliferation and viability, which were not altered in alloys oxidized at 500°C, (iii) several markers of osteoblastic differentiation such as procollagen

I peptide, alkaline phosphatase, osteocalcin, and mineralized nodule formation were found either unaffected or differentially increased by alloys treated either at 500° or 700°C, and (iv) thermal oxidation at 700°C also increased osteoprotegerin secretion. It was therefore concluded that thermal oxidation treatments at 500° or 700°C for 1 h improve the *in vitro* biocompatibility of Ti-6Al-4V [7-98].

Osteoblast response to Ti implants depends not only on the chemistry of the implant but also on the physical properties of the implant surface, such as micro-topography and roughness, as has been reviewed in the above. Early changes in cell morphology and gene expression during the early phase of osteoblast interaction with Ti-6Al-4V surfaces of two different roughnesses were examined. MG63 osteoblast-like cells were cultured for 2, 6, 24, and 72 h on smooth (R_a=0.18 μm) and rough (R_a=2.95 μm) Ti-6Al-4V surfaces. Changes in gene expression for extracellular signal-regulated kinase 2 (Erk2), type I collagen, phospholipase C-γ2 (Plc-γ2), and β-actin were measured by RT-PCR (Reverse Transcription Polymerase Chain Reaction) after 6 and 24 h in culture. It was found that cell number was significantly higher on the smooth surface. Of the genes examined, only Erk2 and β-actin showed a change in expression with surface roughness. Both genes were upregulated on the rough surface at 6 h. Hence, it was indicated that Ti-6Al-4V surface roughness affects osteoblast proliferation, morphology, and gene expression, and that these effects can be measured after periods as short as 2–6 h [7-99].

Stainless steel, CpTi, and Ti-6Al-7Nb are frequently used metals in orthopedic internal fracture fixation. Particularly, clinical uses of Ti-6Al-7Nb are becoming more popular due to its excellent wear resistance. The *in vitro* soft-tissue compatibility of Ti-6Al-7Nb in comparison to stainless steel and CpTi, using a human fibroblast model, was conducted by Meredith *et al.* [7-100]. It was found that (i) cell morphology and growth rate were similar for stainless steel, electro-polished CpTi, and Ti-6Al-7Nb, with cells well spread-out and forming a confluent mono-layer after 10 days, (ii) cell growth on standard CpTi was similar to the electropol-ished samples; however, they showed a less spread-out morphology, with more filopodia and surface ruffling present, and (iii) the endocytosis of α-phase particles originating from the standard Ti-6Al-7Nb surface was noted [7-100].

Polished Ti disks were implanted with high (1×10^{17} ions/cm^2), medium (1×10^{16} ions/cm^2), and low (1×10^{15} ions/cm^2) doses of Ca ions at 40 keV to pro-mote an enhanced osseointegration. The effects of different levels of Ca implanta-tion on morphology, attachment, and spreading of MG-63 cells seeded on the surface of the control (nonimplanted) Ti and Ca-Ti disks were examined. The effects of pre-immersion of the Ca (high)-Ti in tissue culture medium on cell attachment were measured and correlated with specific chemical changes at the Ti surface. It was reported that (i) although cell adhesion on high-dosed Ca-Ti was initially reduced, it nevertheless was not only restored, but substantially increased with progressing culture times, (ii) a significantly enhanced cell spreading, formation of focal adhesion plaques, and expression of integrins were measured on this particular surface, (iii) in contrast, no marked differences were observed in cell behavior on low- and medium-dosed Ca-Ti, (iv) pre-immersion studies indicated that the decrease in cell attachment to high-dosed Ca-Ti at early time periods may be linked to the presence of Ca- and P-rich particles on the surface, and (v) the absence of these particles at 24 h was consistent with a significant increase in cell attachment [7-101].

A Ca-deficient carbonate apatite coating on titanium was prepared by pre-calcifying titanium in a saturated $Ca(OH)_2$ solution, and then immersing in a supersaturated calcium phosphate solution by Feng *et al.* [7-102]. The interaction of the protein with the apatite coating on Ti was investigated by SME with X-ray energy dispersion spectroscopy, X-ray photoelectron spectroscopy, X-ray diffraction, and Fourier transform infrared spectroscopy. During immersion of the coating in BSA solution, it was observed that calcium and phosphate ions dissolved and re-precipitated, resulting in the formation of the coating contamination BSA from the surface to subsurface layers. The adsorption modified the structure and morphology of the apatite and changed the protein configuration. It was also found that the protein chemically adsorbed onto surfaces containing calcium or phosphorous, showing that both Ca and P on the apatite coating were the protein binding sites [7-102].

7.4. ROUGHNESS AND CELLULAR RESPONSE TO BIOMATERIALS

Eriksson *et al.* [7-29] pointed out the importance of surface topography on cellular reactions. Surface plays a crucial role in biological interactions for four reasons. First, the surface of a biomaterial is the only part in contact with the bioenvironment. Second, the surface region of a biomaterial is almost always different in morphology and composition from the bulk. Differences arise from molecular rearrangement, surface reaction, and contamination. Third, for biomaterials that neither release nor leak biologically active or toxic substances, the characteristics of the surface govern the biological response. And fourth, some surface properties, such as topography, affect the mechanical stability of the implant/tissue interface [7-103]. Biomaterials used in a living organism may come into contact with cells in the related tissue for a long period of time. For this reason, they should naturally be harmless to the organism, and their mechanical properties should be suited to the purpose and compatible with receiving tissue surfaces. Furthermore, they should possess a biological effect capable of providing favorable circumstances for the properties and functions of the cells at the implant site. For example, materials used in the construction of an artificial heart or heart valve must provide for antithrombogenesis, which for a dental or bone implant must be suitable for cell attachment, because both the connective and epithelial cells (with which these materials mainly come into contact) are anchorage-dependent, and therefore need a cell attachment scaffold for cell division and cell differentiation to be conducted. It is therefore important to investigate attachability of the cells to the materials, which is one of the parameters in the evaluation of biomaterials.

It has been shown that methods of implant surface preparation can significantly affect the resultant properties of the surface and subsequently the biological responses that occur at the surface [7-104–7-106]. Recent efforts have shown that the success or failure of dental implants can be related not only to the chemical properties of the implant surface, but also to its micromorphologic nature [7-107–7-109]. From an *in vitro* standpoint, the response of cells and tissues at implant interfaces can be affected by surface topography or geometry on a macroscopic basis [7-107, 7-108], as well as by surface morphology or roughness on a microscopic level [7-109, 7-110].

Figure 7-1 shows a distribution of surface roughness of implants, which have been clinically used and evaluated as successful implants by the original author(s)'s judgment. It was found that regardless of the type of implant materials used, if the surface roughness is manipulated to have a range from 1 to 50 μm, the implant's survival rate is excellent. This finding leads toward a morphological compatibility, which is the third requirement for successful and biofunctional implant systems [7-111].

These characteristics undoubtedly affect how cells and tissues respond to various types of biomaterials. Of all the cellular responses, it has been suggested that cellular adhesion is considered the most important response necessary for developing a rigid structural and functional integrity at the bone/implant interface [7-112]. Cellular adhesion alters the entire tissue response to biomaterials [7-113].

On a macroscopic level (roughness > 10 μm) roughness influences the mechanical properties of the titanium–bone interface, the mechanical interlocking of the interface, and the biocompatibility of the material [7-114, 7-115]. Surface roughness in the range from 10 nm to 10 μm may also influence the interfacial biology, since it is the same order as the size of the cells and large biomolecules [7-52]. Microroughness at this level includes material defects, such as grain boundaries, steps and kinks, and vacancies that are active sites for adsorption, and therefore influence the bonding of biomolecules to the implant surface [7-116]. Microrough surfaces promote significantly better bone apposition than smooth surfaces, resulting in a higher percentage of bone in contact with the implant. Microrough surfaces may influence the mechanical properties of the interface, stress distribution, and bone remodeling [7-117]. Increased contact area and mechanical interlocking

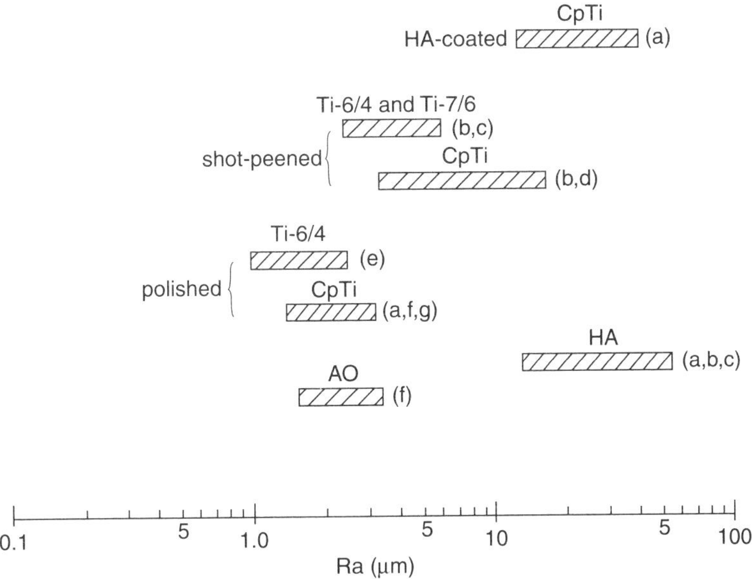

Figure 7-1. Distribution of surface roughness of successfully implanted and clinically biofunctioning materials.

(a) Buser D, Schenk RK, Steinemann SG, Fiorellini JP, Fox CH, Stich H. Influence of surface characteristics on bone integration of titanium implants: a histomorphometric study in miniature pigs. J Biomed Mater Res 1991;25: 889−902.

(b) Jamsen JA, van der Waerden JPCM, Wolke JGC. Histologic investigation of the biologic behavior of different hydroxyapatite plasma-coatings in rabbits. J Biomed Mater Res 1993;27:603−610.

(c) Wang BC, Lee TM, Chang E, Yang CY. The shear strength and the failure mode of plasma-sprayed hydroxyapatite coating to bone: the effect of coating thickness. J Biomed Mater Res 1993;27:1315−1327.

(d) Steinemann SG, Eulenberger J, Maeusli PA, Schoeder A. Biological and bio-mechanical performances of biomaterials. Amsterdam: Elsevier Science, 1986. pp. 409−414.

(e) Hayashi K, Inadome T, Mashima T, Sugioka Y. Comparison of bone-implant interface shear strength of solid hydroxyapatite and hydroxyapatite-coated titanium implants. J Biomed Mater Res 1993;27:557−563.

(f) Thomas KA, Cook SD. An evaluation of variable influencing implant fixation by direct bone apposition. J Biomed Mater Res 1985;19:875−901.

(g) Li J. Behavior of titanium and titanium-based ceramics in vitro and in vivo. Biomaterials 1993;14:229−232.

of bone to a microrough surface can decrease stress concentrations resulting in decreased bone resorption. Bone resorption takes place shortly after loading smooth-surfaced implants [7-118], resulting in a fibrous connective tissue layer, whereas remodeling occurs on rough surfaces [7-119].

Rich and Harris [7-109] reported that macrophages exhibited a rugophilia, or an affinity for rough surfaces, whereas fibroblasts failed to readily adhere to these same surfaces. This finding may complement previous work indicating that rough surfaces promote inflammation via macrophage attraction, rather than wound healing by connective tissue [7-110]. Studies by Brunette [7-90] and Chehroudi et al. [7-120, 7-121] have demonstrated that cells of differing origins can orient themselves in the grooves of differing micromachined surfaces. It was determined that epithelial cells were more likely to attach to grooves of specific dimension than to adjacent smooth surfaces. Michaels *et al.* [7-122] determined that a higher percentage of osteoblast-like cells attached to rough CpTi surfaces produced by sandblasting than to smoother surfaces polished with 1 μm diamond paste. Keller *et al.* [7-123] later confirmed this.

In order to achieve morphological compatibility [7-111, 7-124], titanium implant surfaces need to be modified. They can be treated by additive methods, such as the titanium beads plasma spray procedure, to increase effective surface area. They have also been modified by subtractive methods such as acid pickling, acid etching, sandblasting, and other small particle-blasting to change the texture, as well as to increase the surface area. The development and use of these surface modifications have been based on that an improved osseointegration can be achieved by increasing the topography or roughness of the implant surface [7-125].

7.5. CELL GROWTH

Titanium implant surfaces with rough microtopography exhibit increased pullout trength *in vivo*, suggesting that the bone-to-implant contact was enhanced. This is supported by *in vitro* studies showing that on surfaces with rough microtopographies, osteoblasts secrete factors that enhance osteoblast differentiation, while decreasing osteoclast formation and activity [7-126]. A novel approach to electrochemical processing of a hydroxyapatite (HA) coating was explored by Hu *et al.* [7-127]. Vinyl acetate was added in the electrolytes of calcium and phosphorous in order to improve the adhesion between HA coating and titanium substrate. X-ray powder diffractometer (XRD) spectra indicated that the vinyl acetate did not interfere with the deposition of HA on the surface of titanium cathodes. Results obtained from XPS and SEM revealed that (i) both vinyl acetate and HA were deposited on the titanium cathodes, (ii) the vinyl acetate changed the HA crystalline morphology in the deposition layer, and (iii) the shape and growth direction of the HA crystals in the coating with vinyl acetate differed from those of HA deposition alone. It was moreover mentioned that a preliminary study of the bioactivity showed that osteoblastic cells exhibited higher cell proliferation potential on the HA/vinyl acetate coating than on that of pure HA [7-127].

Vehof *et al.* [7-128] prepared the osteogenic activity of calcium phosphate (CaP)-coated and noncoated porous titanium (Ti) fiber mesh loaded with cultured syngeneic osteogenic cells after prolonged *in situ* culturing to compare in a syngeneic rat ectopic assay model. Rat bone marrow (RBM) cells were loaded onto the CaP-coated and noncoated Ti scaffolds using either a droplet or a suspension loading method. After loading, the RBM cells were cultured for 8 days *in vitro*. Thereafter, implants were subcutaneously placed in 39 syngeneic rats. The rats were euthanized and the implants retrieved at 2, 4, and 8 weeks post-operatively. Further, in the 8-week group, fluorochrome bone markers were injected at 2, 4, and 6 weeks. It was reported that (i) histological analysis demonstrated that only the CaP-coated meshes supported bone formation, (ii) the amount of newly formed bone varied, between single and multiple spheres, filling a significant part of the mesh porosity, (iii) in the newly formed bone, osteocytes embedded in a mineralized matrix could be clearly observed, but (iv) on the other hand, in the noncoated titanium implants, abundant deposition of calcium-containing material was seen. In CaP-coated implants, the accumulation sequence of the fluorochrome markers showed that bone formation started on the mesh fibers. It was concluded that the combination of a thin CaP coating, Ti-mesh, and rat bone marrow cells can generate ectopic bone formation after prolonged *in vitro* culturing [7-128].

Titanium (Ti) and its alloys continue to serve as successful implant materials for skeletal repair because of their physical properties and biocompatibility. The influence of organoapatite (which is an organic compound in which apatite crystals have been nucleated and grown on a polymer), grown directly onto an L-shaped Ti mesh, on preosteoblastic cellular colonization was investigated. Unseeded mesh samples were placed on subconfluent layers of MC3T3-E1 murine calvaria cells and cultured for up to 2 weeks. It was found that (i) cells demonstrated accelerated colonization of the three-dimensional organoapatite-Ti mesh substrates over bare Ti controls, (ii) cells also showed significantly increased proliferation on the organoapatite-Ti mesh over bare Ti controls, and (iii) cellular differentiation, measured by alkaline phosphatase and osteocalcin expression, was observed at late stages of the experiment with no notable differences between the organoapatite-Ti mesh and bare Ti controls, suggesting that organoapatite grown onto porous Ti substrates is capable of inducing accelerated colonization of unseeded implant structures by osteogenic cells [7-129].

Rodriguez *et al.* [7-130] modified titanium surfaces by anodization in a mixed electrolyte of calcium glycerophosphate and calcium acetate. Hydrothermal treatments were performed on two of the anodized groups for either 2 or 4 h. *In vitro* osteoblast response to anodized oxide and the hydrothermal-treated oxide after anodization was evaluated. It was found that (i) calcium and phosphorus ions were deposited on the Ti oxide during anodization, (ii) anodized surfaces following a 4-h hydrothermal treatment were observed to promote the growth of apatite-like crystals, as compared with anodized surfaces after a 2-h hydrothermal treatment, (iii) cellular function and onset of mineralization, as indicated by protein production and osteocalcin production, respectively, were also observed as enhanced on hydrothermal-treated surfaces. It was thus concluded that calcium phosphate and apatite-like crystals could be deposited on Ti surfaces using anodization and a combination of anodization and hydrothermal treatment, and the phenotypic expression of an osteoblast was enhanced by the presence of calcium phosphate or apatite-like crystals on anodized or hydrothermally treated Ti surfaces [7-130].

The topographic effects of hydroxyapatite (HA) and titanium (Ti) surfaces having identical micropatterns were determined as to whether there was synergistic interaction between surface chemistry and surface topography. Surface microgrooves with six different groove widths (4, 8, 16, 24, 30, and 38 μm) and three different groove depths (2, 4, and 10 μm) were made on single crystalline silicon wafers using microfabrication techniques. Ti and HA thin films were coated on the microgrooves by radio-frequency magnetron sputtering, followed by seeding human osteoblast-like cells and culturing on the microgrooved surfaces for up to 7 days. It was reported that (i) contact guidance and cell shape changes were observed on the HA and Ti microgrooves, and (ii) no difference in orientation angle between HA and Ti microgrooves was found, suggesting that surface chemistry was not a significant influence on cell guidance [7-131].

The homogeneous deposition of calcium phosphate (Ca-P) can be coated on porous implants by immersion in simulated physiologic solution. In addition, various Ca-P phases, such as octacalcium phosphate or bone-like carbonated apatite, can be applied. Barrère *et al.* [7-132] conducted experiments (1) to investigate the osteoinduction of octacalcium phosphate-coated and noncoated porous tantalum cylinders, and of dense titanium alloy cylinders (5 mm in diameter and 10 mm in length) in the back muscle of goats at 12 and 24 weeks, and (2) to compare the osteogenic potentials of bone-like carbonated apatite-coated, octacalcium phosphate-coated, and bare porous tantalum cylinders in a gap of 1 mm created in the femoral condyle of a goat at 12 weeks. It was found that (i) in the goat muscle, after 12 weeks, the octacalium phosphate-coated porous cylinder had induced ectopic bone, as well as bone within the cavity of the octacalcium phosphate-coated dense titanium cylinder, (ii) in the femoral condyle, bone did not fill the gap in any of the porous implants, but (iii) in contrast with the two other groups, octacalcium phosphate-coated porous cylinders exhibited bone formation in the center of the implant [7-132].

Studies examining osteoblast response to controlled surface chemistries indicate that hydrophilic surfaces are osteogenic. The TiO_2 surfaces, however, produced until now exhibit low surface energy because of adsorbed hydrocarbons and carbonates from the ambient atmosphere, or roughness-induced hydrophobicity. Zhao *et al.* [7-133] used hydroxylated/hydrated Ti surfaces in order to retain high surface energy of TiO_2. It was reported that (i) osteoblasts grown on this modified surface exhibited a more differentiated phenotype, characterized by increased alkaline phosphatase activity and osteocalcin, and generated an osteogenic microenvironment through higher production of PGE2 and TGF-β1, and (ii) 1α,25(OH)2D3 increased these effects in a manner that was synergistic with high surface energy, suggesting that increased bone formation observed on modified Ti surfaces *in vivo* was due in part to stimulatory effects of high surface energy on osteoblasts [7-133].

To further understanding the biological mechanisms underlying bone–titanium integration and biomechanical properties of the integrated bone, Takeuchi *et al.* [7-134] cultured rat bone marrow-derived osteoblastic cells on either the machined titanium disk or acid-etched titanium disk. It was found that (i) the modulus of elasticity (MOE) of the mineralized tissue was also 3 times greater on the acid-etched surface than on the machined surface, and (ii) after 28 days of culture, the mineralized nodule area was significantly lower on the acid-etched surface than on the machined surface, while total calcium deposition did not differ between the two surfaces, indicating denser mineral deposition on the acid-etched surface. Scanning electron microscopic images revealed that tissue cultured on the acid-etched titanium exhibited plate-like, compact surface morphology, while the tissue on the machined titanium appeared porous and was covered by fibrous and punctate structures. It was therefore concluded that culturing osteoblasts on rougher titanium surfaces enhances the hardness and elastic modulus of the mineralized tissue associated with the condensed mineralization, accelerated collagen synthesis, and upregulated expression of selected bone-related genes [7-134].

7.6. TISSUE REACTION AND BONE INGROWTH

The oral epithelium and connective tissue was able to form a peri-implant epithelium similar in appearance to that around natural teeth [7-107, 7-135]. In studies of peri-implant mucosa conducted in 9-month-old beagle dogs, placement of one- and two-stage implants demonstrated that the mucosal barrier that formed had a definite epithelial, and the connective tissue component was similar in dimension in both the one- and two-stage implants [8-136]. It was proposed that a certain minimum width of the peri-implant mucosa was required, and that bone resorption would occur if the soft-tissue dimension was not satisfied, being similar to the phenomenon of the biologic width that occurs in natural human dentition [7-137]. The connective tissue lateral to the junctional epithelium is comprised of dense collagenous tissue with few vascular structures and scattered inflammatory cells. The "zone of connective tissue integration" between the bone crest and the junctional epithelium is characterized as a collagen-rich and cell-poor scar-like connective tissue. There was a narrow zone close to the implant that had mainly collagen fibers running parallel to the surface of the implant and attaching to the periosteum of the bone [7-136]. Using light microscopy, it was noted that the inner peri-implant epithelium ended 1–2 mm from the bone [7-138]. Using transmission electron microscopic studies, it was noted that the outermost epithelial cells close to the implant exhibited condensed structures resembling hemidesmosomes [7-139]. Good adhesion between the implant surface and the soft tissue is extremely important to assure the survival of dental implants [7-140]. Percutaneous and permucosal implants require a strong support from the underlying connective tissue if an epithelial downgrowth and subsequent failure are to be avoided [7-141]. Systematic and continuous monitoring of peri-implant tissues during maintenance care is recommended for the early diagnosis of peri-implant disease [7-90, 7-142]. Tissue reaction has been shown to vary with micrometer range changes in the substratum [7-143]. Studies have shown that changes in the range of 130 nm are adequate to produce changes in contact guidance of the cells [7-144]. Surface roughness may have had a significant effect on the cell density and proliferation. Some studies concluded that surface microtexture influences the attachment and growth of human gingival fibroblasts [7-145, 7-146], while others suggested that the surface roughness does not influence soft-tissue attachment [7-147]. Significantly greater numbers of cells were attached to rougher milled and sintered surfaces than to polished surfaces, possibly indicating higher proliferation capacity [7-148].

The distribution of released metal in the surrounding tissue is important information to estimate the relation between the dissolution content and tissue response. However, the distributions of released metal have not been studied well because of the difficulty in detecting the very low concentration of rarely contained metal elements in the surrounding tissue. Soft tissues implanted with Cu, Ni, Fe, Ag, Ti, NiTi, 304 (18Cr-8Ni) and 316 (18Cr-8Ni-2Mo) stainless-steel wires were investigated with XSAM (X-ray scanning analytical microscope), and compared with histological observation. The relationship between the distribution metal elements and the tissue response was evaluated. Severe damage was observed around Ni and Cu implants, while fibrous connective tissue was formed around the Fe implant. The dissolved concentration was approximately 10–20 mM for Ni and Cu, and was considered to be on the order of 10 times higher in Fe. It was concluded that (i) the toxicity at the same concentration followed the sequence (from greater toxicity to least): Ni > Cu + Fe, and (ii) for Ag, Ti, NiTi, 304 and 316 stainless-steel implants, significant dissolution and severe tissue damage were not observed [7-149].

Sundgren *et al.* [7-150] used Auger electron spectroscopy to study the interface between human tissue and implants of Ti and stainless steel. Both the thickness and the nature of the oxide layers on the implant have been found to change during the time of implantation. The stainless-steel implants have a surface oxide about 50 Å thick prior to implantation. The metal atoms in the oxide are mainly Cr, with which Cr_2O_3 may be formed. For implants located in cortical bone, the thickness of the interfacial oxide layer remains unaffected, while it increases by a factor of 3–4 on samples located in bone marrow. In both these cases, Ca and P are incorporated in the oxides. Implants located in soft tissue have an interfacial oxide with a thickness of about one and a half times that of an unimplanted sample. On these samples, C and P are not incorporated in the oxide layer. Also, for Ti an increase in oxide thickness and an incorporation of Ca and P are found. In the cases with Ti implants, it is also demonstrated that the oxidation process occurs over a long period of time, up to several years. For the elements incorporated in these interfacial oxide layers on both stainless-steel and Ti implants, it is found that P is strongly bound to oxygen, suggesting the presence of phosphate groups in the oxide. Other species that can oxidize the implants are H_2O_2 and free radicals, such as O_2^- or OH^-. Such species are normally produced during the inflammatory reactions [7-150, Chapter 4].

The immuno-inflammatory responses to stainless-steel (21 implants in 20 patients) and Ti plates (22 implants in 20 patients) used in the treatment of long bone fractures were studied. All fractures healed without complications. It was reported that (i) in the soft tissue adjacent to the surface of the implants, a dark discoloration of the tissue was visible in 18 out of 21 stainless-steel implants and 20 out of 22 Ti plates, and (ii) tissue specimens of all patients contained positive staining for macrophages (CD68-positive cells), demonstrating clearly the presence of a marked inflammation and tissue reaction in the soft tissue covering stainless-steel and Ti plates which were used for internal fixation of bones, independently from the material [7-151].

Successful clinical performance of machine/turned CpTi implants has resulted in a wide-spread usage of them. However, in bone of poor quality and quantity, the results have not always been so good, motivating the development of "novel types of osseointegrated implants." The development of Ti implants has depended on new surface processing technologies. Recently developed clinical oral implants have been focused on topographical changes of implant surfaces, rather than alterations

          Bioscience and Bioengineering of Titanium Materials

of chemical properties [7-152–7-156, Figure 7-1]. These attempts may have been based on the concept that mechanical interlocking between tissue and implant materials relies on surface irregularities in the nanometer to micrometer level. Recently, published *in vivo* investigations have shown significantly improved bone tissue reactions by modification of the surface oxide properties of Ti implants [7-157–7-164]. It was found that in animal studies, bone tissue reactions were strongly reinforced with oxidized titanium implants, characterized by a titanium oxide layer thicker than 600 nm, a porous surface structure and an anatase type of Ti oxide with large surface roughness compared with turned implants [7-162, 7-163]. This was later supported by work done by Lim *et al.* [7-165], who found that the alkali-treated CpTi surface was covered mainly with anatase type TiO_2, and exhibited hydrophilicity, whereas the acid-treated CpTi was covered with rutile-type TiO_2 with hydrophobicity. Besides this characteristic crystalline structure of TiO_2, it was mentioned that good osseointegration, bony apposition, and cell attachment of Ti implant systems [7-166, 7-167] are partially due to the fact that the oxide layer, with unusually high dielectric constant of 50–117, depending on the TiO_2 concentration, may be the responsible feature [7-52, 7-168].

Turned (as-machined implants) with amorphous TiO_2, anatase TiO_2 with S, amorphous TiO_2 with P, and anatase TiO_2 with Ca were prepared by the micro-arc oxidation method. Eighty implants were inserted in the femora and tibiae of 10 mature New Zealand white rabbits for 6 weeks. It was reported that (i) the removal torque values showed significant differences between S implants and controls, Ca implants and controls, Ca implants and P implants but did not show significant differences between the others, and (ii) the bone to metal contact around the entire implants demonstrated a 186% increase in S implants, 232% increase in P implants, and 272% increase Ca implants when compared to the paired control groups. It was therefore concluded that surface chemistry and topography separately or together play important roles in bone response to the oxidized implants [7-156].

The effects of pure commercial titanium implants on the process of primary mineralization was studied by Kohavi *et al.* [7-169] by insertion of titanium implants into rat tibial bone after ablation. The effects of the titanium were studied through the behavior of extracellular matrix vesicles. It was found that the insertion of titanium implants was followed by an increase in the number of extracellular matrix vesicles, as well as vesicular diameter, and by a decrease in vesicular distance from the calcified front when compared to normal healing. It was therefore suggested that (i) the process of extracellular matrix vesicles maturation around titanium implants was delayed when compared to normal primary bone formation during bone healing, and (ii) the delay in mineralization was compensated by an increase in vesicle production, resulting in an enhancement of primary mineralization by the titanium [7-169].

Jasty *et al.* [7-170] investigated bone ingrowth into uncemented hemispherical canine acetabular components, porous-coated with Co-Cr spheres, and compared them to those porous-coated with titanium fiber mesh. Good bone ingrowth was noted in both types of porous coatings at 6 weeks after surgery, with no difference in the histologic quality of the ingrowth bone. Quantification of bone ingrowth showed that significantly more bone grew into the titanium fiber mesh porous-coated components. It was demonstrated that the mean values of the amount of bone ingrowth, the area density of the ingrowth bone inside the pores, the ingrowth bone penetration depth, and the degree of the periphery of the porous layer with bone ingrowth of titanium mesh exceeded those values with Co-Cr spheres [7-170].

Bony ingrowth to control (nontreated) and heat-treated 316L stainless-steel and Ti-6Al-4V implants, which were placed into the medullary canal of the femur in rats, was studied by mechanical, chemical, and Auger electron spectroscopic microanalysis. Control samples of both materials are autoclaved at 280°C. Heat treatment was done at 550°C for 1.5 h. It was found that, at all time intervals up to 35 days post-implantation, the shear strengths of the heat-treated Ti-6Al-4V and stainless-steel implants were significantly higher (1.6- to 3.4-fold) than in control implants. Using Auger electron spectroscopy depth profiling methods, it was also found that the heat treatment modified the implant surface composition significantly, resulting in a thicker oxide layer and other chemical changes. It is therefore concluded that pre-oxidation of metallic implants prior to their insertion alters their chemical surface property and augments bony ingrowth to them [7-171].

Interstitial bone ingrowth is extremely important for optimum fixation of implanted materials under load-bearing conditions. Three types of biomaterial tests pieces were manufactured in solid and open-pore structures, and implanted into dog femoral condyles. Bone formation and remodeling were observed histologically and roentgenologically for 24 weeks thereafter. It was reported that (i) at 24-week post-implantation, thick fibrous tissue surrounded by corticalized bone was formed around both solid smooth-surfaced alumina and titanium implants, but (ii) on the other hand, with an implant made of an artificial osteochondral composite material, thickening of ingrowth trabeculae could be observed as early as 4 weeks, (iii) bone ingrowth into the titanium fiber mesh was abundant and increased with time after implantation. This interstitial bone ingrowth resulted in the complete integration of the implant and the viable host bone. It was thus suggested that interstitial bone ingrowth has great significance; even new bone formation and remodeling follows Wolff's law [7-172] after the completion of the bonding between the bone and implanted material under load-bearing conditions. The artificial osteochondral composite material could lead to complete integration of the implant and viable bone, suggesting that it is a promising material for joint replacements. Moreover, the tibial joint surface which bore against the polyvinyl alcohol hydrogel surface of this implant, remained intact, suggesting that this composite is a very promising biomaterial for use in joint prostheses [7-173].

The histological study was conducted on the surrounding tissues of two screw-shaped titanium plasma-sprayed implants, retrieved due to the fractured abutment after 18 and 42 months. The microscopical examination showed that a large part of the implant surface on both implants was covered by compact, mature lamellar bone, with the presence of many Haversian systems and osteons. With von Kossa staining, at the interface with the implant, the bone was highly mineralized, and any connective tissue or inflammatory cells were not present at the interface [7-174].

The micromotion chamber for implantation in the rabbit tibia consists of two titanium components that have a 1 mm continuous (not independent) pore for bone ingrowth. The fixed, outer cylinder of the chamber contains a movable inner core that can be manually rotated. A comparison was made between the histological and scintigraphic results of bone ingrowth into chambers having a congruently shaped interface that was moved at 20 cycles/day, with amplitude of either 0.5 or 0.75 mm. It was found that (i) histological sections from both amplitude groups contained extensive new woven and trabecular bone, embedded in a fibrovascular network; however, (ii) the chambers with a larger amplitude of motion yielded less bone ingrowth than those with a smaller amplitude. It was therefore suggested that short, discrete periods of motion can stimulate the formation of fibrous tissue rather than bone, using the parameters chosen in this model [7-175].

Li *et al.* [7-176] investigated the effect of the surface macrostructure of a dimpled CpTi implant on bone ingrowth *in vivo* by means of histological examination and a push-out test. Cylindrical implants were inserted in one femur of each experimental rabbit and the animals were killed at 1.5, 3, and 13 months after implantation. The femur with the implant of each animal was then examined in a push-out test. The fracture surfaces of the bone–implant interface were examined after the push-out test under light and electron microscopy. It was reported that the dimpled CpTi surface-enhanced mechanical retention of the CpTi implant in bone due to interlocking between vital bone and the dimples [7-176].

A nonsubmerged ITI-Bonefit implant, after a loading period of 4 years, was examined. It was reported that the behavior of peri-implant bone was tightly related to the magnitude and concentration of stresses which were transmitted to the implant [7-177]. These peri-implant stresses are subject to several variables: opposing occlusion, bite force, number of implants available to carry the load, position of the implant within the prostheses, rigidity of the prosthesis, and implant geometry [7-177]. Moreover, torque or bending moments can be applied to implants, for example, by excessively cantilevered bridges, and can produce a breakdown at the interface, a resorption of bone, a loosening of the screw, and/or a fracture of the bar/bridge [7-178]. It is also significant that the fracture was associated with a severe and rapid marginal bone height loss, and with an almost 100% bone–implant contact [7-179].

The *in vivo* effect of biomimetic calcium phosphate coating of titanium implants on peri-implant bone formation and bone/implant contact was investigated by Schliephake *et al.* [7-180]. Five types of implants were used: Ti-6Al-4V implants with a polished surface; Ti-6Al-4V implants with collagen coating; Ti-6Al-4V implants with a mineralized collagen layer; Ti-6Al-4V implants with a sequential coating of hydroxyapatite (HA) and collagen; and Ti-6Al-4V implants with HA coating only. All implants were press fit inserted into 4.6 mm trephine burr holes in the mandibles of 10 beagle dogs. The implants of five animals each were evaluated after a healing period of 1 and 3 months, respectively, during which time sequential fluorochrome labeling of bone formation had been performed. Bone formation was evaluated by morphometric measurement of the newly formed bone around the implants and the percentage of implant bone contact. It was reported that (i) after 1 month, there was a significantly higher percentage of mean bone/implant contact in the HA-coated implants compared to those with polished surfaces and those with the collagen-coated surface, but (ii) after 3 months, these differences were not present anymore, and (iii) bone apposition was significantly higher next to implants with sequential HA/collagen coating compared to the polished surfaces and mineralized collagen layer. It is concluded that the HA coating to titanium implants has increased bone–implant contact in the early ingrowth period, and the addition of collagen to an HA coating layer may hold some promise when used as a sequential HA/collagen coating with mineralized collagen as the surface layer [7-180].

A major consideration in designing dental implants is the creation of a surface that provides strong attachment between the implant and bone, connective tissue, or epithelium. In addition, to maintain plaque-free implants, it is important to inhibit the adherence of oral bacteria on titanium surfaces exposed to the oral cavity. Therefore, Groessner-Schreiber *et al.* [7-181] examined antimicrobial characteristic TiN. Mouse fibroblasts were cultured on smooth titanium disks that were either magnetron-sputtered with a thin layer of titanium nitride or thermal oxidized, or modified with laser radiation (using an Nd:YAG laser). It was reported that (i) fibroblasts on oxidized titanium surfaces showed a more spherical shape, whereas cells on laser-treated titanium and on TiN appeared intimately adherent to the surface, and (ii) when compared to all other surface modifications, fibroblast metabolism activity and total protein were significantly increased in the fibroblasts cultured on titanium surfaces coated with TiN, suggesting that a titanium nitride TiN coating might be suitable to support tissue growth on implant surfaces [7-181].

Osteoconductive apatite coatings represent an established technology for enhancing the integration of orthopedic Ti implants with living bone. The *in vivo* evaluation of a biomimetic apatite coating was performed. A dense and substantially pure ceramic coating with a crystal size of less than 1 μm was fabricated by using a mixed solution including all major inorganic ions (Na^+, K^+, Mg^{2+}, Ca^{2-}, Cl^-, HPO_4^{2-}, HCO_3^- and SO_4^{2-}). The biomimetic apatite coating was grown on Ti-6Al-4V at 45°C for 4 days. Three types of measurements were taken on linear ingrowth percentage, area ingrowth percentage, and continuous bone apposition percentage. It was demonstrated that (i) under controlled conditions, the apatite coating appears to resorb in 8 weeks, and stimulated early osseointegration with the metal surface with a reduction in fibrous tissue encapsulation, and (ii) this coating may, therefore, be useful in facilitating early bone ingrowth into porous surfaces without the potential for coating debris, macrophage infiltration, fibrous tissue encapsulation, and eventual coating failure, as may occur with current plasma-sprayed hydroxyapatite coating techniques [7-182].

Roughened titanium surfaces have been favorably fabricated and widely used for dental implants, as has been seen above. In recent years, there has been the tendency to replace Ti plasma-sprayed surfaces by sandblasted and acid-etched surfaces in order to enhance osseous apposition. Another approach has been the utilization of hydroxyapatite (HA)-coated implants. Knabe *et al.* [7-183] examined the effect of two roughened Ti dental implant surfaces on the osteoblastic phenotype of human bone-derived cells, and compared this behavior to that for cells on an HA-coated surface. Test materials were an acid-etched and sandblasted Ti surface, a porous Ti plasma-sprayed coating, and a plasma-sprayed porous HA coating (HA). Smooth Ti machined surfaces served as the control. Human bone-derived cells were grown on the substrata for 3, 7, 14, and 21 days, counted and probed for various bone-related mRNAs and proteins (type I collagen, osteocalcin, osteopontin, osteonectin, alkaline phosphatase, and bone sialoprotein). It was reported that (i) dental implant surfaces significantly affected cellular growth and the temporal expression of an array of bone-related genes and proteins, (ii) HA-coated Ti had the most effect on osteoblastic differentiation, inducing a greater expression of an array of osteogenic markers than recorded for cells grown on acid-etched and sandblasted Ti and porous Ti plasma-sprayed Ti, thus suggesting that the HA-coated surface may possess a higher potency to enhance osteogenesis, and (iii) acid-etched and sandblasted Ti surfaces induced greater osteoblast proliferation and differentiation than porous Ti plasma-sprayed Ti [7-183].

Lu *et al.* [7-184] evaluated the osteoconduction of Ti-6Al-4V surfaces under various conditions, including micro-patterned, alkali-treated, micro-patterned plus alkali-treated, and surfaces without any treatment as the control. The *in vitro* calcium phosphate formation on titanium surfaces was in static and dynamic simulated body fluid (SBF). An *in vivo* comparison was conducted in the medullary cavity of a dog femur, which could provide the same physiological environment for specimens with different surface conditions. It was reported that (i) there was good conduction of calcium phosphate on the alkali-treated surfaces, and also better calcium phosphate deposition on the micro-hole surface than on the flat surfaces in the dynamic SBF were found, and (ii) the beneficial effect of alkaline treatment on osteoconduction was confirmed. The *in vivo* experiments also indicate a synergistic effect of the alkaline treatment and the topographic pattern on osteoconduction [7-184], as suggested by basic studies done by Lim *et al.* [7-165].

Early bone ongrowth is known to increase primary implant fixation and reduce the risk of early implant failure. Arg-Gly-Asp (arginine-glycine-aspartate, RGD) peptide has been identified as playing a key role in osteoblast adhesion and proliferation on various surfaces. Elmengaard *et al.* [7-185, 7-186] evaluated the effect of RGD peptide coating on the bony fixation of orthopedic implants. Sixteen unloaded cylindrical plasma-sprayed Ti-6Al-4V implants, coated with cyclic RGD peptide, were press-fit inserted in the proximal tibia of 8 mongrel dogs for 4 weeks. Uncoated control implants were inserted in the contralateral tibia. Results were evaluated by histomorphometry and mechanical push-out tests. It was shown that (i) a significant two-fold increase was observed in bone ongrowth for the RGD-coated implants, (ii) fibrous tissue ongrowth was significantly reduced for the RGD-coated implants, (iii) bone volume was significantly increased in a $0-100$ μm zone around the implant; the increased bony anchorage resulted in moderate increases in mechanical fixation, as the apparent shear stiffness was significantly higher for RGD-coated implants, and (iv) increases in median ultimate shear strength and energy to failure were also observed. It was hence suggested that cyclic RGD coating increases early bony fixation of unloaded press-fit titanium implants [7-185, 7-186].

7.7. OSSEOINTEGRATION AND BONE/IMPLANT INTERFACE

Broadly speaking, two types of anchorage mechanisms have been described: biomechanical and biochemical. Biomechanical binding is when bone ingrowth occurs into micrometer-sized surface irregularities. The term osseointegration is probably, realistically, this biomechanical phenomenon. Biochemical bonding may occur with certain bioactive materials where there is primarily a chemical bonding, with possible supplemental biomechanical interlocking. The distinct advantage with the biochemical bonding is that the anchorage is accomplished within a relatively short period of time, while biomechanical anchorage takes weeks to develop. This would clinically translate into the possibility of earlier restorative loading of implants. Most commercially available implants depend on biomechanical interlocking for anchorage. All implants must exhibit biomechanical as well as morphological compatibility [7-8, 7-124]. As we have seen previously, Ti implants coated with calcium phosphate, or inorganic or organic bone-like apatite possess both anchorage mechanisms, because (1) coated surfaces are normally rough which can facilitate the biomechanical anchorage (morphological compatibility), and (2) coated materials *per se* accommodate biochemical bonding (biological compatibility).

Bone fusing to titanium was first reported in 1940 by Bothe *et al.* [7-187]. Brånemark began extensive experimental studies in 1952 on the microscopic circulation of bone marrow healing. These studies led to dental implant application in early 1960, 10-year implant integration was established in dogs without significant adverse reactions to hard or soft tissues. Studies in humans began in 1965, were followed for 10 years, and reported in 1977 [7-188]. Osseointegration, as first defined by Brånemark, denotes at least some direct contact between vital bone with the surface of an implant at the light microscopic level of magnification [7-189]. The percentage of direct bone−implant contact is variable. To determine osseointegration, the implant must be removed and evaluated under a microscope. Rigid fixation defines the clinical aspect of this microscopic bone contact with an implant, and is the absence of mobility with $1-500$ g force applied in a vertical or horizontal direction. Rigid fixation is the clinical result of a direct bone interface, but has also been reported with fibrous tissue interfaces [7-189]. Osseointegration was originally defined as a direct structural and functional connection between ordered living bone and the surface of a load-carrying artificial implant (which is typically made of titanium materials). It is now said that an implant is regarded as osseointegrated when there is no progressive relative movement between the implant and the bone with which it has direct contact. In practice, this means that in osseointegration there is an anchorage mechanism, whereby nonvital components can be reliably and predictably incorporated into living bone and that this anchorage can persist under all normal conditions of loading. Bioactive (or biochemical) retention can be achieved in cases where the implant is coated with bioactive materials, like hydroxyapatite. These bioactive materials stimulate bone formation leading, to a physio-chemical bond. If it is recognized that the implant is ankylosed with the bone, it is sometimes called a biointegration instead of osseointegration [7-190].

For profound understanding of the osseointegration mechanism, there are several works reported. Fourteen titanium dental implants (TioblastTM) were implanted singly in the proximal tibia of New Zealand rabbits for 120 days. A bone defect was surgically produced and filled with Bio-Oss® (which is natural, osteoconductive bone substitute that promotes bone growth in periodontal and maxillofacial osseous defects) around six of these implants. After the animals were sacrificed and their organs harvested, bone segments were fixed, and methacrylate embedded after the push-in test had been performed. The results showed that (i) the implants were apically and coronally surrounded by bone, whether Bio-Oss® was used or not, (ii) fractures were evident through the newly formed bone and between the preexisting and newly formed bone, and (iii) detachment between the implant and the bone occurred at the coronal extremity of the implants and along its cervical region. Based on these findings, it was concluded that the bone−titanium interface has a high resistance to loading, exhibiting a greater resistance than the newly formed bone [7-191].

The effect of a dual treatment of titanium implants and the subsequent bone response after implantation were investigated by Rønold *et al.* [7-192]. Coin-shaped CpTi implants were placed into the tibias of 12 rabbits. The implant was dually blasted with TiO_2 particles of two different sizes (i.e., finer particles of $22-28$ μm in size, and coarser particles of $180-220$ μm in size). It was found that (i) the Ti surface blasted with coarse particles showed a significantly better functional attachment than the fine surface, and (ii) the Ti surface blasted with fine particles showed lower retention in bone, indicating that there is a positive correlation between the topographical and mechanical evaluation of the surfaces [7-192].

In cementless fixation systems, surface character is an important factor. Alkali and heat treatments of titanium have been shown to produce strong bonding to bone and a higher ongrowth rate. With regard to the cementless hip stem, Nishiguchi *et al.* [7-193] examined the effect of alkali and heat treatments on titanium rods in an intramedullary rabbit femur model. Half of the implants were immersed in 5 mol/L sodium hydroxide solution and heated at 600°C for 1 h, and the other half were untreated. The bone−implant interfaces were evaluated at 3, 6, and 12 weeks after implantation. Pullout tests showed that the treated implants had a significantly higher bonding strength to bone than the untreated implants at each time point. As post-operative time elapsed, histological examination revealed that new bone formed on the surface of both types of implants, but significantly more bone made direct contact with the surface of the treated implants. At 12 weeks, approximately 56% of the whole surface of the treated implants was covered with the bone. Based on aforementioned results, it was concluded that (i) alkali- and heat-treated titanium create strong bone bonding and a high affinity to bone, as opposed to a conventional mechanical interlocking mechanism, and (ii) alkali and heat treatments of titanium may be suitable surface treatments for cementless joint replacement implants [7-193].

Several factors influence the healing process and the long-term mechanical stability of cementless fixed implants, such as bone remodeling and mineralization processes. Histomorphometric and bone hardness measurements were taken in implants inserted in sheep femoral cortical bone at different times to compare the *in vivo* osseointegration of titanium screws with the following surface treatments: machined (Ti-MA); acid-etched with 25 vol% of HNO_3 solution for 1 h (Ti-HF); hydroxyapatite vacuum plasma spray (Ti-HA); and Ca-P anodization (with 0.06 ml/L β-glycerophosphate plus 0.3 ml/L calcium acetate at 275 V and 50 mA/cm^2), followed by a 300°C autoclave-hydrothermal treatment for 2 h (Ti-AM/HA). It was reported that (i) the bone-to-implant contact of Ti-HF was lower than that of the other surface treatments at both experimental times, (ii) significant differences in MAR (mineral apposition rate) were also found between the different experimental times for Ti-MA and Ti-HF, demonstrating that bone growth had slowed inside the screw threads of Ti-HA and Ti-AM/HA after 12 weeks, (iii) no bone microhardness changes in preexisting host bone were found, while Ti-MA showed the lowest value for the inner thread area at 8 weeks. Based on these findings, it was confirmed that osseointegration may be accelerated by adequate surface roughness and a bioactive ceramic coating, such as Ca-P anodization followed by a hydrothermal treatment, which enhances bone interlocking and mineralization [7-194].

Fujibayashi *et al.* [7-195] demonstrated that bone formation took place after 12-month implantation in the muscles of beagle dogs, and reported that chemically and thermally treated plasma-sprayed porous titanium surfaces exhibited an intrinsic osteoinductivity. Takemoto *et al.* [7-196] tested on porous plasma-sprayed Ti blocks with the following characteristics: porosity of 41%, pore size of a ranging between 300 and 500 μm, and yield compressive strength of 85.2 MPa. Three types of surface treatments were applied (1) alkali (in NaOH solution at 60°C for 24 h) and heat treatment (600°C for 1 h, followed by a furnace cooling), (2) alkali, hot water, and heat treatment, and (3) alkali, dilute 40°C HCl for 24 h, hot water, and heat treatment. The osteoinductivity of the materials implanted in the back muscles of adult beagle dogs was examined at 3, 6, and 12 months. It was found that (i) the alkali-acid/heat-treated porous bioactive titanium implant had the highest osteoinductivity, with induction of a large amount of bone formation within 3 months, (ii) the dilute HCl treatment was considered to give both chemical (titania formation and sodium removal) and topographic (etching) effects on the titanium surface, and (iii) adding the dilute HCl treatment to the conventional chemical and thermal treatments is a promising candidate for advanced surface treatment of porous titanium implants [7-196, 7-197].

A number of experimental and clinical data on the so-called oxidized implants have reported promising outcomes, as we have previously reviewed. However, little has been investigated on the role of the surface oxide properties and osseointegration mechanism of the oxidized implant. Sul *et al.* [7-198] recently proposed two action mechanisms for osseointegration of oxidized implants, i.e., (1) mechanical interlocking through bone growth in pores/other surface irregularities, and (2) biochemical bonding. Two groups of oxidized implants were prepared using a micro arc oxidation process, and were then inserted into rabbit bone. One group consisted of magnesium ion incorporated implants, and the other consisted of TiO_2 stoichiometry implants. It was reported that (i) after 6 weeks of follow up, the mean peak values for removal torque of the Mg-incorporated Ti implants dominated significantly over the TiO_2/Ti implants, (ii) bonding failure generally occurred in the bone away from the bone to implant interface for the Mg-incorporated Ti implant and mainly occurred at the bone to implant interface for the TiO_2/Ti implant which consisted mainly of TiO_2 chemistry and significantly rougher surfaces, as compared to the Mg-incorporated Ti implant, and (iii) between bone and the Mg-incorporated-Ti implant surface, ionic movements and ion concentrations gradient were detected. It was therefore concluded that the surface chemistry-mediated biochemical bonding can be theorized for explaining the oxidized bioactive implants [7-198]. Lim and Oshida [7-165] investigated the surface characteristics of acid- and alkali-treated CpTi. It was reported that (i) acid-treated CpTi was covered with rutile-type TiO_2 with high hydrophobicity (in other words, less surface activity), whereas (ii) oxide film formed on the alkali-treated CPTi surface was identified to be an anatase-type dominantly with a trace amount of rutile, which exhibited high hydrophilicity (i.e., high surface energy).

For the past 15 years, orthopedic implants have been coated with hydroxyapatite (HA) to improve implant fixation. The osteoconductive effect of HA coatings has been demonstrated in experimental and clinical studies. However, there are ongoing developments to improve the quality of HA coatings. Aebli *et al.* [7-199] investigated whether a rough and highly crystalline HA coating applied by vacuum plasma spraying had a positive effect on the osseointegration of special, high-grade titanium (Ti) implants with the same surface roughness. Ti alloy implants were vacuum plasma spray-coated with high grade Ti or HA. The osseointegration of the implants was evaluated by light microscopy or pullout tests after 1, 2, and 4 weeks of unloaded implantation in the cancellous bone of 18 sheep. It was found that (i) the interface shear strength increased significantly over all time intervals, (ii) by 4 weeks, values had reached ~10 N/mm^2, but (iii) the difference between the coatings was not significant at any time interval, and (iv) direct bone−implant contact was significantly different between the coatings after 2 and 4 weeks, and reached 46% for Ti and 68% for HA implants by 4 weeks. Therefore, it was indicated that the use of a rough and highly crystalline HA coating, applied by the vacuum plasma spray method, enhances early osseointegration [7-199].

Stewart *et al.* [7-200] investigated the effects of a plasma-sprayed hydroxyapatite/tricalcium phosphate (HA/TCP) coating on osseointegration of plasma-sprayed titanium alloy implants in a lapine, distal femoral intramedullary model. The effects of the HA/TCP coating were assessed at 1, 3, and 6 months after implant placement. The HA/TCP coating significantly increased new bone apposition onto the implant surfaces at all time points. The ceramic coating also stimulated intramedullary bone formation at the middle and distal levels of the implants. There was no associated increase in pullout strength at either 3 or 6 months; however, post-pullout evaluation of the implants indicated that the HA/TCP coating itself was not the primary site of construct failure. Rather, failure was most commonly observed through the peri-prosthetic osseous struts that bridged the medullary cavity. It was therefore concluded that the osteoconductive activity of HA/TCP coating on plasma-sprayed titanium alloy implant surfaces may have considerable clinical relevance to early host−implant interactions, by accelerating the establishment of a stable prosthesis−bone interface [7-200].

The *in vivo* behavior of a porous Ti-6Al-4V material that was produced by a positive replica technique, with and without an octacalcium phosphate coating, has been studied both in the back muscle and femur of goats by Habibovic *et al.* [7-201]. Macro and microporous biphasic calcium phosphate ceramic, known to be both osteoconductive and able to induce ectopic bone formation, was used for comparison. The three groups of materials (Ti-6Al-4V, octacalcium phosphate coated-Ti-6Al-4V and biphasic calcium phosphate ceramic) were implanted transcortically and intramuscularly for 6 and 12 weeks in 10 adult Dutch milk goats in order to study their osteointegration and osteoinductive potential. It was found that (i), in femoral defects, both octacalcium phosphate-coated Ti-6Al-4V and biphasic calcium phosphate ceramic performed better than the uncoated Ti-6Al-4V at both time points, (ii) biphasic calcium phosphate ceramic showed a higher bone amount than octacalcium phosphate-coated Ti-6Al-4V after 6 weeks of implantation, while after 12 weeks, this difference was no longer significant, (iii) ectopic bone formation was found in both octacalcium phosphate coated-Ti-6Al-4V and biphasic calcium phosphate ceramic implants after 6 and 12 weeks, and (iv) ectopic bone formation was not found in uncoated titanium alloy implants, suggesting that the presence of calcium phosphate (CaP) is important for bone induction. [7-201].

The osseointegration of copper vapor laser-superfinished titanium alloy (Ti-6Al-4V) implants with pore sizes of 25, 50, and 200 μm was evaluated in a rabbit intramedullary model by Stangl *et al.* [7-202]. Control implants were prepared by corundum blasting. Each animal received all four different implants in both femora and humeri. Using static and dynamic histomorphometry, the bone−implant interface and the peri-implant bone tissue were examined 3, 6, and 12 weeks post-implantation. Among the laser-superfinished implants, it was reported that total bone−implant contact was smallest for the 25-μm pores, and was similar for 50 and 200-μm pore sizes at all time points, and however, all laser-superfinished surfaces were inferior to corundum-blasted control implants in terms of bone−implant contact. Implants with 25-μm pores showed the highest amount of peri-implant bone volume at all time points, indicating that the amount of peri-implant bone was not correlated with the quality of the bone−implant interface. Accordingly, it was concluded that although laser-superfinished implants were not superior to the corundum-blasted control implants in terms of osseointegration, understanding of the mechanisms of bone remodeling within pores of various sizes was advanced, and optimal implant surfaces can be further developed with the help of modern laser technology [7-202].

There is another work using the laser technology for surface modification. Götz *et al.* [7-203] examined the osseointegration of laser-textured Ti-6Al-4V implants with pore sizes of 100, 200, and 300 μm, specifically comparing 200-μm implants with polished and corundum-blasted surfaces in a rabbit transcortical model. Using a distal and proximal implantation site in the distal femoral cortex, each animal received all four different implants in both femora. The bone/implant interface and the newly formed bone tissue within the pores and in peri-implant bone tissue were examined 3, 6, and 12 weeks post-implantation by static and dynamic histomorphometry. It was shown that (i) additional surface blasting of laser-textured Ti-6Al-4V implants with 200-μm pores resulted in a profound improvement in osseointegration at 12-week post-implantation, (ii) although lamellar bone formation was found in pores of all sizes, the amount of lamellar bone within the pores was linearly related to the pore size, (iii) in 100-μm pores, bone remodeling occurred with a pronounced time lag relative to larger pores, and (iv) implants with 300-μm pores showed a delayed osseointegration compared with 200-μm pores. Therefore, it was concluded that 200 μm may be the optimal pore size for laser-textured Ti-6Al-4V implants, and that laser treating in combination with surface blasting may be a very interesting technology for the structuring of implant surfaces [7-203].

7.8. SOME ADVERSE FACTORS FOR LOOSENING IMPLANTS

Early implant instability has been proposed as a critical factor in the onset and progression of aseptic loosening and periprosthetic osteolysis in total joint arthroplasties. As was seen in Chapter 6, macrophages stimulated with cyclic mechanical strain release inflammatory mediators. But, little is known about the response of these cells to mechanical strain with particles, which is often a component of the physical environment of the cell. Fujishiro *et al.* [7-204] studied the production of prostaglandin E2 (PGE2), an important mediator in aseptic loosening and periprosthetic osteolysis in total joint arthroplasties, for human macrophages treated with mechanical stretch alone, titanium particles alone, and mechanical stretch and particles combined. A combination of mechanical stretch and titanium particles resulted in a statistically synergistic elevation of levels of PGE2 compared with the levels found with either stretch or particles alone. It was hence suggested that, while mechanical strain may be one of the primary factors responsible for macrophage activation and periprosthetic osteolysis, mechanical strain with particles load may contribute significantly to the osteolytic potential of macrophages *in vitro* [7-204].

Corrosion of implant alloys releasing metal ions has the potential to cause adverse tissue reactions and implant failure. Lin and Bumgardner [7-205] hypothesized that macrophage cells and their released reactive chemical species affect the corrosion property of Ti-6Al-4V. It was reported that (i) there was no difference in the charge transfer in the presence and absence of cells, (ii) the alloy had the lowest charge transfer and metal ion release with activated cells was as follows: Ti <10 ppb, V < 2 ppb, attributing to an enhancement of the surface oxides by the reactive chemical species. It was concluded that macrophage cells and reactive chemical species reduced the corrosion of Ti-6Al-4V alloys [7-205].

Prosthetic and osteosynthetic implants from metal alloys will be indispensable in orthopedic surgery, as long as tissue engineering and biodegradable bone substitutes do not lead to products that will be applied in clinical routine for the repair of bone, cartilage, and joint defects. Therefore, the elucidation of the interactions between the peri-prosthetic tissues and the implant remains of clinical relevance, and several factors are known to affect the longevity of implants. Sommer *et al.* [7-206] studied the effects of metal particles and surface topography on the recruitment of osteoclasts *in vitro* in a co-culture of osteoblasts and bone marrow cells. It was reported that (i) the cells were grown in the presence of particles of different sizes and chemical compositions, or on metal disks with polished or sandblasted surfaces, respectively, (ii) at the end of the culture, newly formed osteoclasts were counted, (iii) osteoclastogenesis was reduced when particles were added directly to the co-culture, and (iv) in co-cultures grown on sand-blasted surfaces, osteoclasts developed at higher rates than they did in cultures on polished surfaces. It was therefore summarized that the wear particles and implant surfaces affect osteoclastogenesis, and thus may be involved in the induction of local bone resorption and the formation of osteolytic lesions, leading eventually to the loosening of orthopedic implants [7-206].

REFERENCES

[7-1] Misch CE. editor. Implant dentistry. 2nd ed. St. Loius: Mosby, An Affiliate of Elsevier. 1999.

[7-2] Burden DJ, Eedy DJ. Orthodontic headgear related to allergic contact dermatitis: a case report. Br Dent J 1991;170:447−448.

[7-3] Veien NK, Borchorst E, Hattcel T, Laurberg G. Stomatitis or systemically induced contact dermatitis from metal wire in orthodontic materials. Contact Dermatitis 1994;30:210−213.

[7-4] www.medibix.com.

[7-5] www.bsi-global.com. BS ISO 7206 Implants for surgery. Partial and total hip joint prostheses. Determination of resistance to static load of modular femoral heads.

[7-6] Kasemo B, Lausmaa J. Biomaterials and implant surfaces: a surface science approach. Int J Oral Maxillofac Implants 1988;3:247−259.

[7-7] Albrektsson T, Brånemark PI, Hansson HA, Lindstrom J. Osseointegrated titanium implants. Requirements for ensuring a long-lasting, direct bone-to-implant anchorage in man. Acta Orthop Scand 1981;52:155−170.

[7-8] Albrektsson T. Direct bone anchorage of dental implants. J. Prosthet Dent 1983;50:255−261.

[7-9] Davies JE. Understanding peri-implant endosseous healing. J Dent Edu 2003;67:932−949.

[7-10] Eriksson C, Broberg M, Nygren H, Öster L. Novel *in vivo* method for evaluation of healing around implants in bone. J Biomed Mater Res 2003;66A:662−668.

[7-11] Osborn JF, Kovacs E, Kallenberger A. Hydroxylapatit Keramik − Entwicklung lines neuen Bioverkstoffes und erste tierexperimentelle Ergebnisse. Dtsch Zahnärztl Z 1980;5:54−56.

[7-12] Steinemann S, Werkstoff Titan. In: Orale implantologie. Schroeder A, Sutter F, Krekeler G, editors. Stuttgart: Thieme. 1988.

[7-13] Sugarawa A. Self-healing calcium phosphates − the properties and the applications for bone regeneration. Jpn J Maxillo Facial Implants 2005;4:41−49.

[7-14] Scarano A, Piattelli M, Vrespa G, Petrone G, Iezzi G, Piattelli A. Bone healing around titanium and titanium nitride-coated dental implants with three surfaces: an experimental study in rats. Clin Implant Dent Relat Res 2003;5:103−111.

[7-15] Takemoto S, Yamamoto T, Tsuru K, Hayakawa S, Osaka A, Takashima S. Platelet adhesion on titanium oxide gels: effect of surface oxidation. Biomaterials 2004;25:3485−3492.

[7-16] Huang N, Yang P, Leng YX, Chen JY, Sun H, Wang J, Wang JG, Ding PD, Xi TF, Leng Y. Hemocompatibility of titanium oxide films. Biomaterials 2003;24:2177−2187.

[7-17] Dion I, Baquey C, Monties JR, Havlik P. Harmocompatibility of Ti6Al4V alloy, Biomaterials 1993;14:122−126.

[7-18] Eriksson AS, Bjursten LM, Ericson LE, Thomsen P. Hollow implants in soft tissue allowing quantitative studies of cells and fluid at the implant interface. Biomaterials 1988;9:86−90.

[7-19] Eriksson AS, Thomsen P. Leukotriene B4, interleukin 1 and leucocyte accumulation in titanium and PTFE chambers after implantation in the rat abdominal wall. Biomaterials 1991;12:827−830.

[7-20] Elwing H, Tengvall P, Askendal A, Lunstroem I. Lens on surface: a versatile method for the investigation of plasma protein exchange reactions on solid surfaces. J Biomater Sci Polmer Educ 1991;3:7−16.

[7-21] Graham TR, Dasse K, Coumbe A, Salih V, Marrinan MT, Frazier OH, Lewis CT. Neo-intimal development on textured biomaterial surfaces during clinical use of an implantable left ventricular assist device. Eur J Cardio-Thorac Surg 1990;4:182−190.

[7-22] Ellingen JE. A study on the mechanism of protein adsorption to TiO_2. Biomaterials 1991;12:593− 596.

[7-23] Sunny MC, Sharma CP. Titanium−protein interaction changes with oxide layer thickness. J Biomaterial Appl 1991;6:89−98.

[7-24] Merritt K, Edwards CR, Brown SA. Use of an enzyme linked immunosorbent assay (ELISA) for quantification of proteins on the surface of materials. J Biomed Mater Res 1988;22:99−109.

[7-25] Muramatsu K, Uchida M, Kim H-M, Fujisawa A, Kokubo T. Thromboresistance of alkali − and heat-treated titanium metal formed with apatite. J Biomed Mater Res 2003;65A:409−416.

[7-26] Suska F, Gretzer C, Esposito M, Tengvall P, Thomsen P. Monocyte viability on titanium and copper coated titanium. Biomaterials 2005;26: 5942−5950.

[7-27] Maitz MF, Shevchenko N. Plasma-immersion ion-implanted nitinol surface with depressed nickel concentration for implants in blood. J Biomed Mater Res 2006;76A:356−365.

[7-28] Martinesi M, Bruni S, Stio M, Treves C, Borgioli F. *In vitro* interaction between surface-treated Ti-6Al-4V titanium alloy and human peripheral blood mononuclear cells. J Biomed Mater Res 2005;74A:197−207.

[7-29] Eriksson C, Lausmaa J, Nygren H. Intercations between human whole blood and modified TiO_2-surfaces: influence of surface topography and oxide thickness on leukocyte adhesion and activation. Biomaterials 2001;22:1987−1996.

[7-30] Kasemo B, Lausmaa J. Surfaces science aspects on inorganic biomaterials. CRC Crit Rev Biocomp 1986;2:335−380.

[7-31] http:www.medterms.com.

[7-32] http.alan.kennedy.name/crohns.

[7-33] Masuda T, Salvi GE, Offenbacher S, Fleton D, Cooper LF, Cell and matrix reactions at titanium implants in surgically prepared rat tibiae. Int J Oral Maxillofac Implants 1997;1:472−485.

[7-34] Rosengren A, Johansson BR, Danielsen N, Thomsen P, Ericson LE. Immuno-histochemical studies on the distribution of albumin, fibrigen, fibronectin, IgG and collagen around PTFE and titanium implants. Biomaterials 1996;17:1779−1786.

[7-35] Chehroudi B, McDonnel D, Brunette DM. The effects of micromachined surfaces on formation of bonelike tissue on subcutaneous implants as assessed by radiography and computer image processing. J Biomed Mater Res 1997;34:279−290.

[7-36] Larsson C, Thomsen P, Lausmaa J, Rodahl M, Kasemo B, Ericson LE. Bone response to surface modified implants: studies on electropolsihed implants with different oxide thickness and morphology. Biomaterials 1994;15:1062−1074.

[7-37] Wälivaara B, Askendal A, Lundström I, Tengvall P. Blood protein interaction with titanium surfaces. J Biomater Sci Polym Edn 1996;8:41−48.

[7-38] Kanagaraja S, Lundström I, Nygren H, Tengvall P. Platelet binding and protein adsorption to titanium and gold after short time exposure to heparinized plasma and whole blood. Biomaterials 1996;17: 2225−2232.

[7-39] Nygren H, Eriksson C, Lausmaa J. Adhesion and activation of platelets and polymorpho-nuclear granulocyte cells at TiO_2 surfaces. J Lab Clin Med 1997;129:35−46.

[7-40] Tan P, Luscinskas FW, Homer-Vanniasinkam S. Cellular and molecular mechanisms of inflammation and thrombosis. Eur J Vasc Endovasc Surg 1999;17:373−389.

[7-41] Brown KK, Henson PM, Maclouf J, Moyle M, Ely JA, Worthen GS. Neutrophil−platelet adhesion: relative roles of platelet P-selection and neutrophil $\beta 2$ (CD18) integrins. Am J Respir Cell Mol Biol 1998;18: 100−110.

[7-42] De La Cruz, Haimovich B, Greco RS. Immobilized IgG and fibrinogen differentially affect the cytoskeletal organization and bactericidal function of adherent neutrophils. J Surg Res 1998;80:28−34.

[7-43] Berton G, Yan SR, Fumagalli L, Lowell CA, Neutrophil activation by adhesion: mechanisms and pathophysiological implications. Int J Clin Lab Res 1996;26:160−177.

[7-44] Fröhlich D, Spertini O, Moser R. The Fcγ receptor-mediated respiratory burst of rolling neutrophils to cytokine-activated, immune complex-bearing endothelial cells depends on L-selection but not E-selection. Blood 1998;91:2558−2568.

[7-45] Hansson H-A, Albreksson T, Brånemark P-I. Structural aspects of the interface between tissue and titanium implants. J Prosthet Dent 1983;50: 108−113.

[7-46] Albreksson T, Brånemark P-I, Hansson H-A, Kasemo B, Larsson K, Lundström I, McQueen D.H, Skalak R. The interface zone of inorganic implants *in vivo*: titanium implants in bone. Ann Biomed Eng 1983;11:1−27.

[7-47] Linder L, Albreksson T, Bråemark P-I, Hansson H-A, Ivarsson B, Jönsson U, Lundström I. Electron microscopic analysis of the bone−titanium interface. Acta Orthop Scand 1983;54:45−52.

[7-48] Kempien B, Schulte W, Kleineikenscheidt H, Linder K, Scareyka R, Heimke G. Lichtoptische und rasterelectronen-mikroskopische Untersuchungen an der grenzfläche von Implantaten aus Aluminium-oxidkeramik im Unterkierferknochen von Hunden. Dtsch Zahnärtzl Z 1978;33:332−340.

[7-49] Albrektsson T, Hansson H-A, Ivarsson B. Interface analysis of titanium and zirconium bone implants. Biomaterials 1985;6:97−101.

[7-50] Albreksson T, Hansson H-A. An ultrastructural characterization of the interface between bone and sputtered titanium or stainless steel surfaces. Biomaterials 1986;7:201−205.

[7-51] Parsegian VA. Molecular forces governing tight contact between cellular surfaces and substrates. J Prosthet Dent 1983;49:838−842.

[7-52] Kasemo B. Biocompatibility of titanium implants: surface science aspects. J Prosthet Dent 1983;49:832−837.

[7-53] McQueen D, Sundgren J-E, Ivarsson B, Lundstrøm I, af Ekenstam B, Svensson A, Bränemark P-I, Albreksson T. Auger electroscopic studies of titanium implants. Clinical applications of biomaterials. Lee AJC, Abreksson T, Bränemark P-I, editors. Wiley: Chichester, UK. 1982. pp. 179−185.

[7-54] Abe M. Oxides and hydrous oxides of multivalent metals as inorganic ion exchangers, inorganic ion exchange materilas. Clearfield A, editor. CRC Press: Boca Raton, FL, 1982. pp. 161−273.

[7-55] Bernardi G, Kawasaki T. Chromatography of polypeptides and proteins on hydroxyl-apatite columns. Biochim Biophys Acta 1968;160:301−310.

[7-56] Embery G, Rølla G. Interaction between sulphated macromolecules and hydroxyapatite studied by infrared spectroscopy. Acta Odontol Scand 1980;38:105−108.

[7-57] Rosengren A, Johansson BR, Thomsen P, Ericson LE. A method for immunolocalization of extracellular proteins in association with the implant−soft tissue interface. Biomaterials 1994;15:17−24.

[7-58] Röstlund T, Thomsen P, Bjursten LM, Ericson LE. Difference in tissue response to nitrogen-ion implanted titanium and c.p. titanium in the abdominal wall of the rat. J Biomed Mater Res 1990;24:847−860.

[7-59] Eriksson AS, Lindblad R, Ericson LE, Thomsen P. Distribution of cells around implants in soft tissue in the rat. J Mater Sci: Mater Med 1994;5:269−278.

[7-60] Johansson CB, Albrektsson T, Ericson LE, Thomsen P. A quantitative comparison of the cell response to commercially pure titanium and Ti6Al4V implants in the abdominal wall of the rat. J Mater Sci: Mater Med 1992;3: 126−136.

[7-61] Anderson JM, Miller KM. Biomaterial biocompatibility and the macrophase. Biomaterials 1984;5:5−10.

[7-62] Miller KM, Anderson JM. *In vitro* stimulation of fibroblast activity by factors generated from human monocytes activated by biomedical polymers. J Biomed Mater Res 1989;23:911−930.

[7-63] Bonfield TL, Colton E, Anderson JM. Plasma protein adsorbed biomedical polymers: activation of human monocytes and induction of interleukin. J Biomed Mater Res 1989;23:535−548.

[7-64] Healy KE, Ducheyne P. Hydration and preferential molecular adsorption on titanium *in vitro*. Biomaterials 1992;13:553−561.

[7-65] Zeng H, Chittur KK, Lacefield WR. Analysis of bovine serum albumin adsorption on calcium phosphate and titanium surfaces. Biomaterials 1999;20:377−384.

[7-66] MacDonald DE, Deo N, Msrkovic B, Stranick M, Somasundaran P. Adsorption and dissolution behavior of human plasma fibronectin on thermally and chemically modified titanium dioxide particles. Biomaterials 2002;23:1269−1279.

[7-67] Deligianni DD, Katsala N, Ladas S, Sotiropoulou D, Amedee J, Missirlis YF. Effect of surface roughness of the titanium alloy Ti-6Al-4V on human bone marrow cell response and on protein adsorption. Biomaterials 2001;22:1241−1251.

[7-68] Yang Y, Cavin R, Ong LJ. Protein adsorption on titanium surfaces and their effect on osteoblast attachment. J Biomed Mater Res 2003;67A:344−349.

[7-69] Schenk RK, Buser D. Osseointegration: a reality. Periodontology 1998;17: 22−35.

[7-70] Larsson C, Esposito M, Liao H, Thomsen P. The titanim−bone interface *in vivo*. In: Titanium in medicine: materials science, surface science, engineering, biological responses, and medical applications. Brunette DM, Tengvall P, Textor M, Thomsen, editors. New York: Springer. 2001. pp. 587−648.

[7-71] Masuda T, Yliheikkaila PK, Fleton DA. Cooper LF. Generalizations regarding the process and phenomenon of osseointegration. Part I. *In vivo* studies. Int I Oral Maxillofac Implant 1998;13:17−29.

[7-72] Oshida Y, Sachdeva R, Miyazaki S. Microanalytical characterization and surface modification of NiTi orthodontic archwires. J Biomed Mater Eng 1992;2:51−69.

[7-73] Zinger O, Anselme K, Denzer A, Habersetzer P, Wieland M, Jeanfils J, Hardouin P, Landolt D. Time-dependent morphology and adhesion of osteoblastic cells on titanium model surfaces featuring scale-resolved topography. Biomaterials 2004;25:2695−2711.

[7-74] Leven RM, Virdi AS, Sumner DR. Patterns of gene expression in rat bone marrow stromal cells cultured on titanium alloy discs of different roughness. J Biomed Mater Res 2004;70A:391−401.

[7-75] Li JP, Li SH, Van Blitterswijk CA, de Groot K. A novel porous Ti6Al4V: characterization and cell attachment. J Biomed Mater Res 2005;73A: 223−233.

[7-76] Hughes WDT, Embery G. Adsorption of bovine serum albumin onto titanium powder. Biomaterials 1996;17:859−864.

[7-77] Webster TJ, Ergun C, Doremus RH, Lanford WA. Increased osteoblast adhesion on titanium-coated hydroxylapatite that forms $CaTiO_3$. J Biomed Mater Res 2003;67A: 975−980.

[7-78] MacDonald DE, Rapuano BE, Deo N, Stranick M, Somasundaran P, Boskey AL. Thermal and chemical modification of titanium−aluminum−vanadium implant materials: effects on surface properties, glycoprotein adsorption, and MG63 cell attachment. Biomaterials 2004;25:3135−3146.

[7-79] Oshida Y, Sachdeva R, Miyazaki S. Changes in contact angles as a function of time on some pre-oxidized biomaterials. J Mater Sci: Mater Med 1992;3:306−312.

[7-80] Webster TJ, Ejiofor JU. Increased osteoblast adhesion on nanophase metals: Ti, Ti6Al4V, and CoCrMo. Biomaterials 2004;25:4731−4739.

[7-81] Jain R, von Recum AF. Fibroblast attachment to smooth and microtextured PET and thin cp-Ti films. J Biomed Mater Res 2004;68A:296−304.

[7-82] Zhang YM, Bataillon-Linez P, Huang P, Zhao YM, Han Y, Traisnel M, Xu KW, Hildebrand HF. Surface analyses of micro-arc oxidized and hydrothermally treated titanium and effect on osteoblast behavior. J Biomed Mater Res 2004;68A:383−391.

[7-83] Nayab SN, Jones FH, Olsen I. Human alveolar bone cell adhesion and growth on ion-implanted titanium. J Biomed Mater Res 2004;69A: 651−657.

[7-84] Maitz MF, Poon RWY, Liu XY, Pham M-T, Chu PK. Bioactivity of titanium following sodium plasma immersion ion implantation and deposition. Biomaterials 2005;26:5465−5473.

[7-85] Muhonen V, Heikkinen R, Danilov A, Jämsä T, Ilvesaro J, Tuukkanen J. The phase state of NiTi implant material affects osteoclastic attachment. J Biomed Mater Res 2005;75A:681−688.

[7-86] Walboomers XF, Elder SE, Bumgardner JD, Jansen JA. Hydrodynamic compression of young and adult rat osteoblast-like cells on titanium fiber mesh. J Biomed Mater Res 2006;76A:16−24.

[7-87] Clem WC, Konovalov VK, Chowdhury S, Vohra YK, Catledge SA, Bellis SL. Mesenchymal stem cell adhesion and spreading on microwave plasma-nitrided titanium alloy. J Biomed Mater Res 2006;76A: 279−287.

[7-88] Advincula MC, Rahemtulla FG, Advincula RC, Ada ET, Lemons JE, Bellis SL. Osteo-blast adhesion and matrix mineralization on sol−gel-derived titanium oxide. Biomaterials 2006;27:2201−2212.

[7-89] Wälivaara B, Aronsson B-O, Rodahl M, Lausmaa J, Tengvall P. Titanium with different oxides: *in vitro* studies of protein adsorption and contact activation. Biomaterials 1994;15:827−834.

[7-90] Brunette DM. The effects of implant surface topography on the behavior of cells. Int J Oral Maxillofac Implants 1988;3:231−246.

[7-91] Meenaghan MA, Bielat KL, Budd TW, Scaaf NG, Nagahara K. Nature of metallic implant surfaces and the importance of their preparation. Oral Maxillofac Surg Clin 1991;3:765−773.

[7-92] Meyle J, Wolburg H, von Recum AF. Surface micromorphology and cellular inter-actions. J Biomater Appl 1993;7:362−374.

[7-93] Bowers KT, Keller JC, Randolph BA, Wick DG, Michaels CM. Optimization of surface micromorphology for enhanced osteoblast responses *in vitro*. Int J Oral Maxillofac Implants 1992;7:302−310.

[7-94] Schwartz Z, Martin JY, Dean DD, Simpson J, Cochran DL, Boyan BD. Effect of titanium surface roughness on chondrocyte proliferation, matrix production, and differentiation depends on the state of cell maturation. J Biomed Mater Res 1996;30:145−155.

[7-95] Ahmad M, Gawronski D, Blum J, Goldberg J, Gronowicz G. Differential response of human osteoblast-like cells to commercially pure (cp) titanium grades 1 and 4. J Bimed Mater Res 1999;46:121−131.

[7-96] Khakbaznejad A, Chehroudi B, Brunette DM. Effects of titanium-coated micromachined grooved substrata on orienting layers of osteoblast-like cells and collagen fibers in culture. J Biomed Mater Res 2004;70A: 206−218.

[7-97] Zhu X, Chen J, Scheideler L, Reichl R, Geis-Gerstorfer J. Effects of topography and composition of titanium surface oxides on osteoblast responses. Biomaterials 2004;25:4087−4103.

[7-98] Saldaña L, Vilaboa N, Vallés G, González-Cabrero J, Munuera L. Osteoblast response to thermally oxidized Ti6Al4V alloy. J Biomed Mater Res 2005;73A:97−107.

[7-99] Kim HJ, Kim SH, Kim MS, Lee EJ, Oh HG, Oh WM, Park SW, Kim WJ, Lee GJ, Choi NG, Koh JT, Dinh DB, Hardin RR, Johnson K, Sylvia VL, Schmitz JP, Dean DD. Varying Ti-6Al-4V surface roughness induces different early morphologic and molecular responses in MG63 osteoblast-like cells. J Biomed Mater Res 2005;74A:366−373.

[7-100] Meredith DS, Eschbach L, Wood MA, Riehle MO, Curtis ASG, Richards RG. Human fibroblast reactions to standard and electropolished titanium and Ti-6Al-7Nb, and electropolished stainless steel. J Biomed Mater Res 2005;75A:541−555.

[7-101] Nayab SN, Jones FH, Olsen I. Effects of calcium ion implantation on human bone cell interaction with titanium. Biomaterials 2005;26:4717−4727.

[7-102] Feng B, Chen J, Zhang X. Interaction of calcium and phosphate in apatite coating on titanium with serum albumin. Biomaterials 2002;23:2499−2597.

[7-103] Wen X, Wang X, Zhang N. Microsurface of metallic biomaterials: a literature review. J Biomed Mater Eng 1996;6:173−189.

[7-104] Keller JC, Dougherty WJ, Grotendorst GR, Wrightman JP. *In vitro* cell attachment to characterized cpTi surfaces. J Adhesion 1989;28:115−133.

[7-105] Keller JC, Wrightman JP, Dougherty WJ. Characterization of acid passivated cpTi surfaces. J Dent Res 1989;68:872.

[7-106] Keller JC, Draughn RA, Wrightman JP, Dougherty WJ. Characterization of sterilized CP titanium implant surfaces. Int J Oral Maxillofac Implants 1990;5:360−369.

[7-107] Schroeder A, Van der Zypen E, Stich H, Sutter F. The reactions of bone, connective tissue and epithelium to endosteal implants with titanium sprayed surfaces. J Maxillofac Surg 1981;9:15−25.

[7-108] Buser D, Schenk RK, Steinemann S, Fiorellinni JP, Fox CH, Stich H. Influence of surface characteristics on bone integration of titanium implants. A histomorphometric study in miniature pigs. J Biomed Mater Res 1991;25:889−902.

[7-109] Rich A, Harris AK. Anomalous preferences of cultured macrophages for hydrophobic and roughended substrata. J Cell Sci 1981;50:1−7.

[7-110] Murray DW, Rae T, Rushton N. The influence of the surface energy and roughness of implants on bone resorption. J Bone Joint Surg 1989;71B: 632−637.

[7-111] Oshida Y. Requirements for successful, biofunctional implants. In: International symposium on advanced biomaterials. Yahia L'H, editor. Montréal, Canada. 2000. p. 5.

[7-112] Cherhoudi B, Gould TR, Brunette DM. Effects of a grooved epoxy substratum on epithelial behavior *in vivo* and *in vitro*. J Biomed Mater Res 988;22:459−477.

[7-113] von Recum AF. New aspects of biocompatibility: motion at the interface. In: Clinical implant materials. Heimke G, Soltesz U, Lee AJC, editors. Amsterdam: Elsevier Science Publication. 1990. pp. 297−302.

[7-114] Ratner BD. Surface characterization of biomaterials by electron spectroscopy for chemical analysis. Ann Biomed Eng 1983;11:313−336.

[7-115] Baro AM, Garcia N, Miranda R, Vázquez L, Aparicio C, Olivé J, Lausmaa J. Characterization of surface roughness in titanium dental implants measured with scanning tunneling microscopy at atmospheric pressure. Biomaterilas 1986;7:463−466.

[7-116] Moroni A, Caja VL, Egger EL, Trinchese L, Chao EY. Histomorphometry of hydroxyl-apatite coated and uncoated porous titanium bone implants. Biomaterials 1994;15:926−930.

[7-117] Keller JC, Young FA, Natiella JR. Quantitative bone remolding resulting from the use of porous dental implants. J Biomed Mater Res 1987;21: 305−319.

[7-118] Pilliar RM, Deporter DA. Watson PA, Valiquette N. Dental implant design-effect on bone remodeling. J Biomed Mater Res 1991;25:467−483.

[7-119] Gilbert JL, Berkery CA. Electrochemical reaction to mechanical disruption of titanium oxide films. J Dent Res 1995;74:92−96.

[7-120] Chehroudi B, Gould TRL, Brunette DM. Tiatnium-coated micromachined grooves of different dimensions affect epithelial and connective tissue cells differently *in vitro*. J Biomed Mater Res 1990;24:1206−1219.

[7-121] Chehroudi B, Gould TRL, Brunette DM. Effects of a grooved titanium coated implant surface on epithelial cell behavior, *in vitro* and *in vivo*. J Biomed Mater Res 1990;24:1067−1085.

[7-122] Michaels CM, Keller JC, Stanford CM, Solursh M. Invitro cell attachment of osteblast-like cells to titanium. J Dent Res 1989;68:276 (Abstract No. 321).

[7-123] Keller JC, Stanford CM, Wightman JP, Draughn RA, Zaharias R. Characterizations of titanium implant surfaces (Pt. 3) J Biomed Mater Res 1994;28:939−946.

[7-124] Oshida Y, Hashem A, Nishihara T, Yapchulay MV. Fractal dimension analysis of mandibular bones: toward a morphological compatibility of implants. J Biomed Mater Eng 1994;4:397−407.

[7-125] Klokkevold PR, Nishimura RD, Adachi M, Caputo A. Osseointegration enhanced by chemical etching of the titanium surface. A torque removal study in the rabbit. Clin Oral Implants Res 1997;8:442−447.

[7-126] Lossdörfer S, Schwartz Z, Wang L, Lohmann CH, Turner JD, Wieland M, Cochran DL, Boyan BD. Microrough implant surface topographies increase osteogenesis by reducing osteoclast formation and activity. J Biomed Mater Res 2004;70A:361−369.

[7-127] Hu H, Lin C, Lui PPY, Leng Y. Electrochemical deposition of hydroxyapatite with vinyl acetate on titanium implants. J Biomed Mater Res 2003;65A:24−29.

[7-128] Vehof JWM, van den Dolder J, de Ruijter JE, Spauwen PHM, Jansen JA. Bone formation in CaP-coated and noncoated titanium fiber mesh. J Biomed Mater Res 2003;64A:417−426.

[7-129] Spoerke ED, Stupp SI. Colonization of organoapatite-titanium mesh by preosteoblastic cells. J Biomed Mater Res 2003;67A:960−969.

[7-130] Rodriguez R, Kim K, Ong JL. *In vitro* osteoblast response to anodized titanium and anodized titanium followed by hydrothermal treatment. J Biomed Mater Res 2003;65A: 352−358.

[7-131] Lu X, Leng Y. Quantitative analysis of osteoblast behavior on microgrooved hydroxyapatite and titanium substrata. J Biomed Mater Res 2003;66A:677−687.

[7-132] Barrère F, van der Valk CM, Dalmeijer RAJ, Meijer G, van Blitterswijk CA, de Groot K, Layrolle P. Osteogenecity of octacalcium phosphate coatings applied on porous metal implants. J Biomed Mater Res 2003; 66A:779−788.

[7-133] Zhao G, Schwartz Z, Wieland M, Rupp F, Geis-Gerstorfer J, Cochran DL, Boyan BD. High surface energy enhances cell response to titanium substrate microstructure. J Biomed Mater Res 2005;74A:49−58.

[7-134] Takeuchi K, Saruwatari L, Nakamura HK, Yang J-M, Ogawa T. Enhanced intrinsic biomechanical properties of osteoblastic mineralized tissue on roughened titanium surface. J Biomed Mater Res 2005;72A: 296−305.

[7-135] Buser D, Weber HP, Donath K, Fiorellini JP, Paquette DW, Williams RC. Soft tissue reactions to non-submerged unloaded titanium implants in beagle dogs. J Periodontol 1992;63:225−235.

[7-136] Abrahamsson I, Berglundh T, Wennstrom J, Lindhe J. The peri-implant hard and soft tissues at different implant systems. A comparative study in the dog. Clin Oral Implants Res 1996;7:212−219.

[7-137] Gargiulo AW, Wentz FM, Orban B. Dimensions and relations of the dentogingival junction in humans. J Periodontol 1961;32:261−267.

[7-138] Arvidson K, Fartash B, Hilliges M, Kondell PA. Histological characteristics of peri-implant mucosa around Branemark and single-crystal sapphire implants. Clin Oral Implants Res 1996;7:1−10.

[7-139] Berglundh T, Lindhe J, Ericsson I, Marinello CP, Liljenberg B, Thomsen P. The soft tissue barrier at implants and teeth. Clin Oral Implants Res 1991;2:81−90.

[7-140] Eisenbarth E, Meyle J, Nachtigall W, Breme J. Influence of the surface structure of titanium materials on the adhesion of fibroblasts. Biomaterials 1996;17:1399−1403.

[7-141] Jansen JA, van der Waerden JP, van der Lubbe HB, de Groot K. Tissue response to percutaneous implants in rabbits. J Biomed Mater Res 1990;24:295−307.

[7-142] Hormia M, Kononen M, Kivilahti J, Virtanen I. Immunolocalization of proteins specific for adherents junctions in humans gingival epithelial cells growth o differently processed titanium surfaces. J Periodontal Res 1991;26:491−497.

[7-143] Mustafa K, Silva Lopez B, Hultenby K, Wennerberg A, Arvidson K. Attachment and proliferation of human oral fibroblasts to titanium surfaces with TiO_2 particles. A scanning electron microscopic and histomorphometric analysis. Clin Oral Implants Res 1998;9:195−207.

[7-144] Kononen M, Hormia M, Kivilahti J, Hautaniemi J, Thesleff I. Effect of surface processing on the attachment, orientation, and proliferation of human gingival fibroblasts on titanium. J Biomed Mater Res 1992;26:1325−1341.

[7-145] Abrahamsson I, Zitzmann NU, Bergundh T, Linder E, Wennerberg A, Lindhe J. The mucosal attachment to titanium implants with different surface characteristics: an experimental study in dogs. J Clin Periodotol 2002;29:448−455.

[7-146] Barboza EP, Caula AL, Carvalho WR. Crestal bone loss around submerged and exposed unloaded dental implants: a radiographic and microbiological descriptive study. Implant Dent 2002;11:162−169.

[7-147] Raghoebar GM, Friberg B, Grunert I, Hobkirk JA, Tepper G. Wendelhag I. 3-Year prospective multicenter study on one-stage implant surgery and early loading in the edentulous mandible. Clin Implant Dent Relat Res 2003;5:39−46.

[7-148] Chou L, Firth JD, Uitto VJ, Brunette DM. Substratum surface topography alters cell shape and regulates fibronectin mRNA level, mRNA stability, secretion and assembly in human fibroblasts. J Cell Sci 1995;108: 1563−1573.

[7-149] Uo M, Watari F, Yokoyama A, Matsuno H, Kawasaki T. Tissue reaction around metal implant observed by X-ray scanning analytical microscopy. Biomaterials 2001;22:677−685.

[7-150] Sundgren J-E, Bodö P, Lundström I. Auger electron spectroscopic studies of the interface between human tissue and implants of titanium and stainless steel. J Collid Interf Sci 1986;110:9−20.

[7-151] Voggenreiter G, Leiting S, Brauer H, Leiting P, Majetschak M, Bardenheuer M, Obertacke U. Immuno-inflammatory tissue reaction to stainless-steel and titanium plates used for internal fixation of long bones. Biomaterials 2003;24:247−254.

[7-152] Deporter D, Watson P, Pharoah M, Levy D, Todescan R. Five-to six-year results of a prospective clinical trial using the ENDOPORE dental implant and a mandibular overdenture. Clin Oral Implants Res 1999;10: 95−102.

[7-153] Buser D, Nydegger T, Oxland T, Cochran DL, Schenk RK, Hirt HP, Sneitivy D, Nolte LP. Interface shear strength of titanium implants with a sandblasted and acid-etched surface: a biomechanically study in the maxilla of miniature pigs. J Biomed Mater Res 1999;45:75−83.

[7-154] Palmer RM, Plamer PJ, Smith BJ, A 5-year prospective study of astra single tooth implants. Clin Oral Implants Res 2000;11:179−182.

[7-155] Testori T, Wiseman L, Woolfe S, Porter SS. A prospective multicenter clinical study of the osseotite implant: four-years interim report. Int J Oral Maxillofac Implants 2001;16:193−200.

[7-156] Sul Y-T. The significance of the surface properties of oxidized titanium to the bone response: special emphasis on potential biochemical bonding of oxidized titanium implant. Biomaterials 2003;24:3893−3907.

[7-157] Ishizawa H, Fujiino M, Ogino M. Mechanical and histological investigation of hydrothermally treated and untreated anodic titanium oxide films containing Ca and P. J Biomed Mater Res 1995;29:1459−1468.

[7-158] Larsson C, Emanuelsson L, Thomsen P, Ericson L, Aronsson B, Rodahl M, Kasemo B, Lausmaa J. Bone response to surface-modified titanium implants: studies on the tissue response after one year to machined and electropolished implants with different oxide thicknesses. J Mater Sci: Mater Med 1997;8:721−729.

[7-159] Skripitz R, Aspenberg P, Tensile bond between bone and titanium. Acta Orthop Scand 1998;69:2−6.

[7-160] Fini M, Cigada A, Rondelli G, Chiesa R, Giardino R, Giavaresi G, Aldini N, Torricelli P, Vicentini B. *In vitro* and *in vivo* behaviour of Ca and P-enriched anodized titanium. Biomaterials 1999;20:1587−1594.

[7-161] Henry P, Tan AE, Allan BP. Removal torque comparison of TiUnite and turned implants in the Greyhound dog mandible. Appl Osseointegration Res 2000;1:15−17.

[7-162] Sul Y-T, Johansson CB, Jeong Y, Wennerberg A, Albrektson T. Resonance frequency and removal torque analysis of implants with turned and anodized surface oxide. Clin Oral Implants Res 2002;13:252−259.

[7-163] Sul Y-T, Johansson CB, Jeong Y. Röser K, Wennerberg A, Albreksson T. Oxidized implants and their influence on the bone response. J Mater Sci: Mater Med 2001;12:1025−1031.

[7-164] Sul Y-T, Johansson CB, Albreksson T. Oxidized titanium screws coated with calcium ions and their performance in rabbit bone. Int J Oral Maxillofac Implants 2002;17:625−634.

[7-165] Lim YJ, Oshida Y, Andres CJ, Barco MT. Surface characterizations of variously treated titanium materials. Int J Oral Maxillofac Imaplants 2001;16:333−342.

[7-166] Meachim G, Williams DF. Changes in nonosseous tissue adjacent to titanium implants. J Biomed Mater Res 1973;7:555−571.

[7-167] Albrektsson T, Jacobsson M. Bone-metal interface in osseointegration. J Prosthet Dent 1987;57:5−10.

[7-168] Lausmaa J. Chemical composition and morphology of titanium surface oxides. Mat Res Soc Symp Proc 1986;55:351−359.

[7-169] Kohavi D, Schwartz Z, Amir D, Muller Mai C, Gross U, Sela J. Effect of titanium implants on primary mineralization following 6 and 14 days of rat tibial healing, Biomaterials 1992;13:255−260.

[7-170] Jasty M, Bragdon CR, Haire T, Mulroy RD, Harris WH. Comparison of bone ingrowth into cobalt chrome sphere and titanium fiber mesh porous coated cementless canine acetabular components. J. Biomed Mater Res 1993;27:639−644.

[7-171] Hazan R, Brener R, Oron U. Bone growth to metal implants is regulated by their surface chemical properties. Biomaterials 1993;14:570−574.

[7-172] http://www.teambone.com/wolff.html.

[7-173] Changd YS, Oka M, Kobayashi M, Gu HO, Li ZL, Nakamura T, Ikada Y. Significance of interstitial bone ingrowth under load-bearing conditions: a comparison between solid and porous implant materials. Biomaterials 1996;17:1141−1148.

[7-174] Piattelli A, Corigliano M, Scarano A. Microscopical observations of the osseous responses in early loaded human titanium implants: a report of two cases. Biomaterials1996;17:1333−1337.

[7-175] Goodman S, Aspenberg P. Effect of amplitude of micromotion on bone ongrowth into titanium chambers implanted in the rabbit tibia. Biomaterials 1992;13:944−948.

[7-176] Li J, Liao H, Fartash B, Hermansson L, Johnsson T. Surface-dimpled commercially pure titanium implant and bone ingrowth. Biomaterials 1997;18:691−696.

[7-177] French AA. Bowles CQ, Parharn PL, Eick JD, Killoy WJ, Cobb CM. Comparison of peri-implant stresses transmitted by four commercially available osseointegrated implants. Int J Periodont Rest Dent 1989;9: 221−230.

[7-178] Bidez MW, Misch CE. Force transfer in implant dentistry. Basic concepts and principles. J Oral Implantology 1992;18:264−274.

[7-179] Adell R, Lekholm U, Rockler B, Brånemark P-I. A 15-year study of osseointegrated implants in the treatment of the edentulous jaws. Int J Oral Surg 1981;10:387−416.

[7-180] Schliephake H, Scharnweber D, Dard M, Rößler S, Sewing A, Hüttmann C. Biological performance of biomimetic calcium phosphate coating of titanium implants in the dog mandible. J Biomed Mater Res 2003;64A:225−234.

[7-181] Groessner-Schreiber B, Neubert A, Müller W-D, Hopp M, Griepentrog M, Lange KP. Fibroblast growth on surface-modified dental implants: An *in vitro* study. J Biomed Mater Res 2003;64A:591−599.

[7-182] Vasudev DP, Ricci JL, Sabatino C, Li P, Parsons JR. *In vivo* evaluation of a biomimetic apatite coating grown on titanium surfaces. J Biomed Mater Res 2004;69A:629−636.

[7-183] Knabe C, Howlett CR, Klar F, Zreiqat H. The effect of different titanium and hydroxyapatite-coated dental implant surfaces on phenotypic expression of human bone-derived cells. J Biomed Mater Res 2004;71A: 98−107.

[7-184] Lu X, Leng Y, Zhang X, Xu J, Qin L, Chan C-W. Comparative study of osteoconduction on micromachined and alkali-treated titanium alloy surfaces *in vitro* and *in vivo*. Biomaterials 2005;26:1793−1801.

[7-185] Elmengaard B, Bechtold JE, Søballe K. *In vivo* effects of RGD-coated titanium implants inserted in two bone-gap models. J Biomed Mater Res 2005;75A:249−255.

[7-186] Elmengaard B, Bechtold JE, Søballe K. *In vivo* study of the effect of RGD treatment on bone ongrowth on press-fit titanium alloy implants. Biomaterials 2005;26:3521−3526.

[7-187] Bothe RT, Beaton LE, Davenport HA. Reaction of bone to multiple metallic implants. Surg Gynecol Obstet 1940;71:598−602.

[7-188] Brånemark P-I, Hansson BO, Adell R, Briene U, Lindstrom J, Hallen O, Ohman A. Osseointegrated implants in the treatment of the edentulous jaw. Experience from a 10-year period. Scand J Plast Reconstr Surg Suppl 1977;16:1−132.

[7-189] Misch CE, Misch CM. Generic root form component terminology. In: Implant dentistry. Misch CE, editor. Mosby: An Affiliate of Elsevier. 1999. pp. 13−15.

[7-190] Brånemark R, Brånemark P-I, Rydevik B, Myers RR. Osseointegration in skeletal reconstruction and rehabilitation. J Rehabil Res Dev 2001;38: 175−182.

[7-191] Zaffe D, Baena RR, Rizzo S, Brusotti C, Soncini M, Pietrabissa R, Cavani F, Quaglini V. Behavior of the bone−titanium interface after push-in testing: a morphological study. J Biomed Mater Res 2003;64A: 365−371.

[7-192] Rønold HJ, Lyngstadaas SP, Ellingsen JE. A study on the effect of dual blasting with TiO_2 on titanium implant surfaces on functional attachment in bone. J Biomed Mater Res 2003;67A:524−530.

[7-193] Nishiguchi S, Fujibayashi S, Kim H-M, Kokubo T, Nakamura T. Biology of alkali- and heat-treated titanium implants. J Biomed Mater Res 2003;67A:26−35.

[7-194] Giavaresi G, Fini M, Cigada A, Chiesa R, Rondelli G, Rimondini L, Aldini NN, Martini L, Giardino R. Histomorphometric and microhardness assessments of sheep cortical bone surrounding titanium implants with different surface treatments. J Biomed Mater Res 2003;67A: 112−120.

[7-195] Fujibayashi S, Neo M, Kim HM, Kokubo T, Namamura T. Osteoinduction of porous bioactive titanium metal. Biomaterials 2004;25:443−450.

[7-196] Takemoto M, Fujubayashi S, Neo M, Suzuki J, Kokubo T, Nakamura T. Mechanical properties and osteoinductivity of porous bioactive titanium. Biomaterials 2005;26:6014−6023.

[7-197] Takemoto M, Fujibayashi S, Neo M, Suzuki J, Matsushita T, Kokubo T, Nakamura T. Osteoinductive porous titanium implants: effect of sodium removal by dilute HCl treatment. Biomaterials 2006:27;2682−2691.

[7-198] Sul Y-T, Johansson C, Byon E, Albrektsson T. The bone response of oxidized bioactive and non-bioactive titanium implants. Biomaterials 2005;26:6720−6730.

[7-199] Aebli N, Krebs J, Stich H, Schawalder P, Walton M, Schwenke D, Gruner H, Gasser B, Theis J-C. *In vivo* comparison of the osseointegration of vacuum plasma sprayed titanium- and hydroxyapatite-coated implants. J Biomed Mater Res 2003;66A:356−363.

[7-200] Stewart M, Welter JF, Goldberg VM. Effect of hydroxyapatite/tricalcium-phosphate coating on osseointegration of plasma-sprayed titanium alloy implants. J Biomed Mater Res 2004;69A:1−10.

[7-201] Habibovic P, Li J, van der Valk CM, Meijer G, Layrolle P, van Blitterswijk CACA, de Groot K. Biological performance of uncoated and octacalcium phosphate-coated Ti6Al4V. Biomaterials 2005;26:23−36.

[7-202] Stangl R, Pries A, Loos B, Müller M, Erben RG. Influence of pores created by laser superfinishing on osseointegration of titanium alloy implants. J Biomed Mater Res 2004;69A:444−453.

[7-203] Götz HE, Müller M, Emmel A. Holzwarth U, Erben RG, Stangl R. Effect of surface finish on the osseointegration of laser-treated titanium alloy implants. Biomaterials 2004;25:4057−4064.

[7-204] Fujishiro T, Nishikawa T, Shibanuma N, Akisue T, Takikawa S, Yamamoto T, Yoshiya S, Kurosaka M. Effect of cyclic mechanical stretch and titanium particles on prostagland in E_2 production by human macrophages *in vitro*. J Biomed Mater Res 2004;68A:531−536.

[7-205] Lin H-Y, Bumgardner JD. *In vitro* biocorrosion of Ti-6Al-4V implant alloy by a mouse macrophage cell line. J Biomed Mater Res 2004;68A:717−724.

[7-206] Sommer B, Felix R, Sprecher C, Leunig M, Ganz R, Hofstetter W. Wear particles and surface topographies are modulators of osteoclastogenesis *in vitro*. J Biomed Mater Res 2005;72A:67−76.

Chapter 8
Implant Application

8.1.   General   217
8.2.   Clinical Reports   222
8.3.   Surface and Interface Characterization   227
8.4.   Reactions in Chemical and Mechanical Environments   230
8.5.   Reaction in Biological Environment   235
References   246

Chapter 8
Implant Application

In the past, a metallic post was either cemented or screwed into the canals of teeth that had lost their crowns, but still had their roots. Such teeth are "root-treated" to remove their nerves and blood supplies, and onto the posts ceramic or ceramic metal crowns are themselves cemented. Although many attempts have been made to replace missing roots with all sorts on metallic implants, the satisfactory use of a screwed in implant was found only in the mid-1980s. In practice the gum is slit, and a hole is cut slowly in the bone, and then a CpTi screw is slowly inserted and covered with gum tissue for 6 months. During this time, the bone grows into intimate contact with the passive oxide layer on the titanium and it is said to be osseointegrated. The gum tissue is cut once more, and a titanium sleeve is screwed onto the implant. This will ultimately pass through the healed gum. Onto these sleeves a metallic superstructure can be screwed, and this can support, for example, a polymeric denture base and artificial teeth. For many years, these superstructures have been cast in gold alloys, and getting them to sit perfectly on the titanium sleeves has been a challenge of the highest order. However, titanium frameworks are currently being investigated, particularly those constructed by alternative routes to casting, and considerable promise is being shown by those made by superplastic forming. Titanium implants, for example, have had their surfaces coated with hydroxyapatite to promote the osseointegration [8-1].

The service conditions in the mouth are hostile, both corrosively and mechanically. All parts are continuously bathed in saliva, an aerated aqueous solution of about 0.1 N chloride, with varying amounts of Na, K, Ca, PO_4, CO_2, sulfur compounds and mucin. The pH value is normally in the range of 5.5–7.5, but under plaque deposits it can be as low as 2. Temperatures can vary $\pm 36.5°C$, and a variety of food and drink concentrations apply for short periods. Loads may be up to 1000 N, sometimes at impact speeds. Trapped food debris may decompose, and sulfuric compounds discolor [8-2].

8.1. GENERAL

Direct bony interface promised more predictability and longevity than previously used systems; hence, oral implantology gained significant additional momentum. A carefully planned full rehabilitation of the mouth using "state of the art" methods can free the patient of dental problems for decades. However, this can only be achieved with the complete cooperation of the patient, accompanied by regular supervision and care on the part of the dental surgeon and his assistants (e.g., the dental hygienist). Accordingly, there are four major factors which should be met to accomplish the successful implant: (1) correct indication and favorable anatomic conditions (bone and mucosa), (2) good operative technique, (3) patient cooperation (oral hygiene), and (4) adequate superstructure. According to Binon [8-3], there are approximately 25 dental implants manufacturing companies, producing about 100 different dental-implant systems with a variety of diameters, lengths, surfaces, platforms, interfaces, and body designs. The most logical differentiation and distinctions are based on the implant/abutment interface, the body shape, and the implant-to-bone surface.

There are two ways to retain the implant restorations: cementation or screw-tightening. Implant superstructures are usually made by the lost-wax casting method, and are normally gold alloy beams that fit directly onto metallic components called abutments that protrude through the soft tissues of the mouth, and which are themselves attached to machined titanium components known as dental implants. The dental implants are sited in the hard bone tissues where intimate contact occurs between the metallic component and the natural tissues. The superstructure is screwed or cemented to the abutments, and teeth are then placed along the beam in an aesthetic fashion. Patients prefer this form of fixed bridgework as it is more reminiscent of their natural teeth than a denture. For the whole prosthesis to have the optimal chance of success there must be no misfit between the beam and the abutments. Mainstream philosophy in the design and restoration of implant prostheses from the 1980s through the early 1990s reflected a strong preference for screw-retained over cement-retained restorations. This preference most likely was the result of how osseointegrated implants and their restoration were introduced. Although retrievability remains the primary advantage of a screw-retained approach, cemented implant restorations are attractive for several reasons. First, the preclusion of the interference of screw access holes with esthetic results or with the occlusion of the restoration. Second, cement-retained approaches substantially reduce restoration costs. With some systems, screw-retained restoration involves nearly four times the component cost of cemented restoration. Third, if no occlusal screw is present, its loosening is not a possible complication of implant prosthodontics. Fourth, a cemented restoration is more likely to achieve passive fit than a screw-retained restoration. The argument for increased passivity rests on the assumption that tightening a screw-retained restoration may create substantial strain within the restoration, implant, and investing bone complex. The use of cemented restorations also avoids the clamping effect of multiple screws drawing an ill-fitting prosthesis into place against the support. Fifth, the implant industry's desire to extend routine implant use into the general dental practices has stimulated interest in simplifying implant restorative procedures. Cementation of fixed-implant restorations more closely follow the procedures routinely performed on natural teeth [8-4, 8-5].

The implantation of devices for the maintenance or restoration of a body function imposes extraordinary requirements on the materials of construction. Foremost among these is an issue of biocompatibility, as we had discussed in previous Chapters 6 and 7. There are interactions between the material and the surrounding living tissue, fluid, and blood elements. Some of these are simply adaptive. Others constitute a hazard, both short and long term, to the survival of the living system [8-3, 8-6–8-9]. There are mechanical and physical properties which the material must provide. Some of these govern the ability of the device to provide its intended function from a purely engineering viewpoint. Others, such as, tribology including wear and friction, corrosion and mechanical compliance relate significantly to the biocompatibility issue. Human implantation applications impose more stringent requirements on reliability than any other engineering task. In most applications an implanted device is expected to function for the life of the patient. It is necessary to think in terms of reliability of performance of thousands of devices for the lifetime of a patient and a tolerable expectation of failure of perhaps not more than one in a thousand. Most available data in handbooks, publications, and product brochures represent mean values and supplementary statistical information, and, therefore, comparable reliability is also in question [8-7–8-9]. However, it is surprising to note that there is no safety factor concept in the dental and medical area, whereas it is a very important concept in engineering, particularly designing crucial structures and structural components.

As for orthopedic implants, natural synovial joints, e.g., hip, knee, or shoulder joints, are complex and delicate structures capable of functioning under critical conditions. Unfortunately, human joints are prone to degenerative and inflammatory diseases that result in pain and joint stiffness. Primary or secondary osteoarthritis (osteoarthrosis), and to a lesser extent rheumatoid arthritis (inflammation of the synovial membrane) and chondromalacia (softening of cartilage), are, apart from normal aging of articular cartilage, the most common degenerative processes affecting synovial joints. In fact, 90% of the population over the age of 40 suffers from some degree of degenerative joint disease. Premature joint degeneration may arise from deficiencies in joint biomaterial properties, from excessive loading conditions, or from failure of normal repair processes, where the explicit degenerative processes are not yet completely understood [8-10, 8-11].

The use of Ti materials for medical and dental-implant applications has increased tremendously over the past few decades. One of the prime reasons for this increased use is the design of implants around its unique properties. In the case of high load bearing and fatigues, for devices such as hip prostheses,

Ti should be the metal of choice because of its combination of strength, corrosion resistance, light weight, good biotribological property, and biocompatibility. Advances in Ti materials development, casting technology, and powder metallurgy will be offering the surgeon and patient benefits. Casting of Ti alloys has arrived as a technology, which can be utilized by the implant industry. Hip prostheses made from cast and hot isostatically pressed (HIPed) Ti-6Al-4V are currently being used. Advanced powder metallurgy – metal injection molding (MIM), – can be used to fabricate a whole implant body or surface layer of them. In addition to these technologies, the idea to use superplastic forming started when a dental practitioner was looking for a method of producing dental-implant superstructures with passive fit [8-12, 8-13, Chapter 10].

Traditionally, the majority of geriatric dental services provided to the elderly have been provided to healthy, independent, older adults requiring few special considerations other than normal physiologic age-related changes and their potential impact on the oral cavity. This is no longer an acceptable approach. Over 80% of the geriatric population has at least one chronic disease, and many elderly have several such conditions simultaneously. The rapid growth of the elderly population will have a dramatic impact on the practice of dentistry [8-14]. In 1984, 27.9 million, or one in nine adults in the US, were 65 years or older. It is anticipated that those over 65 will account for 70.2 million, or 20% of the US population by the year 2030 [8-15]. It was mentioned that (1) age should not exclude patients from implant treatment, (2) dental implants and implant-retained and/or supported prostheses are valuable treatment options for geriatric patients, (3) early-implant intervention is strongly recommended when the patient feels able and is willing to undergo dental and prosthetic therapy, (4) diminished levels of oral hygiene that often accompany aging are not a contraindication to implant treatment, and (5) clinicians should be aware of potential risks, possible medical complications, and psychosocial issues in geriatric patients, and how these conditions can affect the implant prognosis [8-14].

There are at least three major required compatibilities for placed implants to exhibit biointegration to receiving hard tissue and biofunctionality thereafter. They include (1) biological compatibility (in other words, biocompatibility), (2) mechanical compatibility (or mechanocompatibility), and (3) morphological compatibility [8-16, 8-17]. One of many universal requirements of implants, wherever they are used in the body, is the ability to form a suitably stable mechanical unit with the neighboring hard or soft tissues. A loose (or unstable) implant may function less efficiently or cease functioning completely, or it may induce an excessive tissue response. In either case, it may cause the patient discomfort and pain. In several situations, a loose implant was deemed to have failed and needed to be surgically removed. For a long time it has been recognized that any types of implants (for both dental implants and orthopedic implants), should possess a biological compatibility against an implant receiving surrounding hard/soft tissues. Accordingly, the material choice for implants is limited to certain types of materials, including titanium materials, stainless steels, or some ceramic materials. The dental or orthopedic prostheses, particularly the surface zone thereof, should respond to the loading transmitting function. The placed implant and receiving tissues establish a unique stress-strain field. Between them, there should be an interfacial layer. During the loading, the strain-field continuity should be held, although the stress-field is obviously in a discrete manner due to different values of modulus of elasticity (MOE) between host tissue and foreign-implant material. If the magnitude of the difference in MOE is large, then the interfacial stress, accordingly, will be so large that the placed implant system will face a risky failure situation. Therefore, materials for implant or surface zone of implants should be mechanically compatible to mechanical properties of receiving tissues, so that the interfacial discrete stress can be minimized. This is the second compatibility and is called as the mechanical compatibility. In a scientific article [8-17], it was found that surface morphology of successful implants has upper and lower limitation in average roughness (1–50 μm) [8-16, 8-17, Figure 7.1 in Chapter 7] and average particle size (10–500 μm) [8-16, 8-17], regardless of types of implant materials (metallic, ceramics, or polymeric materials). If a particle size is smaller than 10 μm, the surface will be more toxic to fibroblastic cells and have an adverse influence on cells due to their physical presence independent of any chemical toxic effects. If the pore is larger than 500 μm, the surface does not maintain sufficient structural integrity because it is too coarse. This is the third compatibility – morphological compatibility [8-16, 8-17].

The criteria for clinical success of osseointegration and the conditions of functional compatibility are dependent on the control of several factors include (1) biocompatible-implant material using commercially pure titanium, (2) design of the fixture: a threaded design is advocated, creating a larger surface per unit volume, as well as evenly distributing loading forces, (3) the provision of optimal prosthodontic design and implant maintenance to achieve ongoing osseointegration, (4) a specific aseptic surgical technique and a subsequent healing protocol which are reconcilable with the principles of bone physiology; this would incorporate a low heat/low trauma regimen, a precise fit and the two-stage surgery program, (5) a favorable status of host-implant site from a health and morphologic standpoint, (6) non-loading of the implant during healing is the basic principle for successful osseointegration, and (7) the defined macro/microscopic surface of the implant as it relates to the host tissue [8-18].

8.2. CLINICAL REPORTS

Before we go into further detail reviewing surface characterization and implant reactions in various environments, it would be worthy to know what is actually happening in clinical fields. Although, as we will observe later, dental implants are enjoying their success rates, there is still much research to be done. This initiative will encourage clinical research on osseointegrated dental implants regarding (1) outcomes when using various surgical and prosthetic protocols (e.g., immediate placement vs. delayed placement; implant placement following osseous grafting and/or sinus augmentation; immediate vs. delayed prosthetic loading), (2) the role of systemic diseases in the success rate of osseointegrated implants, (3) quality of life and patient preferences for dental implants compared to other prosthetic methods for restoring the dentition, and (4) needs in children who have congenitally missing teeth or suffer from developmental disabilities. Osseointegrated dental implants are one of the options available for replacing missing teeth. The success of implants has been attributed to osseointegration or direct contact of the implant surface and bone without a fibrous connective tissue interface. Varying surgical protocols are used for placement of implants, and a wide variety of implant designs are in current use. Implant failure is reported to be higher in smokers than in non-smokers. It is obvious that smoking is one of the contraindications for receiving dental implants, as we discussed in Chapter 7. While there are reports of similar failure rates between people with well-controlled diabetes and those without diabetes, individuals with type 2 diabetes may have a slightly higher failure rate. No studies have documented the success of implants in individuals who develop diabetes, osteoporosis, or other systemic conditions after placement. In general, the relationship between systemic health and implant success is not well documented. The use of surgical procedures such as maxillary sinus lifts, inferior alveolar nerve transposition, and guided bone regeneration have expanded the use of implants [8-19].

Adell *et al.* [8-19] reported the clinical review on the long-term outcome of prostheses and fixtures (titanium implants) in 759 totally edentulous jaws of 700 patients. A total of 4636 standard implants were placed and followed according to the osseointegration method for a maximum of 24 years by the original team at the University of Göteborg. Standardized annual clinical and radiographic examinations were conducted as far as possible. A life-table approach was applied for statistical analysis. Sufficient numbers of implants and prostheses for a detailed statistical analysis were present for observation times up to 15 years. More than 95% of maxillae had continuous prosthesis stability at 5 and 10 years, and at least 92% at 15 years. The figure for mandibles was 99% at all time intervals. Calculated from the time of implant placement, the estimated survival rates for individual implants in the maxilla were 84, 89, and 92% at 5 years; 81 and 82% at 10 years; and 78% at 15 years. In the mandible, they were 91, 98, and 99% at 5 years; 89 and 90% at 10 years; and 86% at 15 years [8-19, 8-20].

Åstrand *et al.* [8-21] reported the following clinical studies. Twenty-three patients with Kennedy Class I mandibular dentition were supplied with prostheses in the posterior parts of the mandible. On one side they were given a prosthesis supported by two implants (type I) and on the other side they received a prosthesis supported by one implant and one natural tooth (type II). Sixty-nine fixtures (titanium implants) were inserted and 46 prostheses constructed. Eight of the implants were lost during the observation period. The failure rate of the implants was about the same in the two types of prostheses: five implants belonged to prostheses type I (10.9%) and two implants belonged to prostheses type II (8.7%), while one implant was lost prior to loading. From a theoretical point of view, the combination of a tooth and an osseointegrated implant should encounter problems with regard to the difference in bone anchorage, and there should be a risk of biomechanical complications. However, the results of this study did not indicate any disadvantages in connecting teeth and titanium implants in the same restoration [8-21].

The structural and functional bonding of load-carrying titanium implants to living bone has been recognized, indicating that the implant was successfully accepted. The effect of CpTi implants on the process of primary mineralization was studied by Kohavi *et al.* [8-22] by insertion of Ti implants into rat bone after ablation. The effects of the Ti were studied through the behavior of extracellular matrix vesicles. It was found that (i) the insertion of Ti implants was followed by an increase in the number of extracellular matrix vesicles as well as vesicular diameter, and by a decrease in vesicular distance from the calcified front when compared to normal healing, suggesting that the process of extracellular matrix vesicles maturation around Ti implants was delayed when compared to normal primary bone formation during bone healing, and (ii) the delay in mineralization was compensated by an increase in vesicular production, resulting in an enhancement of primary mineralization by the titanium [8-22]. Piattelli *et al.* [8-23] reported on the histological picture of the surrounding of two screw-shaped Ti plasma-sprayed implants retrieved for a fracture of the abutment after 18 and 42 months. These two implants were loaded after two months. The microscopic examination showed that both implants were covered in a large part of the implant surface by compact, mature lamellar bone with the presence of many Haversian systems and osteons. With von Kossa staining it was possible to see that the bone at the interface with the implant was highly mineralized. No connective tissue or inflammatory cells were present at the interface [8-23].

Saadown *et al.* [8-24] presented the 8-year study, showing an increased success rate for Steri-Oss osseointegrated implants in full, partial, and single-tooth edentation, compared to previous 6-year studies. A total of 1499 titanium implants placed in 389 women and 216 men had an average success rate of 96.1%. Of the 716 implants placed in the maxilla, 697 were exposed and only 31 failed – a success rate of 95.6%. Of 783 placed in the mandible, 750 were exposed and 25 failed – a success rate of 96%. Performance comparisons of Steri-Oss Ti threaded, HA-coated, cylindrical implants and Ti plasma-sprayed cylindrical implants showed a higher success rate for the coated and sprayed implants, especially for short implants placed in soft maxillary bone [8-24].

A histological analysis of 19 Brånemark titanium implants retrieved for different causes were reported: (1) four implants were removed for abutment fracture, (2) one for dental nerve dysesthesia, (3) two for bone overheating, (4) two for peri-implantitis, (5) nine for mobility, and (6) one for unknown causes. It was mentioned that, in the implants removed for fracture, a high bone–implant contact percentage was present (72%), with compact, mature bone at the interface. It was also reported that, in peri-implantitis, an inflammatory infiltrate was observed in the preimplant tissues: a dense fibrous connective tissue was present around implants failed because of its mobility [8-25].

Successful endosseous implant therapy requires integration of the implant with bone, soft connective tissue, and epithelium. There were several important observations made by Cochran [8-26] as follows. Light and electron microscopy reveal that epithelial structures similar to teeth are found around the implants. A connective tissue zone exists between the apical extension of the junctional epithelium and the alveolar bone. This connective tissue comprises a dense circular avascular zone of connective tissue fibers surrounded by a loose vascular connective tissue. The histologic dimensions of the epithelium and connective tissue comprising the biologic width are similar to the same tissues around teeth. Straumann titanium implants have an endosseous portion that is either coated with a well-characterized and well-documented titanium plasma-sprayed surface or is sand-blasted and acid-etched. Both surfaces have been shown to have advantages for osseous integration compared to as-machined and other smoother-implant surfaces, as we have seen in Chapter 7. These advantages include greater amounts of bone-to-implant contact (mechanical anchorage), more rapid integration with bone tissue (biochemical anchorage), and higher removal torque values (as a result of both). Component connection at the alveolar crest, as seen with submerged implants, results in micro-bial contamination, crestal bone loss, and a more apical epithelial location [8-26].

Leonhardt *et al.* [8-27] followed up longitudinally osseointegrated titanium implants in partially dentate patients by clinical, radiographic, and microbiological parameters in order to evaluate possible changes in the peri-implant health over time. Fifteen individuals treated with titanium implants and followed for 10 years were included in the study. Before implant placement 10 years prior, the individuals had been treated for advanced periodontal disease and thereafter been included in a maintenance care program. The survival rate of the implants after 10 years was 94.7%. Of the individuals, 50% were positive for plaque at the implants. Bleeding on sulcus probing was present at 61% of the implant surfaces. The results of the present study suggest that the presence of these putative periodontal pathogens at implants may not be associated with an impaired-implant treatment. These species are most likely part of the normal resident microbiota of most individuals, and may therefore be found at random at both stable and progressing peri-implant sites [8-27].

ITI® dental implants are available with two bone-anchoring surfaces, a titanium plasma-sprayed surface, and a recently introduced sand-blasted and acid-etched surface. Cell culture and animal tests demonstrate that the sand-blasted and acid-etched surface stimulates bone cell differentiation and protein production, has large amounts of bone-to-implant contact, and results in large removal torque values in functional testing of the bone contact. The 5-year follow-up revealed that 110 patients with 326 implants have completed the 1-year post-loading recall visit, while 47 patients with 138 implants have completed the 2-year recall. Three implants were lost prior to abutment connection. Prosthetic restoration was commenced after shortened healing times on 307 implants. The success rate for these implants, as judged by abutment placement, was 99.3% (with an average healing time of 49 days). It was further analyzed by the life table analyses that there was an implant success rate was 99.1%, both for 329 implants at 1 year and for 138 implants at 2 years. It was also reported that, under defined conditions, solid-screw implants with a sand-blasted and acid-etched endosseous surface can be restored after approximately 6 weeks of healing with a high predictability of success, defined by abutment placement at 35 N cm without counter-torque, and with subsequent implant success rates of greater than 99% 2 years after restoration [8-28].

Meijer *et al.* [8-29] conducted the prospective randomized controlled clinical trial to evaluate the clinical outcomes and prosthetic aftercare of edentulous patients with a mandibular overdenture retained by two IMZ titanium implants or two Brånemark titanium implants during a 10-year period. Patients were allocated to the IMZ group (n=29) or the Brånemark group (n=32). In the IMZ group, four implants were lost during the 10-year follow-up (survival rate: 93%). In the Brånemark group, nine implants were lost (survival rate: 86%). All patients were re-operated successfully. Multiple prosthetic revisions were necessary in both groups. In particular, the precision attachment system in the overdenture (23% of the total number of revisions) and the denture base and teeth (26% of the total number of revisions) were subject to frequent fracture. It was concluded that both the IMZ implant and the Brånemark implant systems supporting an overdenture are functioning well after 10 years of follow-up. There are no indications of a worsening of clinical or radiographical state after 10 years [8-29].

Ichinose *et al.* [8-30] demonstrated that a myriad of fine particles produced by the abrasion of both Co-Cr-Mo and Ti-6Al-4V alloys accumulate in the synovial cells next to surgical implants made from these alloys. The metallic particles were of various sizes, and were observed within the lysosomes. Energy-dispersive X-ray spectroscopy studies revealed that the fine spherical particles consisted solely of Cr, and that other larger particles were composed of the Co-Cr-Mo alloy. It was reported that most of fine particles were 10–15 nm in diameter. Eighty percent of the large particles were 30–35 nm in length and 20–25 nm in width. In addition, the energy-dispersive X-ray spectroscopy analysis clarified that all of the fine particles of the Ti-6Al-4V were composed of that alloy. For Ti-6Al-4V, when discounting the large particles, the fine metal deposits were 20–25 nm in length and 10–15 nm in width. It was concluded that (i) Co-Cr-Mo alloy is easily corroded and that Co is released from the cells, but (ii) Ti-6Al-4V alloy is very stable and does not corrode, although the Ti-6Al-4V alloy does produce particles that are smaller than those produced by the Co-Cr-Mo alloy [8-30].

Goodacre *et al.* [8-31] reviewed articles to identify the types of complications that have been reported in conjunction with endosseous root form implants and associated implant prostheses. The searches focused on Medline publications (which started in 1981) that contained clinical data regarding success/failure/complications. The complications were divided into the following six categories: surgical, implant loss, bone loss, peri-implant soft tissue, mechanical, and esthetic/phonetic. The raw data were combined from multiple studies and means calculated to identify trends noted in the incidences of complications. It was reported that the most common implant complications (those with a greater than a 15% incidence) were loosening of the overdenture retentive mechanism (33%), implant loss in irradiated maxillae (25%), hemorrhage-related complications (24%), resin veneer fracture with fixed partial dentures (22%), implant loss with maxillary overdentures (21%), overdentures needing to be relined (19%), implant loss in type IV bone (16%), and overdenture clip/attachment fracture (16%) [8-31].

Evidence of the successful use of osseointegrated titanium dental implants for the restoration of individual teeth has been reported for anterior teeth more frequently than for posterior teeth. Simon [8-32] collected data from the charts of patients provided with implant–supported single crowns in posterior quadrants in a prosthodontic practice in southern California. It was reported that 49 patients with 126 implants restored with molar or premolar crowns were recalled for examination after periods ranging from 6 months to 10 years. The implant failure rate was 4.6% with complications of abutment screw loosening (7%) and loss of cement bond (22%) [8-32].

8.3. SURFACE AND INTERFACE CHARACTERIZATION

Defining the nature of biomaterial surfaces is crucial for understanding interactions with biological systems. Surface analysis requires special techniques and instruments considering the analysis of a 50 Å thick region in a 1 mm^2 area on the surface of a specimen that is 1 mm in total thickness. Devices intended to be implanted or interfaced intimately with living tissue may be composed of a variety of materials. Understanding the biological performance and efficacy of these biomaterials requires a thorough knowledge of the nature of their surfaces. The nature of the surface can be described in terms of surface chemistry, surface energy, and morphology [8-33][1].

Biomaterials and implant research is a challenging, complex, and extremely cross-disciplinary field. It involves clinical surgery and essentially all other areas of medical science, as well as biology, chemistry, physics, and the technical sciences. Claims for osseointegration, firoosteal integration, bone bonding, and bony ankylosis abound, but there is little precise knowledge of the actual interface between implant and tissue, and of the factors which influence host response and the long-term integrity of the implant system. It is well known that the surface chemistry, surface energy, and surface topography govern the biological response to an implanted material. The tissue response to a dental (or surgical) implant may involve physical factors such as size, shape, surface texture, and relative interfacial movement, as well as chemical factors associated with the composition and surface structure [8-34–8-36].

Ti material is one of the most commonly used biomaterials for dental and orthopedic applications. Its excellent tissue compatibility is mainly due to the properties of the stable oxide layer which is present on the surface. Lausmaa *et al.* [8-37] characterized the surface composition of unalloyed Ti implant materials, prepared according to procedures commonly used in clinical practice (machining, ultrasonic cleaning and sterilization). The surface of the implants is found to consist of a thin surface oxide which is covered by a carbon-dominated contamination layer. The thickness of the surface oxides is 2–6 nm, depending on the method of sterilization. The surface contamination layer is found to vary considerably from sample to sample, and consists of mainly hydrocarbons with trace amounts of calcium, nitrogen, sulfur, phosphorous, and chlorine. Some differences in surface composition between directly prepared surfaces, and some possible contamination sources, are identified [8-37].

---

[1] Suppose we have a cubic with 1 cm on each face. The ratio of molecules involved in the bulk (interior) to that of surface can be estimated as follow. Let the size of molecule be a^3 (in cm^3). The ratio = number of bulk molecules/number of surface molecules = $(1/a^3)/\{6 \times (1/a)^2\} = 1/6a$. If a = 5×10^{-8} cm or 50 nm, the ratio = $1/(6 \times 5 \times 10^{-8}) = 1/(3 \times 10^{-7})$, indicating that surface microanalysis needs at least 3×10^7 times higher sensitivity than the bulk analysis requires.

According to Sutherlan *et al.* [8-38], there are three major reasons for the apparent success of Ti implants. First, Ti, although a highly reactive metal, forms a dense, coherent passive oxide film, preventing the ingress of corrosion products into the surrounding tissue. Second, bone is in a dynamic state, continually resorbing old bone and laying down new. One implant variable in the rate of bone remodeling is the difference in elastic moduli between bone and its replacement. Normally the larger this difference, the more rapidly bone changes take place. Ti is particularly good in this regard because of its lower elastic moduli. The third factor tending to promote osseointegration is the biochemical properties of the oxide coating. The surface is also known to have at least two types of hydroxyl groups attached to it. TiO_2 is non-conducting, but electrons can tunnel through the layer. Thin oxide layers can allow the passage of electrons, leading to conformational changes and denaturing of proteins. Medical grade Ti samples were examined using XPS before and after immersion in various proteins. Additionally, an implant removed from a patient following clinical failure was examined using scanning ion and electron microscopy. The surface of the as-received samples was found to be mainly TiO_2, with contamination of H_2O/OH^-, Ca, and N, which remained after autoclaving. The immersed proteins adhered to the Ti surface, possibly via a Ca-O link. The failed clinical sample was found to be partially fibrously encapsulated with evidence of calcification. Small amounts of $TiO·OH$ were detected at the fibrous periphery [8-38], supporting the theory of Tengvall and co-workers [8-39] that *in vivo* Ti implants are covered in a gel of this material.

The chemical state of the surfaces of three types of osseointegrated Ti implants was analyzed by Sawase *et al.* [8-40]. It was reported that moderately rough surfaces show stronger bone responses than smoother or rougher surfaces. This is supported by a review done by Abrektsson *et al.* [8-41, 8-42], indicating that the majority of currently marketed implants are moderately rough. Oral implants permit bone ingrowth into minor surface irregularities – biomechanical bonding or osseointegration. Four- and seven-year implanted blade-type shape memory effect (SME) NiTi dental implants were microanalyzed to characterize the surfaces by Oshida *et al.* [8-43]. It was found that the entire surfaces of SME NiTi implants were covered with many bubbles and craters which were also found by James [8-44]. These surface irregularities have averaged sizes of 2–10 μm in diameter. Furthermore, the 3D FEM stress analysis showed that several characteristics portions, for example, the tip of the head, neck, and bending portions of the implant, where more bubbles and craters were formed than the other portions of the blades [8-43], were subject to the highest von Mises stress levels [8-45]. Hence, it is suggested that stress-assisted dissolution of nickel would take place at the interface between the NiTi implant and the living hard tissue [8-43].

Fox *et al.* [8-46] characterized the *in vitro* and *in vivo* electrochemical behavior of CpTi using a specialized osseous implant in conjunction with the electrochemical impedance spectroscopy (EIS) measurement technique. Studies performed *in vitro* were used to verify the operation of transducer and develop methods of deconvoluting EIS data. This method was subsequently used to describe an electrochemical equivalent circuit model of the surface oxide and electrical double-layer capacitance of CpTi in the endogenous electrolyte found in the medullary compartment of a baboon tibia. Kinetic profiles of the double-layer capacitance and the polarization resistance were constructed from multiple *in vitro* and *in vivo* EIS measurements performed over 60 min at 0 V (vs. Ag/AgCl) conditioning potential. The profiles demonstrated that the growth of surface oxides was biphasic, with rapid decrease in the double-layer capacitance occurring within 20 min and reaching steady-state conditions at approximately 40 min. These data suggested that a passive, stable biofilm formed on the CpTi surface *in vivo* and *in vitro* [8-46]. There is another study using EIS technique. Surfaces of CpTi and Ti-6Al-4V were subjected to simultaneous polarization/impedance testing and *in situ* electrochemical atomic force microscopy imaging to evaluate how the structure and properties of the passive oxide film is affected by varying potential and hydration. Simultaneous AFM (atomic force microscope) imaging of dry surfaces, initially hydrated surfaces, and surfaces immersed and changing with potential revealed that all sample surfaces were covered with protective Ti oxide domes that grew in area and coalesced due to hydration and as a function of increasing applied voltage and time [8-47].

In order to enhance the tissue-adhering strength onto titanium implants, various physical and chemical treatments have been proposed and conducted. However, the majority of evaluations on such bonding strengths are normally found in bone-bonding strength, tissue bonding strength, or interfacial observation on the order of several mm to mm depth. Interaction between biomaterials and living tissue is strongly influenced by the surface characteristics and surface adsorption of protein or extracellular matrices in the near-surface area, which is more important. Such interaction areas are on the order of nm scale. In order to reach this level of microanalysis, the surface plasmon resonance (SPR) method has been proposed to analyze the real-time interaction between titanium surfaces and biomolecules. The surface plasmon wave is a dense wave of electrons existing on metallic surface. Hirata [8-48] developed a unique SPR sensor accompanied with a titanium which was oxidized in air. It was reported that (i) although the titanium dioxide SPR sensor is effective in optics and industrial fields, since the biofunctionality and bioactivity of biomolecules on preoxidized metallic titanium as artificial bone or dental implant are more important, metallic titanium SPR sensor is more effective, and (ii) this SPR sensor is more useful to investigate the biocompatibility in nanometer regions.

8.4. REACTIONS IN CHEMICAL AND MECHANICAL ENVIRONMENTS

Biocompatibility of metallic materials essentially equates to corrosion resistance because it is thought that alloying elements can only enter the surrounding organic system and develop toxic effects by conversion to ions through chemical or electrochemical process. There are a number of alloys whose tolerance in implanted forms can be as good or as adequate as the ceramic and polymeric materials used in biomedical applications. After implant placement, initial healing of the bony compartment is characterized by formation of blood clots at the wound site, protein adsorption and adherence of polymorphonuclear leukocyte. Then approximately 2 days after placement of the implant, fibroblasts proliferate into the blood clot, and organization begins, and an extracellular matrix is produced. Approximately a week after the implant is placed, appearance of osteoblast-like cells and new bone is seen. New bone reaches the implant surface by osseoconduction (through growth of bone over the surface and migration of bone cells over the implant surfaces) [8-49].

In the cells of living organisms, a delicate equilibrium exists between the quantities of some metals needed in catalytic processes and the level at which these same metals become toxic. Some other metals have no part in the biological cycle, but might be the present in implants used in surgery. Chemical interaction between implants and tissue requires exchange of ions between the solid metal and the biological structure, e.g., by complex formation with proteins. There should be at least three factors: (1) Corrosion frees metal ions, and an immediate conclusion is that slow dissolution of the metal, even if it contains highly toxic elements, produces a weak interaction. (2) The ions can enter the body chemistry or react with water to form surface hydroxides, or oxides. If stable hydroxides or oxides are formed, the dissolved metal concentration will be small and again a possible interaction with the biological structure will be less likely. But solubility of these inorganic compounds can be altered in tissue fluids, and the complex formation of such compounds might even redissolve. (3) A third factor that determined a local metal concentration is related to transport, essentially by chemical diffusion. Ions and any dissolved electrolyte components created at the corroding metal surface move away in a concentration gradient, and if diffusion is slow the local concentration of such electrolyte components will become high. It should also be mentioned that this diffusion, together with an element's solubility, is necessary for systemic effects [8-49]. Corrosion is one of the major processes that cause problems when metals are used as implants in the body. Their proper application to minimize such problems requires that one has an understanding of principles underlying the important degradative process of corrosion. To have such an understanding will result in proper application, better design, choice of appropriate test methods to develop better designs, and the possibility of determining the origin of failures encountered in practice [8-50, 8-51].

All oral and maxillofacial implants are meant to support forces *in vivo*, so it is obvious that biomechanics plays a major role in implant design. Cook [8-52] and Brunski [8-53] pointed out the following biomechanical issues as the most important: (1) what are the *in vivo* loadings that dental implants must be designed to resist, (2) what factors are most important in controlling how the *in vivo* loads are transmitted to interfacial tissues, what are the stress and strain states in bone around the implant, and how can they be controlled, (3) what are safe versus dangerous levels of stress and strain in interfacial bone, and (4) what biomechanical factors contribute most to implants success or failure. It was reported that the potential for long-term implants fixation and the ability to treat younger, more active patients are part of the appeal of porous-coated implants. However, because porous implants eliminate poly(methylmethacrylate) bone cement, initial fixation of the implant is no longer guaranteed, and this may compromise the initial success and long-term stability of the device.

Biomechanics involved in implantology should include at least (1) the nature of the biting forces on the implants, (2) transferring of the biting forces to the interfacial tissues, and (3) the interfacial tissues reaction, biologically, to stress transfer conditions. Interfacial stress transfer and interfacial biology represent more difficult, interrelated problems. Hence, many engineering variables such as implant shape, elastic modulus, extent of bonding between implant and bone, etc., can affect the stress transfer conditions. The successful clinical results achieved with osseointegrated dental implants underscore the fact that such implants easily withstand considerable masticatory loads. In fact, one study showed that bite forces in patients with these implants were comparable to those in patients with natural dentitions. A critical aspect affecting the success or failure of an implant is the manner in which mechanical stresses are transferred from the implant to bone. It is essential that neither implant nor bone be stressed beyond the long-term fatigue capacity. It is also necessary to avoid any relative motion that can produce abrasion of the bone or progressive loosening of the implants. An osseointegrated implant provides a direct and relatively rigid connection of the implant to the bone. This is an advantage because it provides a durable interface without any substantial change in form or duration. There is a mismatch of the mechanical properties and mechanical impedance at the interface of Ti and bone that would be evident at ultrasonic frequencies. It is interesting to observe that from a mechanical standpoint, the shock-absorbing action would be the same if the soft layer were between the metal implant and the bone. In the natural tooth, the periodontum, which forms a shock-absorbing layer, is in this position between the tooth and jaw bone [8-54]. Natural teeth and implants have different force transmission characteristics to bone. Compressive strains were induced around natural teeth and implants as a result of static axial loading, whereas combinations of compressive and tensile strains were observed during lateral dynamic loading. Strains around the natural tooth were significantly lower than the opposing implant and occluding implants in the contralateral side for most regions under all loading conditions. There was a general tendency for increased strains around the implant opposing natural tooth under higher loads and particularly under lateral dynamic loads [8-55].

The above statement is further understood if a natural dentition and implant system are compared, as seen in Figure 8-1. It can be pointed out that the most distinct difference between these two is the fact that the natural dentition has a periodontal membrane, functioning as a mechanical shock absorber, as well as a solid bonding between tooth root surface and surrounding tissue.

The interface mechanical characteristics and histology of CpTi and HA-coated CpTi were studied by Cook *et al.* [8-55]. It was reported that (i) histologic evaluation in all cases revealed mineralization of interface bone directly onto the HA-coated implant surface, becoming part of the implant composite system, and (ii) the uncoated implants had a thin fibrous inter-positional layer present in most areas, with projections of bone in apposition to the implant surface in a limited number of locations, indicating that the HA-coated Ti system may be attractive for use in endosseous dental implant applications [8-56]. Masticatory forces acting on dental implants can result in undesirable stress in adjacent bone, which in turn can cause defects and the eventual failure of implants. It was reported that (i) the maximum stress areas were located around the implant neck, (ii) the decrease in stress was the greatest (31.5%) for implants with a diameter ranging from 3.6 to 4.2 mm, and

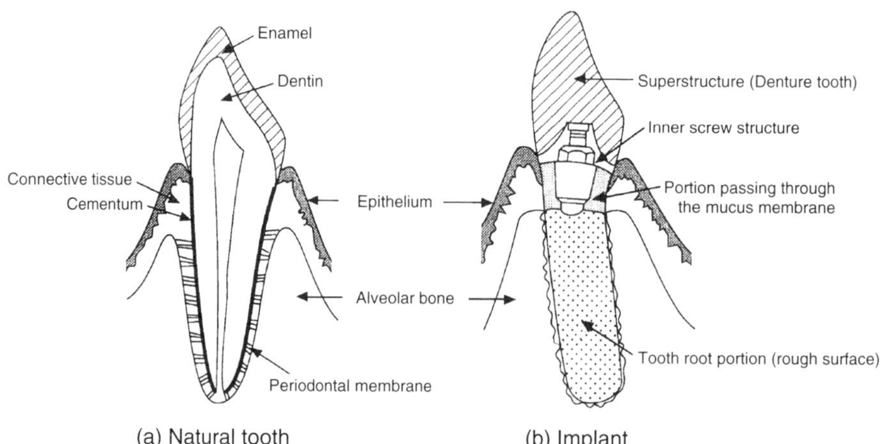

Figure 8-1. Schematic comparison between natural tooth and implant tooth.

(iii) an increase in the implant length also led to a decrease in the maximum von Mises equivalent stress values; the influence of implant length, however, was not as pronounced as that of implant diameter. It was further mentioned that an increase in the implant diameter decreased the maximum von Mises equivalent stress around the implant neck more than an increase in the implant length, as a result of a more favorable distribution of the simulated masticatory forces applied [8-57]. At present, load-bearing implants are designed on a deterministic basis in which the structural strength and applied loading are given fixed values, and global safety factors are applied to cover any uncertainties in these quantities, and to design against failure of the component [8-58]. This approach will become increasingly inappropriate as younger and more active patients demand more exactness, and as devices become more complex. Browne *et al.* [8-58] described a preliminary investigation in which a scientific and probabilistic technique is applied to assess the structural integrity of the knee tibial tray. It was envisaged that by applying such a technique to other load-bearing biomedical devices, reliability theory may aid in future lifting procedures and materials/design optimization.

Although one of the most common procedures performed in implant dentistry is a single tooth replacement, some cases offer two implants for three crowns – three unit implant. The influence of implant number and cantilever design on stress distribution on bone has not been sufficiently assessed for the mandibular overdenture. Sadowsky and Caputo [8-59] reported that while all four prostheses demonstrated low stress transfer to the implant, the plunger-retained prosthesis caused more uniform stress distribution to the ipsilateral terminal abutment compared to the slip-retained prosthesis, and provided retention security under tested loads. The plunger-retained prosthesis retained by two implants provided better load sharing from the ipsilateral edentulous ridge than the clip-retained prosthesis retained by three implants, and lower resultant stresses were seen on the implants [8-59].

By means of finite element analysis method (FEM), stress-distribution in bone around implants was calculated with and without stress-absorbing element [8-60]. A freestanding implant (i.e., single unit) and an implant connected with a natural tooth were simulated. It was reported that (i) for the freestanding implant, the variation in the MOE of the stress-absorbing element had no effect on the stresses in bone; changing the shape of the stress-absorbing element had little effect on the stresses in cortical bone, and (ii) for the implant connected with a natural tooth, a more uniform stress was obtained around the implant with a low MOE of the stress-absorbing element. It was also found that the bone surrounding the natural tooth showed a decrease in the height of the peak stresses [8-60]. Mechanical *in vitro* tests of the Brånemark implant indicated that the screw joint which attaches the prosthetic gold cylinder and the transmucosal abutment to the fixture forms a flexible system. Calculations of vertical load distribution based on measured flexibility data demonstrated that the forces are shared almost equally between the tooth and the implant, even without taking the flexibility of the surrounding bone or the prosthesis into account. The therapy of a single Brånemark implant connected to a natural tooth should be considered without any additional element of a flexible nature. Mechanical tests and theoretical considerations, however, indicate that the transverse mobility of the connected tooth should be limited and that the attachment of the prosthesis to the tooth should be a rigid design to avoid gold-screw loosening [8-61]. The stress distribution pattern clearly demonstrated a transfer of preload force from the screw to the implant during tightening. A preload of 75% of the yield strength of the abutment screw was not established using the recommended tightening torques. Using FEM, a torque of 32 N cm applied to the abutment screws in the implant assemblies was studied in the presence of a coefficient of friction of 0.26, and resulted in a lower than optimum preload for the abutment screws. It was then mentioned that in order to reach the desired preload of 75% of the yield strength, using the 32 N cm torque applied to the abutment screws in the implant assemblies studied, the coefficient of friction between the implant components should be 0.12 [8-62].

The fractured surface of a retrieved Ti screw and metallurgical structures of a dental implant system were analyzed by Yokoyama *et al.* [8-63]. The outer surface of the retrieved screw had a structure different from that of the as-received screw. It was confirmed that a shear crack initiated at the root of the thread and propagated into the inner section of the screw. Gas chromatography revealed that the retrieved screw had absorbed a higher amount of hydrogen than the as-received sample, suggesting that the fracture is, to some extent, related to the hydrogen embrittlement. The grain structure of a Ti screw, immersed in a solution known to induce hydrogen absorption showed features similar to those of the retrieved screw. It was concluded that Ti in a biological environment absorbs hydrogen and this may be the reason for delayed fracture of a Ti implant [8-63]. It is known that the high-cycle fatigue strength of porous-coated Ti-6Al-4V is approximately 75% less than the fatigue strength of uncoated Ti-6Al-4V [8-64]. It was mentioned that (i) the fatigue strength of smooth-surfaced Ti-6Al-4V subjected to hydrogen treatments is 643–669 MPa, significantly greater than that of beta-annealed Ti-6Al-4V (497 MPa), and also greater than that of preannealed, equiaxed Ti-6Al-4V (590 MPa), and (ii) the fatigue strength of porous coated Ti-6Al-4V, however, is independent of microstructure, suggesting that the notch effect of the surface porosity does not allow the material to take advantage of the superior fatigue crack initiation resistance of refined alpha-grain size [8-64]. Bone tissue ingrowth into porous-metal-coated implants is often the preferred method of prosthetic fixation in younger and more active patients [8-52, 8-53]. The short-term clinical results of porous-coated implants have been good. The long-term success of any implants is determined, in part, by the ability of the material to withstand repetitive loading. Kohan *et al.* [8-65] utilized acoustic emission (AE) events and event intensities (e.g., event amplitude, counts, duration, and energy counts) to analyze the fatigue process of uncoated and porous-coated Ti-6Al-4V. AE provides the ability to spatially and temporally locate multiple fatigue cracks in real time. Fatigue of porous-coated Ti-6Al-4V is governed by a sequential, multimode fracture process of: transverse fracture in the porous coating; sphere/sphere and sphere/substrate debonding; substrate fatigue crack initiation; and slow and rapid substrate fatigue crack propagation. Because of the porosity of the coating, the different stages of fracture within the coating occur in a discontinuous manner. Changes in the AE event rate also correspond to changes in crack-extension rate, and may therefore be used to predict failure. It was reported that (i) intergranular fracture and microvoid coalescence generated the highest AE event amplitudes (100 dB), whereas plastic flow and friction generated the lowest AE event amplitudes (55–65 dB), and (ii) fractures in the porous coating were characterized by AE event amplitudes of less than 80 dB [8-65].

8.5. REACTION IN BIOLOGICAL ENVIRONMENT

A goal of biomaterials research has been, and continues to be, the development of implant materials which are predictable, controlled, guided, with rapid healing of the interfacial hard and soft tissues. The performance of biomaterials can be classified in terms of the response of the host to the implant, and the behavior of the material in the host. This is actually related to which side we are looking at the host (vital tissue)/foreign materials (implant) interface. The event that occurs almost immediately upon implantation of metals, as with other biomaterials, is adsorption of proteins. These proteins come first from blood and tissue fluids at the wound site and later from cellular activity in the interfacial region. Once on the surface, proteins can desorb (undenatured or denatured, intact or fragment), remain, or mediate tissue–implant interaction [8-66]. The host response to implants placed in bone involves a series of cell and matrix events, ideally culminating in tissue healing that is as normal as possible and that ultimately leads to intimate apposition of bone to the biomaterials, namely the definition of osseointegration. For this intimate contact to occur, gaps that initially exist between the bone and implant at surgery must be filled initially by a blood clot, and bone damaged during preparation of the implant site must be repaired. During this time, unfavorable conditions, e.g., micromotion (a biomechanical factor) will disrupt the newly forming tissue, leading to formation a fibrous capsule [8-7]. The criteria for clinical success of osseointegration are based on functionality and compatibility, which, depend on the control of several factors including: (1) biocompatible implant material using commercially pure titanium, (2) design of the fixture; a threaded design is advocated creating a larger surface per unit volume as well as evenly distribution loading forces, (3) the provision of optimal prosthodontic design and implant maintenance to achieve ongoing osseointegration, (4) specific aseptic surgical techniques and a subsequent healing protocol which are reconcilable with the principles of bone physiology, which would incorporate a low heat/trauma regimen, a precise fit and the two-stage surgery program, (5) a favorable status of host-implant site from a health and morphologic standpoint, (6) non-loading of the implant during healing is a basic tenet of osseointegration, and (7) the defined macro-microscopic surface of the implant as it relates to the host tissue [8-67]. The implantation of any foreign material in soft tissue initiates an inflammatory response. The cellular intensity and duration of the response is controlled by a variety of mediators and determined by the size and nature of the implanted material, site of implantation and reactive capacity of the host [8-68].

Dental implants vary markedly in the topography of the surfaces that contact cells. According to Brunette [8-69], there are four principles of cell behavior observed in cell culture to explain to some extent the interactions of cells and implants. (1) Contact guidance aligns cells and collagen fibers with fine grooves (known as machine tool-marks). (2) Rugophilia describes the tendency of macrophages to prefer rough surfaces. (3) Two-center effect can explain the orientation of soft connective tissue cells and fibers attached to porous surfaces. (4) Haptotaxis may be involved in the formation of capsules around implants with low-energy surfaces [8-69]. Surface roughness has been shown to be an influencing parameter for cell response. Bigerele *et al.* [8-70] compared the effect of roughness organization of Ti-6Al-4V or CpTi on human osteoblast response (proliferation and adhesion). Surface roughness is extensively analyzed at scales above the cell size (macroroughness) or below the cell size (microroughness) by calculation of relevant classic amplitude parameters and original frequency parameters. It was found that (i) the human osteoblast response on electro-erosion Ti-6Al-4V surfaces or CpTi surface was largely increased when compared to polished or machine-tooled surfaces after 21 days or culture, and (ii) the polygonal morphology of human osteoblast on these electro-erosion surfaces was very close to the aspects of human osteoblast *in vivo* on human-bone trabeculae. On the basis of these findings, it was concluded that electro-erosion technique for creating a rough surface is a promising method for preparation of bone implant surfaces, as it could be applied to the preparation of most biomaterials with complex geometries [8-70].

Ti oxide films were synthesized on Ti, Co alloy, and low-temperature isotropic pryolytic carbon by the ion beam-enhanced deposition technique [8-71], by which the amorphous non-stoichiometrical Ti oxide films (TiO_{2-x}) were obtained. Blood compatibility of the films was evaluated by clotting time measurement, platelet adhesion investigation, and hemoplysis analysis. It was found that (i) the blood compatibility of the material was improved by the coating of Ti oxide films, (ii) the non-stoichiometric TiO_{2-x} has n-type semiconductive properties because very few cavities exist in the valence band of TiO_{2-x}; charge transfer is difficult from the valence band of fibrinogen into the material, but (iii) on the other hand, the n-type semiconductive TiO_{2-x} with a higher Fermi level can decrease the work function of the film, which makes electrons move out from the film easily. As a result, it was concluded that the deposition of fibrinogen can be inhibited and blood compatibility improved [8-71].

Favorable wound-healing responses around metallic implants depend on critical control of the surgical and restorative approaches used in dental implant treatments. One critical parameter that has not been biologically studied is the role of a clean, sterile oxide surface on an implant. This oxide surface can alter the cellular healing responses, and potentially the bone remodeling process, depending on the history of how that surface was milled, cleaned, and sterilized prior to placement. On the basis of this background, Stanford *et al.* [8-72] evaluated the phenotypic responses of rat calvarial osetoblast-like cells on CpTi surfaces. These surfaces were prepared to three different clinically relevant surface preparations (1 um, 600 grit, and 50 μm grit sand-blasting), followed by sterilization with either ultraviolet light, ethylene oxide, argon plasma cleaning, or routine clinical autoclaving. It was found that (i) osteocalcin and alkaline phosphatase, but not collagen expression, were significantly affected by surface roughness when these surfaces were altered by argon plasma cleaning, and (ii) on a per-cell basis, levels of the bone-specific protein, osteocalcin, and enzymatic activity of alkaline phosphatase were highest on the smooth 1 um polished surface, and lowest on the roughest surface for the plasma-cleaned CpTi [8-72].

Carlsson *et al.* [8-73] investigated the glow-discharged Ti implants, with a presumed high surface energy, and conventionally prepared and sterilized CpTi implants which were inserted in the rabbit tibia and femur. The removal torque and histology were compared after 6 weeks *in situ*. It was reported that (i) no qualitative or quantitative differences were detected for implants with different preoperative preparation, and (ii) the conventional implant treatment described is sufficient to give a surface condition with similar early healing response as those observed with glow-discharge-treated implants. Buser *et al.* [8-74] treated CpTi surface by sand-blasting, acid-treatment in HCl/H_2SO_4, and the HA-coating. It was reported that (i) rough implant surfaces generally demonstrated an increase in bone apposition compared to polished or fine-structured surfaces, (ii) the acid-treated CpTi implants had an additional stimulating influence on bone apposition, (iii) the HA-coated implants showed the highest extent of bone–implant interface, and (iv) the HA-coating consistently revealed signs of resorption [8-74].

Generally, roughened surfaces have been used as the endosseous area of a dental implant in order to increase the effective total area available for osseo-apposition. However, there is still considerable controversy concerning the optimal surface geometry and physicochemical properties for the ideal endosseous portion of a dental implant. Knabe *et al.* [8-75] used rat bone marrow cells to evaluate different Ti and HA dental implant surfaces. The implant surfaces were a Ti surface having a porous Ti plasma-sprayed coating, a Ti surface with a deep profile structure, an uncoated Ti substrate with a machined surface, and a machined Ti substrate with a porous HA plasma-sprayed coating. Rat bone marrow cells were cultured on the disk-shaped test substrates for 14 days. The culture medium was changed daily and examined for Ca and P concentration. It was reported that (i) all tested substrates facilitated rat bone marrow cell growth of extracellular matrix formation, (ii) Ti surfaces with a deep profile structure and with porous Ti plasma-sprayed coating to the highest degree, followed by machined Ti and Ti with porous HA plasma-sprayed coating, (iii) Ti surfaces with a deep profile structure and with porous Ti plasma-sprayed coating displayed the highest cell density, and thus seems to be well suited for the endosseous portion of dental implants, and (iv) the rat bone marrow cells cultured on Ti with porous HA plasma-sprayed coating showed a delayed growth pattern due to high phosphate ion release [8-75].

The modern range of medical devices presents contrasting requirements for adhesion in biological environments. For artificial blood vessels, the minimum adhesion of blood is mandatory, whereas the maximum blood cell adhesion is required at placed implant surface. Strong bio-adhesion is desired in many circumstances to assure device retention and immobility. Minimal adhesion is absolutely essential in others, where thrombosis or bacteria adhesion would destroy the utility of the implants. In every case, primary attention must be given to the qualities of the first interfacial conditioning films of bio-macromolecules deposited from the living systems. For instance, fibrinogen deposits from blood may assume different configurations on surfaces of different initial energies, and thus trigger different physiological events. Standard surface modification techniques, such as siliconization, when properly quality controlled, can yield improved blood-compatible devices like substitute blood vessels and artificial heart sacs [8-76, 8-77].

Sunny *et al.* [8-78] showed that the Ti oxide film on Ti affects the adsorption rate of albumin/fibrinogen significantly. Multinucleated giant cells have been observed at interfaces between bone marrow and Ti implants in mouse femurs, suggesting that macrophage-derived factors might perturb local lymphomopoiesis, possibly even predisposing to neoplasia in the B lymphocyte lineage. It has been found that (i) an implant–marrow interface with associated giant cells persists for at least 15 years, (ii) precursor B cells show early increase in number and proliferative activity; however (iii) at later intervals they do not differ significantly from controls. Rahal *et al.* [8-79] mentioned, in mice study, that following initial marrow regeneration and fluctuating precursor B cell activity, and despite the presence of giant cells, Ti implants apparently become well-tolerated by directly apposed bone marrow cells in a lasting state of the so-called myelointegration.

Sukenik *et al.* [8-80] modified the surface of Ti by covalent attachment of an organic monolayers anchored by a siloxane network. This coating completely covers the metal and allows controlled modifications of surface properties by the exposed chemical end-groups of the monolayer forming surfactant. The attachment of such a film allows different bulk materials (e.g., glass and Ti) to have identical surface properties, and this can be used in regulating cell adhesion responses. It was found that (i) this control over surface functionality can modulate the functions of fibronectin in regulating attachment and neurite formation by neuronal cells, and (ii) the effect on bacteria adherence that is achieved by using such monolayers to vary surface hydrophilicity is also assessed [8-80].

Cell adhesion is involved in various natural phenomena such as embryogenesis, maintenance of tissue structure, wound healing, immune response, and metastasis, as well as tissue integration of biomaterial [8-81]. The biocompatibility of biomaterials is very closely related to cell behavior on contact with them, and particularly to cell adhesion to their surface. Surface characteristics of materials (whether their topography, chemistry or surface energy) plays an essential part in osteoblast adhesion on biomaterials. Thus attachment, adhesion and spreading belong to the first phase of cell/material interactions, and the quality of this first phase will influence the cell's capacity to proliferate and to differentiate itself on contact with the implant. It is essential for the efficacy of orthopedic or dental implants to establish a mechanically solid interface with complete fusion between the materials' surface and the bone tissue with no fibrous interface, as mentioned previously. Moreover, the recent development of tissue engineering in the field of orthopedic research makes it possible to envisage the association of autologous cells and/or proteins that promote cell adhesion with osteoconductive material to create osteoinductive materials or hybrid materials. Thus, a complete understanding of the cell adhesion, and particularly osteoblast adhesion on materials is now essential to optimize the bone/biomaterial interface at the heart of these hybrid materials [8-81].

Dmytryk *et al.* [8-82] examined the ability of tissue culture fibroblasts to attach and colonize on the surface of CpTi dental implants following instrumentation of the implant surface with curettes of dissimilar composition. CpTi dental implants were scaled with a plastic, Ti-6Al-4V alloy, or stainless-steel curette and then immersed in a cell suspension of 3T3 fibroblasts. Counts of attached cells were made at 24 and 72 h; the implants were then processed for SEM. At 24 h, only surfaces scaled with a stainless-steel curette showed a significant reduction in number of attached cells relative to untreated control surfaces. It was found that, at 72 h, both stainless steel and Ti-6Al-4V alloy curette instrument surfaces showed significantly fewer attached cells than untreated control surfaces, with the greatest reduction in cell attachment observed on stainless-steel curette instrument surfaces [8-82]. Ryhänen *et al.* [8-83] evaluated the new bone formation, modeling, and cell–material interface responses induced by the NiTi SME alloy after periosteal implantation. The regional acceleratory phenomenon (RAP) model was employed, in which a periosteal contact stimulus provokes an adaptive modeling response. The test implant was placed in contact with the intact femur periosteum, but it was not fixed inside the bone. It was found that (i) a typical peri-implant bone wall modulation was seen due to the normal RAP, (ii) the maximum new woven bone formation started earlier (2 weeks) in the Ti-6Al-4V group than the NiTi group, but also decreased earlier, and at 8 weeks the NiTi and stainless steel groups had greater cortical bone width, and (iii) at 12 and 26 weeks no statistical significances were seen in the histomorphometric values. On the basis of these findings, it was concluded that NiTi had no negative effect on total new bone formation or normal RAP after periosteal implantation during a 26-week follow-up [8-83]. The bone formation around CpTi implants with varied surface properties was tested by Larsson *et al.* [8-84]. Machined and electropolished CpTi samples with and without anodically formed surface oxides were prepared and inserted into the cortical bone of rabbits (1, 3, and 6 weeks). It was reported that (i) the smooth, electropolished implants, irrespective of anodic oxidation, were surrounded by less bone than the machined implants after 1 week, (ii) after 6 weeks, the bone volume as well as the bone–implant contact, was lower for the merely electropolished implants than for other three groups; however (iii) the result with the smooth (electropolished) implants indicates that a reduction of surface roughness, in the initial phase, decreases the rate of bone formation in rabbit cortical bone [8-84]. Larsson *et al.* [8-85], in another experimental study, analyzed the bone formation around systematically modified CpTi implants which were machined, electropolished, and anodically oxidized, and were inserted in the cortical bone of rabbits (7 and 12 weeks). It was found that (i) light microscopy revealed that all implants were in contact with bone, and a large proportion of bone within the threads, and (ii) the smooth, electropolished implants were surrounded by less bone than the machined implants with similar oxide thickness (4–5 nm) and the anodically oxidized implants with 21–180 nm after 7 weeks. It was further reported that a high degree of bone contact and bone formation could be achieved with CpTi implants which are modified with respect to oxide thickness and surface topography; however, it appears that a reduction of surface roughness may influence the rate of bone formation in rabbit cortical bone [8-85]. Since interactions among cell/matrix/substrate, which are associated with cell signaling, occur in the nanoscale dimension, Oliveira and Nanci [8-86] evaluated the influence of nanotexturing of Ti-based surfaces (CpTi and Ti-6Al-4V, treated by H_2SO_4 and 30% H_2O_2) on the expression of matrix proteins by cultures osteogenic cells at initial time points. It was reported that (i) after 6 h, nanotextured surfaces exhibited up to a nine-fold increase in the proportion of cells with peripheral osteopontin labeling, (ii) at day 3, the proportion of osteopontin and bone sialoprotein labeled cells was higher, and the intensity of immunoreactivity dramatically increased, and (iii) no significant differences were observed in the expression pattern and the proportion of cells immunoreactive for fironectin [8-86].

Biological performance of dental and orthopedic implants depends on factors such as composition, design, surface topography, sterilization, and surgical technique. Common dental implants are made of different grades of CpTi. Ahmad *et al.* [8-87] conducted *in vitro* tests to determine if human osteoblast-like cells, Saos-2, would respond differently when plates on disks of CpTi grades 1 and 4, with glass disks serving as controls. It was reported that (i) rhodamine phalloidin fluorescence microscopy showed variations in the actin-based cytoskeleton between grades 1 and 4 CpTi in cell spreading, shape, and the organization of stress fibers, (ii) immuno-fluorescent staining showed differential expression of vinculin, a focal adhesion protein, on the substrates, (iii) at 24 h, the percent of collagen synthesized was significantly more on grade 1 than on grade 4 and on glass, and (iv) the calcium content was significantly higher on grade 1 than on grade 4 and on glass at 24 h and at 4 weeks. It was, thus, concluded that commonly used CpTi induced differential morphologic and phenotypic changes in human osteoblast-like cells depending on the grade of the material [8-87]. The spreading and orientation of human gingival fibroblasts in relation to the rim of smooth-surfaced and porous-coated Ti disks was *in vitro* investigated. It was found that (i) the cells migrated from the multilayer onto the smooth-surfaced disks forming bridges between them, and orientated along parallel circumferential grooves in the rim of the disks, and (ii) cellular bridges were also formed between the porous-coated disks and the multilayer, but because the cells that migrated onto and between the sphere of the porous-coated showed no preferred orientation, the bridges retained their orientation at right angles to surface of the rim. This suggests that the geometrical configuration of the surface of the implants could influence whether a capsule or an orientated fibrous attachment is developed in relation to implants *in vivo* [8-88].

Proliferation and adhesion of mouse osteoblastic cells and primary osteoblastic cells were carried out on Ti-6Al-4V samples with varied surface roughness [8-89]. Mechanically or manually polished surfaces were prepared to produce, respectively, non-orientated or orientated polishing grooves. Sand-blasted surfaces were prepared using 500 μm or 3 mm alumina particles. It was found that (i) surface roughness parameters showed a negative correlation in comparison to proliferation and adhesion parameters, and (ii) X-ray microprobe chemical microanalysis showed complete disturbance of the surface element composition of the Ti-6Al-4V following sand-blasting treatment, on which an AlO_x-enriched layer was observed. This may lead to the suspicion that the concomitant effect of surface roughness amplitude and AlO_x surface concentration has an effect on osteoblastic cell proliferation and adhesion. It was demonstrated that usual surface preparation for increasing the bone integration of Ti-6Al-4V surgical implants (i.e., sand-blasting) may extensively transform their chemical surface composition. Therefore, the expected bone integration of implants may be extensively impaired by surface treatments [8-89]. The possible contamination by used blasting media on Ti surfaces will be discussed in Chapter 11. Mustafa *et al.* [8-90] analyzed variations in the oxide films on CpTi surfaces blasted with TiO_2 particles of various sizes after cultures with cells derived from human mandibular bone. Turned Ti surfaces, and sand-blasted with 63–90, 106–180, and 180–300 μm TiO_2 particles were cultured with osteoblast-like cells. The surfaces were characterized before and after the cell culture with electrochemical impedance spectroscopy (EIS). It was found that (i) EIS revealed, with respect to the turned surfaces, the effective surface area was about 5, 6, and 4 times larger on the surfaces blasted with 63–90, 106–180, and 180–300 μm particles, respectively, (ii) after 28 days of the cell culture, the corrosion resistance on all sample types was unaffected; the impedance characteristics suggest a considerable effect of ion incorporation and precipitation during culturing, (iii) XPS revealed that before the cell culture, a typical surface layer consisted of TiO_2, and (iv) after the culture, the surface oxide film contained both P and Ca, along with large amounts of oxidized carbon (carbonate) and nitrogen [8-90]. Lee *et al.* [8-91] examined the cell attachment and proliferation of neonatal rat calvarial osteoblasts on Ti-6Al-4V as affected by the surface modifications (as polished, heated in vacuum at 970°C for 8 h to have 300 μm thick brazed, at 970°C for 8 h to have 600 μm brazed). During the brazing, a foil of Ti-15Cu-15Ni with melting point of 934°C was placed top of the Ti-6Al-4V. All samples were passivated by boiling them in distilled water for 24 h, followed by autoclaving at 121°C for 15 min. The modifications could alter simultaneously the surface chemistries of the alloy, as well as the surface characteristics of oxides on the metal. It was found that a slight change of the surface characteristics of oxides as a result of the modifications neither alters the *in vitro* capacity of Ca and P ion adsorption nor affects the metal ion dissolution behavior of the alloy, suggesting that any influence on the cytocompatibilities of the materials should only be correlated to the effect of surface chemistries of the alloy and the associated metal ion dissolution behavior of the alloy. It was therefore concluded that the cell response of neonatal rat calvarial osteoblasts on the Ti-6Al-4V alloy should not be affected by the variation of surface chemistries of the alloy in a range studied [8-91]. To investigate the roles of compositions and properties of Ti surface oxides in cellular behavior of osteoblasts, the surface oxides of Ti were modified in compositions and topography by anodic oxidation in two kinds of electrolytes, 0.1 M H_3PO_4 and a mixture of 0.03 M calcium glycerophosphate and 0.15 M calcium acetate. It was found that (i) phosphorous (P: ~ 10 at%) or both Ca (1–6 at%) and P (3–6 at%) were incorporated into the anodized surfaces in the form of phosphate and calcium phosphate, and (ii) contact angles of all the anodized oxides were in the range of 60–90°, indicating relatively high hydrophobicity. It was also mentioned that cell culture experiments demonstrated absence of cytotoxicity and an increase of osteoblast adhesion and proliferation by the anodic oxides [8-92].

Bony ingrowth to control (non-treated) and heat-treated (autoclaved at 280°C for 3 h) stainless-steel and Ti-6Al-4V implants into the medullary canal of the femur in rats was studied by Hazan *et al.* [8-93]. It was found that (i) at all time intervals up to 3 days post-implantation, the shear strengths of heat-treated Ti-6Al-4V and stainless-steel implants were significantly higher (1.6- and 3.4-fold) than the controls, and (ii) the heat treatment of metal implants prior to their insertion alters their chemical surface properties and augments bony ingrowth to them [8-93]. Characteristics of the porous surfaces may be important in improving the bone ingrowth into the porous coatings. Quicker and more mature interstitial bone formation was obtained using a porous rather than a solid structure, due to differences of penetration by some growth factors and bone marrow cells [8-94–8-96]. Li *et al.* [8-97] examined the effect of the surface macrostructure of a dimpled CpTi implant on bone ingrowth *in vivo* by means of histological examination and a push-out test. Dimples had diameters of 100, 140, and 160 μm, and distance between dimple centers of about 400 μm. Cylindrical implants were inserted in rabbits for 1.5, 3, and 13 months. The femur with the implant of each animal was then examined in a push-out test. It was found that the dimpled CpTi surface results in an increased retention of the implant in bone due to interlocking between vital bone and the dimples [8-97].

NiTi SME alloy makes it possible to prepare functional implants that apply a continuous benign force to the bone. Kujala *et al.* [8-98] investigated if bone modeling can be controlled with a functional intramedullary NiTi nail. Preshaped NiTi nail with a curvature radius of 25–37 mm were implanted in the cooled martensite phase form in the medullary cavity of the right femur in 8 rats, where they restored their austenite form, causing a bending force. After 12 weeks, the operated femurs were compared with their non-operated contralateral counterpairs. It was found that (i) quantitative densitometry showed a significant increase in the average cross-sectional cortical area and cortical thickness, which were most obvious in the mid-diaphyseal area, (ii) cortical bone mineral density increased in the proximal part of the bone and decreased in the distal part, and (iii) polarized light microscopy of the histological samples revealed that the new bone induced by the functional intramedullary nail was mainly woven bone. It was therefore concluded that bone modeling can be controlled with a functional intramedullary nail made of NiTi SME alloy [8-98]. The biocompatibility of NiTi as a potential implant material was investigated through *in vivo* studies on beagles. A high-purity alloy was fabricated into prototype bone plates and implanted into the femurs of beagles. Commercial Co-Cr alloy bone plates served as reference controls. The bone plates were removed from the animals and examined after exposures of 3, 6, 12, and 17 months. It was reported that (i) there was no evidence of localized or generalized corrosion on the surfaces of the bone plates and screws, (ii) gross clinical, radiological, and morphological observations of the tissue at the implantation sites during the autopsies uncovered no signs of adverse tissue reactions resulting from the implants, and (iii) a histological analysis was performed on samples of muscle and bone adjacent to the implantation sites and of tissues removed from such organs as the liver, spleen, brain, and kidneys. It was further found that there was no metallic contamination in the organs due to the implants; however, there does appear to be some chromium contamination from the Co-Cr alloy implants in the adjacent bone, and concluding that NiTi alloy is sufficiently compatible with dog tissue to warrant further investigation of its potential as a biomaterial [8-99].

Nishiguchi *et al.* [8-100] evaluated the bone-bonding ability of three differently treated samples of CpTi: (1) as a smooth surface control, (2) treated in 5 M NaOH solution for 24 h at 60°C, and (3) plus heated at 600°C for 1 h. The plates were inserted transcortically into the proximal metaphyses of bilateral rabbit tibiae. The tensile failure loads between implants and bones were measured at two intervals using a detaching test. It was reported that (i) the tensile failure loads of the alkali- and alkali-heat-treated group were 27 and 40 MPa, at 8 and 16 weeks, and significantly higher than those of the other Ti groups, (ii) histological examination revealed that alkali- and heat-treated Ti was in direct contact with bone, but the other Ti groups had a thin intervening fibrous tissue, and (iii) the alkali-treated Ti without heat treatment had no bone-bonding ability due to the unstable reactive surface layer of alkali-treated Ti. It was therefore concluded that both alkali and heat treatment are essential for preparing bioactive Ti and this bioactive Ti is thought to be useful for orthopedic implants with cementless fixation [8-100].

There is an increasing attention given to the influence of surface condition, and in particular on methods to control and improve titanium implant surfaces; such techniques should include laser surface treatment [8-101], strain hardening [8-102] and chemical passivation methods [8-103]. From histological examinations, it has been found that implant loosening is generally associated with the formation of fibrous tissue at the bone–implant interface. To improve implant biocompatibility and osseointegration, Ti-6Al-4V alloy was treated by passivation by immersion in 30% HNO_3 for 1 h, or passivation in boiling distilled water for 10 h [8-104], resulting in an increase in the oxide layer thickness of Ti-6Al-4V alloy as orthopedic implant material. The kinetics of gene expression over 120 h was followed for 58 genes to quantify the effect of he developed surface treatment. Twenty eight genes were further selected to compare the effects of surface treatments on osteoblasts. On the basis of the genes studies, a general pathway for the cell reaction was proposed according to the surface treatment used: (1) metal ion release changes the time course of gene expression in the focal adhesion kinase pathway; (2) once the accumulation of metal ions released from the Ti surface exceeds a threshold value, cell growth is diminished and apoptosis may be activated; (3) protein tyrosine kinase upregulation is also induced by metal ion release; (4) expression of the Bcl-2 family and Bax may suggest that metal ions induce apoptosis [8-104].

The cleanliness of Ti dental implant surfaces is considered to be important for achieving osseointegration, and it has been hypothesized that the presence of inorganic contaminants could lead to a lack of clinical success. As Anselme *et al.* [8-89] pointed out that aluminum ions are suspected of impairing bone formation by a possible competitive action to calcium. Piattelli *et al.* [8-105] investigated the effects of residual Al_2O_3 particles on the implant surface on the integration of Ti dental implants as compared to decontaminated implants in a rabbit experimental model. Threaded screw-shaped machined grade 3 CpTi dental implants were used. The implants were sand-blasted with 100–120 μm Al_2O_3 particles at 1.5 atm pressure for 1 min, then 24 implants (control implants) underwent the ASTM 86-68 decontamination process in an ultrasonic bath. The other 24 implants (test implants) were washed in saline solution for 15 min. Both test and control implants were air-dried and sterilized at $120°C$ for 30 min. After sterilization, the implants were inserted into the tibiae of New Zealand white mature rabbits. All animals were euthanized after 4 weeks. A total of 48 implants were retrieved. It was found that (i) percentage of the implant surface covered by inorganic particles was significantly higher than that of control implants, (ii) no statistically significant differences were found in the bone-implant contact percentage of test and control implants, and (iii) no differences were found in the number of multinucleated cells and osteoblasts in contact with the implant surface. Hence, it was concluded that histological results do not provide evidence to support the hypothesis that the residual aluminum oxide particles on the implant surface could affect the osseointegration of Ti dental implants [8-105].

Primary stability and an optimized load transfer are assumed to account for an undistributed osseointegration process of implants. Immediate-loading type newly designed Ti dental implants inserted in the mandible of minipigs, were used for the characterization of the interface area between the implant surface and the surrounding bone tissue during the early healing phase. Histological and SEM studies were done. Two different load regimes were applied to test the load-related tissue reaction. It was shown that (i) a direct bone apposition on the implant surfaces, as well as the attachment of cells and matrix protein in the early loading phase, and (ii) a striking finding of the ultrastructural immunocytochemical investigations was the synthesis and deposition of bone proteins (osteonection, fibronectin, fibronectin receptor) by osteoblasts from day one of bone/biomaterials interaction. It was therefore suggested that loadings of specially designed implants can be performed immediately after insertion without disturbing the biological osseointegration process [8-106]. Apatite formation on implants is important in achieving a direct bonding to bone tissue. It was shown that a Ti surface which was chemically treated with a H_2O_2 solution containing tantalum chloride had the ability to form a HA layer in simulated body fluid (SBF) that had an inorganic ion composition similar to human blood plasma [8-107]. A CpTi cylinder treated with a H_2O_2 solution with tantalum chloride was inserted into a hole in a rabbit's tibia for 16 weeks. It was found that (i) 8 weeks after surgery, the shearing force of the treated Ti implanted in the 4.2 mm hole was significantly higher than that of the non-treated Ti, although the surface roughness was not changed after the treatments, and (ii) the shearing force was higher for the implanted sample in the 4.0 mm hole than that in the 4.2 mm hole [8-107].

REFERENCES

[8-1] Brown D. Dental implants. Mater World 1996;4:259–261.

[8-2] Brockhurst PJ. Dental materials: new territories for materials science. In: Metals forum. Australian Inst Metals 1980;3:200–210.

[8-3] Binon PP. Implants and components: entering the new millennium. Int J Oral Maxillofac Implants 2000;15:76–94.

[8-4] Carr D, Oshida Y, Platt, JA, Barco MT, Hovijitra S, Andres CJ. Effects of pre-fatigue damage on remaining cement bond strength in a cement-retained Cera-One implant system. In: Medical Device Materials. Proceedings of materials and processes for medical devices conference Shrivastava S, editor. Material Park OH: ASM International. 2003. pp. 463–466.

[8-5] Carr D, Oshida Y, Platt JA, Barco MT, Andres CJ. The effect of fatigue damage on the force required to remove a restoration in cement-retained implant system. J Prosthodont 2006, September issue, to be published.

[8-6] Rostoker W, Galante JO. The influence of titanium surface treatments on the wear of medical grade polyethylene. Biomaterials 1981;2:221–224.

[8-7] Bannon BP, Mild EE. Titanium alloys for biomaterial application: an overview. In: Titanium alloys in surgical implants. Luckey HA, Kubli F, editors. ASTM STP 796, 1983. pp. 7–15.

[8-8] Long M, Rack HJ. Review: titanium alloys in total joint replacement – a materials science perspective. Biomaterials 1998;19:1621–1639.

[8-9] Brunski JB, Puleo DA, Nanci A. Biomaterials and biomechanics of oral and maxillofacial implants: current status and future developments. Int J Oral Maxillofac Implants 2000;15:15–46.

[8-10] http://www.jointpainremedy.com.

[8-11] Carlos CS. Osteoarthritis/Osteoarthrosis concepts. http://www.cvm.umn.edu/academics.

[8-12] Curtis R. Superplastic forming of dental implants. Mater World 1999;10: 607–609.

[8-13] Oshida Y, Barco MT. Superplastically-formed prosthetic components, and equipment for same. US Patent 6,116,070 (2003).

[8-14] Al Jabbari Y, Nagy WW, Iacopino AM. Implant dentistry for geriatric patients: a review of the literature. Quintessence Int 2003;34:281–285.

[8-15] Ship JA, Chavez EM. Management of systemic diseases and chronic impairments in older adults: oral health consideration. Gen Dent 2000; 48:557–558.

[8-16] Oshida Y. Requirements for successful biofunctional implants. Int Sym Advanced Biomaterials 2000;5–10.

[8-17] Oshida Y, Hashem A, Nishihara T, Yapchulay MV. Fractal dimension analysis of mandibular bones: toward a morphological compatibility of implants. J Biomed Mater Eng 1994;4:397–407.

[8-18] Edwards BN, Gold BR. Analysis of surface cleanliness of three commercial dental implants. Biomaterials 1992;13:775–780.

[8-19] Adell R, Eriksson B, Lekholm U, Brånemark P-I, Jemt T. A long-term follow-up study of osseointegrated implants in the treatment of totally edentulous jaws. Int J Oral Maxillofac Implants 1990;5:347–359.

[8-20] http://ocw.tufts.edu/courses/10/content/2446.58; Tufts Open Course Ware.

[8-21] Åstrand P, Borg K, Gunne J, Olsson M. Combination of natural teeth and osseointegrated implants as prosthesis abutments: a 2-year longitudinal study. Int J Oral Maxillofac Implants 1991;6:305–312.

[8-22] Kohavi D, Schwartz Z, Amir D, Mai CM, Gross U, Sela J. Effect of titanium implants on primary mineralization following 6 and 14 days of rat tibial healing. Biomaterials 1992;13:255–260.

[8-23] Piattelli A, Corigliano M, Scarano A. Microscopical observations of the osseous responses in early loaded human titanium implants: a report of two cases. Biomaterials 1996;17:1333–1337.

[8-24] Saadoun AP, Le Gall MG. An 8-year complication of clinical results obtained with Steri-osseoendosseous implants. Compendium 1996;17: 669–680.

[8-25] Piattelli A, Scarano A, Nora AD, De Bona G, Favero GA. Microscopical features in retrieved human Brånemark implants: a report of 19 cases. Biomaterials 1998;19:643–649.

[8-26] Cochran DL. The scientific basis for and clinical experiences with Straumann implants including the ITI® Dental Implant System: a consensus report. Clin Oral Implants Res 2000;11:33–58.

[8-27] Leonhardt Å, Gröndahl K, Bergström C, Lekholm U. Long-term follow-up of osseo-integrated titanium implants using clinical, radiographic and microbiological parameters. Clin Oral Implants Res 2002;13:127–132.

[8-28] Cochran DL, Buser D, ten Bruggenkate CM, Weingart D, Taylor TM, Bernard JP, Peters F, Simpson JP. The use of reduced healing times on ITI® implants with a sandblasted and acid-etched (SLA) surface: early results from clinical trials on ITI® SLA implants. Clin Oral Implants Res 2002;13:144–153.

[8-29] Meijer HJA, Raghoebar GM, Van't Hof MA, Visser A. A controlled clinical trial of implant-retained mandibular overdentures: 10 years' results of clinical aspects and aftercare of IMZ implants and Brånemark implants. Clin Oral Implants Res 2004;15:421–427.

[8-30] Ichinose S, Muneta T, Aoki H, Tagami M. TEM observation of seven retrieved total knee joints made of Co-Cr-Mo and Ti-6Al-4V alloys. J Biomed Mater Eng 2003;13:125–134.

[8-31] Goodacre CJ, Bernal B, Rungcharassaeng K, Kan J. Clinical complications with implants and implant prostheses. J Prosthet Dent 2003;90: 121–132.

[8-32] Simon RL. Single implant-supported molar and premolar crowns: A ten-year retrospective clinical report. J Prosthet Dent 2003;90:517–521.

[8-33] Ratner BD, Johnston AB, Lenk TJ. Biomaterial surfaces. J Biomed Mater Res:Applied Biomater 1987;21:59–90.

[8-34] Smith DC, Pilliar RM, Chernecky R. Dental implant materials. I. Some effects of preparative procedures on surface topography. J Biomed Mater Res 1991;25:1045–1068.

[8-35] Smith DC, Pilliar RM, Metson JB, McIntyre NS. Dental implant materials. II. Preparative procedures and surface spectroscopic studies. J Biomed Mater Res 1991;25:1069–1084.

[8-36] Kasemo B, Lausmaa J. Biomaterial and implant surfaces: a surface science approach. Int J Oral Maxillofac Implants 1988;3:247–259.

[8-37] Lausmaa J, Kasemo B. Surface spectroscopic characterization of titanium implant materials. Applied Surf Sci 1990;44:133–146.

[8-38] Sutherlan DS, Forshaw PD, Allen GC, Brown IT, Williams KR. Surface analysis of titanium implants. Biomaterials 1993;14:893–899.

[8-39] Tengvall P, Lundstrom I. Physio-chemical considerations of titanium as a biomaterial. Clin Mater 1992;9:115–134.

[8-40] Sawase T, Hai K, Yoshida K, Baba K, Hatada R, Atsuta M. Spectroscopic studies of three osseointegrated implants. J Dent 1998;26:119–124.

[8-41] Albrektsson T, Wennerberg A. Oral implant surfaces: Part 1 – review focusing on topographic and chemical properties of different surfaces and *in vitro* responses to them. Int J Prosthodont 2004;17:536–543.

[8-42] Albrektsson T, Wennerberg A. Oral implant surfaces: Part 2 – review focusing on clinical knowledge of different surfaces. Int J Prosthodont 2004;17:544–564.

[8-43] Oshida Y. Surface characterization and microanalytical studies on TiNi shape memory dental devices. In: Perio root implant and medical application of shape memory alloy. Fukuyo S, Sachdeva R, editors. Tokyo: Japan Medical Culture Center. 1992. pp. 17–30.

[8-44] James RA. The support system and the pergingival defense mechanism of oral implants. Oral Implantol 1975;6:270–285.

[8-45] Weiss CM. Tissue integration of dental endosseous implants: description and comparative analysis of the fibro-osseous integration and osseous integration system. J Oral Implantol 1986;12:169–214.

[8-46] Fox WC, Miller MA. Osseo implant for studies of biomaterials using an *in vivo* electro-chemical transducer. J Biomed Mater Res 1993;27:763–773.

[8-47] Bearinger JP, Orme CA, Gilbert JL. *In situ* imaging and impedance measurements of titanium surfaces using AFM and SPIS. Biomaterials 2003;24:1837–1852.

[8-48] Hirata I. Development of titanium biosensor for surface plasma resonance analysis system. J Dent Eng 2005;No.152:33–35.

[8-49] Steinmann SG. Corrosion of surgical implants – *in vivo* and *in vitro* tests. In: Evaluation of Biomaterials. Winter GD, Leray JL, de Groot K, editors. New York: Wiley, 1980. pp. 1–34.

[8-50] Kruger J. Fundamental aspects of the corrosion of metallic implants. ASTM STP 684. In: Corrosion and degradation of implant materials. Syrett B.C, Acharya A. editors. Philadelphia, PA: American Society for Testing and Materials. 1979. pp. 107–127.

[8-51] Greene ND. Corrosion of surgical implant alloys: a few basic ideas. ASTM STP 859; In: Corrosion and degradation of implant materials: second symposium. Fraker AC, Griffin CD, editors. 1983. pp. 5–10.

[8-52] Cook SD. Interface mechanics of porous metal-coated implants. In: Implant prosthodontics: surgical and prosthetic techniques for dental implants. Fagan MJ, editor. Chicago: Year Book Medical Pub. 1990. pp. 305–309.

[8-53] Brunski JB. Biomechanical factors affecting the bone–dental implant interface. Clin Mater 1992;10:153–201.

[8-54] Skalak R. Biomechanical considerations in osseointegrated prostheses. J Prosthet Dent 1983;49:843–848.

[8-55] Hekimoglu C, Anil N, Cehreli MC. Analysis of strain around endosseous implants opposing natural teeth or implants. J Prosthet Dent 2004;92:441–446.

[8-56] Cook SD, Kay JF, Thomas KA, Jarcho M. Interface mechanics and histology of titanium and hydroxyapatite-coated titanium for dental implant applications. Int J Oral Maxillofac Implants 1987;2:15–22.

[8-57] Himmlová L, Dostálová T, Kácovký A, Konvičková S. Influence of implant length and diameter on stress distribution: a finite element analysis. J Prosthet Dent 2004;91:20–25.

[8-58] Browne M, Langley RS, Gregson PI. Reliability theory for load bearing biomedical implants. Biomaterials 1999;20:1285–1292.

[8-59] Sadowky SJ, Caputo AA. Stress transfer of four mandibular implant overdenture cantilever designs. J Prosthet Dent 2004;92:328–336.

[8-60] van Rossen IP, Braak LH, de Putter C, de Groot K. Stress-absorbing elements in dental implants. J Prosthet Dent 1990;64:198–205.

[8-61] Rangert B, Gunne J, Sullivan DY. Mechanical aspects of a Brånemark implant connected to a natural tooth: an *in vitro* study. Int J Oral Maxillofac Implants 1991;6:177–186.

[8-62] Lang LA, Kang B, Wang R-F, Lang BR. Finite element analysis to determine implant preload. J Prosthet Dent 2003;90:539–546.

[8-63] Yokoyama K, Ichikawa T, Murakami H, Miyamoto Y, Asaoka K. Fracture mechanisms of retrieved titanium screw thread in dental implant. Biomaterials 2002;23:2459–2465.

[8-64] Kohn DH, Ducheyne P. A parametric study of the factors affecting the fatigue strength of porous coated Ti-6Al-4V implant alloy. J Biomed Mater Res 1990;24:1483–1501.

[8-65] Kohn DH, Ducheyne P, Awerbuch J. Acoustic emission during fatigue of porous-coated Ti-6Al-4V implant alloy. J Biomed Mater Res 1992;26: 19–38.

[8-66] Bruck SD. Biostability of materials and implants. Long-Term Effects Med Imp 1991;1:89–106.

[8-67] Adell R, Lekholm U, Rockler B, Brånemark PI. A 15 year study of osseointegrated implant in the treatment of the edentulous jaw. J Oral Surg 1981;10:387–416.

[8-68] Marchant RE. Cell attachment and interactions with biomaterials. J Adhesion 1986;20:211–215.

[8-69] Brunette DM. The effects of implant surface topography on the behavior of cells. Int J Oral Maxillofac Implants 1988;3:231–246.

[8-70] Bigerele M, Anselme K, Noël B, Ruderman I, Hardouin P, Iost A. Improvement in the morphology of Ti-based surfaces: a new process to increase *in vitro* human osteoblast response. Biomaterials 2002;23: 1563–1577.

[8-71] Nan H, Ping Y, Xuan C, Yongxang L, Xiaolan Z, Guangjun C, Zihong Z, Feng Z, Yuanru C, Zianghuai L, Tingfei X. Blood compatibility of amorphous titanium oxide films synthesized by ion beam enhanced deposition. Biomaterials 1998;19:771–776.

[8-72] Stanford CM, Keller JC, Solursh M. Bone cell expression on titanium surfaces is altered by sterilization treatments. J Dent Res 1994;73: 1061–1071.

[8-73] Carlsson LJ, Albrektsson A, Berman C. Bone response to plasma-cleaned titanium implants. Int J Oral Maxillofac Implants 1989;4:199–204.

[8-74] Buser D, Schenk RK, Steinmann S, Fiorellini JP, Fox CH, Stich H. Influence of surface characteristics on bone integration of titanium implants. A histomorphometric study in miniature pigs. J Biomed Mater Res 1991;25:889–902.

[8-75] Knabe C, Klar F, Fitzner R, Radlanski RJ, Gross U. *In vitro* investigation of titanium and hydroxyapatite dental implant surfaces using a rat bone marrow stromal cell culture system. Biomaterials 2002;23: 3235–3245.

[8-76] Baier RE. Modification of surfaces to meet bioadhesive design goals: a review. J Adhesion 1986;20:171–186.

[8-77] Glantz P-O, Baier RE. Recent studies on nonspecific aspects of intraoral adhesion. J Adhesion 1986;20:227–244.

[8-78] Sunny MC, Sharma CP. Titanium-protein interaction: change with oxide layer thickness. J Biomater Appl 1991;5:89–98.

[8-79] Rahal MD, Delorme D, Brånemark P-I, Osmond DG. Myelointegration of titanium implants: B lymphopoiesis and hemopoietic cell proliferation in mouse bone marrow exposed to titanium implants. Int J Oral Maxillofac Implants 2000;15:175–184.

[8-80] Sukenik CN, Balachander N, Culp LA, Lewandowska K, Merritt K. Modulation of cell adhesion by modification of titanium surfaces with covalently attached self-assembled monolayers. J Biomed Mater Res 1990;24:1307–1323.

[8-81] Anselme K. Osteoblast adhesion on biomaterials. Biomaterials 2000;21:667–681.

[8-82] Dmytryk JJ, Fox SC, Moriarty JD. The effects of scaling titanium implant surfaces with metal and plastic instruments on cell attachment. J Periodontol 1990;61:491–496.

[8-83] Ryhänen J, Kallioinen M, Tuukkanen J, Lehenkari P, Junila J, Niemelä E, Sandvik P, Serlo W. Bone modeling and cell-attachment interface responses induced by nickel-titanium shape memory alloy after periosteal implantation. Biomaterials 1999;20:1309–1317.

[8-84] Larsson C, Thomsen P, Aronsson B-O, Rodahl M, Lausmaa J, Kasemo B, Ericson LE. Bone response to surface-modified titanium implants: studies on the early tissue response to machined and electropolished implants with different oxide thickness. Biomaterials 1996;17: 605–616.

[8-85] Larsson C, Thomsen P, Lausmaa J, Rodahl M, Kasemo B, Ericson LE. Bone response to surface modified titanium implants: studies on electropolished implants with different oxide thickness and morphology. Biomaterials 1994;15:1062–1074.

[8-86] de Oliveira PT, Nanci A. Nanotexturing of titanium-based surfaces upregulates expression of bone sialoprotein and osteopontin by cultured osteogenic cells. Biomaterials 2004;25:403–413.

[8-87] Ahmad M, Gawronski D, Blum J, Goldberg J, Gronowicz G. Differential response of human osteoblast-like cells to commercially pure (cp) titanium grades 1 and 4. J Biomed Mater Res 1999;46:121–131.

[8-88] Inoue T, Cox JE, Pilliar RM, Melcher AH. Effect of the surface geometry of smooth and porous-coated titanium alloy on the orientation of fibroblasts *in vitro*. J Biomed Mater Res 1987;21:107–126.

[8-89] Anselme K, Linez P, Bigerelle M, Le Maguer D, Le Maguer A, Hardouin P, Hildebrand HF, Iost A, Leroy JM. The relative influence of the topography and chemistry of TiAl6V4 surfaces on osteoblastic cell behaviour. Biomaterials 2000;21:1567–1577.

[8-90] Mustafa K, Pan J, Wroblewski J, Leygraf C, Arvidson K. Electrochemical impedance spectroscopy and X-ray photoelectron spectroscopy analysis of titanium surfaces cultured with osteoblast-like cells derived from human mandibular bone. J Biomed Mater Res 2002;59:655–664.

[8-91] Lee TM, Chang E, Yang CY. Attachment and proliferation of neonatal rat calvarial osteoblasts on Ti6Al4V: effect of surface chemistries of the alloy. Biomaterials 2004;25:23–32.

[8-92] Zhu X, Chen J, Scheideler L, Reichl R, Geis-Gerstorfer J. Effects of topography and composition of titanium surface oxides on osteoblast responses. Biomaterials 2004;25:4087–4103.

[8-93] Hazan R, Brener R, Oon U. Bone growth to metal implants is regulated by their surface chemical properties. Biomaterials 1993;14:570–574.

[8-94] Jasty M, Bragdon CR, Haire T, Mulroy RO, Harris WH. Comparison of bone ingrowth into cobalt chrome sphere and titanium fiber mesh porous coated cementless canine acetabular components. J Biomed Mater Res 1993;27:639–644.

[8-95] Simske SJ, Sachdeva R. Cranial bone apposition and ingrowth in a porous nickel-titanium implant. J Biomed Mater Res 1995;29:527–533.

[8-96] Chang Y-S, Oka M, Kobayashi M, Gu H-O, Li Z-L, Nakamura T, Ikada Y. Significance of interstitial bone ingrowth under load-bearing conditions: a comparison between solid and porous implant materials. Biomaterials 1996;17:1141–1148.

[8-97] Li J, Liao H, Fartash B, Hermansson L. Surface-dimpled commercially pure titanium implant and bone ingrowth. Biomaterials 1997;18:691–695.

[8-98] Kujala S, Ryhänen J, Jämsä T, Danilov A, Saaranenm J, Pramila A, Tuukkanen J. Bone modeling controlled by a nickel-titanium shape memory alloy intramedullary nail. Biomaterials 2002;23:2535–2543.

[8-99] Castleman LS, Motzkin SM, Alicandri FP, Bonawit VL. Biocompatibility of Nitinol alloy as an implant material. J Biomed Mater Res 1976;10:695–731.

[8-100] Nishiguchi S, Nakamura T, Kobayashi M, Kim H-M, Miyaji F, Kokubo T. The effect of heat treatment on bone-bonding ability of alkali-treated titanium. Biomaterials 1999;20:491–500.

[8-101] Villermaux F, Tabrizian M, Yahia L'H, Meunier M, Piron DL. Excimer laser treatment of NiTi shape memory alloy biomaterials. Apply Surf Sci 1997;109/110:62–66.

[8-102] Montero-Ocampo C, Loprez H, Salinas Rodriguez A. Effect of compressive straining on corrosion resistance of a shape memory Ni-Ti alloy in ringer's solution. J Biomed Mater Res 1996;32:583–591.

[8-103] Trépanier C, Tabrizian M, Yahia L'H, Bilodeu L, Piron DL. Effect of modification of oxide layer on NiTi stent corrosion resistance. J Biomed Mater Res (Apply Biomater) 1998;43;433–440.

[8-104] Ku C-H, Browne M, Gregson PI, Corbeil J, Pioletti DP. Large-scale gene expression analysis of osteoblasts cultured on three different Ti-6Al-4V surface treatments. Biomaterials 2002;23:4193–4202.

[8-105] Piattelli A, Degidi M, Paolantonio M, Mangano C, Scarano A. Residual aluminum oxide on the surface of titanium implants has no effect on osseointegration. Biomaterials 2003;24:4081–4089.

[8-106] Meyer U, Joos U, Mythili J, Stamm T, Hohoff A, Fillies T, Stratmann U, Wiesmann HP. Ultrastructural characterization of the implant/bone interface of immediately loaded dental implants. Biomaterials 2004;25:1959–1967.

[8-107] Kaneko S, Tsuru K, Hayakawa S, Takemoto S, Ohtsuki C, Ozaki T, Inoue H, Osaka A. *In vivo* evaluation of bone-bonding of titanium metal chemically treated with a hydrogen peroxide solution containing tantalum chloride. Biomaterials 2001;22:875–881.

Chapter 9
Other Applications

9.1.	Denture Bases	257
9.2.	Crowns and Bridges	259
9.3.	Clasps	265
9.4.	Posts and Cores	266
9.5.	Titanium Fiber and Titanium Oxide Powder as Reinforcement for Bone Cement	267
9.6.	Sealer Material	268
9.7.	Shape-Memory Dental Implant	268
9.8.	Orthodontic Appliances	269
9.9.	Endodontic Files and Reamers	271
9.10.	Clamps and Staples	273
References		274

Chapter 9

Other Applications

This section will describe a variety of titanium applications other than the implants discussed in the previous chapter. Some applications are realized owing to advanced and modified manufacturing technologies, and other applications are developed and promoted due to material's uniqueness, such as superelasticity and/or SME associated with NiTi alloys.

9.1. DENTURE BASES

Denture materials are materials, which require precise fabrication (or these prostheses should be manufactured in a near-net-shape form). Traditionally, various types of polymers have been used, including polymethylmethacrylate (PMMA), polycarbonate (PC), polystylene, and polysulfone. However, any polymeric resins have drawbacks. These include weakening of mechanical properties due to water-sorption, which also causes a dimensional instability, and tissue irritation due to unavoidable residual unpolymerized monomer.

CpTi and Ti-based alloys are generally used as fixtures for dental implantation [9-1, 9-2]. When these metals are used as restoratives for anterior teeth, the metal prostheses are generally covered with esthetic materials, such as porcelain or resin. Veneering resin is more popular than porcelain because it is less expensive and more resistant to abrasion [9-3]. When constructing removable dentures, implant supported prostheses and most especially resin-bonded bridges, the interfacial bond between metal and acrylic resin has been wholly reliable or durable [9-4]. Numerous investigations have been carried out to measure metal/acrylic resin bond strengths between materials that have surface or chemical modifications that are reported to enhance retention [9-4−9-6]. These have included surface treatments of the alloys by macro-, micromechanical and chemical methods, or by the addition of chemical bonding agents to the acrylic [9-6, 9-7]. PMMA resins have a linear thermal coefficient of expansion of 81×10^{-6} m/m·°C, whereas noble and base metals range from 7 to 19×10^{-6} m/m·°C [9−8]. The lack of adhesion and differences in the thermal expansion of the materials places severe demands on the metal/acrylic resin surfaces. Ban [9-1] prepared CpTi (grade 1), Ti-6Al-4V, experimental β-Ti alloy (54Ti-29Nb-13Ta-4.6Zr), and 12Au-20Pd-52Ag-15Cu-1Zn alloy plates which were sand-blasted with alumina Al_2O_3 powder (100−250 μm), water-washed and dried. CpTi plates were further oxidized in-air at 600°C for 1 h.

They were also treated in 5 M LiOH, 5 M KOH, and 5 M NaOH at 25 or 60°C for 20 h, followed by in-air oxidation at 600°C for 1 h. The bonding strengths of such treated CpTi plates were measured using a pull-shear bonding method after immersion in physiological saline solution at 37°C for 24 h. It was reported that (i) the bonds of the standard alkaline-treated CpTi and two Ti alloys to resin were 1.5−1.9 times stronger than those of sand-blasted specimens, but no significant effects of the alkaline treatment were observed on 12Au-20Pd-52Ag-15Cu-1Zn alloy, (ii) the greatest bonding strengths were found for CpTi treated with NaOH and KOH and the heated at 600°C, and (iii) alkaline treatment is a simple, effective surface modification of Ti that improves bonding to veneering resin [9-1].

The recognition that a chemical linkage between acrylic resin and metal would lead to an enhanced bond led to the development and use of an intermediate layer based upon silicon oxide. The so-called "SiO-C" layer was introduced as an intermediate layer between the metal substrate and polymer as a means for enhancing the strength and durability of the bond [9-9]. Ti-6Al-4V, after bonding ($n = 3$), was stored at 37°C in distilled water for 7 days, followed by thermocyclings between 5 and 55°C for 500 cycles (for 21 days), and then was subjected to the 4-point bending at a crosshead speed of 0.5 mm/m. The bond strength $\sigma = F(L - \ell)/\pi R^3$, where σ is in N/m^2, F the load at failure (N), L the distance between specimen support (m) = 0.024 m, ℓ = a distance between loading bars (m) = 0.008 m, and R is a radius of specimen (m) = 0.0035 m. It was concluded that the most durable surface treatment was the silicoating process after sand-blasting of the metal surface with 250 μm alumina for heat-cured acrylic resin materials [9-10].

Denture bases are now mainly manufactured by a casting process of Co-Cr or Ni-Cr alloys. A novel superplastic fabrication process for dentures, as well as implant systems, has been developed [9-11]. Ti-6Al-4V alloy is known as a biocompatible metal, and with both high strength and low density, it seems to be the best material for denture bases. The Ti-6Al-4V also has superplastic properties at high temperature, forming very simple processes by argon pressure (10 kg/cm^2) at 800−900°C. Superplastically formed (SPFed) denture bases of Ti-6Al-4V showed some merits compared with conventional cast or cold-pressed denture base plates, namely, good fitness, light weight, and low producing cost using a smaller number of manufacturing processes [9-12]. A SPFed Ti-6Al-4V removable denture framework was fabricated. Clinical application in short-term follow-up periods from 6 months to 3 years was evaluated. It was reported that (i) the dentures functioned well and did not cause any major clinical difficulties, and (ii) the patients have expressed satisfaction with the dentures at regular recall appointments [9-13].

Srimaneepong *et al.* [9-14] cast Ti-6Al-7Nb alloy into three differently designed, removable partial denture frameworks: palatal strap, anterior–posterior bar, and horseshoe-shaped bar. The vertical displacement and local strain of

Ti-6Al-7Nb alloy frameworks were investigated to compare against those of Co-Cr alloy frameworks. A vertical loading force of 19.6 N was applied at two locations, 10 and 20 mm, from the distal end of the frameworks. It was found that (i) although higher vertical displacement and local strain were observed for Ti-6Al-7Nb alloy frameworks than those for Co-Cr alloy frameworks, the palatal strap framework appeared to show the least deformation, (ii) the strain at 10 mm location was higher than that at 20 mm location for anterior–posterior bar and horseshoe-shaped frameworks, and (iii) the palatal strap design was evaluated to be a suitable design for the removable denture framework with Ti-6Al-7Nb alloy [9-14].

9.2. CROWNS AND BRIDGES

Dental porcelains provide dentistry with biologically acceptable and aesthetic (tooth-colored restoration) methods of repairing diseased and fractured teeth [9-15]. Modern dental porcelains are pigmented glass-crystal composites that were evolved from traditional ternary whiteware compositions, namely clay-feldspar-quartz. Most of the modern dental porcelains used for the fabrication of crowns and bridges require only short, low-temperature sintering cycles. Porcelains have been fluxed so that their thermal expansions match precious or non-precious alloys. Such alloys are manufactured into frameworks for the crowns or bridges, and the porcelains are bonded to them. The metal crown or bridge is faced or veneered with porcelain. The bond between precious alloys and the porcelain is more reliable than the bond between non-precious alloys and the porcelain. The presence (by alloying in the original coping materials or coating on the coping materials) of small amounts of tin (Sn) or indium (In) in the precious coping alloys makes the chemical bonding possible in these cases. It is generally believed that oxygen in InO and SnO shares with oxygen in the silica component in porcelain, resulting in enhanced bonding strengths. All these porcelains have been opacified as to cover their dark metal-oxide background. They are highly reflective and are inferior aesthetically for use in anterior teeth. Porcelains formulated for metal-ceramic prostheses have a tendency to densify during long and successive firing cycles [9-15]. porcelain-fused-to-metal (PFM) is sometimes called porcelain-bonded-to-metal (PBM), porcelain-to-metal (PTM), ceramometal, or metal–ceramics. According to fusing temperature, porcelain can be classified into (1) high fusing (1300°C) for denture tooth fabrication, (2) medium fusing (1100–1300°C) for denture tooth fabrication, (3) low fusing (850−1100°C) for crown/bridge construction, and (4) ultra-low fusing (<850°C) for crown/bridge construction. As for metallic coping materials, they are also classified into (1) high noble (Au-Pt-Pd, Au-Pd-Ag, Au-Pd), (2) noble (Pd-Au, Pd-Au-Ag, Pd-Ag, Pd-Cu, Pd-Co, Pd-Ga-Ag), and (3) base metal (CpTi, Ti-Al-V, Ni-Cr-Mo-Be, Ni-Cr-Mo, Co-Cr-Mo, Co-Cr-W).

There are several requirements and characteristics of metals applicable to the metal-ceramic systems as follows.

(1) The metal should have a high melting point so that it does not melt during the porcelain application procedure.
(2) The metal should have a similar coefficient of thermal expansion as those values in porcelain. The coefficients of thermal expansion (linear), α (in $\times 10^{-6}$ m/m·°C, or simply in ppm), for major materials are reported as follows: conventional high-fusing porcelain (13.7−15.5 ppm), conventional low-fusing porcelain (14.1−15.1 ppm), low fusing porcelain for Ti veneering (7.7–8.9 ppm), Au-Pd (13.9 ppm), high Pd (14.7 ppm), Ni-Cr base metal (13.9 ppm), CpTi (9.7 ppm) [9-16].[1] However, a slight extent of thermal mismatching is favored by $\alpha_{METAL} > \alpha_{PORCELAIN}$, so that the porcelain will be placed under the compression on cooling.
(3) The metal should have excellent castability (however, casting is not only a fabrication method, but it should also include CAD/CAM or superplastic forming − SPF).
(4) Both materials should have similar values of modulus of elasticity to establish the mechanical compatibility [9-17].

Moreover, since little or no residual stress due to thermal mismatch should exist in the final Ti/porcelain-bond interface, the significant discrepancies already noted in their thermal coefficients of expansion will have to be modified to more closely match. It appears that an interfacial oxide layer, some 100–1000 nm (10–100Å) thick, forms during firing, and the thicker this layer becomes, the weaker the bonding strength between the porcelain and the titanium [9-18].

To have a reasonable function of Ti/porcelain system, the rigid bonding is crucial in terms of in-service longevity. There are several theories to postulate bonding mechanism of porcelain.

(1) *Chemical bonding.* A chemical bonding, existing through the formation of a tin (or indium) oxide layer during porcelain firing and that the oxide formed provides the mechanism of bonding between the porcelain and the metal substrate. Watanabe *et al.* [9-19] characterized the interface reaction between porcelain and pure titanium during porcelain firing, using an electron–probe microanalyzer (EPMA), X-ray diffraction method, and Fourier transform

---

[1] When the volumetric expansion, γ, needs to be calculated, it follows as: $\gamma = (1 + \alpha)^3 - 1 = (1 + 3\alpha + 3\alpha^2 + \alpha^3) - 1 \cong 3\alpha$, since α^2 (in order of 10^{-12}) and α^3 (in order of 10^{-18}) are small enough to be neglected.

infrared spectroscopy. It was reported that (i) separation occurred partly at the interface between titanium and porcelain and partly slightly inside the porcelain after firing three times at 760°C, (ii) in the surface layer of the titanium side, only oxygen was detected as a diffusing element, (iii) X-ray diffraction indicated a decrease in SnO_2 and revealed β-Sn (metallic Sn), and (iv) a considerable amount of TiO_2 was formed at the interface when fired at 760°C for 2 min [9-20].

(2) *Mechanical retention through various surface modifications.* Although chemical bonding is generally regarded to be responsible for metal–porcelain adherence, evidence exists that mechanical interlocking provides the principal bonding. Surface concave texturing (resulting in remarkable increase in an effective surface area) will enhance the porcelain bond strength. While potential contamination of sand-blasting and shot peening media (mostly, 50 μm alumina powder) weakens the porcelain bond strength, recently different techniques produced similar effects, such as a contamination-free indirect laser peening, and chemical treatment (in either acid or alkaline).

(3) *Compressive forces.* In order to strengthen the fused porcelain body (in other words, to retard the crack propagation), there are three major methods available: (i) thermal tempering by rapid cooling the surface of the objects while it is hot and in its molten state, (ii) ion exchange by substituting small ion (e.g., Na) with a larger ion (e.g., K), and (iii) thermal mismatching by $\alpha_{METAL} > \alpha_{PORCELAIN}$, where the porcelain will be placed under the compression on cooling.

(4) *Bonding agent application.* In order to enhance the bond strength or improve the wettability of porcelain, a bonding agent or metal conditioner is applied. In some cases, hydrofluoric acid (HF) treatment is recommended to remove silicate from the investing materials.

(5) *Physical bonding.* X-ray microanalysis on CpTi/porcelain and Ti-6Al-7Nb/porcelain gave some evidence of diffusion of some elements, particularly of the porcelain into the metal, which may assist bonding during firing [9-20]. Diffusion of elements in porcelain into the titanium (i.e., titanium oxide) during firing process is also significant in porcelain-fused-to-titanium systems because bonding occurs between the two components by the mutual diffusion of elements. Hanawa *et al.* [9-21] investigated the diffusion of elements of commercial porcelain for titanium into titanium oxide during heating. Titanium was deposited on three kinds of disk-shaped porcelains by vacuum-vaporization and the porcelains were then heated. A thin titanium oxide film was formed on the porcelains by the heating. X-ray photoelectron spectroscopy was used to characterize the surfaces of the porcelain with and without titanium oxide. It was found that (i) only sodium, potassium, and barium diffused into titanium oxide during heating, where they formed a complex oxide with

titanium, (ii) the diffusion of these elements may be involved in the bonding of porcelain to titanium, and (iii) after heating, the titanium oxide layer mainly consisted of a titanium oxide, whose valence was between trivalent and tetravalent, and contained small amounts of sodium, potassium, and barium [9-21].

If titanium's distinct advantages (such as light-weight, high strength, good corrosion resistance) are to be used for aesthetic crowns and bridges, the ability to apply a porcelain veneer becomes important. Because of titanium's high affinity for oxygen, porcelain firing should take place below 800°C (which is below the 885°C $\alpha \leftrightarrow \beta$ transformation temperature) in order to prevent excess oxide formation [9-18, 9-22]. There are two major problems in the Ti/Porcelain system, as mentioned above. The first one is related to the coefficient of thermal expansion. The coefficient of thermal expansion of titanium is not same as that of porcelain. Hence, this requires the development of a new titanium material, which has an equivalent thermal coefficient value. Alternatively, a new porcelain system must be developed to exhibit the thermal compatibility. The second concern deals with the rapid oxidation of titanium. The conventional firing temperature is in a range from 700°C to 800°C, at which the titanium substrate will be easily oxidized to form a relatively thick oxide film, causing a weakening of the bond strength. As mentioned above, it is obvious that the heating temperature should not exceed the β-transus temperature (i.e., 885°C for pure titanium), since the oxidation rate will accelerate if the process temperature is above the β-transus. Besides, there are differences in physical properties between the α-phase (low-temperature HCP structure) and the β-phase (high-temperature BCC structure), including lattice constants and thermal expansion coefficients. Therefore, several low firing porcelain systems have been developed (including Procera porcelain, Vita titanium porcelain, Duceratin, Noritake super porcelain titan, and Ohara titan bond). Alternatively, the porcelain firing process should be operated under a high vacuum condition or an argon-gas atmosphere.

Various casting processes have been developed for titanium in dentistry, some of which are used commercially. From the success of these casting procedures, the bonding of porcelain to titanium has been attempted, and several porcelains for titanium are now available commercially [9-21]. The bonding strength is primarily governed by a reaction layer existing between porcelain and titanium that is formed by the mutual diffusion of elements. In this regard, porcelain-fused-to-titanium systems have been studied. Togaya *et al.* [9-23] investigated the feasibility of using titanium in a metal-porcelain system, and demonstrated properties of titanium at high temperatures. Adachi *et al.* [9-24] reported the relationship between the oxidation of titanium and Ti-6Al-4V and bonding strength; bonding strength is low when the thickness of the oxide layer reaches 1 μm. Kimura *et al.* [9-25] also examined the effects of oxidation on porcelain–titanium interface reactions and bonding strength, and found a layer at the interface that is formed by the diffusion of titanium into the porcelain. Nagayama *et al.* [9-26] characterized the interface between titanium and porcelain using EPMA and XPS (X-ray photoelectron spectroscopy), and revealed that titanium atoms diffused into the porcelain, and that a reaction layer was formed, possibly corresponding to that found by Kimura *et al.* [9-25]. During the inter-diffusion process, a reaction layer between two bodies normally forms through the fact that titanium atoms apparently diffuse into the porcelain during the firing process because only commercial pure titanium is used in porcelain-fused-to-titanium systems at present. The three studies described above [9-23−9-25] show that titanium is easily oxidized during the firing process. In a porcelain-fused-to-titanium system, Oshida *et al.* [9-27, 9-28] examined the oxidation kinetics of titanium during the porcelain firing process. Historically, there are several attempts and ideas proposed to improve the porcelain bond strengths. Such modifications should include bonding agents improvements [9-29, 9-30, 9-31], titanium substrate modification (such as nitrizing) to prevent excess oxidation during the firing process [9-27, 9-28] sand-blasting [9-32] or chemical texturing CpTi surface [9-33], firing process alternation in argon process [9-34], improving porcelain materials [9-35, 9-36] (see also Chapter 11). To improve the bonding strength between Ti substrate and fired porcelain, a bonding agent is sometimes applied [9-21]. Bonding porcelain to titanium and its alloys is problematic if exposure of titanium to temperatures greater than 800°C, causing the formation of thick titanium oxide (TiO_2) layers. At approximately 1 μm in thickness, the oxide layer will spontaneously delaminate from the surface due to induced stresses caused by the volume differences between titanium and its oxides. While the use of fusing temperatures below 800°C has been found to increase porcelain to titanium bond strength, it does not eliminate the oxidation process. During firing, oxygen diffusing from the atmosphere and oxygen obtained from the reduction of porcelain oxides continue to react with titanium. A recently introduced bonding agent and low fusing porcelain/titanium system is claimed by the Nobelpharma to overcome these technical difficulties and provide adequate porcelain to metal bonding. A bonding agent has recently been introduced that prepares the surface of milled titanium copings (by CAD/CAM machining for special cases) for bonding to low fusing dental porcelains. Gilbert *et al.* [9-29] evaluated the effectiveness of the bonding agent by comparing the shear and 3-point bending strengths of specimens made with three combinations of materials: milled titanium/porcelain with bonding agent, the same milled titanium/porcelain without bonding agent, and cast high palladium/conventional porcelain. It was found that (i) when a titanium bonding agent was used, porcelain to titanium bond strength was found greater than the porcelain to high palladium bond strength in both shear test (28.3 MPa) and 3-point bend test (300 MPa), and (ii) without the bonding agent, the shear strength (18.7 MPa) of porcelain to titanium was significantly decreased. On the basis of these findings, it was therefore concluded that (i) the elimination of the bonding agent application resulted in a significant decrease in the bond strength of porcelain to titanium as measured by the shear test, but showed no significant effect in the flexure tests, and (ii) sand-blasting the titanium metal surfaces resulted in a significant amount of alumina becoming embedded in the titanium surface [9-29]. We will discuss the contamination issue in more detail in Chapter 11.

There is a concern with evaluation methods of bond strength of fired porcelain. It is known that a 3-point bending test produces internal shear stress, which does not exist between the two central loading points in a 4-point bending test [9-37]. In Ti-porcelain systems, under the 3-point bending conditions, there are at least two possible fracture processes: (1) a cohesive fracture (from external maximum-tensile side toward the interfacial zone) within a porcelain body, followed by the debonding the porcelain from titanium substrate, and (2) delamination of porcelain from the titanium substrate, then followed by the cohesive fracture of the porcelain body. When a thin plate of double-layered structure is subjected to the 3-point flexure tests, unlike the mono-layer case, the central (neutral) axis may not necessarily be on the geometrical center line of the beam. It is located rather far from the geometrical center line due to a large difference in values of modulus of elasticity between CpTi and the used porcelain. The distance from the top of the beam to the neutral axis, c, can be calculated by; $c = \{[(E_2/E_1)\, h\, (h + 2t)]\} + t^2\}/\{2[(E_2/E_1)\, h + t]\}$, where E_2 is the modulus of elasticity (MOE) of porcelain, E_1 is MOE of CpTi, h the layer thickness of fired porcelain, and t the layer thickness of CpTi substrate [9-38, 9-39]. White *et al.* [9-38] conducted a 3-point flexural test on porcelain-titanium system. The strength of layered porcelain-ceramic beams was limited by the cohesive tensile or compressive strengths of the porcelain, not by the porcelain−titanium interfacial bond, namely, the porcelain failure occurred at lower loads than did failure of the porcelain−titanium interface. It was reported that (i) the porcelain-titanium bond strength was at least 26 MPa, and (ii) SEM and energy-dispersive X-ray spectroscopy demonstrated that the bond was limited by delamination of a thin Ti−Ti oxide interface. It was therefore suggested that, from curves of relative layer and total thickness vs. force-bearing capacity of model beam, porcelain−titanium prostheses should be made as thick as is practical, but that the relative thickness of the porcelain and titanium layers would be less important [9-38]. The Ti−porcelain system can be considered as a double-layer structure, comprising of at least titanium substrate and porcelain body, including bonding agent. Stress distribution pattern of such a double-layer structure is not necessarily same as that of a single beam under the 3-point bending testing mode. Previously tested porcelain-fired commercially pure titanium samples ($n = 285$) [9-28, 9-33] were re-evaluated. It was found that all higher maximum bond strength data were in an area where $c>t$, or in other words, that the calculated neutral axis is located within the porcelain layer (actually in the bonding agent layer), while lower bond strengths were found when $c<t$ [9-40]. These results support conclusions by White *et al.* [9-38].

In contrast with most dental alloys for metal−ceramic restorations, titanium cannot be veneered with conventional feldspathic dental porcelains for several reasons [9-16]. As discussed above, at temperatures above 800°C, which are required for firing conventional dental porcelains (about 950°C), titanium oxidizes rapidly, producing a thick oxide layer with a rather weak bond to the underlying titanium, resulting in inadequate metal-ceramic bond strength [9-24, 9-25]. Although low-fusing dental porcelains (firing temperature <800°C) are available for veneering dental alloys, the use of these porcelains on titanium substrates is limited owing to the great mismatch in the coefficients of thermal expansion, since CpTi (grades $1-4$) has a much smaller thermal expansion coefficient (in units of 10^{-6} m/m·°C, or ppm) than does conventional porcelain. Zinelis *et al.* [9-16] studied various Ti-based alloys with CpTi as a control. The tested Ti materials included Ti-18Sn, Ti-12Ga, Ti-18In, Ti-12Co and combinations of Ga+Sn, Co+Ga, In+Al, and Sn+Al as alloying mixtures. It was reported that the addition of Al, Ga, In, Sn, Co, and Mg increased the coefficient of Ti to $10.1-13.1$ ppm ($25-500$°C).

Eriksson *et al.* [9-41] employed a powder metallurgy-forming process to fabricate titanium copings. Titanium powder was pressed at $300-900$ MPa, milled to the right shape, and sintered at 1200°C for a 2 h holding time. It was reported that the titanium in the sintered copings was ductile, dense ($97-99\%+$), and free from open porosity after sintering, suggesting that the fabrication of copings with this sintering method could have a potential for becoming a very cost-efficient method of producing dental crowns.

9.3. CLASPS

Removable partial denture (RPD) retentive clasp arms must be capable of placing and returning to original form and should satisfactorily retain a prosthesis [9-42]. In addition, clasps should not unduly stress abutment teeth or be permanently distorted during service. In this respect, gold alloys have been favored as clasps compared with base meal alloys because of high yield strengths and low moduli of elasticity [9-43]. However, gold alloys have not been regarded as the ideal metal for clasp construction due to the relatively high cost and unavoidable permanent deformation after long-term service. On the other hand, if clasps are too flexible, they may not provide adequate RPD retention when placed in shallow retentive undercuts. Accordingly, available materials to meet such requirements are limited. Co-Cr alloy, type IV gold alloy, and NiTi alloy are normally employed. NiTi clasps were made to engage one of two retentive undercut depths. The force required to remove the clasps was measured using a universal testing machine with a crosshead speed of 10 mm/min. It was concluded that (i) the retentive force of all tested clasps, except NiTi alloy clasps, significantly decreased during cycling sequences simulating clasp placement and removal for 500 times, (ii) the end-point retention for all clasps was nearly identical, and (iii) NiTi alloy clasps might be suitable for removable prosthodontic restorations because the loss of retention during simulated service was significantly lower than losses recorded for other clasp materials studied [9-42].

Vallittu *et al.* [9-44] determined the fatigue resistance of the cast clasps of RPDs using Co-Cr, CpTi, Ti-6Al-4V, and gold alloy. The test method used was a constant-deflection fatigue test in which the force required to deflect the clasp for 0.6 mm and the number of loading cycles required to fracture the clasp were determined. The results revealed that a fatigue fracture occurred in the 63Co-29Cr-5Mo alloy clasp after about 25 000 cycles, in CpTi after 4500 cycles, in Ti-6Al-4V after 20 000 cycles, and the 72Au-3Pt-11Ag-13Cu alloy clasp 21 000 cycles. It was then suggested that significant differences exist in the fatigue resistance of removable denture clasps made from different commercial metals, which may cause the loss of retention of the RPD and clasp failures [9-44].

9.4. POSTS AND CORES

Restorations through post crowns have revolutionized the treatment of badly damaged teeth to be made functional over the past century. Posts and cores are often inserted in endodontically treated teeth to provide retention and stability for a crown or a fixed partial denture [9-45]. Posts are normally available for either a pre-fabricated and retained type post (actively or passively) or are a custom-made type post. Posts are an essential component of the armamentarium involved with the retention of crowns and bridges. A standard post is cast gold with a zinc phosphate cement. Failure of the post, usually by loosening, will lead to catastrophic failure of the crows and bridges. Concerns about possible toxic effects of the metals used in posts and the adverse stress concentration due to the mismatch in physical properties between the post, luting cement, and tooth have led to the development of ceramic and composite posts using a variety of filler reinforcements with carbon and quartz fibers [9-46]. Among a number of factors influencing the long-term clinical performance of pre-fabricated posts and cores, retention of the core material is one of the most significant of these [9-47−9-53]. Most *in vitro* studies have been performed to evaluate the posts and cores subjected to tensile [9-54, 9-55], compressive [9-55, 9-56], and shear forces [9-57−9-59] as well as trauma and fatigue tests [9-60]. Clinically, however, posts are also subjected to torsional or rotational forces produced by functional tooth contacts [9-61, 9-62]. Failures of posts

and cores usually have serious consequences. Reasons for failure are loss of retention, with risk of caries development in the root canal and fracture of the root, the
post, or the core. Some of these failures are related to the diameter, length, and
mechanical properties − such as stiffness and flexural strength − of the post. Loss
of retention of posts has been found to be the most frequent mode of failure, while
root fracture is the most serious type of failure [9-63−9-67].

Akisli *et al.* [9-47] evaluated eight different core materials. Silica coating/silaniza-
tion, and acrylization were treated on CpTi posts. Following thermocyclings (5
↔55°C, dwell time of 30 s, for 5000 cycles), maximum torsional forces were determined with an electronic torque movement key. It was reported that (i) the resistance
to torsional forces for the core materials on Ti posts increased with the use of chemical surface-conditioning techniques, and varied in accordance with the opaque type,
and (ii) the type of core materials also significantly influenced the resistance after
thermocycling [9-47]. Kozono *et al.* [9-68] examined CpTi prefabricated post and its
compatibility with cast metal core in comparison with those of the commercially
available prefabricated posts made of Ni-Cr alloy and 304 (18Cr-8Ni-Fe) stainless
steel. It was found that, although Ni-Cr and 304 stainless-steel were subjected to
annealing during the casting process of the core, showing marked decrease in hardness and tensile strength, the CpTi post had already annealed during the working
process, and showed no significant changes in properties when heat-treated up to
1000°C [9-68].

9.5. TITANIUM FIBER AND TITANIUM OXIDE POWDER AS REINFORCEMENT FOR BONE CEMENT

Poly(methyl methacrylate) (PMMA) bone cement is used for maintaining the fixation of artificial joint components and for transferring loads from the artificial joint
to the bone [9-69]. For artificial hips or knee, the bone cement must survive in a
fatigue loading environment, and gross *in vivo* failure of bone cement occurs predominantly by fatigue fracture [9-70−9-73]. Even when the cause of loosening or
failure of an artificial joint cannot be linked to obvious bone cement fracture, undetected fatigue microcracking can reduce the mechanical integrity of the PMMA,
affect the load transfer capabilities, and alter the stress distributions within the prosthesis [9-74]. Methods to increase the *in vivo* fatigue life of currently used PMMA-
based bone cements include improved surgical cementing techniques, porosity
reduction by centrifugation or vacuum mixing and fracture resisting or toughening
additives. One of the problems encountered in previous attempts to create an effec-
tive-fiber reinforced bone cement, for example, was that the cement viscosity
increased with the addition of fibers, and mixing and delivery were difficult.
Oshida *et al.* [9-75−9-77] investigated the best type of additives (ZrO_2, TiO_2,

Al$_2$O$_3$, MgO, CaO and their mixture) as well as additive percentage to enhance mechanical properties of PMMA cement. It was found that adding titania alone or a mixture with zirconia (up to 1 vol.%) to conventional bone cement polymeric material (PMMA cement) enhanced the mechanical strengths, as well as fracture strength. Such enhancement was mainly due to crack branching strengthening mechanisms.

9.6. SEALER MATERIAL

It is noticed that zinc is considered as an allergic element. Hence, a new type of zinc-free sealer material for root canal treatments has been developed. The new type is a titanium oxide system sealer, and satisfies ISO 6876 specifications [9-78].

The following applications are mainly realized using NiTi's physical and mechanical uniqueness — shape memory effect (SME) and superelasticity (SE). By SME, a material first undergoes a martensitic transformation. After deformation in the martensitic condition the apparently permanent strain is recovered when the specimen is heated to cause the reverse martensitic transformation. Upon cooling, it does not return to its deformed shape. When SME alloys are deformed in the temperature regime a little above the temperature at which martensite normally forms during cooling, a stress-induced martensite is formed. This martensite disappears when the stress is released, giving rise to a superelastic stress−strain loop with some stress hysteresis; this is the other unique property associated with NiTi - superelasticity (SE). There is a term called psuedoelasticity, which is a more generic term which encompasses both superelastic and rubberlike behavior. The following is a short list of important functions of NiTi medical devices [9-79, 9-80]: (1) biomechanical compatibility, (2) thermal deployment (SME), (3) elastic deployment (SE), (4) hysteresis/biased stiffness, (5) kink resistance, (6) constancy of stress/temperature dependency, (7) fatigue resistance, (8) dynamic interference, (9) MR (magnetic resonance) compatibility, and (10) biocompatibility.

9.7. SHAPE-MEMORY DENTAL IMPLANT

Fukuyo *et al.* [9-81] successfully introduced dental implant systems of perio root type, as well as blade type, using NiTi shape memory alloy. It was reported that (i) the perio root implant demonstrated excellent biocompatibility, and provided for rigid and sustained bone compression in order to encourage bone adhesion, and (ii) this implant system had the potential for use in craniofacial surgery and other allied medical fields. Shape memory NiTi alloy has also been successfully utilized in an open blade-type dental endosseous implant to enhance the fixation force and to exhibit a strong ability of mastication compared with ordinary non-opening blade type implants.

9.8. ORTHODONTIC APPLIANCES

For orthodontic applications, wires and brackets can be easily named. Between them, research and developments on wires are dominant over those involved with brackets. NiTi has a unique property which is of practical use to the orthodontist. That property is its extreme elasticity when it is drawn into high-strength wire. NiTi wire is much more difficult to deform during handling and seating in bracket slots than stainless-steel wire. The wire can be used for longer periods of time without changing it, and it can shorten treatment time needed in leveling the dentition, due to its extreme elasticity [9-82]. Yoneyama *et al.* [9-83] investigated the effect of heat treatment temperature on bending properties and transformation temperatures of a SE NiTi alloy wire. The heat treatment process was at 440°C for 1800 s and between 400 and 540°C for 1800 s. A 3-point bending test and differential scanning calorimetry were performed. It was found that (i) the transformation temperatures of the wires were lowered with increasing the heat treatment temperature, (ii) the reverse transformation finishing temperature was below the body temperature with the treatment above 480°C, (iii) residual deflection of the NiTi wire after bending was small with the secondary heat treatment above 460°C, and (iv) the load in the unloading process was less changeable and increased with the treatment temperature between 460 and 540°C. Hence, it was concluded that secondary heat treatment in this range was suitable for using SE in expansion arch appliances [9-83].

There is evidence that orthodontic archwires can be recycled and re-used among different patients [9-84]. Special attention should be placed on this practice because chemical degradation and mechanical damage of the archwire can accumulate as the frequency of recycling increases [9-85, 9-86]. It is likely that nickel release from the NiTi might be accelerated, and some individuals may be sensitized to this reaction or may develop a hypersensitivity reaction. Furthermore, nickel induced toxicity and carcinogenicity is well documented in the literature [9-87−9-89]. It has been reported that the rate of nickel allergy is increasing rapidly in the Western world [9-86]. On the basis of this background, Oshida *et al.* [9-90] microanalyzed orthodontic SE NiTi archwires of both used (4 weeks) and unused conditions. They were subjected to immersion and polarization corrosion tests in 0.9% NaCl solution. On the basis of results, surfaces of NiTi archwires were further electrochemically treated to etch away Ni selectively and reform the surface morphology to uniform and porous surface layers. It was concluded that (i) surface layers of used wires were covered contaminants causing the discoloration, and the contaminants were identified as mainly KCl crystals, (ii) surfaces of both used and unused wires were observed to have irregular features characterized by lengthy island-like structure, where Ni was selectively dissolved, (iii) corrosion tests in 0.9% NaCl solution immersion and polarization methods indicated that by increasing temperature from $3-60°C$ and acidity from pH 11 to 3, calculated corrosion rates increased, and (iv) surface layers of NiTi wires can be electrochemically modified to selectively etch Ni away, leaving a Ti-enriched surface layer and forming a uniformly distributed porous surface that may reduce the coefficient of friction against the orthodontic brackets [9-90].

Physio-mechanical properties of NiTi wire materials were investigated by Kusy *et al.* [9-91]. Tested materials were eight straight-wire materials, including four stable alloys: one TiMo (beta phase), three NiTi (marteniste phase), and four active alloys: NiTi (martensite phase), NiTi (austenite phase), NiTi (biphase), and NiTi (hybrid martensite phase). With a tensile mode, each sample was scanned from - $120-200°C$ at $2°C/min$. From the data, plots of log storage modulus, log tan delta, and percent change in length vs. temperature were generated. It was shown that (i) the dynamic mechanical properties of the alloy within this Ti system are quite different, (ii) the TiMo alloy was invariant with temperature, having a modulus of 7.30×10^{11} dyn/cm^2, and (iii) the three cold-worked alloys appeared to be similar, with 5.74×10^{11} [9-91].

Yanaru *et al.* [9-92] investigated the orthodontic forces of NiTi wires under the retained condition on the dental arch model, and evaluated with the changes in temperature and deflection. The tested specimens were a commercially available SE NiTi wire and two SME NiTi wires with their nominal A_f (austenitic transformation finish) points were 35 and 40°C. It was shown that (i) typical SE hysteresis loops under the restraint condition was at 40°C, (ii) the recovery forces in the plateau region at 1.0 mm deflection were much larger than desired in the clinical guidelines around oral temperatures, and (iii) in the SME wire with A_f of 40°C, the recovery force rapidly decreased to zero by a small reduction of the deflection from its maximum, whereas the wire exerted the force with the remaining permanent deflection by temperatures rising [9-92].

Kwon *et al.* [9-93] examined extensively the effect of acidic fluoride solution on NiTi arch wires by testing crystal structure, tensile strength, morphology after fracture, and element release from wire under four different test solutions after 1 or 3 day immersion. It was found that (i) three-day immersion in a 0.2%NaF with pH 4 solution did not form any new crystal structure; however (ii) tensile strength after immersion was changed compared to the as-received wires. It was further found that (iii) element release in the test solution increased as NaF concentration and the period of immersion increased, and as pH valued decreased, and (iv) wires immersed in a 0.2% NaF with pH 4 solution released a several-fold greater amount of elements than wire in a 0.05% NaF with pH 4 solution [9-93].

The shape transformation behavior in three commercial NiTi orthodontic wires having different transformation temperatures was studied by micro X-ray diffraction by Iijima *et al.* [9-94]. MicroXRD spectra were obtained at three different NiTi wires included bending angles (135°, 146° and 157°) and three different temperatures (25, 37, and 60°C). The regions analyzed by microXRD were within the separate areas of a given wire specimen that experienced only tensile or compressive strain. The diffraction intensity ratio (M002/A110) between the plane (002) peak in martensite phase and the plane (110) peak in austenite phase of NiTi was used as the index to the proportional of the martensite and austenite phases. It was reported that the ratio of martensite to austenite increased in all three Ni-Ti wires with decreasing included bending angle (greater permanent bending deformation), and was lower within the compression area for all wires at all bending angles than within the tension area [9-94].

Orthodontic brackets made of CpTi have been proposed [9-95], which can be fabricated through the metal injection mold (MIM) technique, as will be described in Chapter 10 [9-96].

9.9. ENDODONTIC FILES AND REAMERS

During the instrumentation of curved canals, inadvertent procedural errors, such as ledging, zip formation, and separation of instruments, occasionally occur [9-97]. When such problems occur, the root canal morphology is adversely altered. This violates a basic endodontic principle, which is that endodontic preparation should conform as closely as possible to the general configuration of the original canal [9-98, 9-99]. The American National Standards Institute/American Dental Association (ANSI/ADA) specification No.28 lists certain torsional property requirements for endodontic files and reamers [9-100]. Instrument separation is a serious concern in endodontics. Because stainless-steel instruments usually deform before they separate, dentists can inspect them for visible signs of instrument deformation. A deformed instrument usually shows severe bending or unwinding of the flutes, indicating that the elastic limit of the metal has been exceeded, and that the instrument should be discarded. NiTi endodontic instruments were introduced to facilitate instrumentation of curved canals. NiTi instruments are superelastic and will flex far more than stainless-steel instruments before exceeding their elastic limit. This flexibility is an important property that allows preparation of curved canals while minimizing transportation. Despite this increased flexibility, separation is still a concern with NiTi instruments, and they have been reported to undergo unexpected fracture [9-101, 9-102]. Separation can occur without any visible signs of previous permanent deformation, apparently within the elastic limit of the instrument [9-102]. This indicates that the NiTi instruments could be deteriorated by low-cycle (or high-stress) fatigue damage.

The torsional properties of stainless-steel endodontic files and NiTi superelastic endodontic files were compared by Rowan *et al.* [9-97]. File sizes 15, 25, 35, 45, and 55 were subjected to torsional loads in both the clockwise (CW) and counterclockwise (CCW) directions independently. Results showed that stainless steel files had a significantly greater rotation to failure in the CW direction, whereas the NiTi files had a significantly greater rotation to failure in the CCW direction. Despite these differences in rotation to fracture, it was reported that there was essentially no difference between the stainless steel and NiTi instruments in the torque that it took to cause failure in both the CW and CCW directions. Therefore, it was concluded that whereas the number of CW and CCW rotations to failure differed for the two instruments, the actual force that it took to cause that failure was the same [9-97]. Pruett *et al.* [9-102] studied cyclic fatigue of NiTi engine-driven instruments by determining the effect of canal curvature and operating speed on the breakage of light-speed instruments. A new method of canal curvature evaluation that addressed both angle and abruptness of curvature was introduced. Canal curvature was simulated by constructing six curved stainless-steel guide tubes with angles of curvature of 30, 45, or 60°, and radii of curvature of 2 or 5 mm. A simulated operating load of 10 g cm was applied. Instruments were able to rotate freely in the test apparatus at speeds of 750, 1300, and 2000 rpm until separation occurred. It was found that (i) cycles to failure were not affected by rotation speeds, (ii) instruments did not separate at the head, but rather at the point of maximum flexure of the shaft, corresponding to the midpoint of curvature within the guide tube, and (iii) cycles to failure significantly decreased as the radius of curvature decreased from 5 to 2 mm, and as the angle of curvature increased greater than 30°. On the basis of these findings, it was concluded that, for NiTi engine-drive rotary instruments, the radius of curvature, angle of curvature, and instrument size are more important than operating speed for predicting separation [9-102].

Instrumentation of the curved canal remains a challenging task to maintain the curvature while avoiding apical transportation and strip perforation. Tucker *et al.* [9-103] quantified the canal wall planning achieved by hand instrumentation using stainless steel files, compared with engine-driven NiTi files in curved canals. Twenty-two mesial roots of extracted human mandibular molars were divided into two groups based on root curvature and length. The mesiolingual canals were instrumented using either stainless steel files in a step-back anticurvature filing method, or they were instrumented with engine-driven 0.02 taper NiTi files. Ground sections were prepared at 1-, 2.5-, and 5-mm levels from the working length. The mesiobuccal canal was used as an uninstrumented control for predentin character. It was found that there was no significant difference in overall canal wall planning between the two groups and no significant difference at each of the three levels [9-103]. Coleman *et al.* [9-104] compared step-back preparations in curved canals of resin blocks using NiTi K-files and stainless-steel K-files [9-104]. Forty canals in resin blocks were cross-sectioned at 3 levels; $1-2$ mm from the apical foramen, middle of the curve, and coronal. Direct digital computer images were recorded before and after instrumentation, and superimposition of the images combined with digital subtraction computer software allowed direct measurement of area instrumented, distance of transportation, and shape analysis. It was found that (i) area removed by instrumentation was significantly greater for stainless-steel files at the middle level and (ii) it took significantly longer to prepare a canal with NiTi files in resin blocks compared to stainless steel.

There are various methods to assess the progressive damage due to fatigue cycling by either destructive method or non-destructive method. They might include subsequent measurements on hardness, ductility, electrical resistance, density, modulus of elasticity [9-105], and dislocation density measurements by X-ray diffraction technique [9-106]. Files are the most efficient instruments in endodontics for the removal of hard tissue and canal enlargement. If the file is evaluated as non-damaged, it will be sterilized and re-used (or recycled) on the next patient. Although the breakage and abandonment of a portion of the endodontic file in the root canal may be considered an option, it may cause many problems to certain type of patients. If the patient is hypersensitive to nickel, an allergic reaction may occur. In a more serious case, the dissolved nickel element will act. Accordingly, it can be speculated that the dissolution of nickel out of broken NiTi is likely to be much greater than in the intraoral environment. Even with the materials containing potentially harmful elements, there are no scientific standards to assess the cumulative damage on NiTi before mechanical failure occurs. Oshida *et al.* [9-107] investigated progressive fatigue damage on two NiTi files (0.25 and 0.35 mm in diameter) under fatigue cycling in a push-pull mode inside the root canal having three different canal curvatures ($\alpha = 25$, 40, and 55°). Changes in DC electrical resistance was monitored with the Wheatstone's bridge type galvanometer (4.5 DCV) to an accuracy of 0.001 Ω at room temperature. It was concluded that (i) normalized cycle-to-failure is linearly related to the normalized damage parameter (change in resistance), and (ii) all the fatigue-fractured files demonstrated an electric resistance ranging between 35 and 40 m Ω [9-107].

9.10. CLAMPS AND STAPLES

The forming and clinical tests of osteosynthesis clamps in NiTi alloy were described. It was reported that (i) by application of localized heating a defined change in shape is induced after implantation, which compresses the osteotomy gap, and (ii) histological evaluation of the tissue surrounding the clamp shows that the alloy satisfies tissue compatibility needs [9-108]. The SME NiTi staple with an oval part joining the two prongs provides not only a double-compression permanent effect in the prongs and in the oval part, but also provides better stability of the osteosynthesis [9-109]. Its application in the foot has been studied in 1850 cases since 1986 with few (5%) of the contra-indications. It was reported that (i) they are associated to the compression strength of the staple in very osteoporotic bones and in the arthrodesis with important resection of the subchandral and cancellous bones, (ii) they have been used successfully for shaft osteotomies of the first phalanx, but also for first joint arthrodesis, and for middle or hind foot arthrodesis, and (iii) in each of these indications, the use of SME staples results in a great improvement not only concerning the reliability of the osteosynthesis, but also in healing time due to the double compression effect [9-109].

Coronary stents made out of NiTi should not be omitted here. NiTi stents have excellent mechanical properties to provide strength to artery walls. The perfect stent does not exist, and it will probably never exist. However, several approaches have been or are currently under investigation to increase their clinical performance. One of these is the implanting of biodegradable stents made of non-toxic materials. The major clinical complication with stents is restenosis, defined as the blockage of blood flow by the coagulation of blood inside the stent [9-110]. The Homer Mannalok wire needle localizer was also developed using NiTi's superelastic characteristics [9-80].

REFERENCES

[9-1] Ban S. Effect of alkaline treatment of pure titanium and its alloys on the bonding strength of dental veneering resins. J Biomed Mater Res 2003;66A:138−145.

[9-2] McCraken M. Dental implant materials: commercially pure titanium and titanium alloys. J Prosthodont 1999;8:40−43.

[9-3] Abe Y, Sato Y, Taji T, Akagawa Y, Lambrechts P, Vanherle G. An *in vitro* wear of posteria denture tooth materials on human enamel. J Oral Rehabil 2001;28:407−412.

[9-4] McCaughey AD. Sandblasting and tin-plating-surface treatments to improve bonding with resin cements. Dental Update 1993;20:153−157.

[9-5] Chang J.C., Powers JM, Hart D. Bond strength of composite to alloy treated with bonding systems. J Prosthodont 1993;2:110−114.

[9-6] Yoshida K, Taira Y, Matsumura H, Atsuta M. Effect of adhesive metal primers on bonding a prosthetic composite resin to metals. J Prosthet Dent 1993;69:357−362.

[9-7] Anagnostopoulos T, Eliades G, Palaghias G. Composition, reactivity and surface interactions of three dental saline primers. Dent Mater 1993;9:182−190.

[9-8] Craig RG. Restorative dental materials. St Lois: Mosby. 1993. pp. 517−519.

[9-9] Musil R, Tiller H. The adhesion of dental resins to metal surfaces. In: The Kulzer silicating technique, Kulzer and Co., GmbH., 1984.

[9-10] Mudford L, Curtis RV, Walter JD. An investigation of debonding between heat-cured PMMA and titanium alloy (Ti-6Al-4V). J Dent 1997;25(5): 415−421.

[9-11] Oshida Y, Barco MT. Superplastically-formed prosthetic components, and equipment for same. US Patent 6,116,070 (2000).

[9-12] Okuno O, Nakano T, Hamanaka H, Kato I. Diffusion boding to titanium alloys during superplastic forming. J Jpn Soc Dent Mater Devices 1991; 10:483−491.

[9-13] Wakabayashi N, Ai, M. A short-term clinical follow-up study of super-plastic titanium alloy for major connectors of removable partial dentures. J Prosthet Dent 1997;77:583−587.

[9-14] Srimaneepong V, Yoneyama T, Wakabayashi N, Kobayashi E, Hanawa T, Doi H. Deformation properties of Ti-6Al-7Nb alloy casting for removable partial denture frameworks. Dent Mater J 2004;23:497−503.

[9-15] Southan DE. Restorative Dental Materials. In: Metals forum. Australian Inst Metals 1980;3:222−227.

[9-16] Zinelis S, Tsetsekou A, Papadpoulos T. Thermal expansion and microstructural analysis of experimental metal-ceramic titanium alloys. J Prosthet Dent 2003;90:332−338.

[9-17] Miura I. Titanium alloys for biomaterials, especially dental materials. Titanium and Zirconium 1991;38:17−18.

[9-18] Hautaniemi JA, Hero H, Juhanoja JT. On the bonding of porcelain on titanium. J Mater Sci: Mater Med 1992;3:180−191.

[9-19] Watanabe K, Okawa S, Miyakawa O, Nakano S, Honma H, Shiokawa N, Kobayashi M. Interface reactions between titanium and porcelains during firing. Jpn J Dent Mater 1993:12:620−629.

[9-20] Suansuwan N, Swain MV. Adhesion of porcelain to titanium and a tita-nium alloy. J Dent 2003;31:509−518.

[9-21] Hanawa T, Kon M, Ohkawa S, Asaoka K. Diffusion of elements in porce-lain into titanium oxide. Dent Mater J 1994:13:164−173.

[9-22] Brockhurst PJ. Dental Materials: new territories for materials science. In: Metals forum. Australian Inst Metals 1980;3:200−210.

[9-23] Togaya T, Suzuki M, Tsutsumi S, Ida K. An application of pure titanium to the metal porcelain. Dent Mater J 1983;2:210−219.

[9-24] Adachi M, Mackert JR, Parry EE, Fairhurst CW. Oxide adherence and porcelain bonding to titanium and Ti-6Al-4V alloy. J Dent Res 1990;69: 1230−1235.

[9-25] Kimura H, Horng CJ, Okazaki M, Takahashi J. Oxidation effects on porcelain-titanium interface reactions and bond strength. Dent Mater J 1990;9:9199.

[9-26] Nagayama K, Izumi T, Kikui T, Okazaki M, Noguchi H. Porcelain-fused-to-titanium-system − EPMA, ESCA studies on the titanium surface and

the interface between titanium and fused porcelain. J Dent Mater 1993;12: 224−225.

[9-27] Oshida Y, Hashem A. Titanium-porcelain system Part 1: Oxidation kinetics of nitrided pure titanium, simulated to porcelain firing process. J Biomed Mater Eng 1993;3:185−198.

[9-28] Oshida Y, Fung LW, Isikbay SC. Titanium-porcelain system. Part II: Bond strength of fired porcelain on nitrided pure titanium. J Biomed Mater Eng 1997;7:12−34.

[9-29] Gilbert JL, Covey DA, Lautenschlager EP. Bond characteristics of porcelain fused to milled titanium. Dent Mater 1994;10:134-140.

[9-30] Könönen M, Kivilahti J. Joining of dental ceramics to titanium by metallic interlayers. Proceedings of 7th international meeting on modern ceramic technologies. Vinzencini P, editor. Amsterdam: Elsevier. 1990. pp. 41−47.

[9-31] Dérand T, Herø H. Bond strength of porcelain on cast vs. wrought titanium. Scand J Dent Res 1992;100:84−88.

[9-32] Rammenlsberg P, Aschl I, Pospiech P. Verbundsfestigket neidrigschmelzender Keramiken zu Titan unter Berücksichtigung der Obenflächen Konditionering. Dtsch Zahnärtzl Z 1990;45:200−203.

[9-33] Reyes MJD, Oshida Y, Andres CJ, Barco MT, Hovijitra S, Brown D. Titanium-porcelain system. Part III. Effects of surface modification on bond strengths. J Biomed Mater Eng 2001;11:117−136.

[9-34] Atsu S, Berksun S. Bond strength of three porcelains to two forms of titanium using two firing atmospheres. J Prosthet Dent 2000;84: 567−574.

[9-35] Oka K, Hanawa T, Kon M, Lee HH, Kawano F, Tomotake Y, Mastumoto N, Asaoka K. Effect of barium in porcelain on bonding strength of titanium-porcelain system. Dent Mater J 1996;15:111−120.

[9-36] Nergiz I, Meine H-C, Niedermeyer W. Untersuchung zur Schverbunds festigkeit von titankeramischen Systemen. Dts Zahnärtzl Z 1999;54: 688−690.

[9-37] Newham RC. Mechanical properties of ceramic materials: strength tests for brittle materials. Proc Br Ceram Soc 1975;25:281−293.

[9-38] White SN, Ho L, Caputo AA, Goo E. Strength of porcelain fused to titanium beams. J Prosthet Dent 1996;75:640−648.

[9-39] White SN, Caputo AA, Vidjak FMA, Seghi RR. Moduli of rupture of layered dental ceramics. Dent Mater 1994;10:52−58.

[9-40] Oshida Y, Reyes MJD. Titanium-porcelain system. Part IV. Some mechanistic considerations on porcelain bond strengths. J Biomed Mater Eng 2001;11:137−142.

[9-41] Eriksson M, Andersson M, Carlström E. Titanium dental copings prepared by a powder metallurgy method: A preliminary report. Int J Prosthodont 2004;17:11−16.

[9-42] Kum D, Park C, Yi Y, Cho L, Comparison of cast Ti-Ni alloy clasp retention with conventional removable partial denture clasps. J Prosthet Dent 2004;91:374−382.

[9-43] McGiveney GP, Carr AB. McCracken's removable partial prosthodontics. 10th Ed. Elsevier Science. 1999. pp. 258−265.

[9-44] Vallittu PK, Kokkonen M. Deflection fatigue of cobalt-chromium, titanium, and gold alloy cast denture clasp. J Prosthet Dent 1995;74:412−419.

[9-45] Sahafi A, Peutzfeldt A, Asmussen E, Gotfredsen K. Bond strength of resin cement to dentin and to surface-treated posts of titanium alloy, glass fiber, and zirconia. J Adhes Dent 2003;5:153−162.

[9-46] Hedlund S-O, Johansson NG, Sjögren G. Retention of prefabricated and individually cast root canal posts *in vitro*. Br Dent J 2003;195:155−158.

[9-47] Akisli I, Özcan M, Nergiz I. Resistance of core materials against torsional forces on differently conditioned titanium post. J Prosthet Dent 2002;88:367−374.

[9-48] Standlee JP, Caputo AA. Hanson EC. Retention of endodontic dowels: effect of cement, dowel length, diameter, and design. J Prosthet Dent 1978;39: 400−405.

[9-49] Standlee JP, Caputo AA, Holcomb J, Trabert KC. The retentive and stress-distributing properties of a threaded endodontic dowel. J Prosthet Dent 1980;44:398−404.

[9-50] Kovarik RE, Breeding LC, Caughman WF. Fatigue life of three core materials under simulated chewing conditions. J Prosthet Dent 1992;68: 584−590.

[9-51] Chang WC, Millstein PL. Effect of design of prefabricated post heads on core materials. J Prosthet Dent 1993;69:475−482.

[9-52] Cohen BI, Condos S, Deutsch AS, Musikant BL. Fracture strength of three different core materials in combination with three endodontic posts. Int J Prosthodont 1994;7:178−182.

[9-53] Huysmans MC, van der Varst PG. Mechanical longevity estimation model for post-and-core restorations. Dent Mater 1995;11:252−257.

[9-54] Lepe X, Bales DJ, Johnson GH. Tensile dislodgement evaluation of two experimental prefabricated post systems. Oper Dent 1996;21:209−212.

[9-55] Burgess JO, Summit JB, Robbins JW. The resistance to tensile, compression, and torsional forces provided by four post systems. J Prosthet Dent 1992;68:899−903.

[9-56] Cohen BI, Condos S, Musikant BL, Deutsch AS. Retentive properties of threaded split-shaft posts with titanium-reinforced composite cement. J Prosthet Dent 1992;68:910−912.

[9-57] Levartovsky S, Goldstein GR, Georgescu M. Shear bond strength of several new core materials. J Prosthet Dent 1996;75:154−158.

[9-58] Stockton LW, Williams PT. Retention and shear bond strength of two post systems. Oper Dent 1999;24:210−216.

[9-59] Stockton LW, Williams PT, Clarke CT. Post retention and post/core shear bond strength of four post systems. Oper Dent 2000;25:441−447.

[9-60] Huysmans MC, van der Varst PG, Schafer R, Peters MC, Plasschaert AJ, Soltesz U. Fatigue behavior of direct post-and-core-restored premolars. J Prosthet Dent 1997;71:1145−1150.

[9-61] Cohen BI, Pagnillo MK, Newman I, Musikant BL, Deutsch AS. Effects of three bonding systems on the torsional resistance of titanium-reinforced composite cores supported by two post designs. J Prosthet Dent 1999;81:678−683.

[9-62] Cohen BI, Penugonda B, Pagnillo MK, Schulman A, Hittelman E. Torsional resistance of crowns cemented to composite cores involving three stainless steel endodontic post designs. J Prosthet Dent 2000;83:38−42.

[9-63] Bergman B, Lundquist P, Sjögren U, Sundquist G. Restorative and endodontic results after treatment with cast posts and cores. J Prosthet Dent 1989;61:10−15.

[9-64] Hatzikyriakos AH, Reisis GI, Tsingos N.A 3-year post operative clinical evaluation of posts and cores beneath existing crowns. J Prosthet Dent 1992;67:454−458.

[9-65] Sorensen JA, Martinoff JT. Clinically significant factors in dowel design. J Prosthet Dent 1984;52:28−35.

[9-66] Taira Y, Yanagida H, Matsumura H, Yoshida K, Atsuta M, Suzuki S. Adhesive bonding of titanium with a thione-phosphate dual functional primer and self-curing luting agents. Eur J Oral Sci 2000;108:456−460.

[9-67] Turner CH. Post-retained crown failure: A survey. Dent Update 1982;9:221−234.

[9-68] Kozono Y, Tajima K, Kakigawa H, Hayashi I, Murakami Y. Pure titanium prefabricated post − compatibility with cast metal core. Dent Mater J 1986;5: 237−245.

[9-69] Topoleski LDT, Ducheyne P, Cuckler JM. Flow intrusion characteristics and fracture properties of titanium-fibre-reinforced bone cement. Biomaterials 1998;19:1569−1577.

[9-70] James SP, Hasty M, Davies J, Piehler H, Harris WH. A fractographic investigation of PMMA bone cement focusing on the relationship between porosity reduction and increased fatigue life. J Biomed Mater Res 1992;26:651−662.

[9-71] Rose RM, Litsky AS. Biomechanical considerations in the loosening flip replacement prostheses. Current perspectives on implantable devices. Vol. 1 New York: JAI Press Inc. 1989. pp. 1−45.

[9-72] Topoleski LDT, Ducheyne P, Cuckler JM. A fractographic analysis of *in vivo* poly(methyl methacrylate) bone cement failure mechanisms. J Biomed Mater Res 1998;24:135−154.

[9-73] Topoleski LDT, Ducheyne P, Cuckler JM. The effect of centrifugation and titanium fibre reinforcement on fatigue failure mechanisms in poly(methyl methacrylate) bone cement. J Biomed Mater Res 1995;29:299−307.

[9-74] Sih GC, Berman AT. Fracture toughness concepts applied to methyl methacrylate. J Biomed Mater Res 1980;14:311−324.

[9-75] Zucarri AG, Oshida Y, Moore BK. Reinforcement of acrylic resins for provisional fixed restorations. Part I. Mechanical properties. J Biomed Mater Eng 1997;7:327−334.

[9-76] Panyayong W, Oshida Y, Andres CJ, Barco TM, Brown DT, Hovijitra S. Reinforcement of acrylic resins for provisional fixed restorations. Part III. Effects of addition of titania and zirconia mixtures on some mechanical and physical properties. J Biomed Mater Eng 2002;12:353−366.

[9-77] Chu KT, Oshida Y, Hancock EB, Kowolik MJ, Barco MT, Zunt SL. Hydroxy-apatite/PMMA composites as bone cements. J Biomed Mater Eng 2004;14:87−105.

[9-78] Nakamura A, Wanibe H, Ando K, Iwama A, Kitamura K, Shibata N, Nakata K, Zinno S, Nakano K, Tsuruta S, Kawai T, Nakamura H. Mechanical properties of root canal sealers containing titanium oxide. J Jpn Soc Dent Mater Devices 2005; 24:287−291.

[9-79] Stöckel D. Nitinol medical devices and implants. Min Invas Ther Allied Technol 2000;9:81−88.

[9-80] Pelton AR, Duerig TW, Berg B, Hodgson D, Mertmann M, Mitchell M, Proft J, Wu M, Yang J. Nitinol medical devices. Adv Mater Processes 2005;163:63−65.

[9-81] Fukuyo S, Sachdeva, RCL, Oshida Y, Sairenji E. The perio root implant (PRI). In: Perio root implant and medical application of shape memory alloy. Fukuyo S, Sachdeva RCL, editors. Tokyo: Japan Medical Culture Center. 1992.

[9-82] Andreasen G. A clinical trial of alignment of teeth using a 0.019 inch thermal nitinol wire with a transition temperature range between 31°C and 45°C. Am J Orthod 1980;23:528−537.

[9-83] Yoneyama T, Doi H, Hamanaka H, Yamamoto M, Kuroda T. Bending properties and transformation temperatures of heat treated Ni-Ti alloy wire for orthodontic appliances. J Biomed Mater Res 1993;27:399−402.

[9-84] Edman B, Moeller H. Trends and forecasts for standard allergy in a 12-year old patch test material. Contact Dermatitis 1982;8:95−98.

[9-85] Oshida Y, Sachdeva R, Miyazaki S, Fukuyo S. Biological and chemical evaluation of TiNi alloys. Mater Sci Forum. Trans Tech Pub. 1990;56/58: 705−710.

[9-86] Sachdeva R, Miyazaki S. Superelastic NiTi alloys in orthodontics. In: Engineering aspects of shape memory alloys. Duerig TW, editor. London: Butterworth-Heinemann. 1990. pp. 452−469.

[9-87] Sunderman FW. A review of the metabolism and toxicity of nickel. Ann Clin Lab Sci 1977;7:377−382.

[9-88] Samitz MH, Kleen A. Nickel dematerial harzards from prostheses. J Am Med Assoc 1972;223:1159−1163.

[9-89] Sunderman FW. A review of carcinogenicities of nickel, chromium, and arsenic compounds in man and animals. Prev Med 1976;5:279−283.

[9-90] Oshida Y, Sachdeva RCL, Miyazaki S. Microanalytical characterization and surface modification of TiNi orthodontic archwires. J Biomed Mater Eng 1992;2:51−69.

[9-91] Kusy RP, Wilson TW. Dynamic mechanical properties of straight titanium alloy arch wires. Dent Mater 1990;6:228−236.

[9-92] Yanaru K, Yamaguchi K, Kakigawa H, Kozono Y. Temperature- and deflection-dependences of orthodontic force with Ni-Ti wires. Dent Mater J 2003;22:146−159.

[9-93] Kwon Y-H, Cheon Y-D, Seol H-J, Lee J-H, Kim H-I. Changes on NiTi orthodontic wired due to acidic fluoride solution. Dent Mater J 2004;23: 557−565.

[9-94] Iijima M, Ohno H, Kawashima I, Endo K, Brantley WA, Mizoguchi I. Micro X-ray diffraction study of superelastic nickel-titanium orthodontic wires at different temperatures and stresses. Biomaterials 2002;23: 1769−1774.

[9-95] Sachdeva RCL, Oshida Y. Orthodontic Bracket, US Patent, 5,232,361 (1993).

[9-96] Deguchi T, Ito M, Obata A, Koh Y, Yamagishi T, Oshida Y. Trial production of titanium orthodontic brackets fabricated by sintering metal injection molding methods. J Dent Res 1996;75:1491−1496.

[9-97] Rowan MB, Nicholls JI, Steiner J. Torsional properties of stainless steel and nickel-titanium endodontic files. J Endodon 1996;22:341−345.

[9-98] Weine FS, Kelly RF, Lio PJ. The effect of preparation procedures on original canal shape and on apical foramen shape. J Endodon 1975;1:255−262.

[9-99] Walia H, Brantley WA, Gerstein H. An initial investigation of the bending and torsional properties of Nitinol root canal files. J Endodon 1988;14:346−351.

[9-100] Council on Dental Materials, Instruments and Equipment. Revised American National Standards Institute/American Dental Association Specification No.28 for root canal files and reamers, type K (revised 1988).

[9-101] Glosson CR, Haller RH, Dove SB, del Rio CE. A comparison of root canal preparation using Ni-Ti hand, Ni-Ti engine driven, and K-Flex endodontic instruments. J Endodon 1995;21:146−151.

[9-102] Pruett JP, Clement DJ, Carnes DL. Cyclic fatigue testing of nickel-titanium endodontic instruments. J Endodon 1997;23:77−85.

[9-103] Tucker DM, Wenkus CS, Bentkover SK. Canal wall planning by engine-driven nickel-titanium instruments, compared with stainless steel hand instrumentation. J Endodon 1997;23:170−173.

[9-104] Coleman CL, Svec TA. Analysis of Ni-Ti versus stainless steel instrumentation in resin simulated canals. J Endodon 1999;25:332−335.

[9-105] Pluvinage GC, Raguet MN. Physical and mechanical measurements of damage in low-cycle fatigue: Application for two-level test. In: Fatigue mechanisms. Lankford J, Davidson DL, Morris WL, Wei RP, editors. American Society for Testing and Materials STP811. Philadelphia: ASTM. 1983. pp. 139−150.

[9-106] Weiss V, Oshida Y. Fatigue damage characterization by x-ray diffraction line analysis and failure probability, EPRI AP-4477, EPRI. 1986. pp. 1−18.

[9-107] Oshida Y, Farzin-Nia F. Progressive damage assessment of TiNi endodontic files. In: Shape memory implants. Yahia L. editor. New York: Springer. 2000. pp. 236−249.

[9-108] Bensmann G, Baumgart F. Osetosyntheseklammern aus Nicekltitan Herstellung, Vorversuche und klinischer Einsatz. Tech Mitt Krupp 1982;40:123−134.

[9-109] Barouk LS. The double compressive nickel-titanium shape-memory staple in foot surgery. In: Shape memory implants. Yahia L. editor. New York: Springer. 2000. pp. 162−173.

[9-110] Lévesque J, Dubé D, Fiset M, Mantovani D. Materials and properties for coronary stents. Adv Mater Process 2004;162:45−48.

Chapter 10
Fabrication Technologies

10.1.	Casting	285
10.2.	Machining	288
10.3.	Electro Discharge Machining (EDM) and CAD/CAM	289
10.4.	Isothermal Forging	291
10.5.	Superplastic Forming (SPF)	291
10.6.	Diffusion Bonding (DB)	294
10.7.	Powder Metallurgy	295
10.8.	Metal Injection Molding (MIM)	297
10.9.	Laser Welding/Forming	299
10.10.	Soldering	301
10.11.	Heat Treatment (HT)	302
References		303

Chapter 10
Fabrication Technologies

There are many conventional and advanced technologies available, including casting, milling, spark erosion, forging, powder metallurgy, a combination of milling or spark erosion with laser welding [10-1–10-6] fabricating medical, and dental prostheses and/or devices. This chapter will discuss and review these technologies.

10.1. CASTING

There are unusual features associated with titanium casting. Unlike castings of other metals, (i) there are no commercial titanium alloys developed strictly for casting applications, therefore all titanium castings have compositions based on those of the common wrought alloys, and (ii) titanium castings are equal or nearly equal in strength to their wrought counterparts. The casting of titanium dental appliances was noted in the early 1970s with the works of Waterstratt [10-7], followed by numerous studies in Japan, Europe, and United States toward the precision casting of dental prostheses, and the development of casting machines and suitable investment materials. There are a number of ways to cast titanium. At this moment, there are about 10 different systems available for casting titanium [10-8, 10-9]. Among them, for example, the Ohara (atmospheric argon shielded arc melting followed by centrifugal casting) and the Castmatic (pressurized argon shielded arc melting followed by gravity casting accelerated by two atmospheres argon pressure above and one atmosphere vacuum below) are examples of such machines. These casting machines are designed and manufactured in particular to overcome the challenging parameters such as high activity and low density of titanium. It was found that the dental titanium castings in these machines are susceptible to surface oxygen contamination and possible interaction with the investment, causing a tripling of surface hardness (about 600 Knoop hardness number), whereas the rest of the casing (200 KHN) at 200 μm underneath the surface case [10-10, 10-11].

Evaluation of titanium casting via radiographic means is a technique that has been found useful in detecting internal porosities [10-12]. Sprue design in Ti, as in gold casting is important in the prevention of porosities. The most favorable sprue design was determined to be a rounded or curved sprue [10-13]. While many authors discuss porosity as a challenge when casting Ti, one study found that 97% of the removable partial denture (RPD) framework cast was technically acceptable [10-14]. Cecconi *et al.* [10-14] evaluated the degree of porosity via radiographic examination of 300 cast CpTi (grade 2) RPD frameworks. A single-chamber, gas pressure Titec 205 M casting machine was used. One lab technician cast all frameworks. The frameworks were used clinically. It was reported that (i) no framework was retuned for any reason, (ii) 250 out of 300 frameworks were technically acceptable as cast, and (iii) 41 were technically acceptable after laser welding modifications and nine were technically unacceptable after casting and needed to be redone.

The castability of titanium and its alloys can be metallurgically improved [10-15, 10-16]. Since commercially pure titanium has a relatively high melting point (1670°C), meta-stable β-phase Ti-based alloys have recently been developed to reduce the melting points by adding Cr or Pd (e.g., Ti-20Cr-0.2Si, or Ti-23Pd-5Cr) [10-17, 10-18]. It was reported that Ti casting was done by the argon-arc melting and subsequently argon or vacuum-pressurized casting of Ti-6Al-4V, Ti-15V, Ti-20Cu, and Ti-30Pd [10-19], and NiTi, Ti-Pt, Ti-Cr, and Ti-Zr alloys [10-20], as well as CpTi [10-21, 10-22]. However, special care must always be taken to control casting defects with titanium (which can not be totally avoided when using a casting process). Defects include such things as shrinkage cavities, pinholes, or voids, and they are particularly a problem with denture bases [10-23, 10-24], where the surface area is relatively large compared to the thin thickness. It is important to avoid the creation of an alpha layer (or α-case), eliminate porosity, and achieve a good fit of the casting onto a die. Since titanium is an active metal, particularly with oxygen, the investment casting of Ti is adversely affected by reaction of the melt with the atmosphere in the melting or casting chamber, reaction of the melt with the melting crucible, and reaction of the melt with the mold materials [10-2, 10-25–10-29]. Even when refractories having thermodynamic stability similar to that of the titanium oxide family are used (such as MgO and Al_2O_3), the very strong reducing power of titanium decomposes such common oxides. The oxygen liberated from the oxide diffuses into the surface of the casting to form a hardened layer (about 200 μm thick), depending on the investment used [10-10, 10-11, 10-30].

Such an oxygen-rich layer (α-case) possesses several drawbacks, including (i) unnecessary hard and brittleness [10-31], (ii) reducing ductility and fatigue resistance [10-9, 10-32], and (iii) resulting in inferior titanium/porcelain bond strengths [10-33–10-35]. Several methods have been evaluated to minimize the formation of this reaction layer, including coating the wax pattern with an oxide that is thermodynamically more stable than the titanium oxides before investing [10-36]. When a wax pattern was coated with a slurry containing Y_2O_3 powder, it was reported that (i) the surface hardness (25 μm from the cast surface) of cast commercially pure titanium was found to be remarkably reduced (280 VHN) compared to the surface hardness of the uncoated cast Ti (530 VHN) [10-11], and (ii) use of ZrO_2

reduced the hardness near the cast surface (457 VHN) [10-37]. Miyakawa *et al.* [10-29] investigated reaction layer structures of titanium cast in three molds containing spinel ($MgO \cdot Al_2O_3$). For this spinel mold, the reaction zone consisted of baked product layer and the alpha case, which was divided into two layers. The outside was stabilized and hardened by a solid solution of Al, whereas the inside was not so much hardened without a solid solution of Al. It was found that, for the investment mixed with magnesium acetate solution, the Al-stabilized alpha case with higher hardness value formed on limited surfaces [10-29]. A novel investment material was developed by Nakamura *et al.* [10-38]. Fine particle sizes of alumina (Al_2O_3) and zirconia (ZrO_2) were imbedded among relatively large particle size silica (SiO_2) crystals, so that the direct contact between molten titanium and silica can be avoided to a great extent. Zr metal powder was mixed with magnetia (magnesium oxide: MgO) to provide an expansion-controlling mold material [10-38]. It was also reported that, using a large particle size of colloidal silica as a mixing liquid to phosphate-bonded investment, thermal expansion can be controlled and surface roughness, as well as the surface reaction layer, was improved [10-39]. Cast titanium for dental crowns has been investigated using the investment (consisted essentially of SiO_2, $Mg_2P_2O_7$, $SiO_2 \cdot H_2O$ and Mg_2SiO_4), claiming that formation of α-case is reduced [10-40]. A relatively simple method with an investment casting of titanium was developed with various characteristics (such as arc melting in argon, combination of vacuum and pressured assisted casting, and shell casting with zirconia), and a three-unit crown was successfully cast. It was claimed that the formation of a brittle α-case was avoided [10-41].

There are several unique melting methods. Traditionally, titanium alloys are melted, refined, and cast into round ingots following two or more runs of the vacuum arc remelting process. Multiple vacuum arc remelting runs are used to homogenize the chemical compositions and, further, in an attempt to overcome two types of melt defects – type I hard α inclusions and high-density inclusions – that are a primary quality concern when producing high-performance titanium alloys. These melt defects, when present undetected at sub-surface location, could promote premature fatigue failure in rotating components, shorten the life cycle of a component, and compromise product performance and safety. Sampath [10-42] introduced a competing melting method as plasma arc cold hearth melting and electron beam cold hearth melting, followed by a vacuum arc remelting. Melting Ti-Nb, Ti-Ta, or Ti-Zr binary alloy is not easy because of melting point and density differences between Ti and these alloying elements are large. To overcome this technical problem, a cold crucible levitation melting, using cast molding made of a mixture of alumina cement and calcium zirconate with a lithium carbonate as a liquid was developed, and successful melting without any segregation was reported [10-43].

10.2. MACHINING

Conventionally, research and development of dental materials has been focused on casting, with very few activities in machinability, specifically aiming on CAD/CAM materials. For an ordinary machining process, one should remember that titanium's work hardening rate ($n = 0.05$ and n is in the equation $\sigma = K\varepsilon^n$) is much less than that of 316 (18Cr-8Ni-2Mo) stainless steel ($n = 0.45$), copper ($n = 0.54$), and alpha brasses ($n = 0.49$), but with proper technique it can readily be machined or polished to a very fine finish. At the same time, it should be noticed that Ti is a poor conductor of heat (0.16 cal/cm/s·°C cm² vs. 0.71 for gold). While this allows excellent localized electric arc spot welding, it means that during cutting heat will not dissipate quickly [10-10]. It is also true that the relatively low value of modulus of elasticity (MOE) of titanium causes a tendency for work to move away from the cutting tool. But if one uses a few basic rules - low cutting speeds, high feed rates, copious amounts of cutting fluid, and sharp tools replaced at the first signs of wear – good results are obtainable [10-10]. In general, poor machinability can arise for various reasons, including high tensile strength, high high-temperature strength, high hardness, large ductility, low value of modulus of elasticity, high work-hardenability, low thermal conductivity, large thermal expansion coefficient, and large reactivity with tool materials. Among these reasons, low value of MOE, low thermal conductivity and high reactivity are typified for titanium materials. Titanium materials have unique machining properties. While the cutting forces are only slightly higher than in machining steels, there are other characteristics that make these alloys more difficult to machine than steels having an equivalent hardness. For example, the chip-tool contact area while turning a titanium alloy is only about one-third to one-half as great as that of turning steel. Also, the thermal conductivity of titanium alloy (21.9 W/(m·K)) is about one-fourth of that of steels (80.4 W/(m·K)). This combination of a small contact area and the low thermal conductivity result in very high cutting temperatures. At a cutting speed of 100 ft/min, the temperature developed at the cutting edge of a carbide tool is 537°C when cutting a steel, while on the titanium alloy, the temperature can reach 704°C. Hence, the cutting speeds on titanium alloys must be lower in order to maintain a tool–chip temperature below that which results in short tool life [10-44]. Normally, S, Pb, P, Bi, Te, or rare-earth elements are alloyed to improve the machinability. For Ti materials, it was reported that alloying with Au, Ag, or Cu (as β-eutectoid type) or with Nb or Hf (as β solid solution type) improved the machinability [10-45].

For the final polishing of titanium materials as well as titanium casts, chemical polishing and buff-polishing are normally practiced. As a novel polishing method, an alcohol-based electrolyte was used while stirring; the original average roughness of

CpTi (0.38 μm) was refined to 0.03 μm, and the 0.55 μm original roughness of Ti-6Al-4V was improved to be 0.12 μm [10-46]. To prevent the implant periodontitis, due to plaque accumulation, a certain portion of dental implants need to be finished as smooth as possible to reduce the surface energy. Kyo *et al.* [10-47] developed the electrolytic in-process dressing (ELID) grinding technique to the mirror surface polishing in the neck region of the NiTi alloy in contact with alveolar bone. It was reported that (i) the ten-point average surface roughness was 185 nm and the center line surface roughness was 22 nm, indicating that good surface properties can be obtained, (ii) the results of analysis of the ELID ground surface using AES showed oxygen peaks other than the titanium and nickel elements, and (iii) when compared to the ground surface, the Ti peak was found to change only slightly along the depth, while the oxygen peak decreased and the Ni peak increased, indicating that the ground surface is oxidized during the ELID machining [10-47].

Since all materials have a finite modulus of elasticity, plastic deformation due to machining is followed by some elastic recovery when the load is removed. In bending, the recovery is called springback. Springback in forming titanium and titanium alloys increases as the R/T ratio (R represents the bend radius and T the work-metal thickness) and yield strength increases and the elastic modulus decreases, and inversely with forming temperature [10-48]. Springback in titanium alloys is more difficult to predict than springback in steel, although it depends on the same principles. Differences in the yield strength of various heats in titanium can cause differences in springback; higher stress ratios of yield strength to tensile strength generally result in greater springback. The springback phenomenon is mostly important when orthodontic arch wires are concerned. Orthodontists usually activate orthodontic wires somewhat into the permanent deformation range. Consequently, the practical working range is considered to be the elastic strain at the yield strength of the wire, namely the quotient of yield strength (σ_{YS}) and MOE. This important property (σ_{YS}/MOE) is termed springback [10-48].

10.3. ELECTRO DISCHARGE MACHINING (EDM) AND CAD/CAM

EDM, also known as spark erosion dentistry, has been used in industry for over 40 years as a means of machining hard metals with electrical impulses used to cut metal passively and precisely into various shapes and forms [10-49]. In 1982, Rubeling [10-50] introduced spark erosion and its application to the dental technology. Spark erosion can be defined as a metal removal process using a series of electrical sparks to erode material from a work piece in a liquid medium under carefully controlled conditions. It is a pressure-less process which does not transmit heat to the alloy, making it possible to consistently machine fit in the range of 0.05 mm [10-51]. The procedure consists of a tension current passing through an electrode, which produces a burning-away of minute particles. A graphite or copper electrode acts as a tool that invades the piece of the base metal and erodes a negative form in the shape of the electrode. The process is accompanied by sparks generated between the electrode and the restoration. The sparks melt the alloy by heating it to between 3000 and 5000°C [10-50, 10-52]. The sparks remove small amounts of the metal substrate within microseconds. The entire process takes place in a dielectric liquid bath that prevents the alloy from burning. Advantages of this technique are (1) it is possible to work without pressure, (2) no heat is conducted to the object being treated, so parts of the prosthesis which have already been finished with porcelain or acrylic resin will not be damaged, (3) all parts can be absolutely parallel, (4) very little surface roughness, and (5) no difficulties with any type of material [10-50, 10-52, 10-53].

Anderson and Anderson [10-54] developed a method by which the principles of spark erosion are combined with a method of machine duplication for CpTi fabricating crowns. The Procera process developed by the Nobelpharm company is based on a precision duplication milling through CAD/CAM, combined with spark erosion and laser welding of the metal frame. Procera is designed for use with titanium and densely sintered alumina making it possible to produce metal ceramic crowns or all ceramic crowns [10-55]. An alternative to casting Ti for ceramic applications was the development of CAD/CAM machining and spark erosion of Ti [10-56, 10-57]. The main advantage of this technique is that it overcomes the hardened surface layer that is encountered with Ti castings, therefore providing adequate porcelain–titanium bonding [10-58, 10-59]. It was mentioned that adaptation of the current Procera CAD/CAM method using pressed-powder technology in the fabrication of Ti copings has several advantages compared to the original Procera method [10-60, 10-61]. The original method required at least four milling operations, the outer surface milling of the Ti blank and milling of three electrodes for spark erosion of the inner form. The powder method needs only two milling operations, one for the enlarged refractory die and one for the outer contour of the coping. Using the spark-erosion process, the milling of the outer contour of the coping is performed on a Ti blank. With the pressed-powder method, the powder is compacted onto the surface of the refractory die and then milled before the sintering, which speeds up the process and reduces wear of the milling tools. Although the spark-erosion process requires careful alignment of the carbon electrodes with the Ti blank in fabricating the coping, with powder technology, the alignment is not critical in the same way. The spark-erosion process is a time-consuming and relatively costly operation. The pressed-powder technology could become a very cost-effective operation. The sintering process does take time; however, when the process is used commercially in creating multiple copings at the same time, it could become cost effective. Several approaches have been reported in the literature using powder technology in dentistry [10-60, 10-61].

10.4. ISOTHERMAL FORGING

Isothermal forging differs from conventional forging in that the temperature of the workpiece is held constant during forging by heating the dies. It is found that the microstructure is affected by isothermal forging in various ways, the most important being the production of a fine grained, equiaxed structure, and free of slip banding. Therefore, by careful control of strain rate and temperature, control of the microstructure, and hence the properties, can be achieved. Forging is a common method of producing wrought titanium alloy articles. Forging sequences and subsequent heat treatment as post-forging heat treatment can be used to control the microstructure and resulting properties of the product. Forging is more than just a shape-making process. The key to successful forging and heat treatment is the beta (β) transus temperature (i.e., 885°C). The higher the processing temperature in the $(\alpha + \beta)$ region, the more beta is available to transform on cooling. The solution heat treatment offers a chance to modify or tune the as-forged microstructure, while the age cycle modifies the transformed beta structures to an optimum dispersion [10-62].

Titanium alloys are known to be very susceptible to their metallurgical background, and their properties can vary considerably depending on the alpha (HCP) $\leftrightarrow$ beta (BCC) transformation temperature. Microstructural control is basic to the successful processing of titanium alloys. Undesirable structures (grain-boundary alphas, beta fleck, elongated alpha) can interfere with optimum property development. Titanium ingot structures can carry over to affect the forged product. Beta processing, despite its adverse effects on some mechanical properties, can reduce forging costs, while isothermal forging offers a means of reducing forging pressure and/or improving die fill and part detail. The first step is to heat the tooling to the forging temperature. A good deal of progress has been made in this area and tooling sets are now available that can withstand elevated temperatures. A second factor for the success of isothermal forging is the forgeability of the alloy: the lower the temperature needed to forge the alloy, the easier it will be to form at uniform temperature. This is why isothermal forming is easier with β or near β alloys, which have a much lower β-transus and lower resistance to hot forming [10-62].

10.5. SUPERPLASTIC FORMING (SPF)

The phenomenon of superplasticity was first observed over 70 years ago, but active research in this field did not begin until 1960. Since then, superplastic effects have been reported over 100 material systems [10-63]. The earliest observations of this unique phenomenon were made in the mid-1940s in Russia when a scientist investigated the thermal-fatigue, and discovered an unusual amount of deformation which could not be explained by the ordinal high temperature creep theory. The extensive research activity was stimulated by the researchers including Backhofen, Underwood, Greenwood, Sherby, Weiss, Oelshlägel, de Jong, Oshida, and others, beginning about the 1960s. There are several conference proceedings and review books available [10-64–10-66]. Superplasticity in metals and alloys is a phenomenon which, by its very uniqueness, has attracted scientists and technologists over a considerable number of years. Superplasticity is the deformation process that produces essentially neck-free elongations of many hundreds of percent in metallic materials deformed in tension. High ductility is also encountered in superplastic alloys during torsion, compression, and indentation hardness testing. Superplasticity can be induced both in materials possessing a stable, ultra-fine grain size at the temperature of deformation, and in those subjected to special environment conditions, e.g., thermal cycling through a phase change. The former type of superplasticity is called ultra-fine grain superplasticity, while the latter type is called transformation superplasticity [10-67–10-70]. Superplasticity, a necking-resistant flow which results from a high strain-rate sensitivity of flow stress, is generally associated with elongation of 100% or greater in tensile tests. The essential requirements for the occurrence of superplasticity are (1) a temperature greater than 0.5, so that diffusional processes are dominant; (2) equiaxed grains $<10~\mu$m in size, so that flow stresses are low and the grain-boundary sliding contribution to strain is high, and (3) a two-phase microstructure to retard grain coarsening [10-67, 10-71].

First, there is "micrograin" plasticity, in which very fine-grained materials (grain size: 1–10 μm) exhibit superplastic flow within a narrow elevated temperature range at low strain rates ($10^{-5} \sim 10^{-3}$/s). The first disadvantage of the "micrograin" superplasticity is its prerequisite for the metallic materials to be pre-heat-treated to obtain very fine microstructures. Secondly, during the superplastic forming, the original micrograin structure will be coarsened, hence the SPFed articles will exhibit somewhat less mechanical properties than the raw materials. This is the second disadvantage. Mechanistically, micrograin superplasticity is accompanied with high-temperature creep deformation. However, coarsened grain structures exhibit higher resistance against the creep deformation than fine-grained structures. This is a main reason why the strain rate with micrograin superplasticity is low. This is the third disadvantage associated with the micrograin superplasticity. Although the transformation superplasticity requires the material have a phase transformation temperature, the preparation to have very fine structure is not required. Contrarily, during the superplastic forming the materials subjected to the transformation superplastic forming exhibit the grain refining. Accordingly, the strain rate is much higher ($10^{-3} \sim 10^{-2}$/s) than that for the micrograin superplasticity, and the transformation-SPFed article shows higher mechanical properties than those fabricated by micrograin superplasticity [10-63].

Since ultrafine grain microstructure is a prerequisite for exhibiting superplasticity [10-72], the superplastic materials are not limited to metallic materials, but they can include ceramic materials as well as fine powder particles [10-73, 10-74]. Dense and translucent hydroxyapatite polycrystals ($Ca_{10}(PO_4)_6(OH)_2$) with a grain size of 0.64 μm were obtained by hot isostatic pressing (HIP) at 203 MPa and 1000°C for 2 h in argon. The material exhibited superplastic elongation ($>150\%$) in a tension test at temperatures from 1000 to 1100°C and at strain rates from 7.2×10^{-5} to $3.6 \times 10^{-4}\,s^{-1}$. [10-75].

For the superplastic forging, the material selection for the molds is important from the standpoint of high temperature strength and oxidation resistance. Ni-based alloy can be used as a mold material for superplastic forming and forging of titanium and its alloys. Although ceramics including SiC and Si_3N_4 exhibit very high temperature strength over 1000°C, they are prone to break due to the thermal stress and the stress concentration at particularly corner portions [10-76]. In order to develop a die for superplastic forming of the sheets of titanium and its alloy into dental prostheses, the CaO-ZrO_2-TiO_2 system was investigated. A practical ceramic die, which hardly reacted with titanium at about 850°C under vacuum, was made by the following process. The mixture of 20 wt%CaO, 60 wt%ZrO_2, and 10 wt%$MgCl_2$ was kneaded together with methanol, slipped into a stainless-steel ring fixed on a pattern, and formed under the pressure of about 20 kgf/cm^2 into a green compact. After drying, the green compact was sintered at 1150°C for 6 h in air [10-77]. A lubricant agent should also be properly selected in terms of the ease for the mold-separation and surface flatness of the formed products. In general, a mixture in which a glass components comprising SiO_2, B_2O_3, CaO, FeO and water are mixed with xylene, is used. Sometimes, BN (boron nitride) powder is admixed to the aforementioned mixture [10-76].

Because of their large commercial significance, isothermal forging and superplastic sheet forming of alpha/beta titanium alloys deserve special mention. Both processes rely on the superplastic nature of the fine, equiaxed alpha microstructures developed during substructures deformation processing. These microstructures provide low flow stresses, as well as high values of the strain-rate sensitivity. The superplastic properties of alpha/beta alloys are a result of the fine, two-phase microstructure of equiaxed alpha and beta grains. At appropriate strain rates, these microstructures deform by diffusion-controlled, grain-boundary sliding, thus giving rise to low flow stresses and high values of the strain-rate sensitivity index, as well as deformed microstructures that retain much of their original equiaxed nature.

The presence of two phases helps to limit the growth of either, to a large extent, thereby enabling the retention of the plastic properties to large strains. Denture bases are now mainly manufactured by a casting process of Co-Cr or Ni-Cr alloys. Ti-6Al-4V alloy is known as a biocompatible metal, and also has high strength and low density, so it seems to be the best material for denture bases. Okuno *et al.* [10-78] investigated the feasibility study on the superplastic dentue base forming of Ti-6Al-4V. It was demonstrated that (i) the Ti-6Al-4V denture base is superplastically formed by very simple processes by argon gas pressure (10 kgf/cm^2) at 800–900°C, and (ii) SPFed denture bases of Ti-6Al-4V showed some merits compared with conventionally cast or cold-pressed denture base plates, namely good fitness, light weight, and low producing cost using a smaller number of manufacturing processes [10-78]. Superplastic forming of the Ti-6Al-4V alloy is quite an attractive method of fabricating a full denture base. However, it is not easy to fabricate a partial denture frame by superplastic forming because a retainer-like clasp cannot be fabricated with it. The thickness of the superplastic denture base cannot be controlled to change partially as in casting. Diffusion bonding, which makes it possible to join the retainer or titanium plate to the denture during the superplastic forming of Ti-6Al-4V alloy denture base was attempted. Pure titanium, Ti-6Al-4V, Ti-20Cr-0.2Si successfully joined the Ti-6Al-4V alloy solders by diffusion bonding during the superplastic forming. Co-Cr alloys, which were coated in advance with 14–20 K gold alloy solders, were also joined to the superplastic forming Ti-6Al-4V alloy denture base [10-79].

Applications of superplastic phenomenon can be versatile [10-71, 10-80, 10-81], including forming, bonding, forming/bonding, sintering, heat treatment, and thermo-mechanical treatment. Each application is faced to the conventional technology. For example, SP forming will be faced to sheet metal molding (press forming) and hot forging, SP bonding will be faced to diffusion bonding and soldering/brazing, SP sintering will be faced to powder metallurgy, MIM, and the combustion synthesis method, and the SP heat treatment will be faced to carburization, nitriding, and carbo-nitriding as well.

10.6. DIFFUSION BONDING (DB)

Shaping and joining are two fabrication processes fundamental to the manufacturing of metallic structure. Limitations in the efficiency and cost effectiveness of these basic processes may seriously inhibit the use of an otherwise valuable structural material. Such has been the case with titanium in the aerospace industry. Superplastic forming/diffusion bonding (SPF/DB) is a new fabrication process which provides an attractive answer to this dilemma [10-80–10-82]. Because few systematic studies have been made of the effects of material and process variables on the superplasticity of titanium alloys, the effects of metallurgical and process variables on SPF/DB processes were investigated by Sastry *et al.* [10-67]. The alloy, processing schedules, microstructure, time, and strain-rate dependency of the superplasticity indices were given special emphasis. Generalized relations between metallurgical and process parameters for superplastic forming of titanium alloys were determined from a systematic characterization of the superplasticity indices for several alpha-beta titanium alloys. The generalized relations can be used to adjust superplastic processing parameters without time-consuming forming tests on each alloy lot. Alternatively, if the processing parameter limits are fixed, the generalized relations can be used to specify the grain size and alloy compositional limits for any two-phase titanium alloy [10-67]. It was concluded that techniques for determining the generalized three-dimensional plots of the interrelation between flow stress, strain rate, and strain should thus be valuable for evaluating the relative superplasticity of different titanium alloys [10-67].

The SPF/DB process takes advantage of elevated temperature characteristics inherent in the Ti-6Al-4V alloy. This alloy is endowed with a high degree of superplasticity and excellent characteristics for diffusion bonding. Thus, monolithic structures which combine severely formed configurations and joints of parent metal properties may be fabricated from titanium alloy sheet stock to achieve a new level of structural efficiency and cost effectiveness. Okuno *et al.* [10-78] developed a SPF/DB method by inserting intermediate metal foils such as palladium and gold alloy solders between CpTi or Ti-6Al-4V to Ti-6Al-4V denture base under argon pressure of 4 kg/cm^2 and forming temperature at 950°C for 2 h.

SPF/DB can be considered as a near-net shape (NNS) forming technique. NNS processing offers cost reduction by minimizing machining, reducing part count, and avoiding part distortion from welding. NNS technologies such as flow-forming, casting, forging, superplastic forming, powder metallurgy, metal injection molding, three-dimensional laser deposition, and plasma arc deposition have been explored for potential use in tubular geometries and other shapes of varying complexity. It appears that the reduction in cost of a given titanium product will be maximized by achieving improvements in all of the manufacturing steps, from extraction to finishing [10-83].

10.7. POWDER METALLURGY

Even though titanium alloys offer increased performance, the question of cost must be faced. Of the 'near-net shape' processes recently developed or advanced for the purpose of reducing material input and machining for titanium alloy part production, powder metallurgy exhibits the widest range of process variations and potential applications, particularly in the aerospace industry. Powder metallurgy (P/M) has advantages such as the elimination of casting defects, good yield, less segregation, and short processing. A processing route that results in parts as close as possible to near-net shape forming reduces the overall cost of these parts. P/M is one of these routes, along with isothermal forging and casting. Sintering has a further advantage when dealing with alloys for which problems of segregation are encountered in traditional metallurgical processes involving melting. There are needs to have porosity for titanium implant surface. To provide appropriate porosity to accommodate the bone ingrowth toward a complete osseointegration, titanium powder can be sintered with certain designed porosity. A major problem with the use of titanium is its cost. One of the ways to overcome this problem is by use of a NNS process such as powder metallurgy [10-84]. A typical powder metallurgy process includes blending/mixing sponge fines or master alloy $\rightarrow$ cold isostatic press or mechanical pressing $\rightarrow$ vacuum sintering $\rightarrow$ cold work + anneal, hot isostatic press, forging, coining $\rightarrow$ final machining. Sintering reactions and fine structures of the biocomposites prepared from powder mixtures of titanium (α-Ti), hydroxyapatite (HA) and bioactive glass (BG) ($SiO_2-CaO-P_2O_5-B_2O_3-MgO-TiO_2-CaF_2$) were investigated by X-ray diffraction and transmission electron microscopy. The results showed that complex reactions among the starting materials mainly depended on the initial Ti/HA ratios, as well as the sintering temperatures. And the reaction could be expressed by the following illustrative equation: $Ti + Ca_{10}(PO_4)_6(OH)_2 \rightarrow CaTiO_3 + CaO + Ti_xP_y + (Ti_2O) + (Ca_4P_2O_9) + H_2O$ [10-85].

Nano and conventional TiO_2 powders were deposited onto titanium alloy using atmospheric plasma-spraying technology. As-sprayed titania coatings were treated by H_2SO_4 and HCl solutions at room temperature for 24 h, and the bioactivity was evaluated by simulated body fluid tests. Measured X-ray diffraction patterns indicated that as-sprayed titania coatings obtained from both nano and conventional powders were composed of primary rutile, as well as a small quantity of anatase and Ti_3O_5. The surface of as-sprayed coatings prepared from the conventional powder was rougher than that from nanopowder. After immersion in simulated body fluid for 2 weeks, both the acid-treated nano and titania coatings were induced with carbonate-containing hydroxyapatite to form on the surfaces. However, this phenomenon did not appear on the surface of as-sprayed nano and conventional titania coatings. The results obtained indicated that (i) the bioactivity of plasma-sprayed titania coatings was improved by the acid treatment, and (i) it had nothing to do with the phase composition and particle size of the original TiO_2 powders [10-86].

P/M technology can be applied to modify the surface zone of implants. Porous-coated Ti-6Al-4V alloy implant systems provide a biocompatible interface between implant and bone, resulting in a firm fixation and potential long-term retention via bony ingrowth. In order for an acceptable porous coating structure, the sintering protocol for Ti-6Al-4V alloy systems often requires that the material be heat treated above the β-transus (i.e., 885°C). This transforms the as-received equiaxed microstructure, recommended for surgical implants, to a lamellar alpha–beta distribution, which has been shown to have the worst fatigue properties of the most common structures attainable in Ti-6Al-4V alloy. However, post-sintering heat treatment may be used to improve these properties by producing microstructures more resistant to crack initiation and propagation. Cook *et al.* [10-87] investigated the influence of microstructural variations on the fatigue properties of porous-coated Ti-6Al-4V alloy material. Non-porous-coated and porous-coated Ti-6Al-4V fatigue specimens were subjected to a standard sintering heat treatment to produce a lamellar microstructure, In addition, two post-sintering heat treatments were used to produce coarse and fine acicular microstructures. Rotating beam (reversed bending) fatigue testing was performed, and the endurance limits determined for the non-coated and porous-coated microstructures. The values determined were 680 MPa (non-coated as-received equiaxed), 394 MPa (non-coated lamellar), 488 MPa (non-coated coarse acicular), 494 MPa (non-coated fine acicular), 140 MPa (porous-coated lamellar), 161 MPa (porous-coated coarse acicular), and 162 MPa (porous-coated fine acicular). The non-coated coarse and fine acicular specimens displayed an approximate 25% increase over the non-coated lamellar specimens. The porous coated coarse and fine acicular specimens showed an approximate 15% improvement over the porous coated lamellar specimens [10-87].

Diffusion bonding phenomenon associated with superplasticity can be applied to the bonding powder particles, resulting in superplastic powder metallurgy [10-80]. Superplasticity in metals has received a considerable amount of attention. However, the possibility that ceramic materials also may show anomalous deformational properties coincident with a phase transformation has been neglected. Examples will be given where ready deformation of ceramic bodies may occur with (1) simple allotropic phase transformation, (2) eutectic exsolution, (3) exsolution from solid solution, (4) exsolution by decomposition of mixed crystal systems, and (5) simple decomposition producing one or more solid-phase products and a gas phase; these will be related to the better known processes in metallic systems. It was demonstrated that alumina, magnesia, Mg-Cr spinel ($MgO.Cr_2O_3$), zirconia (ZrO_2), TCP, and HAP can be sintered by using their superplastic characteristics (some cases, being accompanied with HIP process) [10-88, 10-89].

10.8. METAL INJECTION MOLDING (MIM)

MIM technology as an advanced powder metallurgy offers NNS shaping and mass production of products of the same design [10-90–10-92]. MIM has been used successfully in the industrial field, particularly in watch manufacturing, which requires a precise NNS shape forming [10-92]. The recent evolution has been to maximize the content of solid particles and to remove the polymer binder during the sintering. As a consequence, a new powder-forming process has evolved, allowing shape complexity, low cost forming, and high performance properties. Over the last decades, powder metallurgy has made significant contributions as a near-net shape forming technology. This new process, termed metal powder injection molding [10-93], is begun by mixing selected powders and binders. Among its latest spin-offs, injection molding, usually associated with organic polymers, although its use for metal and ceramic components dates back to World War II, has recently received considerable attention [10-94]. The MIM process comprises the following sequences: (1) mixing/kneading metal powder, binder, solvent, and lubricant to form the so-called compound, (2) granulating the compound, (3) injecting the mixed compound into a mold, (4) debinding the binding material, and (5) sintering. Metal powder is made mainly of atomized powder. Normally, in order to provide the compound better mobility and sintered products of high density, fine particle powder (particle size ranges from several μm to several 10 μm) is used. In this technology, the use of microsized powders, allowing for greater packing density, results in higher final densities than those generally achievable by conventional powder metallurgy [10-94].

The binder is added (volume % of 40–50 of organic binding substances) and consists of thermoplastic resins such as polyethylene or polypropylene, wax, plasticizer, and lubricant. The principle purpose of mixing the binder is to provide sufficient flow mobility when the compound is injected into the mold. The mixture (or compound) is then granulated and injection-molded into a desired shape. The polymer imparts viscous flow characteristics to the mixture to aid forming, die filling, and uniform packing. The molded products are then subjected to debinding by either a heating process or a dissolving process in order to remove the binding material. The atmosphere used in the debinding stage includes nitrogen, air, vacuum, wet hydrogen, and hydrogen with hydrochloric acid [10-95].

The debonded mold is, at the final stage of the MIM process, sintered in either vacuum or inert gas such as argon or nitrogen gas. The maximum sintering temperature depends on the type of metal powder. For example, Fe-based alloys are generally sintered in a range from 1200 to 1400°C. At this moment, the debonded mold is subject to shrinkage. It is reported that the linear shrinkage for Fe-based alloys is 13–20% [10-95]. However, the shrinkage is approximately linear because the molded product is uniformly filled with metal powder [10-92]. The product may then be further densified, heat treated, or machined to complete the fabrication process if required. The application of the MIM method to fabricate the orthodontic bracket was successfully introduced [10-92].

10.9. LASER WELDING/FORMING

Applications of laser (light amplification by stimulated emission of radiation) energy are versatile, including engineering applications (forming, honing, welding, surface modification, cutting), caries detection and oral surgery in dentistry, as well as various microsurgeries through the medical field. With laser welding, it is possible to join parts by the self-welding of the metal parts themselves. Although numerous studies have been made on the use of lasers for welding other dental metals [10-96–10-99], few have been made on the welding of titanium materials [10-100]. Ordinarily the metal parts of prosthetic devices are soldered by the investment method. However, this procedure is relatively complex and time consuming. Since a laser is capable of concentrating high light energy on a small spot, the possibility of using a laser for joining the precise and miniature parts of prosthetic devices was assessed. Welding experiments were carried out by using a normal pulse, Nd:Yag laser processing apparatus to weld titanium, which is being used frequently in dentistry. Laser-welding of titanium must be carried out in an argon atmosphere. When titanium was irradiated in an atmosphere of ordinary air, cracks appeared in the metal surface, welds cracked, and the mechanical properties of the metal were debilitated. Greater tensile strength, yield strength and elongation of titanium were generally observed with increases in laser irradiation power. Even in an argon atmosphere, the intensity of the laser irradiation must be at least 21 J/P to maintain the tensile strength and yield strength of the original metal. Differences in the laser irradiation atmosphere significantly affected the elongation of titanium [10-96, 10-100]. With laser welding, it is possible to join parts by the self-welding of the metal parts. However, the laser welding in same occasions induces volume reduction. In order to minimize the volumetric reduction, thin titanium foil was inserted between the titanium plates. It was found that laser welded titanium plates produced porosity both on the surface and in the cross-sectional areas. Yamakura *et al.* [10-101] evaluated the insertion of a titanium foil to be effective in preventing volume reduction during the welding process. It was reported that (i) no porosity was found at the weldment with thin foil insertions, (ii) greater bending strength was generally observed for self-welding titanium plates using 30 μm titanium foil; however (iii) the bending strength decreased as the thickness of titanium foil was increased, and (iv) laser-welded titanium plates decreased in elasticity, with hardness of the welded area being higher than the original un-welded area [10-101].

With the introduction of new techniques for the fabrication of titanium frameworks for implant-supported prostheses comes the need to understand how the components used compare to those used for conventional cast frameworks. The relationship of measured machining tolerances between conventional implant components and those components use for stereo laser-welded frameworks was determined using a standardized protocol. It was reported that statistically significant differences in the horizontal interface relationship were found between paired implant components, which had a mean range from 23.1 to 51.7 μm [10-102].

The obvious differences in structure and properties of the samples by the welding process can be explained by the differing duration of the welding processes. The finding is that laser welding affects the surrounding material only minimally. However, the optimal result will only be achieved if the most suitable energy and duration of the laser impulse are chosen. The large heat-affected zone (HAZ) seen in plasma-welded samples may be attributed to the long and continuous plasma-welding process as well as its difficult manual execution [10-100]. Roggensack *et al.* [10-103] investigated two different methods of welding. Specimens machined from CpTi rods were fused either by laser welding or plasma welding. Hardness profiles and light microscopy images were taken in the region of the weld. The mechanical properties were tested by an alternating bending fatigue test of up to 3 million cycles. Light microscopy images and hardness profiles showed a larger HAZ after plasma welding compared to laser welding. No significant differences comparing fatigue strength could be found between the two methods of welding. SEM images of the laser-welded joints showed fractures in the welding zone, while the plasma-welded specimens fractured mostly beyond the HAZ. It was therefore concluded that (i) both methods are suitable for welding Ti, and (ii) the laser welding is the more suitable technique in dentistry because of its lower heat affect on the workpieces [10-103].

If a clasp on a RPD is damaged or broken, the entire prosthesis may become unstable, lose retention, and become a source of discomfort to the patient. Clasps may be repaired by soldering with a conventional torch technique. However, there some disadvantages of soldering [10-100, 10-103, 10-104–10-106]: (1) the soldered clasp is likely to break again because of stress concentration, (2) discoloration caused by electric and chemical corrosion may occur, (3) soldering of commercially pure Ti is difficult because of oxidation problems, (4) use of fire and gas may be hazardous, (5) technical skill is necessary, and (6) heating may burn the denture base resin. Laser welding is an attractive alternative method to join dental casting alloys [10-104–10-106]. During the past decade, laser welding has been increasingly used because there is no need for investment and soldering alloy, working time is decreased, lasers are easy to operate, little damage is caused to the denture resin from the pin-point heat, and there are few effects of heating and oxidation [10-103]. The penetration depth of the weld into Ti is significantly greater than into gold alloy [10-106]. Ti has lower thermal conductivity and a greater rate of laser beam absorption compared with gold alloy [10-106]. These properties make it easier to laser weld Ti to repair broken Ti clasps.

Iwasaki *et al.* [10-107] investigated the distortion of laser welded Ti plates for different operating conditions of the laser welding devices, and with different welding parameters (weld points and pre-welding). Nd:Yag laser was used to join CpTi (grade 2) plates. Weld surfaces were sand-blasted with 50 μm alumina under 0.2 MPa pressure. It was found that (i) distortion increased stepwise after each welding point along the welding zone (one-side welding), but decreased consecutively as the welding proceeded on the second side of the weld (two-side welding), (ii) in the case of one-side welding, the dependence of distortion on current and spot diameter presented maxima, (iii) for two-side welding the same parameters exercised little influence on its distortion recovery due to the effect of the solidified weld pool from the first side, and (iv) four-point pre-welding significantly decreased the final distortion for both one and two-side welds [10-107].

Laser energy is not only applied to welding, but also to forming or peening technology, being similar to shot-forming and shot-peening. Laser forming (Lasform) is a laser direct metal deposition process that combines high-power laser cladding technologies with advanced rapid prototyping methods to directly manufacture complex three-dimensional components. Mechanical test data and compositional analysis have shown that the laser-deposited material meets ASTM specifications for CpTi, Ti-6Al-4V, and Ti-5Al-2.5Sn. The system itself is comprised of a large processing chamber, a 14 kW CO_2 laser, and a newly developed powder feed system that provides consistent powder flow at high mass-flow rates. Lasform is an additive process and can create near-net shape geometries requiring minimal post-machining and heat treatment [10-108]. Laser peening (which is characterized by a contamination-free and non-contact peening for surface hardening) is a process in which a laser beam is pulsed upon a metallic surface, producing a planar shockwave that travels through the workpiece and plastically deforms a layer of material. The depth of plastic deformation and resulting compressive residual stress are significantly deeper than possible with most other surface treatments [10-109]. These compressive layers are typically 1.0 mm thick, compared with typical depths of 0.25 mm for standard shot peening. As example of laser peening on Ti-6Al-4V, it was reported that the surface compressive residual stress (which is particularly beneficial for enhancing fatigue strength as well as fatigue life) was 830 MPa, and at even 0.25 mm beneath the surface, remained at half of the surface compressive stress (400 MPa) as a compressive internal stress [10-110].

10.10. SOLDERING

Although there are several disadvantages of soldering [10-100, 10-103, 10-104–10-106], as seen in above, titanium materials are frequently employed in dentistry, and their solder-joining is often conducted. Like as other technologies involving elevated temperatures with titanium materials, the joining portion is appropriately sealed from the ambient atmosphere to avoid unwanted oxidation. Solder strength normally depends on various factors, such as the solder material's mechanical strength, extent and completion of diffusion into base material, penetration and cavity at the soldering interface, and base material's strength. Heat's affect on the base material, as well as the solder, will influence the strength of the prosthesis after the solder process is completed, and corrosion resistance when the prosthesis is used in intraoral environment. Soldering is normally done by electro-resistance soldering, infrared soldering, or plasma soldering, all in an argon atmosphere. It is generally believed that casting and soldering are two major technical difficulties associated with titanium materials, since titanium has a relatively high melting point and high reactivity with oxygen. In industry, there are three types of solders used for titanium materials: silver-based, aluminum-based, and eutectic solders. If appropriate measures are taken in order to control the excessive oxidation, titanium materials can be successfully solder-joined in the dental/medical fields. These measures might include material factors (using high activity flux and low melting point solders) and environment factors (non-oxidizing atmosphere and shorter operation time) [10-111].

10.11. HEAT TREATMENT (HT)

For reducing residual stresses (mostly tensile residual stresses) which was developed during fabrication, titanium materials are heat-treated – stress relieving. For improving mechanical properties (ductility), machinability, and structural stability, titanium products are also heat-treated – annealing. For increasing mechanical strength, titanium is heat-treated – solution treating and aging. Alpha and near-alpha titanium alloys can be stress relieved and annealed, but high strength cannot be developed in these alloys by any type of heat treatment. Beta phase in the commercial beta and metastable beta alloys decomposes at selected elevated temperature to cause the material strengthen. Phase compositions, sizes and distributions can be manipulated by heat treatment within certain limits to enhance a specific property or to attain a range of strength levels [10-112]. For a certain type of titanium alloy, post-welding heat-treatment (PWHT) is recommended. Development of maximum properties in alpha-beta alloy weldments requires solution treating followed by rapid quenching and then aging. However, if reduced strength is acceptable or higher ductility is required, workpieces may be annealed only, after welding.

The microstructural evolution and attendant strengthening mechanisms in two novel orthopedic alloy systems, Ti-34Nb-9Zr-8Ta and Ti-13Mo-7Zr-3Fe, have been compared by Banerjee *et al.* [10-113]. In the homogenized condition, both alloys exhibited a microstructure consisting primarily of a beta-phase matrix with grain boundary alpha precipitates and a low-volume fraction of intra-granular alpha precipitates. On aging of the homogenized alloys at 600°C for 4 h, both alloys exhibited the precipitation of refined scale secondary alpha precipitates homogeneously in the beta matrix. However, while the hardness of the Ti-13Mo-7Zr-3Fe alloy marginally increased, that of the Ti-34Nb-9Zr-8Ta alloy decreased substantially as a result of the age-treatment. In order to understand this difference in the mechanical properties after ageing, TEM studies were carried out on both alloys prior to and post aging. It was reported that a metastable B2 ordering in the Ti-34Nb-9Zr-8Ta alloy was found in the homogenized condition which is destroyed by the ageing treatment, consequently leading to a decrease in the hardness [10-113].

REFERENCES

[10-1] Bessing C, Bergman B. The castability of unalloyed titanium in three different casting machines. Swed Dent J 1992;16:109–113.

[10-2] Chai TI, Stein RS. Porosity and accuracy of multiple-unit titanium casting. J Prosthodont 1995;73:534–541.

[10-3] Schädlich-Stubenrauch J, Augthun M, Sahm PR. Untersuchungen zu den mechanischen Eigenschaften und zur Porenbilding von Titan bei verschiedenen Gussverfahren. Dtsch Zahnäatzl Z 1994;49:774–776.

[10-4] Anderson M, Bergman B, Bessing C, Ericsson G, Lundquist P, Nilson H. Clinical results with titanium crowns fabricated with machine duplication and spark erosion. Acta Odontol Scand 1989;47:279–286.

[10-5] Sjögren G, Andersson M, Bergman M. Laser welding of titanium in dentistry. Acta Odontol Scand 1988;46:247–253.

[10-6] Russell MM, Andersson M, Dahlmo K, Razzoog ME, Lang BR. A new computer-assisted method for fabrication of crowns and fixed partial dentures. Quintessence Int 1995;26:757–763.

[10-7] Mueller HJ, Giuseppetti AA, Waterstratt RM. Phosphate-bonded investment materials for titanium casting. J Den Res 1990;69:367 (Abstract No. 2072).

[10-8] Dunigan-Miller J. Electrochemical corrosion behavior of commercially pure titanium materials fabricated by different methods in three electrolytes. Indiana University Master Thesis, 2004.

[10-9] Okabe T, Herø H. The use of titanium in dentistry. Cell Mater 1995;5: 211–230.

[10-10] Lautenschlager EP, Monaghan P. Titanium and titanium alloys as dental materials. Int Dent J 1993;43:245–253.

[10-11] Okabe T, Shimizu H, Woldu M, Brezner M, Carrasco L, Watanabe K. Mechanical properties of titanium cast using yttria face-coating. THER-MEC 2000-Proceedings of international conference on processing and

manufacturing of advanced materials, Vol. 117/3, Section A7, Las Vegas, December CDROM, 2000.

[10-12] Wang RR, Fenton A. Titanium for prosthetic applications; a review of the literature. Quintessence Int 1996;27:401–408.

[10-13] Baltag I, Watanabe K, Kusakari H, Miyakawa O. Internal porosity of cast titanium removable partial dentures: influence of sprue direction on porosity in circumferential clasps of a clinical framework design. J Prosthet Dent 2002;88:151–158.

[10-14] Cecconi BT, Koeppen RG, Phoenix RD, Cecconi ML. Casting titanium partial denture frameworks: a radiographic evaluation. J Prosthet Dent 2002;87:277–280.

[10-15] Yoda M, Konno T, Takada Y, Iijima K, Griggs J, Okuno O, Kimura K, Okabe T. Bond strength of binary titanium alloys to porcelain. Biomaterials 2001;22:1675–1681.

[10-16] Okuno O, Hamanaka H. Application of beta titanium alloys in dentistry. Dentistry Jpn 1989;26:101–104.

[10-17] Okuno I. New beta-titanium alloys. Metal 1990;44:4–9.

[10-18] Ida K, Tsustumi S, Togaya M. Titanium and titanium alloys for dental casting. J Dent Res 1980;59:985 (Abstract No.397).

[10-19] Taira M, Moser JB, Greener EH. Studies of Ti alloys for dental castings. Dent Mater 1989;5:45–50.

[10-20] Ita I. Dental casting technique of titanium. Met Technol 1985;55:4–9.

[10-21] Nakamura S. Ohara's titanium castings for dental use. Titanium and Zirconium 1986;34:154–156.

[10-22] Nakamura S. Casting of titanium for dental use. Met Technol 1988;58:16–19.

[10-23] Eriksson M, Andersson M, Carlström E. Titanium dental coping prepared by a powder metallurgy method: a preliminary report. Int J Prosthodont 2004;17:11–16.

[10-24] Stoll R, Okuno O, Ai M, Stachniss V. Titanium casting technique. Possibilities, problems, and hopes. Position report on pure titanium castings. ZWR 1991;100:38–42.

[10-25] Ida K, Togaya T, Tsutsumi S, Takeuchi M. Effect of magnesia investments in the dental casting of pure titanium and titanium allow. Dent Mater J 1982;1:8–21.

[10-26] Mori T, Jean-Louis M, Yabugami M, Togaya T. The effect of investment type on the fit of cast titanium crowns. Aust Dent J 1994;39:348–352.

[10-27] Okuno O, Yoneyama T, Hamanaka H. Advanced casting of titanium. Iron and Steel 1990;76:1633–1641.

[10-28] Koike M, Chai Z, Fujii H, Brezner M, Okabe T. Corrosion behavior of cast titanium with reduced surface reaction layer made by a face-coating method. Biomaterials 2003;24:4541–4549.

[10-29] Miyakawa O, Watanabe K, Okawa S, Nakano S, Kobayashi M. Reaction layers of titanium cast in molds containing spinel. J Dent Mater 1995;14:560–568.

[10-30] Zhang J, Okazaki M, Takahashi J. Effect of casting methods on surface state of pure titanium casting. J Dent Mater 1994;13:221–227.

[10-31] Bauccio ML. Technical note 9: descaling and special surface treatments. In: Materials properties handbook: titanium alloys. Boyer R, Welsch G, Collings EW, editors, Materials Park, OH: ASM International. 1994. p. 1080.

[10-32] Donachie MJ, Titanium: A technical guide. Metals Park: OH, ASM International, 1988, p. 28.

[10-33] Cai Z, Bunce N, Nunn M. Okabe T. Porcelain adherence to dental cast CP titanium: effects of surface modifications. Biomaterials 2001;22: 979–986.

[10-34] Walter M, Reppelp P, Boning K, Freesmeyer W. Six-year follow-up of titanium and high-gold porcelain-fused-to-metal fixed partial dentures. J Oral Rehabil 1999;26:91–96.

[10-35] Bergman B, Marklund S, Nilson H, Hedlund S. An intraindividual clinical comparison of 2 metal-ceramic systems. Int J Prosthodont 1999;12: 444–447.

[10-36] Hashimoto H, Kuroiwa A, Wada K. Hibino Y, Kouchi H, Hashimoto T, Hasegawa Y, Ando Y, Akaiwa Y. Reaction product on the surface of titanium castings. J Dent Mater 1992;11:603–614.

[10-37] Shafer T, Cai Z, Carrasco L, Okabe T. Porcelain bonding to cast titanium made from oxide face-coated patterns. J Dent Res 2002;81:A-261 (Abstract No. 2003).

[10-38] Nakamura M, Jyohzaki K, Kimura T, Togaya T, Ida K. Ceramic dental molds for the precision casting of titanium and titanium alloys: 1, Relationship between particle size distribution of aggregates and mold expansion. J Am Ceram Soc 1990;73:35–38.

[10-39] Nakamura S, Komasa Y. Titanium casting procedure using thermal expansion-inhibited investment. Part 3: In the condition of the large colloidal silica particles as mixing liquid. J Jpn Prosthedont Soc 1999;38: 753–765.

[10-40] Ferenczi AM, Demri B, Moritz M, Muster S. Casted titanium for dental applications: an XPS and SEM study. Biomaterials 1998;19:1513–1515.

[10-41] Ott D. Gießen von Titan im Dentallabor (Entwicklung eines Verfahrens). Metall 1990; 44:366–369.

[10-42] Sampath K. The use of technical cost modeling for titanium alloy process selection. J Metals 2005;57:25–32.

[10-43] Fukui H, Wei Y, Yamada S, Fujishiro Y, Morita A, Niinomi M. Cold crucible levitation melting of biomedical Ti-30wt%Ta alloy. Dent Mater J 2001;20:156–163.

[10-44] Zlatin N, Field M. Procedures and precautions in machining titanium alloys. Titanium Sci Technol 1973;1:489–504.

[10-45] Ohkubo C, Shimura I, Aoki T, Hanatani S, Hosoi T, Hattori M, Oda Y, Okabe T. Wear resistance of experimental Ti-Cu alloys. Biomaterials 2003;24:3377–3381.

[10-46] Tajima K, Miyawaki A, Nagamatsu Y, Kakigawa H, Kozono Y. Electro-polishing of titanium and its alloys for miller finishing; J Jpn Dent Mater 2005;24:406.

[10-47] Kyo K, Ohmori H, Okamoto Y. Characteristics of mirror surface grinding of dental nickel-titanium alloy using ELID (electrolytic in-process dressing). J Jpn Soc Precision Eng 1998;32:183–187.

[10-48] Kalpakjian S. Manufacturing engineering and technology. New York: Addison-Wesley Pub., 1989. pp. 463–464.

[10-49] Seman GA. Practical guide to electro-discharge machining, Ed. 2, Geneva, Ateliers Des Charmilles SA, 1975, Chapter 2.

[10-50] Rubeling G, Hans-Albert K. Spark erosion in dental technology: possibilities and limitations. Quintessence Dent Tech 1984;8:649–657.

[10-51] Salinas TJ, Finger IM, Thaler JJ, Clark RS. Spark erosion implant supported over-dentures: clinical and laboratory techniques. Implant Dent 1992;1:246–251.

[10-52] Weber H, Frank G. Spark erosion procedure: A method for extensive combined fixed and removable prosthodontic care. J. Prosth Dent 1993;69:222–227.

[10-53] Jemt T, Linden B. Fixed implant-supported prostheses with welded titanium frame- works. Int J Periodont Restotr Dent 1992;12:177–183.

[10-54] Anderson M, Anderson M. Spark erosion. Swedish Patent 8400396-1 (1987).

[10-55] Anderson M, Oden A. A new all ceramic crown. A dense sintered, high purity alumina coping with porcelain. Acta Odontol Scand 1993;51:59–64.

[10-56] Maeda Y, Yamada M, Nokubi T, Tsustusmi S, Urabe T. Clinical application of the DCS precident CAD/CAM system. Quintessence Dent Tech 1996;19:9–20.

[10-57] Andersson M, Carlsson L, Persson M, Bergman B. Accuracy of machine milling and spark erosion with a CAD/CAM system. J Prosthet Dent 1996;76:187–193.

[10-58] Adell R, Lekholm V, Rockler B, Branemark PI. A 15-year study of osseointegrated implants in the treatment of the edentulous jaw. Int J Oral Surg 1981;10:387–416.

[10-59] Dérand T, Herø H. Bonding strength of porcelain on cast vs. wrought titanium. Scan J Dent Res 1992;100:184–188.

[10-60] Setz J, Diehl J. Gingival reaction on crowns with cast and sintered metal margins: a progressive report. J Prosthet Dent 1994;71:442–446.

[10-61] van der Zel JM. Ceramic fused to metal restoration with a new CAD/CAM system. Quintessence Int 1993;24:769–778.

[10-62] Krishna CG, Prasad YVRK, Birla NC, Rao GS. Hot-deformation mechanisms in near-alpha titanium alloy. J Metal 1996;48:56–59.

[10-63] Oshida Y. Current trends in research on superplasticity. J Jpn Soc Plasticity 1986;27:357–363.

[10-64] Padmanabhan KA, Davies GJ. Materials research and engineering 2: superplasticity, New York: Springer-Verlag, 1980.

[10-65] Paton NE, Hamilton CH, editors. Superplastic forming of structural alloys. The metallurgical society of AIME, conference proceedings. 1982.

[10-66] Ghosh AK, Bieler TR, editors. Superplasticity and superplastic forming 1998, TMS, conference proceeding. 1998.

[10-67] Sastry SML, Lederich RL, Mackay TL, Kerr WR. Superplastic forming characterization of titanium alloys. J Metals 1983;135:48–53.

[10-68] Langdon TG, Furukawa M, Horita Z, Nemoto M. Using intense plastic straining for high-strain-rate superplasticity. J Metals 1998;150: 41–45.

[10-69] Mishra RS, McFadden SX, Valiev RZ, Mukherjee AK. Deformation mechanisms and tensile superplasticity in nanocrystalline materials. J Metals 1999;151:37–40.

[10-70] Friedrich HE, Winkler P-J. Near-net shape airframe part. Adv Mater Process 1991;145:16–22.

[10-71] Oshida Y. Transformation plasticity of steels and titanium alloys in compression. Syracuse University Thesis, 1967.

[10-72] Ultrafine-Grain Metals. Proceedings of 6th Sagamore army materials research conference. Vol. 16. Burke JJ, Weiss V, editors. Syracuse: Syracuse University Press. 1970.

[10-73] Ultrafine-Grain Ceramics. Proceedings of 15th Sagamore army materials research conference. Vol. 15. Burke JJ, Reed NL, Weiss V, editors. Syracuse: Syracuse University Press. 1970.

[10-74] Ceramics and Polymers. Proceedings of 20th Sagamore army materials research conference. Vol. 20. Burke JJ, Weiss V, editors. Syracuse: Syracuse University Press. 1975.

[10-75] Wakai F, Kodama Y, Sakaguchi S. Superplasticity of hot isostatically pressed hydroxylapatite. J Am Ceram Soc 1990;73:457–460.

[10-76] Matsusita T, Tsuda T. Superplastic forging of Ti-based alloys and Ni-based alloys. Materials 1990;38:130–136.

[10-77] Kato M, Murakami H. On the dental application of titanium-based alloy, Part 2: ceramic die for superplastic forming. Gov Indust Res Inst 1990;39:309–317.

[10-78] Okuno O, Nakano T, Hamanaka H, Kato I. Diffusion bonding to titanium alloys during superplastic forming. J. Jpn Soc Dental Mater Devices 1991;10:483–491.

[10-79] Wakabayashi N, Ai M. Thickness and accuracy of superplastic Ti-6Al-4V alloy denture frameworks. Int J Prosthodont 1996;9:520–526.

[10-80] Oshida Y. An application of superplasticity to powder metallurgy. J Jpn Powder and Powder Met 1975;22:147–153.

[10-81] Takase S, Oshida Y. On the solid-state bonding in cast irons using dynamic superplastic phenomena. Trans Iron Steel Inst Jpn 1977;17: 506–515.

[10-82] Mitsuya H, Okada M, Kato I. Superplastic forming of titanium alloy for dental use. Titanium and Zirconium 1990;37:19–23.

[10-83] Klug KL, Ucok I, Gungor MN, Guclu M, Kramer LS, Troy Tack W, Nastac L, Martin NR, Dong H. The near-net-shape manufacturing of affordable titanium components for the M777 lightweight Howitzer. J Metals 2004;56:35–42.

[10-84] Froes FH, Moshl SJ, Moxson VS, Hebeisen JC, Daz VA. The technologies of titanium powder metallurgy. J Metals 2004;56:46–48.

[10-85] Ning CQ, Zhou Y. On the microstructure of biocomposites sintered from Ti, HA and bioactive glass Biomaterials 2004;25:3379–3387.

[10-86] Zhao X, Liu X, Ding C. Acid-induced bioactive titania surface. J Biomed Mater Res 2005;75A:888–894.

[10-87] Cook SD, Thongpreda N, Anderson RC, Haddad RJ. The effect of post-sintering heat treatments on the fatigue properties of porous coated Ti-6Al-4V alloy. J Biomed Mater Res 1988;22:287–302.

[10-88] Morgan PED. Superplasticity in ceramics. In: Ultrafine-grain ceramics. Burke JJ, Reed NL, Weiss V, editors. Syracuse NY: Syracuse University Press. 1970. pp. 251–271.

[10-89] Nonami T, Wakai H. Superplastic bio-ceramics. Met Technol 1991;12: 36–41.

[10-90] Bulger M. Metal injection molding. Adv Mater Process 2005;163:39–40.

[10-91] Erickson AR, Amaya HE. Recent development in injection molding of p/m parts. In: Modern developments in powder metallurgy. Aqua EN, editor. New Jersey: Princeton, American Powder Metallurgy Institute 1984. pp. 145–155.

[10-92] Yamagishi T, Ito M, Kou Y, Obata A, Deguchi T, Hayashi S, Igarashi Y. Mechanical properties of sintered titanium using metal injection molding. J Jpn Dent Mater 1995;14:1–7.

[10-93] Wiech RE. Method of making inelastically compressible ductile particulate material article and subsequent working thereof. US. Patent No.4,445,936 (1984).

[10-94] Billiet R. Net-shape full density p/m parts by injection molding. Int J Powder Metall Powder Technol 1985;21:119–129.

[10-95] Wei TS, German RM. Injection molded tungsten heavy alloy. Int J Powder Metall 1988;24:327–335.

[10-96] Yamagishi T, Ito M, Fujimura Y. Mechanical properties of laser welds of titanium in dentistry by pulsed Nd:YAG laser apparatus. J Prosthet Dent 1993;70:264–273.

[10-97] Gordon TE, Smith DL. Laser welding of prostheses – an initial report. J Prosthet Dent 1970;24:472–476.

[10-98] Huling JS, Clark RE. Comparative distortion in three-unit fixed prostheses joined by laser welding, conventional soldering, or casting in one piece. J Dent Res 1977;56:128–134.

[10-99] van Benthem H, Munster JV. Korrosionsversuche an Dentallegierungen vor und nach dem Laserschweiben. 1. Hochgoldhaltige Dentallegierungen. Dtsch Zahnarztl Z 1985;40:286–289.

[10-100] Yamagishi T, Ito M, Masuhara E. Laser-welding of titanium and other dental alloys, Part 1. Mechanical properties of laser welds of titanium by pulsed Nd:YAG laser apparatus. J Jpn Dent Mater 1991;10: 763–772.

[10-101] Yamakura K, Kajima M, Mori K, Yokoyama K, Igarashi T, Onizawa T, Hideka Y, Ito M. Studies on the laser welding in titanium. J Jpn Soc Oral Implant 1997;10:420–425.

[10-102] Rubenstain JE, Ma T. Comparison of interface relationships between implant components for laser-welded titanium frameworks and standard cast frameworks. Int J Oral Maxillofac Implants 1999;14:491–495.

[10-103] Roggensack M, Walter MH, Böning KW. Studies on laser- and plasma-welded titanium. Dent Mater 1993;9:104–107.

[10-104] Suzuki Y, Ohkubo C, Abe M, Hosoi T. Titanium removable partial denture repair using laser welding: A clinical report. J Prosthet Dent 2004; 91:418–420.

[10-105] Berg E, Wanger WC, Davik G, Dootz ER. Mechanical properties of laser-welded cast and wrought titanium. J Prosthet Dent 1995;74:250–257.

[10-106] Watanabe I, Liu J, Atsuta M. Effects of heat treatment on mechanical strength of laser-welded equi-atomic AuCu-6at%Ga alloy. J Dent Res 2001;80:1813–1817.

[10-107] Iwasaki K, Ohkawa S, Rosca ID, Uo M, Akasaka T, Watari F. Distortion of laser welding titanium plates. Dent Mater J 2004;23:593–599.

[10-108] Abbott DH, Arcella FG. Laser Forming Titanium Components. Adv Mater Processes 1998;153:29–30.

[10-109] Hill MR, DeWald AT, Demma AG, Hackel LA, Chen H-L, Specht RC, Harris FB. Laser peening technology. Ad Mater Process 2003;158:65–67.

[10-110] Dane CB, Hackel LA, Daly J, Harrison J. Shot peening with lasers. Ad Mater Process 1998;153:37–38.

[10-111] Miura I, Ida, K. Titanium for dental use, Tokyo: Quintessence. 1988. pp. 123–131.

[10-112] ASM Committee on Titanium and Titanium Alloys. Heat treating of titanium and titanium alloys. Metals Handbook, 1981. pp. 330–341.

[10-113] Banerjee R, Nag S, Stechschulte J, Fraser HL. Strengthening mechanisms in Ti–Nb–Zr–Ta and Ti–Mo–Zr–Fe orthopedic alloys Biomaterials 2004;25:3413–3419.

Chapter 11
Surface Modifications

11.1.	Sandblasting and Surface Texturing	314
11.2.	Shot-Peening and Laser-Peening	317
11.3.	Chemical, Electrochemical, and Thermal Modifications	319
11.4.	Coating	322
	11.4.1 Carbon, Glass, Ceramic Coating	322
	11.4.2 Hydroxyapatite Coating	324
	11.4.3 Ca-P Coating	335
	11.4.4 Composite Coating	342
	11.4.5 TiN Coating	347
	11.4.6 Ti Coating	353
	11.4.7 Titania Film and Coating	357
11.5.	Porosity Controlled Surface and Texturing	360
11.6.	Foamed Metal	363
11.7.	Coloring	364
References		364

Chapter 11
Surface Modifications

Surface modifications have been applied to metallic biomaterials in order to improve mechanical, chemical, and physical properties such as wear resistance, corrosion resistance, biocompatibility, and surface energy, etc. For enhancing the mechanical retention between two surfaces, one or both surfaces are normally modified to increase effective surface area by either sandblasting, shot-peening, or laser-peening method. Another distinct purpose of surface modification is found on implant surfaces for both dental and orthopedic applications to exhibit biological, mechanical, and morphological compatibilities to receiving vital hard/soft tissue, resulting in promoting osseointegration. Such modifications are, in general, divided into two categories: surface concave texturing and surface convex texturing. Surface concave textures can be achieved by either material removal by chemical or electrochemical action, or mechanical indentations (caused by sandblasting, shot-peening, or laser-peening). On the other hand, surface convex textured surfaces can be formed by depositing certain types of particles by one of several physical or chemical depositing techniques (like CVD, PVD, plasma-spraying, etc.) or diffusion bonding. If density and porosity of deposited particles can be appropriately controlled, a porous surface can be achieved, leading to successful bone ingrowth. Surface roughness measurement is one of the most frequently and easily employed methods to characterize the modified surfaces. Hence, alternation of surface roughness should also be discussed in association with surface modifications.

Biological survival, particularly longevity of biological adhesive joints, is often dependent on thin surface films. Surfaces and interfaces behave completely different from bulk properties. The characteristics of a biomaterial surface govern the processes involved in biological response. Surface properties such as surface chemistry, surface energy, and surface morphology may be studied in order to understand the surface region of biomaterials. The surface plays a crucial role in biological interactions for four reasons: (1) the surface of a biomaterials is the only part contacting with the bioenvironment, (2) the surface region of a biomaterial is almost always different in morphology and composition from the bulk, (3) for biomaterials that do not release or leak biologically active or toxic substance, the characteristics of the surface governs the biological response (foreign material vs. host tissue), and (4) some surface properties such as topography affect the mechanical stability of the implant–tissue interface [11-1–11-5, Figure 8-1]. Like the interface, the surface has a certain characteristic thickness, (1) for the case when the interatomic reaction is dominant, such as wetting or adhesion, atoms within a depth of 100 nm (1000 Å) will be important, (2) for the case of the mechanical interaction, such as tribology and surface hardening, since the elasticity due to the surface contact and the plastically deformed layer will be a governing area, the thickness of about 0.1–10 μm will be important, and (3) for the case when mass transfer or corrosion is involved, the effective layer for preventing the diffusion will be within 1–100 μm [11-1–11-5].

Such important surfaces can be further modified or altered in a favorable fashion to accommodate, facilitate, or promote more biofunctionality and bioactivity in mechanical, chemical, electrochemical, thermal, or any combination of these methods.

11.1. SANDBLASTING AND SURFACE TEXTURING

Sandblasting, as well as shot-peening (which will be discussed in the following section), possesses three purposes: (1) cleaning surface contaminants, (2) roughening surfaces to increase effective surface area, and (3) producing beneficial surface compressive residual stress. As a result, such treated surfaces exhibit higher surface energy, indicating higher surface chemical and physical activities, and enhancing fatigue strength as well as fatigue life.

In order to obtain satisfactory fixation and biofunctionality of biotolerated and bioinert materials, some of the mechanical surface alternation such as threaded surface, grooved surface, pored surface, and rough surface have been produced that promote tissue and bone ingrowth. But so far, there is no report on suitable roughness to specific metallic biomaterials (see Figure 7-1). In general, on the macroscopic level (>10 μm), roughness will influence the mechanical properties of the interface, the way stresses are distributed and transmitted, the mechanical interlocking of the interface, and the biocompatibility of biomaterials. On a smaller scale, surface roughness in the range from 10 nm to 10 μm may influence the interface biology, since it is of the same order in size as cells and large biomolecules. Topographic variations of the order of 10 nm and less may become important because microroughness on this scale length consists of material defects such as grain boundaries, steps, and vacancies, which are known to be active sites for adsorption, and thus may influence the bonding of biomolecules to the implant surface. There is evidence that surface roughness on a micron scale allows cellular adhesion that alters the overall tissue response to biomaterials. Microrough surfaces allow early better adhesion of mineral ions or atoms, biomolecules, and cells, form stronger fixation of bone or connective tissue, result in a thinner tissue-reaction layer with inflammatory cells decreased or absent, and prevent microorganism adhesion and plaque accumulation, when compared with the smooth surfaces [11-6].

Piattelli *et al.* [11-7] conducted a histological and histochemical evaluation in rabbits to study the presence of multinucleated giant cells (MGCs) at the interface with machined, sandblasted (with 150 μm alumina media), and plasma-sprayed titanium implants. It was reported that (i) MGCs were not observed at any of the experimental times around machined and sandblasted titanium surfaces; whereas (ii) MGCs were present at the interface with titanium plasma-sprayed (TPS) implants at 2 weeks and at 2 months, (iii) at 4 and 8 weeks these cells tended to decrease in number, and (iv) an inflammatory infiltrate was not present in connection with the MGCs [11-7].

Although alumina (Al_2O_3) or silica (SiO_2) particles are most frequently used as a blasting media, there are several different types of powder particles utilized as media. Surface roughness modulates the osseointegration of orthopedic and dental titanium implants. This process may cause the release of cytotoxic silicium or aluminum ions in the peri-implant tissue. To generate a biocompatible roughened titanium surface, an innovative grid-blasting process using biphasic calcium phosphate (BCP) particles was developed. Ti-6Al-4V disks were either polished, BCP grid-blasted, or left as-machined. BCP grid-blasting created an average surface roughness of 1.57 μm compared to the original machined surface of 0.58 μm. X-ray photoelectron spectroscopy indicated traces of calcium and phosphorus and relatively less aluminum on the BCP grid-blasted surface than on the initial titanium specimen. It was reported that (i) scanning electronic microscopy observations and measurement of mitochondrial activity (MTS assay) showed that osteoblastic MC3T3-E1 cells were viable in contact with the BCP grid-blasted titanium surface, (ii) MC3T3-E1 cells expressed alkaline phosphate (ALP) activity and conserved their responsiveness to bone morphogenetic protein BMP-2, and (iii) the calcium phosphate grid-blasting technique increased the roughness of titanium implants and provided a non-cytotoxic surface with regard to mouse osteoblasts [11-8]. Tribo-chemical treatment has been proposed to enhance the bond strength between titanium crown and resin base [11-9]. Using silica-coated alumina as a blasting media under relatively low pressure, silica layer is expected to remain on the blasted surface so that retention force is enhanced by silan-coupling treatment.

Although the recent development of investment materials and casting machines has enabled clinical applications of titanium in dentistry, there remain several problems to be solved. First of all, efficient finishing techniques are required. Titanium is known to be difficult to grind because of its plasticity, stickiness, low heat conductivity, and chemical reactivity at high temperatures [11-10, 11-11]. Although blasting shows several advantages, there is evidence of adverse effects: (1) surface contamination, depending on type of blasting media, and (2) distortion of blasted workpiece, depending on blasting manner and intensity. Miyakawa *et al.* [11-12]

studied the surface contamination of abraded titanium. Despite low grinding speeds and water cooling, the abraded surfaces were found to be contaminated by abrasive constituent elements. Element analysis and chemical bond state analysis of the contaminants were performed using an electron probe microanalyzer. X-ray diffraction of the abraded surface was performed to identify the contaminants. It was reported that (i) the contamination of titanium is related to its reactivity as well as its hardness, (ii) in spite of water cooling and slow–speed abrading, titanium surfaces were obviously contaminated, (iii) contaminant deposits with dimensions ranging from about 10 to 30 μm occurred throughout the surfaces, and (iv) the contaminant of titanium, although related also to the hardness, resulted primarily from a reaction with abrasive materials, and such contamination could negatively influence titanium's resistance to corrosion and its biocompatibility [11-12].

Normally, fine alumina particles (50 μm Al_2O_3) are recycled within the sandblasting machine. Ceramics such as alumina are brittle in nature, therefore some portions of recycled alumina might be brittle-fractured. If fractured sand blasting particles are involved in the recycling media, it might result in irregular surfaces, as well as potential contamination. Using fractal dimension analysis [11-13–11-15], a sample plate surface was weekly analyzed in terms of topographic changes, as well as chemical analysis of sampled recycled Al_2O_3 particles. It was found that after accumulated use time exceeded 30 min, the fractal dimension (D_F) [11-13] remained a constant value of about 1.4, prior to that it continuously increased from 1.25 to 1.4. By the electron probe microanalysis on collected blasting particles, unused Al_2O_3 contains 100% Al, whereas used (accumulated usage time was about 2400 s) particles contained Al (83.32 wt%), Ti (5.48), Ca (1.68), Ni (1.36), Mo (1.31), S (1.02), Si (0.65), P (0.55), Mn (0.49), K (0.29), Cl (0.26), and V (0.08), strongly indicating that used alumina powder was heavily contaminated, and a high risk for the next material surface to be contaminated. Such contaminants are from previously blasted materials having various chemical compositions, and investing materials as well [11-16].

There are several evidences of surface contamination due to mechanical abrasive actions. As a metallographic preparation, the surface needs to be mechanically polished with a metallographic paper (which is normally SiC-adhered paper) under running water [11-17, 11-18]. It is worth mentioning here that polishing paper should be changed between different types of materials, and particularly when a dissimilar metal-couple is used for galvanic corrosion tests, such couple should not be polished prior to corrosion testing because both materials could become cross-contaminated [11-18]. Hence, there are attempts to use TiO_2 powder for blasting onto titanium material surfaces. It was reported that titanium surfaces were sandblasted using TiO_2 powder (particle size ranging from 45, 45–63, and 63–90 μm) to produce the different surface textures prior to fibroblast cell attachment [11-19].

In the rabbit tibia, CpTi implants, which were sandblasted with 25 μm Al_2O_3 and TiO_2 particles, were inserted in the rabbit tibia for 12 weeks. Even though the amount of Al on the implant surface was higher than for the Al_2O_3-blasted implants compared to implants not blasted with Al_2O_3, any negative effects of the Al element were not detected [11-20], which is in contrast to those reported by Johansson *et al.* [11-21], who reported that Al release from Ti-6Al-4V implants was found to coincide with a poorer bone-to-implant over a 3-month period. It is possible that the lack of differences between TiO_2-blasted and the Al_2O_3-blasted implants depends on lower surface concentrations of toxic Al ions than those reported by Johansson *et al.* [11-21]. Wang [11-22] investigated the effects of various surface modifications on porcelain bond strengths. Such modifications included Al_2O_3 blasting, TiO_2 blasting, HNO_3+HF+H_2O treatment, H_2O_2 treatment, and pre-oxidation in air at 600°C for 10 min. Ti-porcelain couples were subjected to 3-pt. bending tests. It was concluded that TiO_2 air abrasion showed the highest bond strength, which was significantly different from other surface treatments.

Recently, it was reported that sandblasting using alumina as a media caused a remarkable distortion on a Co-Cr alloy and a noble alloy [11-23, 11-24]. It was estimated that the stress causing the deflection exceeded the yield strength of tested materials. It was also suggested that the sandblasting should be done using the lowest air pressure, duration of blasting period, and particle size alumina in order to minimize distortion of crowns and frameworks. To measure distortion, Co-Cr alloy plates (25 mm long, 5 mm wide, and 0.7 mm thick) were sandblasted with Al_2O_3 of 125 μm. Distortion was determined as the deflection of the plates as a distance of 20 mm from the surface. It was reported that (i) the mean deflections varied between 0.37 and 1.72 mm, and (ii) deflection increased by an increase in duration of the blasting, pressure, particle size, and by a decrease in plate thickness [11-23].

11.2. SHOT-PEENING AND LASER-PEENING

Shot peening (which is a similar process to sandblasting, but has more controlled peening power, intensity, and direction) is a cold working process in which the surface of a part is bombarded with small spherical media called shot. Each piece of shot striking the material acts as a tiny hammer, imparting to the surface small indentations or dimples. In order for the dimple to be created, the surface fibers of the material must be yielded in tension. Below the surface, the fibers try to restore the surface to its original shape, thereby producing below the dimple a hemisphere of cold-worked material highly stressed in compression. Overlapping dimples (which are sometimes called forged dimples) develop an even layer of metal in residual compressive stress. It is well known that cracks will not initiate or propagate in a compressively stressed zone due to a tendency of crack-closure. Since nearly all fatigue and stress corrosion cracking failures originate at the surface of a part, compressive stresses induced by shot peening provide considerable increases in part life, since advancing crack-opening is suppressed by pre-existing compressive residual stress. The maximum compressive residual stress produced at or under the surface of a part by shot peening is at least as great as half the yield strength of the material being peened. Many materials will also increase in surface hardness due to the cold working effect of shot peening [11-25–11-27]. Both compressive stresses and cold working effects are used in the application of shot peening in forming metal parts, called "shot forming" [11-26].

Oshida investigated the compressive residual stress distribution in depth underneath the surface of Ti-6Al-4V alloy using the X-ray diffraction. It was found that the compressive residual stress at the surface was −50 MPa, followed by the maximum compressive residual stress of −80 MPa at 20 μm below the surface. Then, the compressive residual stress starts to decrease to show almost zero stress at about 50 μm underneath the surface [11-27]. It was also shown that shot-peening resulted in better bonding strengths between titanium substrate and porcelain than sandblasting [11-28].

The laser peening technology is recently developed, claiming non-contact, no-media, and contamination-free peening method. Before treatment, the workpiece is covered with a protective ablative layer (paint or tape) and a thin layer of water. High-intensity (5–15 GW/cm^2) nanosecond pulses (10–30 ns) of laser light beam (3–5 mm width) striking the ablative layer generate a short-lived plasma which causes a shock wave to travel into the workpiece. The shock wave induces compressive residual stress that penetrates beneath the surface and strengthens the workpiece [11-29–11-32], resulting in improvements in fatigue life and retarding in stress corrosion cracking occurrence. Cho and Jung [11-33] laser-treated CpTi screws and inserted in right tibia metaphysics of white rabbits for 8 weeks. It was reported that (i) SEM of laser-treated implants demonstrated a deep and regular honeycomb pattern with small pores, and (ii) eight weeks implantation, the removal torque was 23.58 N cm for control machined and 62.57 N cm for laser-treated implants. Gaggl *et al.* [11-34] reported that (i) surfaces of laser-treated Ti implants showed a high purity with enough roughness for good osseointegration, and (ii) the laser-treated Ti had regular patterns of micropore with interval of 10–12 μm, diameter of 25 μm, and depth of 20 μm [11-34].

At the end of this section, it is necessary to summarize various techniques to measure and characterize the surface roughness. They include that (1) surface roughness can be measured using a profilometer with sharp edge stylus, which is a contact method, (2) atomic force microscopy can provide non-contact surface topography from which the surface roughness can be indirectly measured, and (3) fractal dimension analysis can be used to present the surface roughness in non-Euclidian dimension [11-13, 11-16]. Recently, Hansson and Hansson [11-35] employed computer simulations to measure surface roughness. The lateral resolution was defined as the pixel size of a profiling system. A surface roughness was simulated by a trigonometric function with random periodicity and amplitude. The function was divided into an array of pixels simulating the pixels of the profiling system. The mean height value for each pixel was used to calculate the surface roughness parameters. It was found that the accuracy of all the surface roughness parameters investigated decreased with increasing pixel size. This tendency was most pronounced for mean slope and developed length ratio, amounting to about 80% of their true values for a pixel size of 20% of the true mean high-spot spacing. It was concluded that the lateral resolution of an instrument/method severely compromises the precision of surface roughness parameters which are measured for roughness features with a mean high-spot spacing less than five times the lateral resolution [11-35].

11.3. CHEMICAL, ELECTROCHEMICAL, AND THERMAL MODIFICATIONS

There are several experimental results on chemical, electrochemical, thermal, and combinations of these with regard to altering the Ti surface to facilitate better surface chemical, mechanical, and biological reactions. Endo [11-36] treated NiTi in 30% NHO_3, heated at 400°C for 0.75 h, and NHO_3 treatment, followed by boiling in water for 6–14 h. The variously treated NiTi surfaces were tested for dissolution resistance in bovine serum. It was found that (i) those stems thermally treated were found to have significantly lower metal ion release due to stable rutile oxide (TiO_2) formation, (ii) human plasma fibronectin (an adhesive protein) was covalently immobilized onto an alkylaminosilane derivate of NiTi substrate with glutaraldehyde, and (iii) the XPS spectra suggested that gamma-aminopropyltriethoxysilane (γ-APS) was bonded to the surface through metallosiloxane bonds (Ti–O–Si) formed via a condensation reaction between the silanol end of γ-APS and the surface of the hydroxyl group, with a highly cross-linked siloxane network formed after heat treatment of the silanized surface at 100°C. Based on these findings, it was concluded that human plasma fibronectin was immobilized at the surface, and significantly promoted fibroblast spreading, suggesting that this chemical modification offers an effective means of controlling metal/cell interactions [11-36]. In a study done by Browne and Gregson [11-37], hip replacement stems manufactured from the Ti-6Al-4V alloy were surface- treated and tested for dissolution resistance in bovine serum. Specimens were degreased in 1,2-dichloroethane vapor and surface treated in one of four ways: (1) 35% nitric acid for 10 min – typical commercial treatment, (2) 35% nitric acid for 16 h and rinsed in distilled water, (3) thermal heating in a furnace for 0.75 h at 400°C, and (4) 35% nitric acid, then aged in boiling distilled water in a silica beaker for various times, 6, 8, 10, and 14 h.

It was found that thermal treatment and aging of the surface oxide encourages the formation of dense rutile structure. This is effective in reducing metal ion dissolution (up to 80%), particularly in the early stages of implantation where the stem surface is equilibrating with its surroundings. This benefit is further enhanced on rough surfaces with an increased surface area. It was therefore concluded that (i) the thermal treatment and aging of the surface oxides are important with respect to cementless and porous implants, and (ii) such treatments could be incorporated in commercial manufacturing procedures and reduce the risk of metal dissolution being a contributory factor toward revision surgery [11-37].

Krozer *et al.* [11-38] investigated the possible influence of an amino-alcohol solution on machined Ti surface properties. Screw-shaped CpTi implants and CpTi studs were used. They were rinsed (1) in running deionized water for 2 min, (2) NaCl solution for 2 min followed by deionized water washing, and (3) rinsed in 5% H_2O_2 for 2 min followed by deionized water washing, and rinsed in deionized water for 2 min. The amino-alcohol solution was supplied to the sample surfaces, and four methods were used in order to remove the adsorbed alcohol molecules. It was shown that (i) rinsing in water, saline solution, and 5% H_2O_2 did not remove the amino-alcohol from the surface; however, (ii) exposure to ozone produced by using a commercial mercury lamp in ambient air resulted in complete removal of the adsorbed amino-alcohol, and (iii) the presence of such a film most likely prevents reintegration to occur at the implant–tissue interface *in vivo* [11-38]. In study done by Rupp *et al.* [11-39], CpTi was first blasted with 354–500 μm large grits, followed by (1) $HCl/HF/HNO_3$ etching, (2) HCl/H_2SO_4 etching, (3) $HCl/H_2SO_4/HF$/oxalic acid + neutralized, and (4) $HCl/H_2SO_4/HF$/oxalic acid + oxidized. It was reported that the Ti modifications which shift very suddenly from a hydrophobic (high-surface contact angle) to a hydrophilic (low surface contact angle) state adsorbed the highest amount of immunologically assayed fibronectin, suggesting that microtexturing greatly influenced both the dynamic wettability of Ti implant surfaces during the initial host contact and the initial biological response of plasma protein adsorption [11-39].

MacDonald *et al.* [11-40] investigated the microstructure, chemical composition, and wettability of thermally, and chemically modified Ti-6Al-4V disks, and correlated the results with the degree of adsorption between the radiolabeled fibronectin and Ti-6Al-4V alloy surface and subsequent adhesion of osteoblast-like cells. It was found that (i) heating either in pure oxygen or atmosphere resulted in an enrichment of Al and V within the surface oxide, (ii) heating (in pure oxygen or atmosphere) and hydrogen peroxide treatment, both followed by butanol treatment, resulted in a reduction in content of V, but not in Al, (iii) heating (oxygen/atm) or hydrogen peroxide treatment resulted in a thicker oxide layer and a more hydrophilic surface when compared with chemically passivated controls (in 40% NHO_3); however, the post-treatment with butanol resulted in a less hydrophilic surface than heating or hydrogen peroxide treatment alone, and (iv) the greatest increases in the adsorption of radiolabeled fibronectin following treatment were observed with hydrogen peroxide/butanol-treated samples followed by hydrogen peroxide/butanol and heat/butanol, although binding was only increased by 20–40% compared to untreated control. These experiments with radiolabeled fibronectin indicated that enhanced adsorption to the glycoprotein was more highly correlated with changes in chemical composition, reflected in V content and decrease in the V/Al ratio, than with changes in wettability. It was therefore concluded that an increase in the absolute content of Al and/or V, and/or in the Al/V ratio is correlated with an increase in the fibronectin-promoted adhesion of an osteoblast-like cell line [11-40]. Li *et al.* [11-41] modified the surface of CpTi (grade 2) implants by the micro-arc oxidation, operated under voltage ranging from 190, 230, 270, 350, 450, and 600 V to form a porous layer. It was found that (i) with increasing voltage, the roughness (from 0.3 to 2.5 μm) and thickness (from 1 to 15 μm) of the film increased, and (ii) the TiO_2 phase changed from anatase to rutile, supporting what was discussed in Chapter 4. The micro-arc oxidation was carried out in an aqueous electrolyte with calcium acetate monohydrate and calcium glycerophosphate in deionized water. During the micro-arc oxidation, it was found that (i) Ca and P ions were incorporated into the oxide layer, (ii) the *in vitro* cell responses were also dependent on the oxidation condition, and (iii) with increasing voltage, the alkaline phosphatase activity increased, while the cell proliferation rate decreased. Preliminary *in vivo* tests of the micro-arc oxidation-treated specimens on rabbits showed a considerable improvement in their osseointegration capacity as compared to the unmodified CpTi implant [11-41].

The surface bioactivity of titanium was investigated after water and hydrogen plasma immersion ion implantation (PI^3) by Xie *et al.* [11-42]. PI^3 method excels in the surface treatment of components possessing a complicated shape such as medical implants. In addition, water and hydrogen plasma immersion ion implantation has been extensively studied as a method to fabricate silicon-on-insulator substrates in the semiconductor industry, and so it is relatively straightforward to transfer the technology to the biomedical field. Water and hydrogen were plasma-implanted into titanium sequentially. It was found that (i) after incubation in simulated body fluids (SBFs) for cytocompatibililty evaluation *in vitro*, bone-like hydroxyapatite was found to precipitate on the (H_2O+H_2) implanted samples, while no apatite was found on titanium samples plasma implanted with water or hydrogen alone, and (ii) human osteoblast cells were cultured on the (H_2O+H_2)-implanted titanium surface and they exhibited good adhesion and growth. It was therefore suggested that plasma immersion ion implantation is a practical means to improve the surface bioactivity and cytocompatibility of medical implants made of titanium [11-42].

Rohanizadeh *et al.* [11-43] investigated methods of preparing different types of titanium oxide (TiO_2) and their effects on apatite deposition and adhesion on titanium surfaces. CpTi disks were subjected to the following treatments: (1) heat treatment at 750°C; (2) oxidation in H_2O_2 solution followed by heat treatment; (3) dipping in rutile/gelatin slurry; and (4) dipping in anatase/gelatin slurry. Surface-treated Ti disks were immersed in a supersaturated calcium phosphate solution to allow apatite deposition. It was shown that (i) the percentage of area covered by deposited apatite was highest in sample disks which were dipped in an anatase/gelatin slurry, compared to the other groups, (ii) apatite deposited on Ti disks pre-treated in H_2O_2 solution demonstrated the highest adhesion to the titanium substrate, and (iii) the surface treatment method affects the type of TiO_2 layer formed (anatase or rutile) and affects apatite deposition and adhesion on the Ti surface [11-43].

11.4. COATING

The coating layer is not only required to exhibit an expected function, depending on its original specific aims, but it is also important to notice that the coating layer is only functional if it adheres well to the metal substrate and if it is strong enough to transfer all loads. Coated substrate possesses at least two layers and one intermediate interface. If such coupled is subjected to stressing, although the strain field should be assumed to be a continuum, the stress field of the couple exhibits a discrete one due to differences in modulus of elasticity. This discrete stress field results in interfacial stress, and if the interfacial stress is higher than the bonding strength, the couple can be debonded or delaminated, causing the structural integrity to no longer be maintained.

11.4.1 Carbon, glass, ceramic coating
The surface of Ti-6Al-4V has been modified by ion beam mixing a thin C film by Demri *et al.* [11-44]. XPS analysis showed that after mixing, the surface film consists essentially of a Ti compound containing (Ti, O, and C), TiO_2, Ti, and C. The composition of the surface modified film determined by Rutherford backscattering spectrometry is approximately $Ti_{0.5}O_{0.3}C_{0.2}$ and its thickness is about 200 μm. It was also reported that after 3 months immersion in an SBF, the growth of calcium phosphate species containing both HPO_4^- and $H_2PO_4^-$ (probably $CaHP_4$ and $Ca(PO_4)_2$) have been observed [11-44]. The corrosion resistance and other surface and biological properties of NiTi were enhanced using carbon plasma immersion ion implantation and deposition (PI^3). Poon *et al.* [11-45]

mentioned that either an ion-mixed amorphous carbon coating fabricated by plasma immersion ion implantation and deposition or direct carbon PI[3] can drastically improve the corrosion resistance and block the out-diffusion on Ni from the metal. The tribological tests showed that the treated surfaces are mechanically more superior and cytotoxicity tests revealed that both sets of plasma-treated samples favored adhesion and proliferation of osteoblasts [11-45]. With regard to potential toxicity of Ni, this is one of methods to prevent or shield the Ni element to diffuse out from NiTi surface. There is another way to achieve the similar outcome by selectively leaching out Ni from the NiTi surface layer by chemically etching the NiTi surface in mixed acid aqueous solution of $HF+HNO_3+H_2O$ (1:1:3 by volume) [11-46].

Bioactive glass (BAG) is a bioactive material with a high potential as implant material. Reactive plasma spraying produces an economically feasible BAG-coating for Ti-6Al-4V oral implants. It was shown that (i) the coating withstands, without any damage, an externally generated tensile stress of 47 MPa, and (ii) adhesion testing after 2 months of *in vitro* reaction in an SBF showed that coating adhesion strength decreased by 10%, but the implant was still adequate for load-bearing applications [11-47].

Saiz *et al.* [11-48] evaluated the *in vitro* response in SBF of silicate glass coating on Ti-6Al-4V. Glasses belonging to the SiO_2-CaO-MgO-Na_2O-K_2O-P_2O_5 system were used to prepare 50–70 μm thick coatings by employing a simple enameling technique. It has been found that (i) coatings with silica content lower than 60 wt% are more susceptible to corrosion and precipitate carbonated HA on their surface during *in vitro* tests; however, (ii) these coatings have a higher thermal expansion than the metal and are under tension, (iii) after 2 months in SBF, crack grows in the coating, reaches the glass/metal interface and initiates delamination, and (iv) glasses with silica content higher than 60 wt% are more resistant to corrosion and have lower thermal expansion, and these coatings do not crack, but such glasses with silica do not precipitate apatite even after 2 months in SBF [11-48]. Lee *et al.* [11-49] prepared calcium-phosphate, apatite–wollastonite ($CaSiO_3$) (1:3 by volume fraction) glass ceramic, apatite–wollastonite (1:1) glass ceramic, and bioactive CaO-SiO_2-B_2O_3 glass ceramic coatings by the dipping method. Coated and uncoated Ti-6Al-4V screws were inserted into the tibia of 18 adult mongrel male dogs for 2, 4, and 8 weeks. It was found that (i) at 2, 4, and 8 weeks, the extraction torque of these ceramic-coated screws was significantly higher than the corresponding insertion torque, and (ii) strong fixation was observed even at 2 weeks in all three coatings except CaO-SiO_2-B_2O_3 glass ceramic coating [11-49].

11.4.2 Hydroxyapatite coating

Enhancement of the osteoconductivity[1] of Ti implants is potentially beneficial to patients since it shortens the medical treatment time and increases the initial stability of the implant. To achieve better osteoconductivity, an apatite [Ca_{10-x} $(HPO_4)_x(PO_4)_{6-x}(OH)_{2-x}$] coating has been commonly employed on the Ti implant surface.

BSE (bovine spongiform encephalopathy) or "mad cow disease" could have been caused by animal-feed contaminated with human remains a controversial theory. Accordingly, although bovine-derived hydroxyapatite was used as a semi-natural HA [11-50], the articles reviewed here are limited to those published before the BSE issue became to be addressed and received public attention.

With the growing demands of bioactive materials for orthopedic as well as max-illofacial surgery, the utilization of hydroxyapatite (HA, with Ca/P=1.67) and tricalcium phosphate (TCP, with Ca/P=1.50) as fillers, spacers, and bone graft substitutes has received great attention during the past two to three decades, pri-marily because of their biocompatibility, bioactivity, and osteoconduction charac-teristics with respect to host tissue. Porous hydroxyapatite granules with controlled porosity, pore size, pore size distribution, and granule size were fabri-cated using a drip-casting process. Granules with a wide range of porosity from 24 to 76 vol%, pore size from 95 to 400 μm, and granular sizes from 0.7 to 4 mm can be obtained. This technique allows the fabrication of porous granules with desir-able porous characters simulating natural bone architecture, and is expected to provide advantages for biomedical purposes [11-51]. Hydroxyapatite is a major mineral component in animal and human bodies. It has been used widely not only

---

[1] *Osteogenesis*: development of bones (bone formation)

Osteoconduction: process in which a graft acts as a frame network or scaffold over which new host bone can grow.

Osteoinduction: processes in which new bone is induced to form through the activation of factors contained within the graft bone, such as proteins of growth factors. Furthermore, there are four dif-ferent types of biomaterials exhibiting osteogenesis:

(1) *Distance osteogenesis*: These types of materials are not in contact with living tissue, but they are not rejected by living tissue either. They are covered with connective tissue and do not connect to bone. Such materials (like Ti and Au) are called biotolerant.

(2) *Contact osteogenesis*: Although it is possible for biomaterials to connect the bone through the connective tissue, they do not connect bone directly. Such materials (like Al_2O_3 or ZrO_2) are called bioinert.

(3) *Bonding osteogenesis*: Bone can grow on biomaterials and can connect to bone directly. Such materials (like high dense sintered hydroxyapatite or bioactive glass) are called bioactive.

(4) *Osteogenesis*: Materials are absorbed by osteoclasts, and bone can grow through the bone remod-eling. Such materials (like low crystalline type hydroxyapatite, tricalcium phosphate) are called bioresorbable.

as a biomedical implant material but also as biological chromatography supports in protein purification and DNA isolation. Spherical HA ceramic beads have recently been developed that show improvements in mechanical properties and physical and chemical stability. These spherical ceramic beads are typically 20–80 μm in size. There are advantages to reducing the granule size of the spherical HA material: (1) the smaller the granule size, the higher the specific surface area and the higher the bonding capacity, (2) theoretically, the specific surface area (i.e., surface area per volume) is proportional to $6/d$, where d is the diameter of the spherical granule, (3) in addition, the mechanical properties of a packed column can be improved by reducing the granule size, resulting in more contacting surface areas, and thereby greater frictional forces between granules, and (4) furthermore, a uniform pack is expected to have a homogeneous pore distribution. The porosity and the specific surface areas of the HA material can be controlled by changing the morphology of the granules, for example, the solid spheres and doughnuts. Hence, Luo *et al.* [11-52] introduced a new method to produce spherical HA granules ranging in size from 1 to 8 μm with controlled morphology. This method involves an initial precipitation followed by a spray-drying process, which is the controlling step, to produce granules with various structures. It was reported that by adjusting the operating parameters (e.g., atomization pressure) and starting slurry (e.g., concentration), the hollow or solid spheres and doughnut-shaped HA were fabricated.

Since HA is normally used as a bone grafting material, it is convenient here to review all types of grafts. *Autologous grafts*: are obtained directly from the patient. Autologous bone grafts possess maximum biocompatibility, and thus are generally considered to be the best material available for bone grafting. *Allogenous grafts*: are obtained from a donor of the same species, and are commonly employed as the "second alternative" when an autograft is not possible. Allogenous bone is also favored when the bone defect is extensive and the quality and quantity of autologous bone is insufficient. The primary problems with allogenous grafts are immunological. *Allografts*: a tissue (bone) graft between individuals of the same species but of non-identical genetic disposition. *Alloplast*: a synthetic bone graft material; a bone graft substitute. *Xenogenous grafts*: are bone materials obtained from another species. Xenogenous bone is primarily harvested from cows. These grafts must be specially treated before use because of genetic incompatibility. *Synthetic implants*: are manufactured from a number of different materials, including plastic polymers (PMMA), ceramics (HA), composites, and metals.

Plasma-sprayed HA-coated devices demonstrated wide variability in the percentage of the HA coating remaining on the stems. Porter *et al.* [11-53] reported that (i) the coating was missing from a substantial portion of a stem after only about 6 months of implantation, and (ii) many ultrastructural features of the bone bonded to the HA coatings on these implants from human subjects were comparable to those on HA-coated devices implanted in a canine model. The geometric design and chemical compositions of an implant surface may have an important part in affecting early implant stabilization and influencing tissue healing.

The influence of different surface characteristics on bone integration of titanium implants was investigated by Buser *et al.* [11-54]. Hollow-cylinder implants with different surfaces were placed in the metaphyses of the tibia and femur in six miniature pigs. After 3 and 6 weeks, the implants with surrounding bone were removed and analyzed in undecalcified transverse sections. The histologic examination revealed direct bone–implant contact for all implants. However, the morphometric analyses demonstrated significant differences in the percentage of bone–implant contact, when measured in cancellous bone. It was reported that (i) electropolished, as well as the sandblasted and acid pickled (medium grit, HF/HNO_3) implant surfaces, had the lowest percentage of bone contact with mean values ranging between 20 and 25%, (ii) sandblasted implants, with a large grit, and TPS implants demonstrated 30–40% mean bone contact, and (iii) the highest extent of bone–implant interface was observed in sandblasted and acid-attacked surfaces (large grit; HCl/H_2SO_4) with mean values of 50–60%, and hydroxyapatite (HA)-coated implants with 60–70%. It was therefore concluded that the extent of the bone–implant interface is positively correlated with an increasing roughness of the implant surface. Moreover the morphometric results indicated that (i) rough implant surfaces generally demonstrated an increase in bone apposition compared to polished or fine structured surfaces, (ii) the acid treatment with HCl/H_2SO_4 used for sandblasted with large grit implants has an additional stimulating influence on bone apposition, (iii) the HA-coated implants showed the highest extent of bone–implant interface, and (iv) the HA coating consistently revealed signs of resorption. It was suggested that sandblasting and chemical etching with HCl/H_2SO_4 as well as HA coating, seemed to be the most promising alternatives to titanium implants with smooth or TPS surfaces [11-54].

Souto *et al.* [11-55] investigated the corrosion behavior of four different preparations of plasma-sprayed hydroxyapatite (HA) coatings (50 and 200 μm) on Ti-6Al-4V substrates in static Hank's balanced salt solution through DC potentiodynamic and AC impedance EIS techniques. Because the coatings are porous, ionic paths between the electrolytic medium and the base material can eventually be produced, resulting in the corrosion of the coated metal. It was concluded that significant differences were found in electrochemical behavior for similar nominal thicknesses of HA coatings obtained under different spraying [11-55]. Filiaggi *et al.* [11-56] reported that evaluations of an HA-coated Ti-6Al-4V implant system using a modified short-bar technique for interfacial fracture toughness determination revealed relatively low fracture toughness values. Using high-resolution electron spectroscopic imaging, evidence of chemical bonding was revealed at the plasma-sprayed HA/Ti-6Al-4V interface, although bonding was primarily due to mechanical interlocking at the interface [11-56]. The modulus of elasticity, residual stress and strain, bonding strength, and microstructure of the plasma-sprayed hydroxyapatite coating were evaluated on Ti-6Al-4V substrate with and without immersion in Hank's balanced salt solution. It was reported that (i) the residual stress and strain, modulus of elasticity, and bonding strength of the plasma-sprayed HA coating after immersion in Hank's solution were substantially decreased, and (ii) the decayed modulus of elasticity and mechanical properties of HA coatings were caused for by the degraded interlamellar or cohesive bonding in the coating due to the increased porosity after immersion that weakens the bonding strength of the coating and substrate system. It was also suggested that the controlled residual stress and strain in the coating might promote the long-term stability of the plasma-sprayed HA-coated implant [11-57]. Yang *et al.* [11-58] investigated the effect of TPS and zirconia (ZrO_2)-coated titanium substrates on the adhesive, compositional, and structural properties of plasma-sprayed HA coatings. Apatite-type and α-tricalcium phosphate phases were observed for all HA coatings. The coating surfaces appeared rough and melted, with surface roughness correlating to the size of the starting powder. It was found that (i) no significant difference in the Ca/P ratio of HA on Ti and TPS-coated Ti substrates was observed, (ii) however, the Ca/P ratio of HA on ZrO_2-coated Ti substrate was significantly increased, (iii) interfaces between all coatings and substrates were observed to be dense and tightly bound, except for HA coatings on TPS-coated Ti substrate interface; however, (iv) an intermediate TPS or ZrO_2 layer between the HA and Ti substrate resulted in a lower adhesive strength as compared to HA on Ti substrate [11-58].

HA can be admixed with Ti powder. 80HA-20Ti powder and 90HA-10Ti powder (by weight) were mechanically mixed and dry-pressed and heat treated at 1100°C for 30 min in vacuum ($<6\times10^{-3}$ Pa). Heat treatment of HA specimens in vacuum resulted in the loss of hydroxyl groups, as well as the formation of a secondary β-tricalcium phosphate phase. It was concluded that the *in vacuo* heat treatment process completely converted the metal-ceramic composites to ceramic composites [11-59]. Moreover, using air plasma spraying and vacuum plasma spraying methods, HA and a mixture of HA+Ti were deposited on Ti-6Al-4V. It was reported that a higher adhesion was obtained with vacuum plasma spraying rather than with air plasma spraying [11-60].

Lee *et al.* [11-61] employed the electron-beam deposition method to obtain a series of fluoridated apatite coatings. The fluoridation of apatite was aimed to improve the stability of the coating and elicit the fluoride effect, which is useful in the dental restoration area. Apatite fluoridated at different levels was used as initial evaporants for the coatings. It was observed that (i) the as-deposited coatings were amorphous, but after heat-treated at 500°C for 1 h, the coatings crystallized well to an apatite phase without forming any cracks, (ii) the adhesion strengths of the as-deposited coatings were about 40 MPa, and after heat treatment at 500°C, it decreased to about 20 MPa; however, the partially fluoridated coating maintained its initial strength. It was also reported that (i) the osteoblast-like cells responded to the coatings in a similar manner to the dissolution behavior, (ii) the cells on the fluoridate coatings showed a lower proliferation level compared to those on the pure HA coating, and (iii) the alkaline phosphatase activity of the cells was slightly lower than of on the pure HA coating [11-61]. Kim *et al.* [11-62] deposited HA and fluoridated HA films on CpTi (grade 2). HA sol was prepared by mixing $P(C_2H_5O)_3$ and distilled water, and fluoridated HA sol was prepared by NH_4F in P containing solution (like HA by replacing the OH group with F ions) and their films were deposited on CpTi (grade 2). The mixture was stirred at room temperature for 72 h. Dipping CpTi into the solutions was performed at 500°C for 1 h. It was concluded that (i) the coatings layers were dense, uniform, and had a thickness of approximately 5 μm after heat treatment at 500°C, (ii) the fluoridated HA layer showed much lower dissolution rate (in 0.9% NaCl as physiological saline solution) than pure HA, suggesting the tailoring of solubility with F-incorporation within the apatite structure, (iii) the osteoblast-like MG63 and HOS cells grew and proliferated favorably on both coatings and pure Ti, and (iv) especially, both coated Ti exhibited higher alkaline phosphatase expression levels as compared to non-coated Ti, confirming the improved activity and functionality of cells on the substrate via the coatings [11-62].

There are several studies done on effects of post-spray coating heat-treatment on mechanical and structural integrities of the coated film. A post-plasma-spray heat treatment (at 960°C for 24 h followed by slow furnace cooling) was performed to enhance chemical bonding at the metal–ceramic interface, and hence improve the mechanical properties. It was found that any improvements realized were lost due to the chemical instability of the coating in a moisture-laden environment, with a concomitant loss in bonding properties, and this deterioration appeared to be related to environmentally assisted crack growth as influenced by processing conditions [11-63]. HA coatings on Ti-6Al-4V were annealed at 400°C in air for 90 h, and were evaluated as post-coating heat treatment. It was found that (i) the oxide species TiO_2, Al_2O_3, V_2O_5, V_2O_3, and VO_2 were present on both as-coated and as heat-treated samples, and (ii) the fatigue resistance of the substrate was found to be significantly reduced by the heat treatment, due to the stress relief [11-64]. Ti-6Al-4V substrate was heated at 25, 160, and 250°C, followed by cooling in air or air$+CO_2$ gas during operating of the HA plasma coating. It was reported that (i) when residual compressive stress was 17 MPa the bonding strength was 9 MPa, while the bonding strength continuously reduced to 3 MPa for a residual compressive stress of about 23 MPa, and (ii) the compressive residual stress weakened the bonding at the interface of the HA and the Ti-substrate [11-65], due to remarkable mechanical mismatching.

As has been reviewed in the above, a plasma spray method has been widely accepted as the apatite coating method because it gives tight adhesion between the apatite coating and Ti. However, the plasma spray method employs extremely high temperatures (10,000–12,000°C) during the coating process. Unfortunately, it results in potentially serious problems including (1) an alteration of structure, (2) formation of apatite with extremely high crystallinity, and (3) long-term dissolution and the accompanying debonding of the coating layer. To reduce these drawbacks associated with the conventional plasma spray method, a high viscosity flame spray method was developed. Although the high viscosity flame spray method employs lower temperature compared with the plasma spray method, it is still 3000°C, which is adequate to alter the crystal structure and formation of apatite with extremely high crystallinity. To avoid shortcomings in the apatite coating methods using high temperatures, many alternative room temperature coating methods were studied extensively including ion beam sputtering, dipping, electrophoretic deposition, and electrochemical deposition [11-66]. Mano *et al.* [11-66] developed a new coating method called blast coating, using common sandblasting equipment at room temperature. Blast-apatite-coated CpTi implants were inserted in tibia of rats for 1, 3, and 6 weeks. It was found that (i) the apatite coating adhered tightly to the Ti surface even after the 6 week implantation, (ii) the implants cause no inflammatory response, showing strong bone response and much better osteoconductivity compared with the uncoated CpTi implant, (iii) the new bone formed on the surface of coated implants was thinner compared with that formed on the surface of Ti implant. Therefore, the blast-apatite implant has a good potential as an osteoconductive implant material [11-66].

A novel method to rapidly deposit bone apatite-like coatings on titanium implants in SBF is proposed by Han and Xu [11-67]. The processing was composed of two steps: micro-arc oxidation of titanium to form titania (TiO_2) films, and UV-light illumination of the titania-coated titanium in SBF. It was reported that (i) the micro-arc oxidation films were porous and nanocrystalline, with pore sizes varying from 1 to 3 μm and grain sizes varying from 10–20 to 70–80 nm; the predominant phase in titania films changed from anatase to rutile, and the bond strength of the films decreased from 43.4 to 32.9 MPa as the applied voltage increased from 250 to 400 V, (ii) after UV-light illumination of the films in SBF for 2 h, bone apatite-like coating was deposited on the micro-arc oxidation film formed at 250 V, and (iii) the bond strength of the apatite/titania bilayer was about 44.2 MPa [11-67].

As we have reviewed the sol method by Kim *et al.* [11-62], the technique was further developed using the sol–gel method to coat Ti surface with hydroxyapatite

(HA) films [11-68]. The coating properties, such as crystallinity and surface roughness, were controlled and their effects on the osteoblast-like cell responses were investigated. The obtained sol–gel films had a dense and homogeneous structure with a thickness of about 1 µm. It was found that (i) the film heat-treated at higher temperature had enhanced crystallinity (600>500>400°C), while retaining similar surface roughness, (ii) when heat-treated rapidly (50°C/min), the film became quite rough, with roughness parameters being much higher (4–6 times) than that obtained at a low heating rate (1°C/min), and (iii) the dissolution rate of the film decreased with increasing crystallinity (400>500>600°C), and the rougher film had slightly higher dissolution rate. The attachment, proliferation, and differentiation behaviors of human osteosarcoma HOS TE85 cells were affected by the properties of these films. It was further reported that (i) on the films with higher crystallinity (heat treated over 500°C), the cells attached and proliferated well, and expressed alkaline phosphatase and osteocalcin to a higher degree as compared to the poorly crystallized film (heat treated at 400°C), and (ii) on the rough film, the cell attachment was enhanced, but the alkaline phosphatase and osteocalcin expression levels were similar as compared to the smooth films [11-68].

The sol–gel method was favored due to the chemical homogeneity and fine grain size of the resultant coating, and the low crystallization temperature and mass-producibility of the process itself. The sol–gel-derived HA and TiO_2 films, with thicknesses of about 800 and 200 nm, adhered tightly to each other and to the CpTi (grade 2) substrate. It was reported that (i) the highest bond strength of the double layer coating was 55 MPa after heat treatment at 500°C due to enhanced chemical affinity of TiO_2 toward the HA layer, as well as toward the Ti substrate, (ii) human osteoblast-like cells, cultured on the HA/TiO_2 coating surface, proliferated in a similar manner to those on the TiO_2 single coating and on the CpTi surfaces; however, (iii) alkaline phosphatase activiy of the cells on the HA/TiO_2 double was expressed to a higher degree than on the TiO_2 single coating and CpTi surfaces, and (iv) the corrosion resistance of Ti was improved by the presence of the TiO_2 coating, as confirmed by a potentiodynamic polarization test [11-69]. Sol–gel-derived TiO_2 coatings are known to promote bone-like hydroxyapatite formation on their surfaces *in vitro* and *in vivo*. Hydroxyapatite integrates into bone tissue. In some clinical applications, the surface of an implant is simultaneously interfaced with soft and hard tissues, so it should match the properties of both. Moritz *et al.* [11-70] introduced a new method for treating the coatings locally in a controlled manner. The local densification of sol–gel-derived titania coatings on titanium substrates with a CO_2 laser was studied in terms of the *in vitro* calcium phosphate-inducting properties. The CO_2-laser-treated multi-layer coating was compared with furnace-fired coating. Additionally, local areas of furnace-fired multilayer coatings were further laser-treated to achieve various properties in the same implant. It was found that (i) calcium phosphate formation can be adjusted locally by laser treatment, (ii) calcium phosphate is a bone-like hydroxyapatite, and (iii) the local treatment of sol–gel-derived coatings with a CO_2 laser is a promising technique for creating implants with various properties to interface different tissues, and a possible way of coating implants that do not tolerate furnace firing [11-70].

Plasma-sprayed HA coatings on commercial orthopedic and dental implants consist of mixtures of calcium phosphate phases, predominantly a crystalline calcium phosphate phase, hydroxyapatite, and an amorphous calcium phosphate with varying ratios. Alternatives to the plasma spray method are being explored because of some of its disadvantages, as mentioned before. Rohanizaded *et al.* [11-71] developed a two-step (immersion and hydrolysis) method to deposit an adherent apatite coating on titanium substrate. First, titanium substrates were immersed in acidic solution of calcium phosphate, resulting in the deposition of a monetite ($CaHPO_4$) coating. Second, the monetite crystals were transformed to apatite by hydrolysis in NaOH solution. Energy-dispersive spectroscopy had revealed the presence of calcium and phosphorous elements on the titanium substrate after removing the coating using tensile or scratching tests. It was reported that (i) the average tensile bond of the coating was 5.2 MPa and cohesion failures were observed more frequently than adhesion failures, (ii) this method has the advantage of producing a coating with homogeneous composition on even implants of complex geometry or porosity, and (iii) this method involves low temperatures and, therefore, can allow the incorporation of growth factors or biogenic molecules [11-71].

Ergun *et al.* [11-72] studied the chemical reactions between hydroxyapatite (HA) and titanium in three different kinds of experiments to increase understanding the bond mechanisms of HA to titanium for implant materials. HA powder was bonded to a titanium rod with hot isostatic pressing. Interdiffusion of the HA elements and titanium was found in concentration profiles measured in the electron microprobe. Titanium was vapor-deposited on sintered HA disks and heated in air; perovskite ($CaTiO_3$) was found on the HA surface with Rutherford backscattering and X-ray diffraction measurements. Powder composites of HA, titanium, and TiO_2 were sintered at 1100°C; again, perovskite was a reaction product, as well as β-$Ca_3(PO_4)_2$, from a decomposition of the HA. Based on these findings, it was reported that (i) chemical reactions and interdiffusion between HA and TiO_2 during sintering resulted in chemical bonding between HA and titanium; thus, cracks and weakness at HA-titanium interfaces probably result from mismatch between the coefficients of thermal expansion of these materials, and (ii) HA composites with other ceramics and different alloys should lead to better thermal matching and better bonding at the interface [11-72].

Piattelli and Trisi [11-73] examined retrieved HA-coated CpTi implants from patients. It was reported that the bone–HA interface presented variable features: in some areas the mineralized bone was directly apposed to the HA surface, while in others unmineralized material, probably osteoid, was interposed [11-73].

The effects of HA coating and macrotexturing of Ti-6Al-4V was tested by Hayashi *et al.* [11-74]. Implants were inserted in dog's femoral condyles. It was demonstrated that when grooved Ti implants were used, the addition of HA coating significantly improved the biological fixation. In addition, a grooved depth of 1 mm was found to give significantly better fixation than 2 mm. When compared to implants with traditional diffusion-bonded bead-coated porous surfaces, HA-coated grooved Ti implants were found to show better fixation at 4 weeks after implantation, but significantly inferior fixation at 12 weeks. Hence, it was concluded that while a groove depth of 1 mm was optimal in HA-coated and further grooved Ti implants, they are still inferior to bead-coated Ti implants with respect to long-term fixation [11-74]. Two different groups of HA coated and uncoated porous Ti implants, 250–350 and 500–700 μm diameter beads, were press-fitted into femoral canine cancellous bone, for 12 weeks. It was reported that (i) the percentage of bone and bone index were higher in the HA-coated implants, when comparing HA coated vs. uncoated implant in the 250–350 μm bead diameter group, (ii) comparing 500–700 μm, bone ingrowth and bone depth penetration were higher in HA- coated samples, and (iii) the HA-coating was an effective method for improving bone formation and ingrowth in the porous implants [11-75].

Several HA-coated and uncoated Ti-6Al-4V femoral endoprostheses were evaluated in adult dogs for 52 weeks. It was found that (i) histologic sections from the uncoated grooved implants showed no direct bone–implant apposition with no fibrous tissue interposition after up to 10 weeks, and (ii) the HA-coated grooved implants demonstrated extensive direct bone-coating apposition after 5 weeks [11-76]. HA-coated cylindrical plugs of CpTi, Cp-Ti screws, and partially HA-coated CpTi screws were inserted in tibia of adult New Zealand white rabbits for 3 months. The histological results demonstrated that although there were no marked differences in bony reaction at the cortical level to the different implant materials, HA-coating appeared to induce more bone formation in the medullary cavity. It was also noted that 3 months after insertion loss of coating thickness had occurred [11-77]. Threaded HA-coated CpTi implants were inserted in the rabbit tibial metaphysis. After 6 weeks and 1 year post-insertion, the semiloaded implants were histomorphometrically analyzed. It was reported that (i) there was more direct bony contact with the HA-coated implants after 6 weeks, and (ii) one year after insertion, there was significantly more direct bone-to-implant contact with the uncoated CpTi controls, suggesting that the HA coating does not necessarily improve implant integration over a long-time period [11-78]. After 6 or 12 weeks, four goats were used for mechanical tests and three for histology. To cope with the severe femoral bone stock loss encountered in revision surgery, impacted trabecular bone grafts were used in combination with an HA-coated Ti stem. In this first experimental study, the results indicated that this technique had a high complication rate. However, it was shown that impacted grafts sustained the loaded stems and that incorporation of the graft occurred with a biomechanically stable implant. The technique allows gradual graft incorporation and stability, but more investigations are needed before its introduction into clinical practice [11-79]. Surface treatments, such as sand-blasting or deposit or rough pure Ti obtained by vacuum plasma spraying, bring about an increase of passive and corrosion current density values as a consequence of the increase of the surface exposed to the aggressive environment. It should, however, be outlined that in the case of rough Ti, only Ti ions are released instead of Al or V out of Ti-6Al-4V. The presence of deposits of HA by vacuum plasma spraying causes an increase of about one order of magnitude in the passive and corrosion current density of the metallic substrate [11-80].

Hayashi *et al.* [11-81] evaluated the bone–implant interface shear strength of HA-coated Ti-6Al-4V and HA-coated Ti-6Al-4V with a rougher surface in a transcortical model in adult dogs for 4 and 12 weeks. It was reported that (i) the bone–implant shear strength of bead-coated porous Ti-6Al-4V was significantly greater than that of both HA-coated implants, (ii) HA coating enhanced early bone ingrowth into the porous surface of the implant, (iii) long-term fixation should depend on bone anchoring to this porous surface, and (iv) HA must be developed which do not obstruct the pores of the surface of the implant [11-81]. Souto *et al.* [11-82] examined HA-coated Ti-6Al-4V (which has a thickness between 50 and 200 μm and is very porous in nature) with EIS (electrochemical impedance spectroscopy) tests in Ringer's solution. It was found that (i) two-layer models describe the electrochemical behavior of the system by considering a porous film on the metallic substrate, (ii) metal dissolution occurs through the pores in the HA-coating, and leads to the precipitation of salts which block the pores, and (iii) the resulting precipitates layer originates an additional barrier to metal dissolution in the system, as it can be followed through a significant increase in the resistance values measured after 20 days. It was also mentioned that after 4-month exposure to Ringer's solution, no evidence was found from the EIS that might indicate a (partial) detachment of the HA film from the substrate, suggesting that the coated system may outstand longer exposure periods in this physiological solution [11-82].

There are several works in regard to mechanical properties of HA and bond strength of HA layer. Cook *et al.* [11-83] conducted interface shear strength tests using a transcortical pushout model in dogs after 3, 5, 6, 10, and 32 weeks on CpTi-coated Ti-6Al-4V and HA-coated Ti-6Al-4V. The mean values for interface shear strength increased up to 7.27 MPa for HA-coated implants after 10 weeks of implantation, while uncoated CpTi had 1.54 MPa [11-83]. Yang *et al.* [11-84] reported that the direction of principal stresses were in proximity to and perpendicular to the spraying direction. The measured modulus of elasticity (MOE) of HA (16 GPa at maximum) was much lower than the theoretical value (i.e., 112 GPa). The denser, well-melted HA exhibited a higher residual stress (compressive, 10–15 MPa at maximum), as compared with the less dense, poor-melting HA. The denser coatings could be affected by higher plasma power and lower powder feed rate. It was also reported that the thicker 200 μm HA exhibited higher residual stress than that of the thinner 50 μm HA [11-84]. Mimura *et al.* [11-85] characterized morphologically and chemically the coating–substrate interface of a commercially available dental implant coated with plasma-sprayed HA, when subjected to mechanical environment. A thin Ti oxide film containing Ca and P was found at the interface on Ti-6Al-4V. When the implant was subjected to mechanical stress, a mixed mode of cohesive and interfacial fractures occurred. The cohesive fracture was due to separation of the oxide film from the substrate, while the interfacial fracture was due to the exfoliation of the coating from the oxide film bonded to the substrate. It was reported that (i) microanalytical results showed diffusion of Ca into the metal substrate, hence indicating the presence of chemical bond at the interface; however, (ii) mechanical interlocking seemed to play the major role in the interfacial bond [11-85]. Lynn and DuQuesnay [11-86] conducted uniaxial fatigue tests ($\sigma_{max}/\sigma_{min}$ stress ratio, $R = -1$, stress amplitude of 620 MPa, frequency of 50 Hz) on blasted-Ti-6Al-4V with HA coating on Ti-6Al-4V with film thickness ranging from 0, 25, 50, 75, 100, and 150 μm. It was found that (i) samples with 150 μm were shown to have significantly decreased fatigue resistance, while coatings of 25–100 μm were found to have no effect on fatigue resistance, and (ii) HA coatings with 25–50 μm show no observable delamination during fatigue tests, while coatings with 75–150 μm thick were observed to spall following but not prior to the initiation of the first fatigue crack in the substrate [11-86].

There are several studies done on apatite-like formation without HA coating. Ti can form a bone-like apatite layer on its surface in SBF when it is treated in NaOH. When pre-treated Ti is exposed to SBF, the alkali ions are released from the surface into the surrounding fluid. The sodium ions increase the degree of supersaturation of the soaking solution with respect to apatite by increasing pH. On the other hand, the released Na^+ causes an increase in external alkalinity that triggers an inflammatory response and leads to cell death. Therefore, it would be beneficial to decrease the release of Na^+ into the surrounding tissue. It was found that (i) the rate of apatite formation was not significantly influenced by a lower amount of Na^+ ion in the surface layer, and (ii) Ti with the lowest content of Na^+ could be more suitable for implantation in the human body (4 at%) [11-87]. Jonášová *et al.* [11-88] pre-treated CpTi surfaces: (1) in 10 M NaOH at 60°C for 24 h, and (2) etched in HCl under inert atmosphere of CO_2 for 2 h, followed by 10 M NaOH treatment. It was found that Ti treated in NaOH can form hydroxycarbonated apatite after exposition in SBF. Generally, Ti is covered with a passive oxide layer. In NaOH this passive film dissolves and an amorphous layer containing alkali ions is formed. When exposed to SBF, the alkali ions are released from the amorphous layer and hydronium ions center into the surface layer, resulting in the formation of Ti-OH groups in the surface. The acid etching of Ti in HCl under inert atmosphere leads to the formation of a micro-roughened surface, which remains after alkali treatment in NaOH. It was shown that the apatite nucleation was uniform and the thickness or precipitated hydroxycarbonated apatite layer increased continuously with time. The treatment of Ti by acid etching in HCl and subsequently in NaOH, is a suitable method for providing the metal implant with bone-bonding ability [11-88]. Wang *et al.* [11-89] reported that bone-like apatite was formed on Ti-6Al-4V surfaces pre-treated in NaOH solution immersed in SBF, while no apatite was formed on untreated Ti-6Al-4V. The increase in electrical resistance (by EIS) in the outermost surface of pre-treated Ti-6Al-4V indicated apatite nucleation. With an increase in immersion time in SBF, islands of apatite were seen to grow and coalesce on pre-treated Ti-6Al-4V under SEM. It was also mentioned that the growth of apatite corresponded to the increase in electrical resistance of the surface layer [11-89].

Titania-based films on Ti were prepared by micro-arc oxidation at various applied voltage (250–500 V) in electrolytic solution containing β-glycerophosphate disodium salt pentahydrate (β-GP) and calcium acetate monohydrate. Such modified Ti samples were examined in SBF and 1.5 times concentrated SBF [11-90]. It was found that (i) macro-porous, Ca- and P-containing titania-based films were formed on the Ti substrates, (ii) in particular, Ca- and P-containing compounds such as $CaTiO_3$, β-$Ca_2P_2O_7$, and β-$Ca_3(PO_4)_2$ were produced at higher voltage (>450 V), (iii) when immersed in SBF, a carbonated HA was induced on the surface at 450 V after 28 days, which is closely related to Ca- and P-containing phases, and (iv) the use of 1.5 concentrated SBF shortened the apatite (AP) induction time, and AP formation was confirmed even on the surface of the films oxidized at 350 V, suggesting that the incorporated Ca and P in the titania films play a similar role to the Ca- and P-containing compounds in the SBF [11-90–11-92].

11.4.3 Ca-P coating

As we have just reviewed the evidence showed a formation of an apatite-like film containing Ca and P ions when Ti was pre-treated in NaOH, followed by immersion in SBF or PBL (phosphate-buffered liquid). Clinically, it was found that for Ti an increase in oxide thickness and an incorporation of Ca and P were found in placed Ti implants which were in the shape of screws, and had been osseointegrated in patients' jaws for times ranging from 0.5 to 8 years [11-93]. These provide "hindsight" knowledge, where later on people will use this knowledge to modify the Ti surfaces.

Formation of calcium-phosphate on alkali- (10 M NaOH at 60°C for 24 h) and heat-treated (alkali-treated then at 600°C for 1 h followed by furnace cooling) CpTi was tested. Samples were immersed in the revised SBF with the same HCO_3^- concentration level as in human blood plasma. It was reported that (i) electron diffraction for the precipitates revealed that octacalcium phosphate (OCP), instead of HA, directly nucleates from amorphous calcium-phosphate, (ii) the OCP crystals continuously grew on the Ti surfaces rather than transforming to AP, and (iii) calcium titanate ($CaTiO_3$) was identified by electron diffraction [11-94]. Hanawa and Ota [11-95] characterized surface films formed on titanium specimens which were immersed in electrolyte solutions (pH 4.5, 5.1, 7.4) at 37°C for 1 h, 1, 30, and 60 days by XPS, FTIR-RAS Fourier transform infrared reflection absorption spectroscopy to understand the reaction between Ti and inorganic ions. For comparison, the surfaces of Ti-6Al-4V and NiTi were also characterized. XPS data revealed that calcium phosphates (CPs) were naturally formed on these specimens. In particular, compared with the CPs formed on the Ti alloys, the CPs formed on Ti immersed for 30 days in the solution with pH 7.4 was more like HA. It was also reported that (i) the compositions of the CPs formed on the specimens changed with the immersion time and the pH value of the solution (the solution with pH 7.4 was Hank's balanced solution without organic species, solution with pH 4.5 was the artificial saliva without sodium sulfide and urea), and (ii) a CP similar to AP is naturally formed on Ti in a neutral electrolyte solution in 30 days [11-95].

Calcium phosphate (Ca-P) coatings have been applied onto titanium alloy prostheses to combine the strength of the metals with the bioactivity of Ca-P. It has been clearly shown in many publications that a Ca-P coating accelerates bone formation around the implant. However, longevity of the Ca-P coating for an optimal bone apposition onto the prosthesis remains controversial. Barrère *et al.* [11-96] evaluated biomimetic bone-like carbonate apatite (BCA) and OCP coatings which were deposited on Ti-6Al-4V samples to evaluate their *in vitro* and *in vivo* dissolution properties. The coated plates were soaked in α-MEM for 1, 2, and 4 weeks, and they were analyzed by backscattering electron microscopy (BSEM) and by Fourier transform infra red spectroscopy (FTIR). Identical coated plates were implanted subcutaneously in Wistar rats for similar periods. BSEM, FTIR, and histomorphometry were performed on the explants. *In vitro* and *in vivo*, a carbonate apatite (CA) formed onto OCP and BCA coatings via a dissolution–precipitation process. It was reported that (i) *in vitro*, both coatings dissolved overtime, whereas *in vivo* BCA calcified and OCP partially dissolved after 1 week, and (ii) thereafter, OCP remained stable. This different *in vivo* behavior can be attributed to (1) different organic compounds that might prevent or enhance Ca-P dissolution, (2) a greater reactivity of OCP due to its large open structure, or (3) different thermodynamic stability between OCP and BCA phases. It was therefore concluded that these structural and compositional differences promote either the progressive loss or calcification of the Ca-P coating, and might lead to different osseointegration of coated implants [11-96]. Barrère *et al.* [11-97] studied the nucleation and growth of a calcium phosphate (Ca-P) coating deposited on titanium implants from SBF, using atomic force microscopy (AFM) and environmental scanning electron microscopy (ESEM). Forty titanium alloy plates were assigned into two groups: a smooth surface group having a maximum roughness R_{max}<0.10 μm, and a rough surface group with an R_{max}<0.25 μm. Titanium samples were immersed in SBF concentrated by five (SBF×5) from 10 min to 5 h and examined by AFM and ESEM. It was observed that scattered Ca-P deposits of approximately 15 nm in diameter appeared after only 10 min of immersion in SBF×5, and (ii) these Ca-P deposits grew up to 60–100 nm after 4 h on both smooth and rough Ti-6Al-4V substrates. It was found that (i) a continuous Ca-P film formed on titanium substrates, (ii) a direct contact between the Ca-P coating and the Ti-6Al-4V surface was observed, and (iii) the Ca-P coating was composed of nanosized deposits and of an interfacial glassy matrix, which might ensure the adhesion between the Ca-P coating and the Ti-6Al-4V substrate. The Ca-P coating detached from the smooth substrate, whereas the Ca-P film extended onto the whole rough titanium surface over time. In the case of rough Ti-6Al-4V, the Ca-P coating evenly covered the substrate after immersion in SBF×5 for 5 h. Accordingly, it was suggested that (i) the heterogeneous nucleation of Ca-P on titanium was immediate and did not depend on the Ti-6Al-4V surface topography, and (ii) the further growth and mechanical attachment of the final Ca-P coating strongly depended on the surface, for which a rough topography was beneficial [11-97].

The bioconductivity of a new biomedical Ti-29Nb-13Ta-6Zr alloy was achieved by a combination of surface oxidation and alkaline treatment [11-98]. It was reported that (i) immersion in a protein-free SBF and fast calcification solution led to the growth of CP phase on the oxidized and alkali-treated (10 M NaOH at 40°C for 24 h) alloy, and the new bioconductive surface was still harder than the substrate, (ii) oxidation at 400°C×24 h led to the formation of a hard layer, (iii) the oxides are mainly composed of TiO_2, with small amounts of Nb_2O_5 and ZrO_2; an oxygen diffusion layer exists beneath the surface oxide layer, and (iv) a titanate layer forms on the pre-oxidized surface after alkali treatment, and growth of a layer of Ca-P has been successfully induced on the titanate layer by immersing in SBF [11-98]. Bogdanski *et al.* [11-99] coated NiTi with CP by dipping in oversaturated CP solution. Since polymorphonuclear neutrophil grannulocytes (PMN) belong to the first cells which will adhere to implant materials, the apoptosis of isolated human PMN after cell culture on non-coated and CP-coated SME NiTi was analyzed by light and scanning electron microscopy and flow cytometry. It was found that (i) in contrast to PMN adherent to non-coated TiNi, the apoptosis of PMN adherent CP-coated samples was inhibited, and (ii) cell culture media obtained from cultured leukocytes with CP-coated were able to transfer the apoptosis inhibiting activities to freshly isolated PMN [1-99]. The compositions of the surface and the interface of calcium phosphate ceramic (CP) coatings electrophoretically deposited and sintered on CpTi and Ti-6Al-4V were evaluated before and after 4 weeks immersion in a simulated physiological solution. In the CP coating–metal interface, it was mentioned that (i) the phosphorus diffused beyond the Ti oxide layer, resulting in the depleted phosphorus in the ceramic adjacent to the metal, and (ii) the surface of the ceramic, however, was substantially unchanged [11-100].

The possible mechanisms of minimization of prosthesis-derived bone growth inhibitors by shielding of the metal and reduction of the associated metal dissolution was investigated by Ducheyne *et al.* [11-101]. Ti, Al, and V release rates were determined *in vitro* for Ti-6Al-4V alloy both with and without a CP coating. Surfaces were passivated in 40% HNO_3 at 55°C for 20 min. Calcium phosphate ceramic was electrophoretically deposited and subsequently vacuum sintered at 925°C for 2 h. Immersion for 1, 2, 4, 8, and 16 weeks in Hank's balanced salt solution, simulated physiological solution with 1.5 mM DS-EDTA. It was found that (i) the CP-coated specimens contained no measurable amounts of Ti, (ii) the Al ion solution around the CP-coated specimen was significantly greater than the concentration around the non-coated specimens; however, (iii) Al did not increase significantly with time, at least up to 4 weeks immersion. The CP coating produced a significant increase of biological fixation, yet at the same time a greater Al release into solution, calling into question the value of calcium phosphate ceramic coating in shielding adverse metal passive dissolution to enhance bone growth [11-101].

Similarly to HA coating, there are various (chemical, electrochemical, and physical) processes for Ca-P coating available. The ions were implanted in sequence, first Ca and then P, both at a dose of 10^{17} ions/cm^2 at a beam energy of 25 keV on CpTi (grade 2). The corrosion tests were done in SBF at 37°C. It was concluded that (i) the CpTi surface subjected to two-step implantation of Ca+P at a dose of 10^{17} ions/cm^2 becomes amorphous, (ii) the implantation of Ca+P ions increases the corrosion resistance in SBF exposure at 37°C for up to 3200 h, (iii) during exposures to SBF, CP precipitates form the implanted as well as non-implanted samples, (iv) the CP precipitates have no effect on the corrosion resistance, and (v) under the conditions of the applied examination, the biocompatibility of Ti subjected to the two-step implantation of Ca+P ions was similar to that of untreated sample [11-102]. CpTi (grade 2) were surface-modified by anodic spark discharge anodization and a thin layer (*ca.* 5 μm) of amorphous TiO_2 containing Ca and P – Ti/AM. Some of the Ti/AM samples were further modified by hydrothermal treatment to produce a thin outermost (*ca.* 1 μm) of HA (Ti/AM/HA). It was reported that (i) non-anodized vacuum annealed and hydrofluoric acid-etched samples, used as control material, showed good bone adsorption producing Ti excellent surface properties, (ii) anodization at a voltage of 275 V that produced a crack-free amorphous TiO_2 film containing Ca and P (Ti/AM) provided good results for cytocompatibility, important morphological characteristics (micropores without crack development) but presented the lowest bone apposition probably due to the degradation of amorphous TiO_2 film, and (iii) hydrothermal treatment at 300°C for 2 h that produced a sub-micrometer layer of HA crystals (*ca.* 1 μm) upon the amorphous TiO_2 film (Ti/AM/HA) gave rise to the highest bone apposition only at 8 weeks [11-103].

Coating by a RF magnetron sputter technique for the production of thin Ca-P coatings can be produced that vary in Ca/P ratio as well as structural appearance. Jansen *et al.* [11-104] studied the effect of non-coated titanium and three different Ca-P sputtered surfaces on the proliferation and differentiation (morphology and matrix production) of osteoblast-like cells. It was found that (i) proliferation of the osteoblast-like cells was significantly higher on non-coated than on Ca-P coated samples, (ii) on the other hand, more mineralized extracellular matrix was formed on the coated surfaces, and (iii) TEM confirmed that the cells on the coated substrates were surrounded by extracellular matrix with collagen fibers embedded in crystallized needle-shaped structures. On the basis of these findings, it was concluded that (i) the investigated Ca-P sputter coatings possess the capacity to activate the differentiation and expression of osteogenic cells, and (ii) bone formation proceeds faster on Ca-P surfaces than on Ti substrates. Further, it was noticed that this bone stimulating effect appeared to be independent of the Ca-P ratio of the deposited coatings [11-104].

Hayakawa *et al.* [11-105] prepared four types of Ti implants: (1) as-blasted with titania (<150 μm), (2) sintered with Ti beads with 50–150 μm diameter, (3) blasted with IBDM (ion beam dynamic mixing) Ca-P coating, and (4) multilayered sintered implants with IBDM Ca-P coating. The Ca-P coating was rapid heat-treated with infrared radiation at 700°C. The implants were inserted into the trabecular bone of the left and right femoral condyles of 16 rabbits, for 2, 3, 4, and 12 weeks. It was reported that (i) histological evaluation revealed new bone formation around different implant materials after already 3 weeks of implantation, (ii) after 12 weeks, mature trabecular bone surrounded all implants, and (iii) the combination of surface geometry and Ca-P coating benefits the implant–bone response during the healing phase [11-105].

Gan *et al.* [11-106] evaluated the interface shear strength of Ca-P thin films applied to Ti-6Al-4V substrates using a substrate straining method – a shear lag model. The Ca-P films were synthesized using sol–gel methods from either an inorganic or organic precursor solution. Strong interface bonding was demonstrated for both film types. It was reported that (i) the films were identified as non-stoichiometric hydroxyapatite, but with different Ca/P ratios, (ii) the Ca-P films were 1–1.5 μm thick, and (iii) the shear strength was approximately 347 and 280 MPa for inorganic and organic route-formed films, respectively [11-106]. NiTi was coated with CP by dipping in oversaturated CP solution, with layer thickness (5–20 μm). The porous nature of the layer makes it mechanically stable enough to withstand both the shape memory transition upon cooling and heating and also strong bending of materials (superelasticity). The adherence of human leukocytes and platelets to the AP layer was analyzed *in vivo*. By comparison, in non-coated SME-NiTi, it was reported that leukocytes and platelets showed a significantly increased adhesion to the coated NiTi [11-107]. Fend *et al.* [11-108] pre-treated CpTi: by immersing in boiling $Ca(OH)_2$ solution for 30–40 min to have only Ca-added pre-calcification, immersing in 20% H_3PO_4 solution at 85–95°C for 30 min to have only P-added pre-phosphatization, and immersed these pre-calcified CpTi in supersaturated CP solution at 37°C for 1 week. It was found that (i) the osteoblast amount and activity on the surfaces containing Ca are higher than those on the surface containing sole P ions, (ii) Ca^{2+} ion sites on the material surfaces favor protein adsorption, such as fibronectin or vitronectin as ligands of osteoblast, onto the surface due to positive electricity, chemical, and biological function, and (iii) on the AP surfaces, Ca^{2+} ions are the active sites of the osteoblast adhesion and also promote the cell adhesion on PO_4^{3-} ion sites [11-108].

Self-assembled monolayer was formed in anhydrous pentane with 5% (v/v) 7-oct-1-enyltrichlorosilane, under argon atmosphere at 25°C for 1 h. CpTi foil was first oxidized in concentrated mixture of H_2SO_4 and H_2O_2. A fast calcium phosphate CP deposition experiment was carried out by mixing Ca^{2+} and $(PO_4)^{3-}$ containing solution in the presence of the surface-modified Ti samples at pH 7.4 at room temperature in order to verify the nucleating abilities of these functional groups. Microanalytical results showed that poorly crystallized HA can be deposited on the self-assembled monolayer surfaces with $-PO_4H_2$ and $-COOH$ functional groups, but not onto the self-assembled monolayer with $-CH=CH_2$ and $-OH$, $-PO_4H_2$ exhibited a stronger nucleating ability than that of $-COOH$. The oxidized Ti sample also showed some CP deposition, but to a lesser extent as compared to self-assembled monolayer surfaces. The pre-deposited HA can rapidly induced biomimetic apatite layer formation after immersion in 1.5 SBF for 18 h regardless of the amount of pre-deposited HA. The results suggested that the pre-deposition of HA onto these functionalized self-assembled monolayer surfaces might be an effective and fast way to prepare biomimetic AP coatings on surgical implants [11-109]. Peng *et al.* [11-110] deposited electrochemically adherent and optically semi-transparent thin calcium phosphate (CaP) films on titanium substrates in a modified SBF at 37°C. It was found that (i) coatings deposited by using periodic pulsed potentials showed better adhesion and better mechanical properties than coatings deposited with use of a constant potential, (ii) the coatings displayed a polydispersed porous structure with pores in the range of a few nanometers to 1 μm, and (iii) cell culture experiments showed that the CaP surfaces are non-toxic to MG63 osteoblastic cells *in vitro* and were able to support cell growth for up to 4 days, outperforming the untreated titanium surface in a direct comparison [11-110].

Biomimetic deposition and electrochemical deposition in solution were used for CP (calcium phosphate) deposition on sintered CpTi bead-cylinders (1300°C for 2 h under vacuum). The supersaturated solution for CP deposition was prepared by dissolving $NaCl$, $CaCl_2$, $Na_2HPO_4 \cdot 2H_2O$ in distilled water buffered to pH 7.4. Biomimetic deposition was done by immersing samples in solution at 37°C for 1, 2, 4, and 6 days, while electrochemical deposition was performed at $-1.8V$ (vs. Hg_2SO_4/Hg reference) for 2 h at 37°C. It was reported that (i) a pre-coating alkaline treatment (5 mol/l NaOH at 60°C for 5 h) is necessary to obtain a uniform coating layer on the inner pore surfaces when the biomimetric deposition is used, (ii) electrochemical deposition is more efficient and less sensitive to the conditions of the Ti surfaces compared to the biomimetric depositions; however, (iii) the electrochemical deposition produces less uniform and thinner coating layers on the inner pore surfaces compared with the biomimetric deposition, and (iv) the crystal structure of the deposited Ca-P is OCP (octacalcium phosphate) regardless of the deposition methods [11-111]. Heimann and Wirth [11-112] deposited hydroxyapatite and duplex hydroxyapatite+titania bond coat layers onto Ti-6Al-4V substrates by atmospheric plasma spraying at moderate plasma enthalpies. From as-sprayed coatings and coatings incubated in SBF, electron-transparent samples were generated by focused ion beam excavation. It was found that (i) adjacent to the metal surface a thin layer of amorphous calcium phosphate (CP) was deposited, (ii) after *in vitro* incubation of duplex coatings for 24 weeks, Ca-deficient defect apatite needles precipitated from amorphous CP, and (iii) during incubation of hydroxyapatite without a bond coat for 1 week diffusion bands were formed within the amorphous CP of 1–2 μm width parallel to the interface metal/coating, presumably by a dissolution–precipitation sequence [11-112]. Aluminoborosilicate glass+HA powder were coated on CpTi at 800–900°C. It was found that (i) the precipitates were Ca-deficient carbonate with low crystallinity, (ii) both morphologies and composition of the precipitates *in vivo* were similar to those *in vitro*, and (iii) the HA particles on the surface of the composite act as nucleation sites for precipitation in physiologic environments, whereas the glass matrix is independent of it [11-113].

For the same purpose as done for the HA-coating, a Ca-P-coated layer is needed to post-coating heat treatment. Lucas and Ong [11-114] investigated the post-deposition heat treatments of ion beam as-sputtered coatings by varying the time and temperature. Ca-P coatings, deposited using a hydroxyapatite–fluorapatite target, received the post-deposition heat treatments as follows: 150, 300, 400, 500, and 600°C for either 30 or 60 min, followed by furnace cooling to room temperature. It was found that (i) the average bond strengths for the coatings heated to 500°C (30 min), 600°C (30 min) and 600°C (1 h) were 40.1, 13.8, and 9.1 MPa, respectively. The 500°C heat treatment for 30 min resulted in an HA-type coating without reducing the tensile bond strength of the coating. Thus, temperature and time are critical parameters in optimizing coating properties [11-114], suggesting that the diffusion process must be significantly involved.

As we have seen above, there are many evidence indicating that, in vitro tests, calcium phosphate is precipitated on such surfaces when they are immersed in a simulated physiological solution, suggesting a main reason for excellent biocompatibility [e.g., 11-80]. On the other hand, there are evidence suggesting that, in vivo, after even 7 day implantation of such modified Ti in rat bone, it was found that Ca/P ratio reduced less than 1 (i.e., Ca-depletion) [11-115].

11.4.4 Composite coating

Bioactive calcium phosphate (CaP) coatings were produced on titanium by using phosphate-based glass (P-glass) and hydroxyapatite (HA), and their feasibility for hard tissue applications was addressed *in vitro* by Kim *et al.* [11-116]. P-glass and HA composite slurries were coated on Ti under mild heat treatment conditions to form a porous thick layer, and then the micropores were filled in with an HA sol–gel precursor to produce a dense layer. The resultant coating product was composed of HA and calcium phosphate glass ceramics, such as tricalcium phosphate (TCP) and calcium pyrophosphate (CPP). It was reported that the coating layer had a thickness of approximately 30–40 μm and adhered to the Ti substrate tightly, (ii) the adhesion strength of the coating layer on Ti was as high as about 30 MPa, (iii) the human osteoblastic cells cultured on the coatings produced by the combined method attached and proliferated favorably, and (iv) the cells on the coatings expressed significantly higher alkaline phosphatase activity than those on pure Ti, suggesting the stimulation of the osteoblastic activity on the coatings [11-116]. Maxian *et al.* [11-117] evaluated the effect of amorphous calcium phosphate and poly-crystallized (60% crystalline) HA coatings on bone fixation of smooth and rough (Ti-6Al-4V powder sprayed) Ti-6Al-4V implants after 4 and 12 weeks of implantation in a rabbit transcortical femoral model. Histological evaluation of amorphous vs. poorly crystallized HA coatings showed significant differences in bone apposition and coating resorption that were increased within cortical compared to cancellous bone. The poorly crystallized HA coatings showed the most degradation and least bone apposition. Mechanical evaluation, however, showed no significant differences in pushout shear strengths. Significant enhancement in interfacial shear strengths for bioceramic coated, as compared to uncoated implants, was seen for smooth-surfaced implants (3.5–5 times greater) but not for rough-surfaced implants at 4 and 12 weeks. Based on these results, it was suggested that once early osteointegration is achieved biodegradation of a bioactive coating should not be detrimental to the bone/coating/implant fixation [11-117]. Plasma-sprayed HA coatings on titanium alloy substrates have been used extensively due to their excellent biocompatibility and osteoconductivity. However, the erratic bond strength between HA and Ti alloy has raised concern over the long-term reliability of the implant. Accordingly, Khor *et al.* [11-118] developed HA/yttria-stabilized-zirconia(YSZ)/Ti-6Al-4V composite coatings that possess superior mechanical properties to conventional plasma-sprayed HA coatings. Ti-6Al-4V powders coated with fine YSZ and HA particles were prepared through a unique ceramic slurry mixing method. The composite powder was employed as feedstock for plasma spraying of the HA/YSZ/Ti-6Al-4V coatings. The influence of net plasma energy, plasma spray standoff distance, post-spray heat treatment on microstructure, phase composition, and mechanical properties were investigated. It was found that (i) coatings prepared with the optimum plasma-sprayed condition showed a well-defined splat structure, (ii) HA/YSZ/Ti-6Al-4V solid solution was formed during plasma spraying, which was beneficial for the improvement of mechanical properties, (iii) the microhardness, modulus of elasticity, fracture toughness, and bond strength increased significantly with the addition of YSZ, and (iv) post-spray heat treatment at 600 and 700°C for up to 12 h was found to further improve the mechanical properties of coatings [11-118]. Yttria-stabilized-zirconia (YSZ) is often used as reinforcement for many ceramics because it has the merits of high strength and enhanced toughening characteristics during crack-particle interactions [11-119–11-122].

Yamada *et al.* [11-123] utilized the Cullet method for which (1) the mixture of HA powder and glass frits are sintered at 900–1000°C for from 5 to 10 min to prepare well- homogenized coating powder, whereas the conventional method is just mixing and not sintering, and (2) the time of etching treatment, through which the bioactive surface is formed using the mixed solution of HNO_3 and HF, is relatively short (within 1 min) compared to the conventional method. Through this method, functionally gradient HA/Ti composite implants were successfully fabricated with higher quality compared with the conventional method [11-123]. Suzuki *et al.* [11-124] coated titanium dioxide onto silicone substrates by radio-frequency sputtering.

It was reported that silicone coating with titanium dioxide enhanced the breakdown of peroxynitrite by 79%. Titanium dioxide-coated silicone inhibited the nitration of 4-hydroxy-phenylacetic acid by 61% compared to aluminum oxide-coated silicone and 55% compared to uncoated silicone. Titanium dioxide-coated silicone exhibited a 55% decrease in superoxide compared to uncoated silicone, and a 165% decrease in superoxide compared to uncoated polystyrene. Titanium dioxide-coated silicone inhibited IL-6 production by 77% compared to uncoated silicone. Based on these findings, it was concluded that the anti-inflammatory properties of titanium dioxide can be transferred to the surfaces of silicone substrates [11-124].

HA coatings with titania addition were produced by the high velocity oxy-fuel spray process by Li *et al.* [11-125]. It was found that (i) the addition of TiO_2 improves the MOE, fracture toughness, and shear strength of high velocity oxy-fuel-sprayed HA-based coatings, (ii) the incorporation of the secondary titania phase is found to have a negative effects on the adhesive strength of high velocity oxy-fuel-sprayed HA coatings, (iii) the titania is found to inhibit the decomposition of HA at elevated temperatures lower than 1410°C, at which point the mutual chemical reaction occurs, and (iv) a small amount of TiO_2 added into high velocity oxy-fuel-sprayed HA coatings with less than 20 vol% is therefore recommended for strengthening of HA-coatings [11-125]. Lu *et al.* [11-126] fabricated a two-layer hydroxyapatite (HA)/HA+TiO_2 bond coat composite coating (HTH coating) on titanium by the plasma spraying technique. The HA+TiO_2 bond coat (HTBC) consists of 50 vol% HA and 50 vol% TiO_2 (HT). The as-sprayed HT coating consists mainly of crystalline HA, rutile TiO_2 and amorphous Ca-P phase, but the post-spray heat treatment at 650°C for 120 min effectively restores the structural integrity of HA by transforming non-HA phases into HA. It was found that there exists interdiffusion of the elements within the HTBC, but no chemical product between HA and TiO_2, such as $CaTiO_3$ was formed. The toughening and strengthening mechanism of HTBC is mainly due to TiO_2 as obstacles resisting cracking, and the reduction of the near-tip stresses resulting from stress-induced microcracking [11-126]. Ng *et al.* [11-127] mimicked biomineralization of bone by applying an initial TiO_2 coating on Ti-6Al-4V by electrochemical anodization in two dissimilar electrolytes, followed by the secondary calcium (CaP) coating, subsequently applied by immersing the substrates in an SBF with three times concentration (SBF×3). Electrochemical impedance spectroscopy (EIS) and DC potentiodynamic polarization assessments in SBF revealed that the anodic TiO_2 layer is compact, exhibiting up to a four-fold improvement in *in vitro* corrosion resistance over unanodized Ti-6Al-4V. X-ray photoelectron spectroscopy analysis indicates that the anodic Ti oxide is thicker than air-formed ones with a mixture of TiO_{2-x} compound between the TiO_2/Ti interfaces. The morphology of the dense CaP film formed, when observed using scanning electron microscopy, is made up of linked globules 0.1–0.5 μm in diameter without observable delamination. It was also found that (i) the calculated Ca:P ratios of all samples (1.14–1.28) are lower than stoichiometric hydroxyapatite (1.67), and (ii) a duplex coating consisting of a compact TiO_2 with enhanced *in vitro* corrosion resistance and bone-like apatite coating can be applied on Ti-6Al-4V by anodization and subsequent immersion in SBF [11-127]. TiO_2 coatings were prepared on NiTi alloy by heat treatment in air at 300, 400, 600, and 800°C. The heat-treated NiTi alloy was subsequently immersed in a SBF for the biomimetic deposition of the apatite layer onto the surface of TiO_2 coating. The apatite coatings, as well as the surface oxide layer on NiTi alloy, were microanalytically characterized by Gu *et al.* [11-128]. Results showed the samples heat-treated at 600°C produced a layer of anatase and rutile TiO_2 on the surface of NiTi. It was surprise to find that no TiO_2 was detected on the surface of NiTi after heat treatment at 300 and 400°C by X-ray diffraction, while rutile was formed on the surface of the 800 C heat-treated sample. It was found that the 600°C heat-treated NiTi induced a layer consisting of microcrystalline carbonate containing hydroxyapatite on its surface most effectively, while 300 and 400°C heat-treated NiTi did not form apatite. This was due to the presence of anatase and/or rutile in the 600 and 800°C heat-treated NiTi, which could provide atomic arrangements in their crystal structures suitable for the epitaxy of apatite crystals, and anatase had better apatite-forming ability than rutile. XPS and Raman results revealed that this apatite layer was a carbonated and non-stoichiometric apatite with Ca/P ratio of 1.53, which was similar to the human bone. It was mentioned that the formation of apatite on 600°C heat-treated NiTi following immersion in SBF for 3 days indicated that the surface modified NiTi possessed excellent bioactivity [11-128].

Chu *et al.* [11-129] characterized microanalytically a bioactive sodium titanate/titania-graded film formed *in situ* on NiTi shape memory alloy by oxidizing in H_2O_2 solution and subsequent NaOH treatment. The bioactivity of the film was investigated using an SBF soaking test. A titania (TiO_2) layer was first found on NiTi substrate after oxidized in H_2O_2 solution, and then a porous sodium titanate (Na_2TiO_3)/titania film with many Ti-OH groups, and a trace of Ni_2O_3 was formed by the reaction of partial TiO_2 phase with NaOH solution. After immersion in SBF for 12 h, AP was observed to nucleate and grow on the film. By prolonging the soaking time, more AP appeared on its surface. Moreover, XPS depth profiles of O, Ni, Ti, and Na show the bioactive film possesses a smooth graded interface structure to NiTi substrate, which is in favor of sufficient mechanical stability of apatite layer by subsequent deposition in SBF [11-129].

Knabe *et al.* [11-130] investigated the effects of novel calcium titanium, calcium, titanium zirconium phosphates suitable for plasma spraying on CpTi substrate on the expression of bone-related genes and proteins of human bone-derived cells, and compared the effects to that on native Ti and HA-coated Ti. Test materials were acid etched and sandblasted, plasma-sprayed HA, and sintered $CaPO_4$ with Ti, Zr, TiO_2, and ZrO_2. Human bone-derived cells were grown on these surfaces for 3, 7, 14, and 21 days, counted and probed for various mRNAs and proteins. It was reported that (i) all surfaces significantly affected cellular growth and the temporal expression of an array of bone-related genes and proteins, (ii) at 14 and 21 days, cells on sintered displayed significantly enhanced expression of all osteogenetic mRNAss, and (iii) surfaces of $55CaO \cdot 20TiO_2 \cdot 31P_2O_5$ and $CaTi_4(PO_4)_6$ had the lost effect on osteoblastic differentiation inducing a greater expression on an array of osteogenetic markers than recorded for cells grown on HA, suggesting that these novel materials may possess a higher potency to enhance osteogenesis [11-130]. Shtansky *et al.* [11-131] performed a comparative investigation of multicomponent thin films based on the systems Ti-Ca-C-O-(N), Ti-Zr-C-O-(N), Ti-Si-Zr-O-(N), and Ti-Nb-C-(N). $TiC_{0.5} + 10\%CaO$, $TiC_{0.5} + 20\%CaO$, $TiC_{0.5} + 10\%ZrO_2$, $TiC_{0.5} + 20\%ZrO_2$, $Ti_5Si_3 + 10\%ZrO_2$, $TiC_{0.5} + 10\%Nb_2C$, and $TiC_{0.5} + 30\%Nb_2C$ composite targets were manufactured by means of self-propagating high-temperature synthesis, followed by DC magnetron sputtering in an atmosphere of argon or in a gaseous mixture of argon and nitrogen. The biocompatibility of the films was evaluated by both *in vitro* and *in vivo* experiments. *In vitro* studies involved the investigation of the proliferation of Rat-1 fibroblasts and IAR-2 epithelial cells on the tested films, morphometric analysis and actin cytoskeleton staining of the cells cultivated on the films. *In vivo* studies were fulfilled by subcutaneous implantation of Teflon plates coated with the tested films in mice and analysis of the population of cells on the surfaces. It was reported that (i) the films showed high hardness in the range of 30–37 GPa, significant reduced modulus of elasticity, low friction coefficient down to 0.1–0.2, and low wear rate in comparison with conventional magnetron-sputtered TiC and TiN films, (ii) no statistically significant differences in the attachment, spreading, and cell shape of cultured IAR-2 and Rat-1 cells on the Ti-Ca-C-O-(N), Ti-Zr-C-O-(N) ($TiC_{0.5} + 10\%ZrO_2$ target), Ti-Si-Zr-O-(N) films and the uncoated substrata was detected, and (iii) the adhesion and proliferation of cultured cells *in vitro* was perfect at all investigated films. Based on these findings, it was concluded that the combination of excellent mechanical properties with non-toxicity and biocompatibility makes Ti-Ca-C-O-N, Ti-Zr-C-O-N, and Ti-Si-Zr-O-N films promising candidates as tribological coatings to be used for various medical applications like total joint prostheses and dental implants [11-131]. von Walter *et al.* [11-132] introduced a porous composite material, named "Ecopore", and described *in vitro* investigation of the material and its modification with fibronectin. The material is a sintered compound of rutile TiO_2 and the volcanic silicate perlite with a macrostructure of interconnecting pores. In an *in vitro* model, human primary osteoblasts were cultured directly on Ecopore. It was reported that human osteoblasts grew on the composite as well as on samples of its single constituents, TiO_2 and perlite glass, and remained vital, but cellular spreading was less than on tissue culture plastic. To enhance cellular attachment and growth, the surface of the composite was modified by etching, functionalization with aminosilane and coupling of fibronectin, resulting in greatly enhanced spreading of human osteoblasts. It was therefore concluded that (i) Ecopore is non-toxic and sustains human osteoblasts growth, cellular spreading being improvable by coating with fibronectin, and (ii) the composite may be usable in the field of bone substitution [11-132].

Different biomaterials have been used as scaffolds for bone tissue engineering. Rodrigues *et al.* [11-133] characterized biomaterial composed of sintered (at 1100°C) and powdered hydroxyapatite and type I collagen (both of bovine origin) designs for osteoconductive and osteoinductive scaffolds. Collagen/HA proportions were 1/2.6 and 1/1 by weight, with particle sizes ranging from 200 to 400 μm. X-ray diffraction and infrared spectroscopy showed that the sintered (1100°C) bone was composed essentially of HA with minimum additional groups as surface calcium hydroxide, surface and crystal water, free carbon dioxide, and possibly brushite. It was reported that osteoblasts adhered and spread on both the HA particle surface and the collagen fibers, which seemed to guide cells between adjacent particles [11-133], suggesting that this biocomposite can be considered as ideal for its use as scaffold for osteoconduction and osteoinduction. Cheng *et al.* [11-134] prepared electrochemically a bovine serum albumin (BSA) protein-containing AP coating on a HA coated Ti-6Al-4V. It was reported that (i) the method resulted in a 70-fold increase in BSA inclusion compared to simple adsorption, and was subsequently released by a slow mechanism (15% loss over 70 h), and (ii) thus, this technique provides an efficient method of protein incorporation at physiological stem, with a potential for sustained release of therapeutic agents, as may be required for metallic implant fixation [11-134].

Redepenning *et al.* [11-135] prepared another type of biocomposite coatings containing brushite ($CaHPO_4 \cdot 2H_2O$) and chitosan by electrochemical deposition. The brushite/chitosan composites were converted to hydroxyapatite/chitosan composites in aqueous solutions of sodium hydroxide. The coatings ranged from about 1 to 15% chitosan by weight. It was mentioned that qualitative assessment of the coatings showed adhesion significantly improved over that observed for electrodeposited coatings of pure HA [11-135].

11.4.5 TiN coating

Mechanical–electrochemical interactions accelerate corrosion in mixed-metal modular hip prostheses. These interactions can be reduced by improving the modular component machining tolerances or by improving the resistance of the components to scratch or fretting damage. Wrought Co-alloy (Co-Cr-Mo) is known to have better tribological properties compared to the Ti-6Al-4V alloy. Thus, improving the tribological properties of this mixed-metal interface should center on improving the tribological properties of Ti-6Al-4V. It was mentioned that (i) the nitrogen-diffusion-hardened Ti alloy samples had a more pronounced grain structure, more nodular surface, and significantly higher mean roughness values than the control Ti-6Al-4V, (ii) the nitrided Ti-6Al-4V samples also exhibited at least equivalent corrosion behavior and a definite increases in surface hardness compared to the control Ti-6Al-4V samples, and (iii) fretting can be reduced by decreasing micromotion or by improving the tribological properties (wear resistance and surface hardness) of the material components at this interface [11-136]. The corrosion behavior of the titanium nitride-coated TiNi alloy was examined in 0.9%NaCl solution by potentiodynamic polarization measurements and a polarization resistance method [11-137]. XPS spectra showed that the titanium nitride film consisted of three layers, a top layer of TiO_2, a middle layer of TiN_x ($x>1$), and an inner layer of TiN, which agreed very well with results obtained by Oshida and Hashem [11-138]. The passive current density for the titanium nitride-coated alloy was approximately two orders of magnitude lower than that of the polished alloy in the potential range from the free corrosion potential to $+500$ mV (vs. Ag/AgCl). Pitting corrosion associated with breakdown of the coated film occurred above this potential. The polarization resistance data also indicated that the corrosion rate of the titanium nitride-coated alloy at the corrosion potential ($+50$ to $+100$ mV) was more than one order of magnitude lower than that for the polished alloy. It was concluded that the corrosion rate of TiNi alloy can be reduced by more than one order of magnitude by titanium nitride coating, unless the alloy is highly polarized anodically *in vivo* [11-137]. Ti-6Al-7Nb alloy was treated at 70 and 100 keV energy using a 150 keV accelerator at different doses between 1×10^{16} and 3×10^{17} ions/cm^2, and was corrosion-tested in Ringer's solution. It was reported that nitrogen ion implantation enhanced the possibility of passivation and reduced the corrosion kinetics of the alloy surface with increasing tendency for repassivation [11-139]. Bordji *et al.* [11-140] prepared Ti alloys treated by (1) glow discharge nitrogen implantation (10^{17} atoms/cm^2), (2) plasma nitriding by plasma diffusion treatment, and (3) deposition of TiN layer by plasma-assisted chemical vapor deposition additionally to plasma diffusion treatment. A considerable improvement was noticed in surface hardness, especially after the two nitriding processes. A cell culture model using human cells was chosen to study the effect of such treatments on the cytocompatibility. The results showed that Ti-5Al-2.5Fe alloy was as cytocompatible as the Ti-6Al-4V alloy, and the same surface treatment led to identical biological consequences on both alloys. It was concluded that (i) after the two nitriding treatments, cell proliferation and viability appeared to be significantly reduced and the SEM revealed somewhat irregular surface states; however, (ii) osteoblast phenotype expression and protein synthesis capability were not affected [11-140]. Goldberg and Gilbert [11-141] utilized the plasma vapor deposition technique, by which the samples were placed into a vacuum chamber and sputtered to remove the oxide film, followed by depositing a 200 nm thick interlayer of titanium to enhance the coating/substrate interface. Alternating layers of TiN and AlN were deposited until a coating thickness of approximately 5 μm was produced. The mechanical and electrochemical behavior of the surface oxides of Co-Cr-Mo and Ti-6Al-4V alloys during fracture and repassivation play an important role in the corrosion of the taper interfaces of modular hip implants. These corrosion properties were investigated in one group of Co-Cr-Mo and Ti-6Al-4V alloy samples passivated with nitric acid and another group coated with TiN/AlN coating. It was found that (i) Co-Cr-Mo had a stronger surface oxide and higher interfacial adhesion strength, making it more resistant to fracture than Ti-6Al-4V, (ii) if undistributed, the oxide on the surface of Ti-6Al-4V significantly reduced dissolution currents at a wider range of potential than Co-Cr-Mo, making Ti-6Al-4V more resistant to corrosion, (iii) the TiN/AlN coating had higher hardness and modulus of elasticity than Co-Cr-Mo and Ti-6Al-4V. It was much less susceptible to fracture, had higher interfacial adhesion strength, and was a better barrier to ionic diffusion than the surface oxides on Co-Cr-Mo and Ti-6Al-4V [11-141].

In spite of their high strength, low density, and good corrosion resistance, the usefulness of Ti alloys in general engineering components is frequently limited by their poor wear resistance. If the alloy surface is subjected to conditions of sliding or fretting, adhesive wear can rapidly lead to catastrophic failure unless appropriate surface engineering is carried out. In order to combat modest contact loads, several surface treatments are commercially available, such as plasma nitriding or PVD coating with TiN. Titanium nitride is known for its high-surface hardness and mechanical strength. It was also reported that the dissolution of Ti ions is very low [11-142]. As for dental implants, they are comprised of various components. The implant abutment part (the mucosa penetration part) is exposed in the oral cavity, and so plaque and dental calculus easily adhere on it. Removal of the plaque and dental calculus is necessary to obtain a good prognosis throughout the long-term maintenance of the implant. Based on this background, Kokubo *et al.* [11-143] prepared CpTi (grade 1) samples by polishing with #2000 grit paper, or buff-polishing with 6 μm diamond emulsion paste, followed by a 0.1% HF acid solution for 10 s to clean the surface, then treated in N_2 atmosphere of 1 atm at 850°C for 7 h (N_2 flow rate: 50 l/min). It was reported that (i) the nitrided layer about 2 μm thick composed of TiN and Ti_2N was formed on Ti by a gas nitriding method, and the dissolved amount of Ti ion in SBF[2] was as low as the detectable limit of ICP-MS (inductively coupled plasma mass spectroscopy), and that the 1% lactic acid showed no significant difference from Ti [11-143].

The most of surface treatments such as plasma nitriding or PVD coating with TiN are, however, carried out in the solid state and the depth of coating or hardening is restricted by low diffusivity. The diffusion coefficient of nitrogen in Ti is more than a thousand times slower than that in steels due to different crystalline structures. The packing factor of Ti (HCP) is 74%, while that of steel (BCC) is 68%, so that steel has more spaces available for diffusing species. In order to achieve the depth of hardening necessary to withstand the subsurface Hertzian stresses induced by heavy rolling contact, it is necessary to alloy the Ti surface in the molten state. The necessary depth of surface hardening can readily be achieved in this way by laser melting the surface in the presence of interstitial alloying elements such as carbon, oxygen, and nitrogen. Of these, nitrogen has been found to provide the best balance between increased hardness and decreased ductility, and can easily be added by laser gas nitriding. The Hertzian compressive stress in the substrate was increased to 1.36 GPa [11-144]. Pure iron has allotropic phase transformations: the first one is $910°C$ between α-BCC and γ-FCC, and the second one is $1390°C$ between γ-FCC and δ-BCC. While investigating the deformation mechanism of transformation super-plasticity, Oshida observed that the transformation front behaves as if semiliquid due to loosing a clear crystalline electron diffraction pattern even both α-BCC and γ-FCC are still in solid state. Based on this finding, the "semiliquid transformation front" model was proposed. As mentioned in Chapter 10, one of various applications using transformation superplasticity is a nitridation of metallic materials if they have an allotropic phase transformation temperature. It was demonstrated that CpTi was successfully nitrided when CpTi was heated and cooled repeatedly passing the β-transus temperature (between 800 and $930°C$) for several times in nitrogen gas filled chamber [11-145]. Ion implantation, diffusion hardening, and coating are surface modification techniques for improving the wear resistance and surface hardness of Ti alloy surfaces [11-146–11-150]. A Ti-6Al-4V sample was diffusion-hardened in a nitrogen atmosphere for 8 h at $566°C$ and argon or nitrogen quenched to room temperature. The nitrogen-diffusion-hardened Ti-6Al-4V had TiN and TiNO complexes at the immediate surface and subsurface layers. The diffusion-hardened samples also had a deeper penetration of oxygen compared to regular Ti-6Al-4V samples [11-151].

Surface topography and chemistry have been shown to be extremely important in determining cell–substrate interactions and influencing cellular properties such

---

[2] SBF: Na^1 (142.0), K^1 (5.0), Mg^{21} (1.5), Ca^{21} (2.5), Cl^- (148.8), HCO_3^- (4.2), HPO_4^- (1.0). H.P. Na (142.0), K (5.0), Mg (1.5), Ca (2.5), Cl (103.0), HCO_3 (13.5), HPO_4 (1.0).

as cell adhesion, cell–cell reactions, and cytoskeletal organization [11-152]. The cell–substrate interaction of primary hippocampal neurones with thin films of TiN was studied *in vitro*. TiN films of different surface chemistries and topographies were deposited by pulsed DC reactive magnetron sputtering and closed field unbalanced magnetron sputter ion plating to result in TiN thin films with similar surface chemistries, but different topographical features. It was reported that (i) primary hippocampal neurones were found to attach and spread to all of the TiN films, (ii) neuronal network morphology appeared to be more preferential on the nitrogen rich TiN films, and also reduced nanotopographical features, (iii) at early time points of 1 and 4 days *in vitro* primary hippocampal neurones respond to the presence of interstitial nitrogen rather than differences in nanotopography; however, (iv) at 7 days more preferential neutronal network morphology is formed on TiN thin films with lower roughness values and decreased size of topographical features [11-152]. Bull and Chalker [11-153] studied the use of thin titanium interlayers to promote the adhesion of TiN coatings on a range of substrates. For thin interlayers, an interstitially hardened titanium layer is formed at the interface, resisting the interfacial crack propagation. However, at a critical interlayer thickness, the surface contaminants are completely dissolved in the interfacial layer, and depositing any further titanium leads to an overall softer interfacial layer which offers less resistance to crack propagation, and delamination can easily take place. For this reason, failure is observed within the interlayer for thick interlayers, whereas it occurs at the interlayer/substrate interface for thinner interlayers. Another contribution to the enhanced adhesion comes from the reduction in coating stresses in the interfacial region due to the presence of a soft compliant layer, which was examined by changing the hardness of the interlayer deposited before coating deposition. It was concluded that (i) softer interlayers do not lead to improved adhesion performance in most cases, and (ii) it appears that the best adhesion results from a hard interlayer that leads to ductile failure at the coating/substrate interface, rather than the brittle failure observed due to the presence of oxide films [11-153].

As briefly described in the above, Oshida *et al.* [11-138, 11-154] applied a TiN coating onto CpTi substrate prior to porcelain firing to develop a new method to control the excessive oxidation. The bonding strength of porcelain to metals depends on the oxide layer between the porcelain and the metal substrate. Oxidation of a metal surface increases the bonding strength, whereas excessive oxidation decreases it. Titanium suffers from its violent reactivity with oxygen at high temperatures that yield an excessively thick layer of TiO_2, and this presents difficulties with porcelain bonding. The oxidation kinetics of titanium simulated to porcelain firing was investigated, and the surface nitridation of CpTi as a process of controlling the oxidation behavior was evaluated. Nitrided samples with the arc ion plating PVD process and un-nitrided control CpTi were subjected to oxidation simulating of Procera porcelain with 550, 700, and 800°C firing temperatures for 10 min in both 1 and 0.1 atmospheric air. The weight difference before and after oxidation was calculated, and the parabolic rate constant, K_p ($mg^2/cm^4/s$), was plotted against inverse absolute temperature (i.e., in an Arrhenius plot). Surface layers of the samples were subjected to X-ray and electron diffraction techniques for phase identifications. Results revealed that both nitrided and un-nitrided samples obey a parabolic rate law with activation energy of 50 kcal/mol. In addition, the study shows that nitrided CpTi had a K_p about 5 times lower than the un-nitrided CpTi, and hence the former needs 2.24 times longer oxidation time to show the same degree of oxidation. Phase identification resulted in confirming the presence of TiO_2 as the oxide film in both groups, but with 1–2 μm thickness for the un-nitrided CpTi and 0.3–0.5 μm thickness for nitrided samples. Therefore, it can be concluded that nitridation of titanium surfaces can be effective in controlling the surface oxide thickness that might ensure satisfactory bonding with porcelain [11-138]. Oshida *et al.* [11-154] evaluated CpTi substrates subjected to porcelain firing and bond strengths under three-point bending mode (span length: 15 mm; crosshead speed: 0.5 mm/min). Experimental variables included surface treatments of CpTi and porcelain-firing schedules. Variables for the surface treatments were: (1) sandblasting, (2) mono- and triple-layered nitridation, and (3) mono-layered chrome-doped nitridation. Variables for the porcelain-firing schedule included (4) bonding agent application, (5) bonding agent plus gold bonding agent application, and (6) Procera porcelain application. Statistically, all of them exhibited no significant differences. Hence, we employed two further criteria: (i) the minimum bond strength should exceed the maximum porcelain strength *per se*, and (ii) the CpTi substrate should not be heated close to the β-transus temperature. After applying these criteria, it was concluded that mono-layered nitridation and mono-layered application of chrome-doped nitridation on both (with and without) sandblasted and non-sandblasted surfaces were the most promising conditions for a successful titanium–porcelain system [11-154]. It seems that an alloy which has the properties of titanium and is relatively inexpensive would be a very good material for surgical purposes. These requirements could be met, for example, by stainless-steel coated with a firmly adhering non-porous titanium film. Głuszek *et al.* [11-155] coated 316L (18Cr-8Ni-2Mo with low carbon content) stainless-steel with Ti or TiN by ion plating. The galvanic effects for the galvanic couples 316L/Ti, 316L/Ti-coated 316L, 316L/TiN-coated 316L were studied in Ringer's solution. It was concluded that (i) both Ti and TiN coatings were non-porous, (ii) Ti served as an anode in the couple 316L/Ti, whereas for the other two couples, the coatings were the cathodes; however, (iii) the dissolution rate of 316L stainless-steel in these couples was smaller than expected owing to a strong polarization of the coatings [11-155].

For hip prostheses, the coupling between the metallic femoral head and the polymeric acetabular cup is normally used. Biotribiological phenomena contribute principally to the clinical failure of the prosthesis. In the metal–polymer coupling, the problems consist of biotribological wear, creep of the UHMWPE (ultra-high molecular weight polyethylene), and fretting corrosion of the metal femoral head. Cyclic stress exceeding the fatigue resistance of UHMWPE produces surface microcracks and particulates that can migrate into the tissue of the host implant. This fretting-wear debris causes local irritation, proliferation of fibrous tissue, and necrosis of bone. Minimizing the wear is critically important for maintaining the long life of the femoral prosthesis. On the other hand, titanium alloys are susceptible to fretting corrosion; this susceptibility can be reduced via surface treatments. UHMWPE was gamma-ray sterilized. This sterilization technique results in a cross-linking of the polymer, which enhances its wear resistance. Tribiological behavior of N_2-implanted and non-implanted titanium alloys coupled with UHMWPE were studied using pin-on-flat tests, according to ASTM F 732-82 in bovine serum. The results show that while the non-implanted titanium alloy and the titanium with N_2 on UHMWPE resulted in high final wear values, titanium implanted with O_2 generates a wear value less than that obtained for polyethylene against 316L stainless-steel. Ti-6Al-4V implanted with chromium exhibited the lowest wear. Hardness values of the implanted material corresponded to the wear rates, which assist in determining optimal elements for implantation. Implantation of certain elements may increase the surface activity, resulting in more adherent oxide layers that also increase wettability [11-3].

11.4.6. Ti coating

Lee *et al.* [11-156] conducted the *in vivo* study to evaluate the behavior and mechanical stability in implants of three surface designs, which were smooth surface Ti, rough Ti surface by plasma spray coating, and alkali and heat-treated. The implants were inserted transversely in a dog thighbone and evaluated at 4 weeks of healing. At 4 weeks of healing after implantation in bone, it was found that (i) the healing tissue was more extensively integrated with an alkali and heat-treated Ti implant than with the implants of smooth surface and/or rough titanium surfaces, (ii) the bone bonding strength (pull-out force) between living bone and implant was observed by a universal testing machine, (iii) the pull-out forces of the smooth surface Ti, plasma spray coated Ti, and alkali and heat-treated Ti implants were 235, 710, and 823 N, respectively, and (iv) histological and mechanical data demonstrated that appropriate surface design selection can improve early bone growth and induce an acceleration of the healing response, thereby improving the potential for implant osseointegration.

In order to improve the biocompatibility of functional titanium-based alloys, Sonoda *et al.* [11-157] investigated pure titanium coatings on the TiNi shape memory alloy and Ti-6Al-4V alloy by sputtering. More high-quality thin film and higher growth rate were obtained by the sputtering with a DC source than with an RF source. After the cleaning method was established, the effect of sputtering on the thickness of the film was investigated with DC sputtering. It was concluded that the growth rate of sputtered titanium film was proportional to the applied electric power, and the orientation of the film highly depended on the heating temperature of the substrate [11-157]. Sonoda *et al.* [11-158] further applied this technique to the complete denture base of the Ti-6Al-4V alloy fabricated by superplastic forming. The base was attached to the substrate holder and cooled by water or heated at 417°C. It was reported that the film deposited on the heated base was superior to that on the cooled one in smoothness, glossiness, uniformity, and covering of the fine cracks [11-158]. Until the late 1970s, PMMA (polymethymethacrylate) bone cement was the predominant means of fixing a joint replacement implant to the skeletal system. At that time, however, bone cement was implicated as a major contributor to implant loosening because of its brittle nature and poor interfacial relationship with the metallic implant and bone tissue [11-159– 11-161]. The use of porous coated implants for long-term biological fixation has been receiving an enthusiastic response, especially when the patients are younger and more active [11-162, 11-163]. The application of TPS coatings to Ti-6Al-4V orthopedic implants results in a dramatic decrease in high-cycle fatigue performance. It was noted that the better bonding of the plasma-sprayed and heat-treated implants results in a lower high-cycle fatigue strength. As with conventional sintered porous coatings, the application of a coating that contains defects serves as the crack initiator of the high-cycle fatigue. It was also mentioned that the addition of the post-coating heat treatment to improve coating bonding strength resulted in a further reduction in the high-cycle fatigue strength, most likely due to a higher frequency of bonding sites between the coating and substrate, and a more intimate metallurgical bond at those sites [11-159].

Reclaru *et al.* [11-164] investigated the corrosion behavior of Co-Cr-Mo implants with rough Ti coatings, applied by different suppliers by either sintering or vacuum plasma spraying, and compared the results with those obtained from uncoated material. The open-circuit potential, corrosion current, and polarization resistance were determined by electrochemical methods in artificial bone fluid. The Co, Cr, and Ti ions released from the samples into the electrolyte during the polarization extraction technique were analyzed using ICP-MS (inductively coupled plasma mass spectroscopy). It was found that (i) among the Ti coated samples, the one with the sintered overcoat turned out to be the most resistant, (ii) yet, on an absolute scale, they all showed a corrosion resistance inferior to that of uncoated Co-Cr-Mo or wrought Ti [11-164]. The degree of the degradation of the implant surface depends on the particular surface treatment employed [11-81, 11-165–11-168]. Anodization is a common surface treatment for Ti implants. More recently, there was a growing interest in TPS overcoats as a viable alternative to sintered bead or diffusion-bonded fiber metal surfaces, since the inherent roughness of such coatings is believed to favor the osteointegration of the bone [11-169, 11-170]. Surface treatment plays an important role in the corrosion resistance of Ti. The cement, in spite of having reduced electrical conductivity in comparison to metal, is an ionic transporter, and therefore capable of participating in the corrosion process. The crevice corrosion at the metal–cement interface was studied by Reclaru *et al.* [11-171] who reported that in the case of plasma spray surfaces, a process of diffusion of Ti particles in the electrolyte could accompany the crevice corrosion.

Xue *et al.* [11-172] modified plasma-sprayed titanium coatings by an alkali treatment. The changes in chemical composition and structure of the coatings were examined by SEM and AES. The results indicated that (i) a net-like microscopic texture feature, which was full of the interconnected fine porosity, appeared on the surface of alkali-modified titanium coatings, (ii) the surface chemical composition was also altered by alkali modification, and (iii) a sodium titanate compound was formed on the surface of the titanium coating and replaced the native passivating oxide layer. The bone bonding ability of titanium coatings were investigated using a canine model. The histological examination and SEM observation demonstrated that more new bone was formed on the surface of alkali-modified implants, and grew more rapidly into the porosity. It was therefore concluded that (i) the alkali-modified implants appose directly to the surrounding bone, (ii) in contrast, a gap was observed at the interface between the as-sprayed implants and bone, (iii) the pushout test showed that alkali-modified implants had higher shear strength than as-sprayed implants after 1 month of implantation, and (iv) an interfacial layer, containing Ti, Ca, and P, was found to form at the interface between bone and the alkali-modified implant [11-172].

Borsari *et al.* [11-173] developed a new implant surface with the purpose of avoiding as much stress shielding as possible, and thus prolong the prosthesis lifespan, and investigated the *in vitro* effect of this new ultra-high roughness and dense Ti (R_a=74 μm) in comparison with medium (R_a=18 μm) and high (R_a=40 μm) roughness and open porous coatings, which were obtained by vacuum plasma spraying. MG63 osteoblast-like cells were seeded on the tested materials and polystyrene, as control, for 3 and 7 days. It was reported that (i) akaline phosphatase activity had lower values for high roughness surfaces than medium and ultra-high rough surfaces, (ii) procollagen-I synthesis reduced with increasing roughness, and the lowest data was found for the ultra-high rough surface, (iii) all tested materials showed significantly higher Interleukin-6 levels than those of polystyrene at both experimental times, and (iv) the new ultra-high roughness and dense coating provided a good biological response, even though, at least *in vitro*, it behaved similarly to the coatings already used in orthopedics [11-173]. The bone response to different TPS implants was evaluated in the trabecular femoral condyles of 10 goats by Vercaigne *et al.* [11-174]. These implants were provided with three different TPS coatings with a R_a of 16.5, 21.4, and 37.9 μm, respectively. An Al_2O_3 grit-blasted implant with a R_a of 4.7 μm was used as a control. After an implantation period of 3 months, the implants were evaluated histologically and histomorphometrically. Only one implant was not recovered after the evaluation period. It was reported that (i) most of the implants showed a different degree of fibrous tissue alternating with direct bone contact, (ii) complete fibrous encapsulation of the implant was observed in some of the sections, and no signs of delamination of the plasma-sprayed coating was visible, (iii) no significant differences in bone contact were measured between the different types of implants, (iv) hismorphometrical analysis revealed significantly higher bone mass close to the implants (0–500 μm) for treated implants placed in medial femoral condyle and implants placed in the lateral condyle, and (v) at a distance of 500–1500 μm no difference in bone mass measurements between the different implants was observed [11-174]. Ong *et al.* [11-175] *in vivo* evaluated the bone interfacial strength and bone contact length at the plasma-sprayed HA and TPS implants. Non-coated Ti implants were used for control. Cylindrical coated or non-coated implants were implanted in the dogs' mandibles. Loading of the implants was performed at 12 weeks after implantation. At 12 weeks after implantation (prior to loading) and 1 year after loading, implants were evaluated for interfacial bone–implant strength and bone–implant contact length. It was found that (i) no significant differences in interfacial bone–implant strength for all groups at 12 weeks after implantation and after 1-year loading in normal bone were found; however, (ii) bone contact length for HA implants was significantly higher than the TPS and Ti implants for both periods tested, and (iii) TPS implants exhibited similar pull-out strength compared to the HA implants [11-175].

Histologically, Ti has been demonstrated to be a highly biocompatible material on account of its good resistance to corrosion, absence of toxic effects on macrophages and fibroblasts, and lack of inflammatory response in peri-implant tissues [11-176–11-179]. Ti endosseous dental screws with different surfaces (smooth Ti, TPS, alumina oxide sandblasted and acid-etched, zirconia sandblasted and acid-etched) were implanted in femura and tibiae of sheep for 14 days to investigate the biological evolution of the peri-implant tissues and detachment of Ti debris from the implant surfaces in early healing. Implants were not loaded. It was reported that (i) all samples showed new bone trabeculae and vascularized medullary spaces in those areas where gaps between the implants and host bone were visible, (ii) in contrast, no osteogenesis was induced in the areas where the implants were initially positioned in close contact with the host bone, (iii) the threads of some screws appeared to be deformed where the host bone showed fractures, and (iv) Ti granules of 3–60 μm were detectable only in the peri-implant tissues of TPS implants both immediately after surgery and after 14 days, thus suggesting that this phenomenon may be related to the friction of the Ti plasma spray coating during surgical insertion [11-180].

11.4.7. Titania film and coating

The high corrosion resistance and good biocompatibility of Ti and its alloys are due to a thin passive film that consists essentially of TiO_2. There is increasing evidence, however, that under certain conditions, extensive Ti release may occur *in vivo*. An ion-beam-assisted sputtering deposition technique has been used to deposit thick and dense TiO_2 films on Ti and stainless-steel surfaces. A higher electrical film resistance, lower passive current density, and lower donor density (in order of 10^{15} cm^{-3}) have been measured for sputter-deposited oxide film on Ti in contrast to the naturally formed passive oxide film on Ti (donor density in the order of 10^{20} cm^{-3}). It was found that (i) the coated surface exhibited improved corrosion resistance in phosphate buffered saline, and (ii) the improved corrosion protection of the sputter-deposited oxide film can be explained by a low defect concentration and, consequently, by a slow mass transport process across the film [11-181]. The National Industrial Research Institute of Nagoya has established technology for forming a functional gradient Ti-O oxide film on Ti-6Al-4V by the reactive DC sputtering vapor deposition method. The overall film thickness in the experiment was 3 μm, and the Vickers hardness of the surface was 1500 (vs. CpTi: 200–300) [11-182].

The conditions for obtaining titanium dioxide from the substrates titanium tetrachloride and oxygen, and applying this to a surgical 316L stainless-steel by PACVD (plasma assisted chemical vapor deposition) were determined. It was established that during the process, Ti dioxide anatase is created. During exposure of the 316L stainless-steel with the Ti dioxide coating, in Ringer's solution, it was found that (i) protective properties of this coating improved, (ii) Ti dioxide covering increased the resistivity of 316L stainless-steel to pitting corrosion and general corrosion, and (iii) any damage or partial removal of the coating did not cause an increased galvanic corrosion of the substrate [11-183].

Wollastonite ($CaSiO_3$) ceramic was studied as a medical material for artificial bones and dental roots because of its good bioactivity and biocompatibility [11-184, 11-185]. Lui and Ding [11-186] prepared wollastonite/TiO_2 composite coating using plasma spraying technology onto Ti-Al6-4V substrate. The composite coating revealed a lamellar structure with alternating wollastonite coatings and

TiO$_2$ coating. In the case of composite coatings, the primarily crystalline phases of the coatings were wollastonite and rutile, indicating wollastonite and TiO$_2$ did not react during the plasma spraying process. It was found that (i) the mean Vickers microhardness of the coatings increased with an increase in the content of TiO$_2$. Wollastonite/TiO$_2$ composite coatings were soaked in SBF to examine their bioactivity, (ii) a carbonate-containing HA layer was formed on the surface of the wollastonite and composite coatings (wollastonite/TiO$_2$: 7/3) soaked in SBF, while a carbonate-containing HA layer was not formed on the surface of the TiO$_2$ and composite (with wollastonite/TiO$_2$: 3/7) coatings after immersion, and (iii) a silica-rich layer appeared at the interface of the carbonate-containing HA and wollastonite and composite (7/3 with wollastonite/TiO$_2$) coatings. The cytocompatibility study of osteoblasts seeded onto the surface indicated that the addition of wollastonite promoted the proliferation of osteoblasts [11-186].

Fraichinger *et al*. [11-187] utilized the anodic plasma-chemical (APC) process to modify CpTi surfaces. This technique allowed for the combined chemical and morphological modifications of Ti surfaces in a single step. The resulting conversion coating, typically several micrometers thick, consisted mainly of Ti oxide and significant amounts of electrolyte constituents. A new electrolyte was developed containing both calcium-stabilized by complexation with EDTA (ethylene diamine tetraacetic acid; as a chelating agent) – and phosphate ions at pH 14. The presence of the Ca-EDTA complex, negatively charged at high pH, favors incorporation of high amounts of Ca into the APC coatings during the anodic (positive) polarization. It was reported that (i) the maximum Ca/P atomic ratio in the coating produced with the new APC electrolyte was approximately 1.3, with higher Ca concentrations than reported in conventional APC coatings, (ii) the coating produced in the new electrolyte system exhibited favorable mechanical stability, and (iii) the new APC technology is believed to be a versatile coating technique to render titanium implant surfaces bioactive [11-187].

Haddow *et al*. [11-188] investigated the effects of dip rate, sintering temperatures, and time on the chemical composition of the titania films, their physical structure and thickness, and adherence to a silica substrate. In order to produce films, the sol–gel method was employed. By this method, films which can be mimicked as closely as possible the natural oxide layer that is found on titanium. CpTi (grade 4) isopropoxide was dissolved in isopropanol to form the starting sol. Owing to the ease of hydrolysis of this sol, a chelating agent, diethanolamine, was added. A small amount of water was added to the solution of alkoxide to partially polymerize the Ti species. Thin surface films of titania have been deposited onto glass substrates. These films are to be used as substrates in an *in vitro* model of osseointegration. If titania has been deposited onto glass substrates, the use of low dipping rates prevents cracking in the films, irrespective of the subsequent firing time or temperature.

Firing at higher temperature (600°C) produces predominantly glass films that mimic closely, in chemistry, the natural oxide layer that is formed on Ti implants. Refinement of the dipping setup, or the use of dilute solutions, may result in the production of thinner films. Thinner films will almost certainly be crack-free after firing, and may result in less organic products being trapped in the film during the firing process. It was mentioned that (i) these data are important to use the titania films to develop an *in vitro* model to study the phenomenon of osseointegration, and (ii) coatings may be deposited onto a wide range of materials; this would be particularly beneficial, enabling the study of osseointegration by TEM and photon-based spectroscopies [11-188]. Sato *et al.* [11-189] used the sol–gel processing to coat Ti substrates with HA, TiO_2, and poly(DL-lactic-glycolic acid). Coatings were evaluated by cytocompatibility testing with osteoblast-like cells (or bone-forming cells). The cytocompatibility of the HA composite coatings was compared to that of a traditional plasma-sprayed HA coating. Results showed that (i) osteoblast-like cell adhesion was promoted on the novel HA sol–gel coating compared to the traditional plasma-sprayed HA coating, and (ii) hydrothermal treatment of the sol–gel coating improved osteoblast-like cell adhesion [11-189]. Wang *et al.* [11-190] treated an amorphous titania gel layer on the CpTi surface after the Ti surface was treated with a H_2O_2/HCl solution at 80°C. The thickness of the gel layer increased almost linearly with the period of the treatment. It was found that (i) a subsequent heat treatment above 300°C transformed gradually the amorphous gels to the anatase crystal structure, and the rutile started to appear after heat treatment at 600°C, and (ii) similar to the sol–gel derived titania gel coatings, titania gel layers exhibited *in vitro* apatite deposition ability after the gel layers exceeded a minimum thickness (0.2 μm) and was subsequently heated in a proper temperature range (400–600°C) [11-190].

Titania (5–20 mol%) was mixed with pure HA or HA containing Ag_2O (10–20 mol%) and was heated at 900°C for 12 h. The sintered samples were found to contain mainly tricalcium phosphate (β-TCP). Enhanced TCP formation with impurity was observed with TiO_2-Ag_2O addition. An *in vitro* solubility study in phosphate buffer at physiological conditions showed the resorbable nature of these materials. It was also mentioned that the gradually functional material structure was formed by spreading a TiO_2-Ag_2O mixture on the surface of the HA green compact and heating at 900°C [11-191].

Apatite is known to show excellent absorbability of bacteria, ammonia, nitrogen oxides, or aldehyde – environment-purification material. One major drawbacks of apatite is that absorbed materials will be sooner or later saturated on apatite surfaces. At the same time, as was discussed in Chapter 4 briefly, it is known that TiO_2 is a photocatalyst. A photocatalyst is a material which absorbs light to bring it to higher energy level, and provides such energy to reacting substances to chemical reaction take place. When the photocatalyst absorbs the light, positive holes are created in the valence electron band to react with electrons. A positive hole of TiO_2 reacts with water or dissolved oxygen to generate OH radicals, which decompose toxic substances. The OH radicals possess stronger oxidation power than chlorine or ozone for sterilization or disinfection. Hence, a multiple-functional ceramic composite can be fabricated to achieve those specific aims. TiO_2 particles or base substrate layer can be partially coated with apatite to make respective materials function individually. Namely, apatite absorbs bacteria or organic materials and TiO_2 can decompose such toxic materials. Since absorbed species on apatite are decomposed by the TiO_2 catalyst, the original absorbability of apatite can always be maintained [11-192, 11-193].

11.5. POROSITY CONTROLLED SURFACE AND TEXTURING

Void metal composite (VMC) is a porous metal developed to fix a prosthesis to bone by tissue ingrowth. The material is made by techniques which produce structures with controlled porosity, density, and physical properties. The ability to produce a range of structures creates porosities to study the effects of pore size, shape, and density on bone–metal interface strength. Ti-6Al-4V is the metal of choice for VMC. It was selected for its corrosion resistance, good mechanical properties, low density, and good tolerance by body tissue. Structures with spherical pore size ranging from 275 to 650 μm, and have been fabricated with up to 80% theoretical densities. The optimum structure for attachment strength seems to be a pore size of 450 μm and 50% theoretical density [11-194]. The control porosity can be achieved by blasting with alumina or titania [11-195].

Petronis *et al.* [11-196] developed a model system for studying cell–surface interactions, based on microfabricated cell culture substrates. Porous surfaces consisting of interconnecting channels with openings of subcellular dimensions are generated on flat, single crystal, silicon substrates. Channel size (width and depth), distribution, and surface coating can be varied independently and used for systematic investigation of how topographical, chemical, and elastic surface properties influence cell or tissue biological responses. Model porous surfaces have been produced by using two different microfabrication methods. Submicron-sized channels with very high depth-to-width aspect ratios (up to 30) have been made by using electron beam lithography and anisotropic reactive ion etching into single-crystal silicon. Another method uses thick-resist photolithography, which can be used to produce channels wider than 1 μm and with depth-to-width aspect ratios below 20 in an epoxy polymer [11-196]. Xiaoxiong and Shizhong [11-197] created pit with controlled pit density and geometry to exhibit porosity controlled surfaces. The pit initiation process on CpTi in bromide solution was investigated by means of surface analysis. The results showed that the titanium surface film formed by anodic polarization in bromide solution was mainly TiO_2. Prior to the pit initiation, Br ions were absorbed and accumulated on localized spots of the TiO_2 film, forming bromine nuclei containing mostly $TiBr_4$. The bromine nuclei grew into the critical nuclei when the film was in the critical state of breaking down. The depth of the critical nuclei was equal to or less than 3 nm. The concentration of bromine in the critical nuclei was the critical surface concentration. It was the requisite condition for pit initiation that the concentration of bromine in bromine nuclei reached critical surface concentration. It was mentioned that, in the system of titanium/bromide solution, the critical surface concentration was 25–35 wt% and was independent of the temperature and concentration of the solution [11-197].

The *in vitro* mineralization of osteoblast-like cells on CpTi with different surface roughness was examined. CpTi disks were polished through 600 grit SiC paper (grooved), polished through 1 μm diamond paste (smooth), or sandblasted (rough). The disks were cleaned, acid passivated and UV sterilized (254 nm, 300 μW/cm^2). Osteoblast-like cells were harvested from rat pups and were cultured. The cultures were grown for 6 or 12 days in media supplemented with 5 mM a-glycerophosphate. It was found that (i) *in vitro* mineralization responses were independent of surface roughness, and (ii) Alizarin red staining indicated small zones of mineralization on all surfaces, indicating that surface topography is known to affect osteoblast-like cell activities [11-195].

Ungersboeck *et al.* [11-198] investigated five types of limited contact dynamic compression plates of CpTi with different surface treatments and electropolished stainless-steel limited contact dynamic compression plates. The surface roughness parameters and chemical surface conditions were determined and checked for probable surface contamination. After an implantation period of 3 months on long sheep bones, the soft tissue adjacent to the plates was evaluated histomorphometrically. The difference in roughness parameters was statistically significant for most surface conditions. It was reported that (i) a correlation was found between the surface roughness of the implants and the thickness of the adjacent soft-tissue layer, (ii) the thinnest soft-tissue reaction layer with a good adhesion to the implant surface was observed for the titanium anodized plates with coarse surface (20% nitric acid at 60°C for 30 min), and (iii) smooth implants, in particular, the electropolished stainless-steel plates, induced statistically significant thicker soft-tissue layers [11-198]. Larsson *et al.* [11-199] investigated the bone formation around titanium implants with varied surface properties. Machined and electropolished samples with and without thick anodically formed surface oxide were prepared and inserted in the cortical bone of rabbits (1, 3, and 6 weeks). It was found that (i) light microscopic morphology and morphometry showed that all implants were in contact with bone and had a large proportion of bone within the threads at

6 weeks, (ii) the electropolished implants, irrespective of anodic oxidation, were surrounded by less bone than the machined implants after 1 week, (iii) after 6 weeks the bone volume, as well as the bone–implant contact, were lower for the merely electropolished implants than for the other three groups, and (iv) a high degree of bone contact and bone formation are achieved with titanium implants which are modified with respect to oxide thickness and surface topography; however, the result with the smooth (electropolished) implants indicates that a reduction of surface roughness, in the initial phase, decreases the rate of bone formation in rabbit cortical bone [11-199].

The topography of titanium implants is of importance with respect to cellular attachment. Chung and McAlarney [11-200] examined the topographies of three as-received implant systems (Nobelpharma, Swede-Vent, and Screw-Vent), followed by thermal (700°C for 240 min) and anodic oxidation (70 V in 1 M acetic acid solution) of the fixtures. Fixtures were self tapped into freshly sacrificed swine rib bone. It was found that (i) thermal and anodic oxidation, as well as implantation shear stress, had no effect on topography, and (ii) the growth of oxides and implantation shear stress had no effect on topography [11-200].

Thelen *et al.* [11-201] investigated mechanics issues related to potential use of a recently developed porous titanium material for load-bearing implants. This material may have advantages over solid Ti for enhancing the bone–implant interface strength by promoting bone and soft-tissue ingrowth, and for reducing the bone–implant modulus mismatch, which can lead to stress shielding. It was mentioned that (i) simple analytic models provide good estimates of the elastic properties of the porous Ti, and (ii) the moduli can be significantly reduced to decrease the mismatch between solid Ti and bone, achieving the mechanical compatibility proposed by Oshida [11-202]. The finite element simulations show that bone ingrowth will dramatically reduce stress concentrations around the pores [11-201]. Takemoto *et al.* [11-203] prepared porous bioactive titanium implants (porosity of 40%) by a plasma spray method and subsequent chemical and thermal treatments of immersion in a 5 M aqueous NaOH solution at 60° C for 24 h, immersion in distilled water at 40°C for 48 h, and heating to 600°C for 1 h. It was reported that compression strength and bending strength were 280 MPa (0.2% offset yield strength 85.2 MPa) and 101 MPa, respectively. For *in vivo* analysis, bioactive and non-treated porous titanium cylinders were implanted into 6 mm diameter holes in rabbit femoral condyles. It was found that (i) the percentage of bone–implant contact (affinity index) of the bioactive implants was significantly larger than for the non-treated implants at all post-implantation times (13.5 vs. 10.5, 16.7 vs. 12.7, 17.7 vs. 10.2, 19.1 vs. 7.8 at 2, 4, 8, and 16 weeks, respectively), and (ii) the percentage of bone area ingrowth showed a significant increase with the bioactive implants, whereas with the non-treated implants it appeared to decrease after 4 weeks (10.7 vs. 9.9, 12.3 vs. 13.1, 15.2 vs. 9.8, 20.6 vs. 8.7 at 2, 4, 8, and 16 weeks, respectively), suggesting that porous bioactive titanium has sufficient mechanical properties and biocompatibility for clinical use under load-bearing conditions [11-203].

The high-cycle fatigue strength of porous coated Ti-6Al-4V is approximately 75% less than the fatigue strength of uncoated Ti-6Al-4V. Kohn and Ducheyne [11-204] designed their study to separate the effects of three parameters thought to be responsible for this reduction: interfacial geometry, microstructure, and surface alterations caused by sintering. To achieve the goal of one parameter variation, hydrogen-alloying treatments, which refined the lamellar microstructure of beta-annealed and porous coated Ti-6Al-4V, were formulated. It was found that (i) Ti-6Al-4V subjected to hydrogen-alloying treatment was is 643–669 MPa, which was significantly greater than the fatigue strength of beta-annealed Ti-6Al-4V (497 MPa), and also greater than the fatigue strength of pre-annealed, equiaxed Ti-6Al-4V (590 MPa); however, (ii) the fatigue strength of porous coated Ti-6Al-4V is independent of microstructure. It is therefore concluded that the notch effect of the surface porosity does not allow the material to take advantage of the superior fatigue crack initiation resistance of a refined α-grain size [11-204].

11.6. FOAMED METAL

The foam-shaped materials exhibit a continuously connected ligamented, reticulated open-cell geometry having a duodecahedronal cell shape. The foam-shaped material is in a density ranging from 1 to 20% theoretical, and a cell density of 10–50 cells per linear inch, with material density and cell density independently variable. Foamed materials are presently produced in a wide range of plastics, metals, and composites having either solid or tubular ligaments. The metals include aluminum, nickel, copper, silver, zinc, leads, tin, magnesium, and stainless-steel [11-205].

Surface modifications of dental and medical implants are very important for various reasons. As numerous reports indicate, the surface texture should be preferably rough for enhanced mechanical retention between the implant surface and newly grown bone structure. Moreover, increased porosity in the modified surface layer will show lower mechanical properties than the bulk material, and their values are close to those of the implant receiving soft/hard vital tissues. This is the so-called mechanical compatibility [11-202]. Besides this and biological compatibility, there is the third compatibility, which was proposed by the present author, and is called morphological compatibility [11-206].

Table 11.1. Summary of interference colors, after in-air oxidation for 30 min.

Material\Temperature	350°C	450°C	550°C	650°C	750°C
CpTi (II)	Yellowish blue	Brown	Blue	Light green	Bright gray
Ti-6Al-4V	Yellow	Violet	Greenish blue	Bright gray	Light gray
TiNi	Metallic color	Light yellow	Violet	Blue	Reddish violet

11.7. COLORING

In order to color titanium, there are basically four different methods available: (1) thermal-oxidation, (2) chemical-oxidation, (3) anodic oxidation, and (4) nitrizing. By either method the color can be well defined by the thickness of the film (mainly TiO_2) formed on the Ti substrate surface. The color is sometimes referred to as the interference color, which can repeatedly appear under the integral of the half-wave length of the light [11-207]. Hence, colors, such as rainbow colors, are not colors *per se*. This is true for the first three methods, but for the nitrizing method, gold is the color of the reaction product (i.e., TiN) and silver is the color of Ti_2N.

As the result of anodization (electrolyte: H_3PO_4:H_2SO_4:H_2O by 8:1:1 ratio), CpTi was covered with anodic oxide film as follows: silver (2 V, 25 Å film thickness), light gold (4 V, 105 Å), gold (12 V, 270 Å), golden violet (20 V, 272 Å), dark blue (24 V, 364 Å), and light blue (30 V, 610 Å). When CpTi (grade 2), Ti-6Al-4V, and TiNi were heated in air for 30 min at different temperatures, different interference colors were obtained as seen in the Table 11.1, due to different oxide film thicknesses (accordingly, different interference index) [11-208].

REFERENCES

[11-1] Kasemo B, Lausmaa J. Biomaterials and implant surfaces: a surface science approach. Int J Oral Maxillofac Implants 1988;3:247–259.

[11-2] Baier RE, Meyer AE. Implant surface preparation. Int J Oral Maxillofac Implants 1988;3:9–20.

[11-3] Reitz WE. The eighth international conference on surface modification technology. J Meatls 1995;47:14–16.

[11-4] Smith DC, Pilliar RM, Chernecky R. Dental implant materials. I. Some effects of preparative procedures on surface topography. J Biomed Mater Res 1991;25:1045–1068.

[11-5] Buddy D, Ratner B, Thomas JL. Biomaterial surfaces. J Biomed Mater Res 1987;21:59–89.

[11-6] Wen X, Wang X, Zhang N. Microrough surface of metallic biomaterials: a literature review. J Biomed Mater Eng 1996;6:173–189.

[11-7] Piattelli A, Scarano A, Coriglano M, Piattelli M. Presence of multinucleated giant cells around machined, sandblasted and plasma-sprayed titanium implants: a histological and histochemical time-course study in rabbit. Biomaterials 1996;17:2053–2058.

[11-8] Citeau A, Guicheux J, Vinatier C, Layrolle P, Nguyen TP, Pilet P, Daculsi G. *In vitro* biological effects of titanium rough surface obtained by calcium phosphate grid blasting. Biomaterials 2005;26: 157–165.

[11-9] Kubo M, Fijishima A, Hotta Y, Kunii J, Shimakura Y, Shimidzu T, Kawawa T, Miyazaki T. Production method of resin-bonded titanium crown by CAD/CAM. The 19th Proceedings of Society Titanium Research. 2006. p. 42

[11-10] Miyakawa O, Watababe K, Okawa S, Shiokawa N, Kobayashi M, Tamura H. Grinding of titanium, Part 1: commercial and experimental wheels made of silicon carbide abrasives. Dent Mater J 1990;9:30–41.

[11-11] Miyakawa O, Watanabe K, Okawa S, Nakano S, Shiokawa N, Kobayashi M, Tamura H. Grinding of titanium, Part 2: commercial vitrified wheels made of alumina abrasives. Dent Mater J 1990;9: 42–52.

[11-12] Miyakawa O, Watanabe K, Okawa S, Kanatani M, Nakano S, Kobayashi M. Surface contamination of titanium by abrading treatment. Dent Mater J 1996;15:11–21.

[11-13] Mandelbrot BB. The fractal geometry of nature. New York: Freeman. 1983. p. 34.

[11-14] Chesters S, Wen HY, Lundin M, Kasper G. Fractal-based characterization of surface texture. Appl Surf Sci 1989;40:185–192.

[11-15] Sayles RS, Thomas TR. Surface topography as a non-stationary random process. Nature 1978;271:431–434.

[11-16] Oshida Y, Munoz CA, Winkler MM, Hashem A, Ito M. Fractal dimension analysis of aluminum oxide particle for sandblasting dental use. J Biomed Mater Eng 1993;3:117–126.

[11-17] Miyakawa O, Okawa S, Kobayashi M, Uematsu K. Surface contamination of titanium by abrading treatment. Dent Jpn 1998;34:90–96.

[11-18] Oshida, Y. Cross-contamination of biomaterials while sample preparation for corrosion tests. to be submitted to Corrosion Science, 2007.

[11-19] Mustafa K, Lopez BS, Hultenby K, Arvison K. Attachment of HGF of titanium surfaces blasted with TiO_2 particles. J Dent Res 1997;76:85 (Abstract No. 573).

[11-20] Wennerberg A, Albrektsson T, Johnsson C, Andersson B. Experimental study of turned and grit-blasted screw-shaped implants with special emphasis on effects of blasting material and surface topography. Biomaterials 1996;17:15–22.

[11-21] Johansson CB, Albrektsson T, Thomsen P, Snnerby I, Lodding A, Odelius H. Tissue reactions to titanium-6aluminum-4vanadium alloy. Eur J Exp Musculoskel Res 1992;1:161–169.

[11-22] Wang C-S. Surface modification of titanium for enhancing titanium–porcelain bond strength. Indiana University Master Degree Thesis. 2004.

[11-23] Peutzfeld A, Asmussen E. Distortion of alloy by sandblasting. Am J Dent 1996;9:65–66.

[11-24] Peutzfeld A, Asmussen E. Distortion of alloys caused by sandblasting. J Dent Res 1996;75:259 (Abstract No. 1162).

[11-25] Oshida Y, Daly J. Fatigue damage evaluation of shot peened high strength aluminum alloy. In: Surface engineering. Meguid SA, editor. New York: Elsevier Applied. 1990. pp. 404–416.

[11-26] Metal Improvement Company, Inc. Shot peening applications 7th ed. 1990.

[11-27] Oshida Y. Beneficial effect of shot-peening on biomaterials. to be submitted to J Biomed Mater Eng.

[11-28] Oshida Y. Enhancing of porcelain bond strengths by application of shot-peening. to be submitted to Dent Mater J.

[11-29] DeWald AT, Rankin JE, Hill MR, Lee MJ, Chen H-L. Assessment of tensile residual stress mitigation in Alloy 22 welds due to laser peening. J Eng Mater Tech, Trans ASME 2004;126:81–89.

[11-30] Fairland BP, Wilcox BA, Gallagher WJ, Williams DN. Laser shock-induced microstructural and mechanical property changes in 7075 aluminum. J Appl Phys 1972;43:3893–3895.

[11-31] Fairland BP, Clauer AH. Laser generation of high amplitude stress waves in materials. J Appl Phys 1979;50:1497–1502.

[11-32] Dane CB, Hackel LA, Daly J, Harrison J. High power laser for peening of metals enabling production technology. Mater Manuf Process 2000;15:81–96.

[11-33] Cho S-A, Jung S-K. A removal torque of the laser-treated titanium implants in rabbit tibia. Biomaterials 2003;24:4859–4863.

[11-34] Gaggl A, Schultes G, Muller WD, Karcher H. Scanning electron microscopical analysis of laser-treated titanium implants surfaces – a comparative study. Biomaterials 2000;21:1067–1073.

[11-35] Hansson S, Hansson KN. The effect of limited lateral resolution in the measurement of implant surface roughness: a computer simulation. J Biomed Mater Res 2005;75A:472–477.

[11-36] Endo K. Chemical modification of metallic implant surfaces with bio-functional proteins Part 1: molecular structure and biological activity of a modified NiTi alloy surface. Dent Mater J 1995;14:185–198.

[11-37] Browne M, Gregson PJ. Surface modification of titanium alloy implants. Biomaterials 1994;15:894–898.

[11-38] Krozer A, Hall J, Ericsson I. Chemical treatment of machined titanium surfaces. Clin Oral Implant Res 1999;10:204–211.

[11-39] Rupp F, Scheideler L, Rehbein D, Axmann D, Geis-Gerstorfer J. Roughness induced dynamic changes of wettability of aid etched titanium implant modifications. Biomaterials 2004;25:1429–1438.

[11-40] MacDonald DE, Rapuano BE, Deo N, Stranick M, Somasundaran P, Boskey AL. Thermal and chemical modification of titanium–aluminum–vanadium implant materials: effects of surface properties, glycoprotein adsorption, and MG63 cell attachment. Biomaterials 2004;25:3135–3146.

[11-41] Li L-H, Kong Y-M, Kim H-W, Kim Y-W, Kim H-E, Heo S-J, Koak J-Y. Improved biological performance of Ti implants due to surface modification by micro-arc oxidation. Biomaterials 2004;15:1867–1875.

[11-42] Xie Y, Liu X, Huang A, Ding C, Chu PK. Improvement of surface bioactivity on titanium by water and hydrogen plasma immersion ion implantation. Biomaterials 2005;26:6129–6135.

[11-43] Rohanizadeh R, Al-Sadeq M, LeGeros RZ. Preparation of different forms of titanium oxide on titanium surface: effects on apatite deposition. J Biomed Mater Res 2004;71A: 343–352.

[11-44] Demri B, Hage-Ali M, Moritz M, Muster D. Surface characterization of C/Ti-6Al-4V coating treated with ion beam. Biomaterials 1997;18: 305–310.

[11-45] Poon RWY, Yeung KWK, Liu XY, Chu PK, Chung CY, Lu WW, Cheung KMC, Chan D. Carbon plasma immersion ion implantation of nickel–titanium shape memory alloys. Biomaterials 2005;26:2265–2272.

[11-46] Oshida Y, Sachdeva RCL, Miyazaki S. Microanalytical characterization and surface modification of TiNi orthodontic archwires. J Biomed Mat Eng 1992;2:51–69.

[11-47] Schrooten J, Helsen JA. Adhesion of bioactive glass coating to Ti6Al4V oral implant. Biomaterials 2000;21:1461–1419.

[11-48] Saiz E, Goldman M, Gomez-Vega JM, Tomsia AP, Marshall GW, Marshall SJ. *In vitro* behavior of silicate glass coatings on Ti6Al4V. Biomaterials 2002;23:3749–3756.

[11-49] Lee JH, Ryu H-S, Lee D-S, Hong KS, Chang B-S, Lee C-K. Biomechanical and histomorphometric study on the bone–screw interface of bioactive ceramic-coated titanium screws. Biomaterials 2005;26: 3249–3257.

[11-50] Chu KT, Oshida Y, Hancock EB, Kowolik MJ, Barco TM, Zunt SL. Hydroxyapatite/PMMA composites as bone cements. J Biomed Mat Eng 2004;14:87–105.

[11-51] Liu DM. Fabrication and characterization of porous hydroxyapatite granules. Biomaterials 1996;17:1955–1957.

[11-52] Luo P, Nieh TG. Preparing hydroxyapatite powders with controlled morphology. Biomaterials 1996;17:1959–1964.

[11-53] Porter AE, Taak P, Hobbs LW, Coathup MJ, Blunn GW, Spector M. Bone bonding to hydroxyapatite and titanium surfaces on femoral stems retrieved from human subjects at autopsy. Biomaterials 2004;25:5 199–5208.

[11-54] Buser D, Schenk RK, Steinemann S, Fiorelline JP, Fox CH, Stich H. Influence of surface characteristics on bone integration of titanium implants. A histomorphometric study in miniature pigs. J Biomed Mater Res 1991;25:889–902.

[11-55] Souto RM, Lemus MM, Reis RL. Electrochemical behavior of different preparations of plasma-sprayed hydroxyapatite coatings on Ti6Al4V substrate. J Biomed Mater Res 2004;70A:59–65.

[11-56] Filiaggi MJ, Coombs NA, Pilliar RM. Characterization of the interface in the plasma-sprayed HA coating/Ti-6Al-4V implant system. J Biomed Mater Res 1991;25:1211–1229.

[11-57] Yang YC, Chang E, Lee ST. Mechanical properties and Young's modulus of plasma-sprayed hydroxyapatite coating on Ti substrate in simulated body fluid. J Biomed Mater Res 2003;67A:886–899.

[11-58] Yang YC, Ong JL. Bond strength, compositional, and structural properties of hydroxyapatite coating on Ti, ZrO_2-coated Ti, and TPS-coated Ti substrate. J Biomed Mater Res 2003;64A:509–516.

[11-59] Yang YC, Kim K-H, Agrawal CM, Ong JL. Interaction of hydroxyapatite-titanium at elevated temperature in vacuum environment. Biomaterials 2004;25:2927–2932.

[11-60] Brossa F, Cigada A, Chiesa R, Paracchini L, Consonni C. Adhesion properties of plasma sprayed hydroxylapatite coating for orthopaedic prostheses. J Biomed Mater Eng 1993;3:127–136.

[11-61] Lee E-J, Lee S-H, Kim H-W, Kong Y-M, Kim H-E, Fluoridated apatite coatings on titanium obtained by electron-beam deposition. Biomaterials 2005;26:3843–3851.

[11-62] Kim H-W, Kim H-E, Knowles LC. Fluor-hydroxyapatite sol–gel coating on titanium substrate for hard tissue implants. Biomaterials 2004;25: 3351–3358.

[11-63] Filiaggi MJ, Pilliar RM, Coombs NA. Post-plasma-spraying heat treatment of the HA coating/Ti-6Al-4V implant system. J Biomed Mater Res 1993;27:191–198.

[11-64] Lynn AK, DuQuesnay DL. Hydroxyapatite-coated Ti-6Al-4V Part 2: the effect of post-deposition heat treatment at low temperatures. Biomaterials 2002;23;1947–1953.

[11-65] Yang YC, Chang E. Influence of residual stress on bonding strength and fracture of plasma-sprayed hydroxyapatite coatings on Ti-6Al-4V substrate. Biomaterials 2001;22:1827–1836.

[11-66] Mano T, Ueyama Y, Ishikawa K, Matsumura T, Suzuki K. Initial tissue response to a titanium implant coated with apatite at room temperature using a blast coating method. Biomaterials 2002;23:1931–1936.

[11-67] Han Y, Xu K. Photoexcited formation of bone apatite-like coatings on micro-arc oxidized titanium. J Biomed Mater Res 2004;71A:608–614.

[11-68] Kim H-W, Kim H-E, Salih V, Knowles JC. Sol–gel-modified titanium with hydroxyl-apatite thin films and effect on osteoblast-like cell responses. J Biomed Mater Res 2005;74A:294–305.

[11-69] Kim H-W, Koh Y-H, Li L-H, Lee S, Kim H-E. Hydroxyapatite coating on titanium substrate with titania buffer layer processed by sol–gel method. Biomaterials 2004;25:2533–2538.

[11-70] Moritz N, Jokinen M, Peltola T, Areva S, Yli-Urpo A. Local induction of calcium phosphate formation on TiO_2 coatings on titanium via surface treatment with a CO_2 laser. J Biomed Mater Res 2003;65A:9–16.

[11-71] Rohanizadeh R, LeGeros RZ, Harsono M, Bendavid A. Adherent apatite coating on titanium substrate using chemical deposition. J Biomed Mater Res 2005;72A:428–438.

[11-72] Ergun C, Doremus R, Lanford W. Hydroxylapatite and titanium: interfacial reactions. J Biomed Mater Res 2003;65A:336–343.

[11-73] Piattelli A, Trisi P. A light and laser scanning microscopy study of bone/hydroxyl-apatite-coated titanium implants interface: histochemical evidence of unmineralized material in humans. J Biomed Mater Res 1994;28:529–536.

[11-74] Hayashi K, Mashima T, Uenoyama K. The effect of hydroxyapatite coating on bony ingrowth into grooved titanium implants. Biomaterials 1999; 20:111–119.

[11-75] Moroni A, Caja VL, Egger EL, Trinchese L, Chao EYS. Histomorphometry of hydroxyapatite coated and uncoated porous titanium bone implant. Biomaterials 1994;15:926–930.

[11-76] Thomas KA, Cook SD, Haddad RJ, Kay JF, Jarcho M. Biologicla response to hydroxyapatite-coated titanium hips. J Arthrop 1989;4:43–53.

[11-77] Jansen JA, van de Waerden JPCM, Wolke JGC, de Groot K. Histologic evaluation of the osseo adaptation to titanium and hydroxyapatite-coated titanium implants. J Biomed Mater Res 1991;25:973–989.

[11-78] Gottlander M, Albrektsson T. Histomorphometric studies of hydroxylapatite-coated and uncoated cp titanium threaded implants in bone. Int J Oral Maxillofac Implants 1991;6:399–404.

[11-79] Schreurs BW, Huiskes R, Buma P, Slooff TJJH. Biomechanical and histological evaluation of a hydroxyapatite-coated titanium femoral stem fixed with an intramedullary morsellized bone grafting technique: an animal experiment on goats. Biomaterials 1996;17:1177–1186.

[11-80] Cabrini M, Cigada A, Rondelli G, Vicentini B. Effect of different surface finishing and of hydroxyapatite coatings on passive and corrosion current of Ti6Al4V alloy in simulated physiological solution. Biomaterials 1997;18:783–787.

[11-81] Hayashi K, Inadome T, Tsumura H, Nakashima Y, Sugioka Y. Effect of surface roughness of hudroxyapatite-coated titanium on the bone–implant interface shear strength. Biomaterials 1994;15:1187–1191.

[11-82] Souto RM, Laz MM, Reis RL. Degradtion characteristics of hydroxyapatite coatings on orthopaedic TiAlV in simulated physiological media investigated by electrochemical impedance spectroscopy. Biomaterials 2003;24:4213–4221.

[11-83] Cook SD, Thomas KA, Kay JF, Jarcho M. Hydroxyapatite-coated titanium for orthopedic implant applications. Clin Orthop 1988;186:225–243.

[11-84] Yang YC, Chang E, Hwang BH, Lee SY. Biaxial residual stress states of plasma-sprayed hydroxyapatite coatings on titanium alloy substrate. Biomaterials 2000;12:1327–1337.

[11-85] Mimura K, Watanabe K, Okawa S, Kobayashi M, Miyakawa O. Morphological and chemical characterizations of the interface of a hydroxyapatite-coated implant. Dent Mater J 2004;23:353–360.

[11-86] Lynn AK, DuQuesnay DL. Hydroxyapatite-coated Ti-6Al-4V Part 1: the effect of coating thickness on mechanical fatigue behaviour. Biomaterials 2002;23:1937–1946.

[11-87] Jonášová L, Müller FA, Helebrant A, Strnad J, Greil P. Hydroxyapatite formation on alkali-treated titanium with different content of Na^+ in the surface layer. Biomaterials 2002;23:3095–3101.

[11-88] Jonášová L, Müller FA, Helebrant A, Strnad J, Greil P. Biomimetic apatite formation on chemically treated titanium. Biomaterials 2004;25: 1187–1194.

[11-89] Wang CX, Wang M, Zhou X. Nucleation and growth of apatite on chemically treated titanium alloy: an electrochemical impedance spectroscopy study. Biomaterials 2003;24:3069–3077.

[11-90] Song W-H, Jun Y-K, Han Y, Hong S-H. Biomimetic apatite coatings on micro-arc oxidized titania. Biomaterials 2004;25:3341–3349.

[11-91] Brinker CJ, Scherer GW. Sol gel science: the physics and chemistry of sol gel processing. San Diego: Academic Press, 1990.

[11-92] Hofacker S, Mechtel M, Mager M, Kraus H. Sol gel: a new tool for coatings chemistry. Prog Organ Coat 2002;45:159–164.

[11-93] Sundgren J-E, Bodö P, Lundström I. Auger electron spectroscopic studies of the interface between human tissue and implants of titanium and stainless-steel. J Colloid Interface Sci 1986;110:9–20.

[11-94] Lu X, Leng Y. TEM study of calcium phosphate precipitation on bioactive titanium surfaces. Biomaterials 2004;25:1779–1786.

[11-95] Hanawa T, Ota M, Calcium phosphate naturally formed on titanium in electrolyte solution. Biomaterials 1991;12;767–774.

[11-96] Barrère F, van der Valk CM, Dalmeijer RAJ, van Blitterswijk CA, de Groot K, Layrolle P. *In vitro* and *in vivo* degradation of biomimetic octacalcium phosphate and carbonate apatite coatings on titanium implants. J Biomed Mater Res 2003;64A:378–387.

[11-97] Barrère F, Snel MME, van Blitterswijk CA, de Groot K, Layrolle P. Nano-scale study of the nucleation and growth of calcium phosphate coating on titanium implants. Biomaterials 2004;25:2901–2910.

[11-98] Li SJ, Yang R, Niinomi M, Hao YL, Cui YY. Formation and growth of calcium phosphate on the surface of oxidized Ti-29Nb-13Ta-4.6Zr alloy. Biomaterials 2004;25:2525–2532.

[11-99] Bogdanski D, Esenwien SA, Prymak O, Epple M, Muhr G, Köller M. Inhibition of PMN apoptosis after adherence to dip-coated calcium pho phate surfaces on a NiTi shape memory alloy. Biomaterials 2004; 25:4627–4623.

[11-100] Kim CS, Ducheyne P. Compositional variations in the surface and interface of calcium phosphate ceramic coatings on Ti and Ti-6Al-4V due to sintering and immersion. Biomaterials 1991;12:461–469.

[11-101] Ducheyne P, Bianco PD, Kim C. Bone tissue growth enhancement of calcium phosphate coatings on porous titanium alloys: effect of shielding metal dissolution product. Biomaterials 1992;13:617–624.

[11-102] Krupa D, Baszkiewicz J, Kozubowski JA, Barcz A, Sobczak JW, Biliński A, Lewandowska-SzumieA M, Rajchel B. Effect of dual ion implantation of calcium and phosphorus on the properties of titanium. Biomaterials 2005;26:2847–2856.

[11-103] Fini M, Cigada A, Rondelli G, Chiesa R, Giardino R, Giavaresi G, Aldini NN, Torricelli P, Vicentini B. *In vitro* and *in vivo* behaviour of Ca- and P-enriched anodized titanium. Biomaterials 1999;20:1587–1594.

[11-104] Jansen JA, Hulshoff JE, Van Dijk K. Biological evaluation of thin calcium-phosphate (Ca-P) coatings. J Dent Res 1997;76:282 (Abstract No. 2150).

[11-105] Hayakawa T, Yoshinari M, Kiba H, Yamamoto H, Nemoto K, Jansen JA. Trabecular bone response to surface roughened and calcium phosphate (Ca-P) coated titanium implants. Biomaterials 2002;23:1025–1031.

[11-106] Gan L, Wang J, Pilliar RM. Evaluating interface strength of calcium phosphate sol–gel-derived thin films to Ti6Al4V substrate. Biomaterials 2005;26:189–196.

[11-107] Choi J, Bogdanski D, Köller M, Esenwein SA, Müller D, Epple M. Calcium phosphate coating of nickel–titanium shape-memory alloys, coating procedure and adherence of leukocytes and platelets. Biomaterials 2003;24:3689–3696.

[11-108] Fend B, Weng J, Yang BC, Qu SX, Zhang XD. Characterization of titanium surfaces with calcium and phosphate and osteoblast adhesion. Biomaterials 2004;25:3421–3428.

[11-109] Lui Q, Ding J, Mante FK, Wunder SL, Baran GR. The role of surface functional groups in calcium phosphate nucleation on titanium foil: a self-assembled monolayer technique. Biomaterials 2002;23:3103–3111.

[11-110] Peng P, Kumar S, Voelcker NH, Szili E, Smart RStC, Griesser HJ. Thin calcium phosphate coatings on titanium by electrochemical deposition in modified simulated body fluid. J Biomed Mater Res 2006;76A:347–355.

[11-111] Zhang Q, Leng Y, Xin R. A comparative study of electrochemical deposition and biomimetic deposition of calcium phosphate on porous titanium. Biomaterials 2005;26:2857–2865.

[11-112] Heimann RB, Wirth R. Formation and transformation of amorphous calcium phosphates on titanium alloy surfaces during atmospheric plasma spraying and their subsequent *in vitro* performance. Biomaterials 2006; 27:823–831.

[11-113] Ban S, Maruno S, Iwata H, Itoh H. Calcium phosphate precipitation on the surface HA-G-Ti composite under physiologic conditions. J Biomed Mater Res 1994;28:65–71.

[11-114] Lucas LC, Ong JL. Post deposition heat treatments of Ca-P coating. J Dent Res 1993;71:228 (Abstract No. 998).

[11-115] Watanabe K, Hashimoto A, Nomura S, Endo MM, Okawa S, Kanatani M, Nakano S, Kobayashi M, Miyakawa O. Surface properties of dental implants Part 2: surface analysis of the titanium implant extracted from a rat bone. J Jpn Dent Mater 2004;23:329–334.

[11-116] Kim H-W, Lee E-J, Jun I-K, Kim H-E. On the feasibility of phosphate glass and hydroxyapatite engineered coating on titanium. J Biomed Mater Res. 2005;75A:656–667.

[11-117] Maxian SH, Zawadsky JP, Dunn MG. Mechanical and histological evaluation of amorphous calcium phosphate and poorly crystallized

hydroxyapatite coatings on titanium implants. J Biomed Mater Res 1993;27:717–728.

[11-118] Khor KA, Gu YW, Pan D, Cheang P. Microstructure and mechanical properties of plasma sprayed HA/YSZ/Ti–6Al–4V composite coatings. Biomaterials 2004;25:4009–4017.

[11-119] Gu YW, Khor KA, Pan D, Cheang P. Activity of plasma sprayed yttria stabilized zirconia reinforced hydroxyapatite/Ti-6Al-4V composite coatings in simulated body fluid. Biomaterials 2004;25:3177–3185.

[11-120] Chou BY, Chang E, Yao SY, Chen JM. Phase transformation during plasma spraying of hydroxyapatite-19wt%-zirconia composite coating. J Am Ceram Soc 2002;85:661–669.

[11-121] Kasugu T, Yoshida M, Ikushima AJ, Tuchiya M, Kusakari H. Stability of zirconia-toughened bioactive glass-ceramics – *in vivo* study using dogs. J Mater Sci: Mater Med 1993;4:36–39.

[11-122] Takagi M, Mochida M, Uchida N, Saito K, Uematsu K. Filter cake forming and hot isostatic pressing for TZP-dispressed hydroxyapatite composite. J Mater Sci: Mater Med 1992;3:199–203.

[11-123] Yamada K, Imamura K, Itoh H, Iwata H, Maruno S. Bone bonding behavior of the hydroxyapatite containing glass–titanium composite prepared by the Cullet method. Biomaterials 2001;22:2207–2214.

[11-124] Suzuki R, Muyco J, McKittrick J, Frangos JA. Reactive oxygen species inhibited by titanium oxide coatings. J Biomed Mater Res 2003;66A: 396–402.

[11-125] Li H, Khor KA, Cheang P. Titanium dioxide reinforced hydroxyapatite coatings deposited by high velocity oxy-fuel (HVOF) spray. Biomaterials 2002;23:85–91.

[11-126] Lu Y-P, Li M-S, Li S-T, Wang Z-G, Zhu R-F. Plasma-sprayed hydroxyapatite + titania composite bond coat for hydroxyapatite coating on titanium substrate. Biomaterials 2004;25:4393–4403.

[11-127] Ng BS, Annergren I, Soutar AM, Khor KA, Jarfors AEW. Characterization of a duplex TiO_2/CaP coatings on Ti6Al4V for hard tissue replacement. Biomaterials 2005;26:1087–1095.

[11-128] Gu YW, Tay BY, Lim CS, Yong MS. Biomimetic deposition of apatite coating on surface-modified NiTi alloy. Biomaterials 2005;26:6916–6923.

[11-129] Chu CL, Chung CY, Zhou J, Pu YP, Lin PH. Fabrication and characteristics of bioactive sodium titanate/titania graded film on NiTi shape memory alloy. J Biomed Mater Res 2005;75A:595–602.

[11-130] Knabe C, Berger G, Gildenhaar R, Klar F, Zreiqat H. The modulation of osteogenesis *in vitro* by calcium titanium phosphate coatings. Biomaterials 2004;25:4911–4919.

[11-131] Shtansky DV, Gloushankova NA, Sheveiko AN, Kharitonova MA, Moizhess TG, Levashov EA, Rossi F. Design, characterization and testing of Ti-based multi-component coatings for load-bearing medical applications. Biomaterials 2005;26:2909–2924.

[11-132] von Walter M, Rüger M, Ragoß C, Steffens GCM, Hollander DA, Paar O, Maier HR, Jahnen-Dechent W, Bosserhoff AK, Erli H-J. *In vitro* behavior of a porous TiO_2/perlite composite and its surface modification with fibronectin. Biomaterials 2005;26:2813–2826.

[11-133] Rodrigues CVM, Serricella P, Linhares ABR, Guerdes RM, Borojevic R, Rossi MA, Duarte MEL, Farina M. Characterization of a bovine collagen-hydroxyapatite composite scaffold for bone tissue engineering. Biomaterials 2003;24:4987–4997.

[11-134] Cheng X, Filiaggi M, Roscoe SG. Electrochemically assisted co-precipitation of protein with calcium phosphate coatings on titanium alloy. Biomaterials 2004;25:5395–5403.

[11-135] Redepenning J, Venkataraman G, Chen J, Stafford N. Electrochemical preparation of chitosan/hydroxyapatite composite coatings on titanium substrates. J Biomed Mater Res 2003;66A:411–416.

[11-136] Venugopalan R, George MA, Weimer JJ, Lucas LC. Surface topography, corrosion and microhardness of nitrogen-diffusion-hardened titanium alloy. Biomaterials 1999;20:1709–1716.

[11-137] Dion I, Rouais F, Trut L, Baquency Ch, Monties JR, Havlik P. TiN coating: surface characterization and haemocompatibility. Biomaterials 1993;14:1230–1235.

[11-138] Oshida Y, Hashem A. Titanium–porcelain system. Part I: oxidation kinetics of nitrided pure titanium, simulated to porcelain firing process. J Boimed Mater Eng 1993;3:185–198.

[11-139] Thair L, Kamachi Mudali U, Bhurvaneswaran N, Nair KGM, Asokamani R, Raj B. Nitrogen ion implantation and *in vitro* corrosion behavior of as-cast Ti-6Al-7Nb alloy. Corr Sci 2002;44:2439–2457.

[11-140] Bordji K, Jouzeau JY, Mainard D, Payan E, Netter P, Rie KT, Stucky T, Hage-Ali M. Cytocompatibility of Ti-6Al-4V and Ti-5Al-2.5Fe alloys according to three surface treatments, using human fibroblasts and osteoblasts. Biomaterials 1996;17:929–940.

[11-141] Goldberg JR, Gilbert JL. The electrochemical and mechanical behavior of passivated and TiN/AlN-coated CoCrMo and Ti6Al4V alloys. Biomaterials 2004;25:851–864.

[11-142] Tamura Y, Yokoyama A, Watari F, Kawasaki T. Surface properties and biocompatibility of nitrided titanium for abrasion resistant implant materials. Dent Mater J 2002;21:355–372.

[11-143] Kokubo T, Ito S, Shigematsu M, Sakka S. Fatigue and life-time of bioactive glass-ceramic A–W containing apatite and wollastonite. J Mater Sci 1987;22:4067–4070.

[11-144] Morton PH, Bell T, Weisheit A, Kroll J, Mordike BL, Sagoo K. Laser gas nitriding of titanium and titanium alloys. In: Surface modification technologies V. Sudarshan TS, Braza JF, editors. London, UK: The Institute of Materials. 1992. pp. 593–609.

[11-145] Oshida Y. Novel nitridation technique – transformation superplasticity application on Ti materials. to be submitted to J Metals.

[11-146] Eberhardt AW, Pandey R, Williams JM, Weimer JJ, Ila D, Zimmerrmann RL. The role of residual stress and surface topography on hardness of ion implanted Ti-6Al-4V. Mater Sci Eng 1997;A229:147–155.

[11-147] Shetty RH. Mechanical and corrosion properties of nitrogen diffusion hardened Ti-6Al-4V alloy. In: Medical applications of titanium and its alloys: the material and biological issues. Brown SA, Lemons JE, editors. ASTM STP 1272. Philadelphia, PA: American Society for Testing and Materials. 1996. pp. 240–251.

[11-148] Semlitsch MF, Weber H, Streicher RM, Schon R. Joint replacement components made of hot-forged ad surface-treated Ti-6Al-7Nb alloy. Biomaterials 1992;13:781–788.

[11-149] Thull R. Semiconductive properties of passivated titanium and titanium based hard coatings on metals for implants – an experimental approach. Med Prog Technol 1990;16:225–234.

[11-150] Wisbey A, Gregson PJ, Tuke M. Application of PVD TiN coating to Co-Cr-Mo-based surgical implants. Biomaterials 1987;8:477–480.

[11-151] Venugopalan R, Weimer JJ, George MA, Lucas LC. The effect of nitrogen diffusion hardening on the surface chemistry and scratch resistance of Ti-6Al-4V alloy. Biomaterials 2000;21:1669–1677.

[11-152] Cyster LA, Parker KG, Parker TL, Grant DM. The effect of surface chemistry and nanotopography of titanium nitride (TiN) films on primary hippocampal neurons. Biomaterials 2004;25:97–107.

[11-153] Bull SJ, Chalker PR. The role of titanium interlayers in the adhesion of titanium nitride thin films. In: Surface modification technologies, Sudarshan TS, Braza JF, editors. The Institute of Materials. Cambridge: The University Press. 1992. pp. 205–215.

[11-154] Oshida Y, Fung LW, Ishikbay SC. Titanium–porcelain system. Part II: bond strength of fired porcelain on nitrided pure titanium. J Biomed Mater Eng 1997;7:13–34.

[11-155] Głuszek J, Jedrokowiak J, Markowski J, Masalski J. Galvanic couples of 316L steel with Ti and ion plated Ti and TiN coatings in Ringer's solutions. Biomaterials 1990;11:330–335.

[11-156] Lee B-H, Kim JK, Kim YD, Choi K, Lee KH. *In vivo* behavior and mechanical stability of surface-modified titanium implants by plasma spray coating and chemical treatments. J Biomed Mater Res 2004;69A: 279–285.

[11-157] Sonoda T, Kotake S, Kakimi H, Yamada M, Naganuma K, Kato M, Saka T, Shimizu T, Katoh K. On the dental application of titanium-base alloy. Part 7: titanium coating by sputtering. Gov Ind Res Inst 1991;40: 291–299.

[11-158] Sonoda T, Kotake S, Kakimi H, Yamada M, Naganuma M, Kato M, Saka T, Shimizu T, Katoh K. On the dental application of titanium-base alloy. Part 8: pure titanium coating on the denture base of Ti-6Al-4V alloy by sputtering. Gov Ind Res Inst 1991;40:300–307.

[11-159] Smith T. The effect of plasma-sprayed coatings on the fatigue of titanium alloy implants. J Metals 1994;46:54–56.

[11-160] Gruen TA, McNeice GM, Amstutz HD. Models of failure: cemented stem-type femoral components – a radiographic analysis of loosening. Clin Orthop 1979;141:17–27.

[11-161] Willert HG, Ludwig J, Semlitsch M. Reaction of bone to methacrylate after hip arthroplasty: a long-term gross, light microscopic and scanning electron microscopic study. J Bone Joint Surg 1974;56A: 1368–1373.

[11-162] Smith TS. Rationale for biological fixation of prosthetic devices. SAMPE J 1985:21:33:14–16.

[11-163] Engh CA, Bobyn JD. Biologic fixation of hip prosthesis: a review of the clinical status and current concepts. Adv Orthop Surg 1984;18:136–149.

[11-164] Reclaru L, Eschler P-Y, Lerf R, Blatter A. Electrochemical and metal ion release from Co-Cr-Mo prosthesis with titanium plasma spray coating. Biomaterials 2005;26:4747–4756.

[11-165] Aparicio C, Gil FJ, Fonseca C, Barbosa M, Planell J. Corrosion behaviour of commercially pure titanium shot blasted with different materials and sizes of shot particles for dental implant applications. Biomaterials 2003;24:263–273.

[11-166] Ku CH, Pioletti DP, Browne M, Gregson PJ. Effect of different Ti-6Al-4V surface treatments on osteoblasts behaviour. Biomaterials 2002;23: 1447–1454.

[11-167] Cai Z, Nakajima H, Woldu M, Berglund A, Bergmann M, Okabe T. *In vitro* corrosion resistance of titanium made using different fabrication methods. Biomaterials 1999;20:183–190.

[11-168] Chen G, Wen Z, Zhang N. Corrosion resistance and ion dissolution of titanium with different surface microroughness. J Biomed Mater Res 1998;8:61–74.

[11-169] Head WC, Mallory TH, Emerson RH. The proximal porous coating alternative for primary total hip arthroplasty. Orthopedics 1999;22:813–815.

[11-170] Bourne RB, Rorabeck CH, Burkart BC, Kirk PG. Ingrowth surfaces-plasma sprayed coating to titanium alloy hip replacements. Clin Orthop Rel Res 1994;298:37–46.

[11-171] Reclaru L, Lerf R, Eschler P-Y, Blatter A, Meyer J-M. Evaluation of corrosion on plasma sprayed and anodized titanium implants, both with and without bone cement. Biomaterials 2003;24:3027–3038.

[11-172] Xue W, Liu X, Zheng XB, Ding C. *In vivo* evaluation of plasma-sprayed titanium coating after alkali modification. Biomaterials 2005;26: 3029–3037.

[11-173] Borsari V, Giavaresi G, Fini M, Torricelli P, Tschon M, Chiesa R, Chiusoli L, Salito A, Volpert A, Giardino R. Comparative *in vitro* study on a ultra-high roughness and dense titanium coating. Biomaterials 2005;26:4948–4955.

[11-174] Vercaigne S, Wolke JBC, Naert I, Jansen JA. The effect of titanium plasma-sprayed implants on trabecular bone healing in the goat. Biomaterials 1998;19:1093–1099.

[11-175] Ong JL, Carnes DL, Bessho K. Evaluation of titanium plasma-sprayed and plasma-sprayed hydroxyapatite implants *in vivo*. Biomaterials 2004;25:4601–4606.

[11-176] Breme J, Steinhäuser E, Paulus G. Commercially pure titanium Steinhäuser plate-screw system for maxillofacial surgery. Biomaterials 1988;9:310–313.

[11-177] Rae T. A study on the effects of particulate metals of orthopaedic interest on murine macrophages *in vitro*. J Bone Joint Surg Br 1975;57:444–450.

[11-178] Rae T. The toxicity of metals used in orthopaedic prostheses. An experimental study using cultured human synovial fibroblasts. J Bone Joint Surg Br 1981;63-B:435–440.

[11-179] Brune D, Evje D, Melson S. Corrosion of gold alloys and titanium in artificial saliva. Scand J Dent Res 1982;90:168–171.

[11-180] Franchi M, Bacchelli B, Martini D, DePasquale V, Orsini E, Ottani V, Fini M, Giavaresi G, Giardino R, Ruggeri A. Early detachment of titanium particles from various different surfaces of endosseous dental implants. Biomaterials 2004;25:2239–2246.

[11-181] Pan J, Leygraf C, Thierry D, Ektessabi AM. Corrosion resistance for biomaterial applications of TiO_2 films deposited on titanium and stainless-steel by ion-beam-assisted sputtering. J Biomed Mater Res 1997; 35:309–318.

[11-182] New Technology Japan. Ti-O coating in Ti-6Al-4V alloy by DC reactive sputtering method. JETR 1994;22:18.

[11-183] Głuszek J, Masalski J, Furman P, Nitsch K. Structural and electrochemical examinations of PACVD TiO_2 films in Ringer solution. Biomaterials 1997;18:789–794.

[11-184] De Aza PN, Guitian F, De Aza S. Reactivity of wollastonite-tricalcium phosphate bioeutectic ceramic in human parotid saliva. Biomaterials 2000;21:1735–1741.

[11-185] Siriphannon P, Hayashi S, Yasumori A, Okada K. Preparation and sintering of $CaSiO_3$ from coprecipitated powder using NaOH as precipitant and its apatite formation in simulated body fluid solution. J Mater Res 1999;14:529–536.

[11-186] Lui X, Ding C. Plasma sprayed wollastonite/TiO_2 composite coatings on titanium alloys. Biomaterials 2002;23:4065–4077.

[11-187] Fraichinger VM, Schlottig F, Gasser B, Textor M. Anodic plasma-chemical treatment of CP titanium surfaces for biomedical applications. Biomaterials 2004;25:593–606.

[11-188] Haddow DB, Kothari S, James PF, Short RD, Hatton PV, van Noort R. Synthetic implant surfaces 1. The formation and characterization of sol–gel titania films. Biomaterials 1996;17:501–507.

[11-189] Sato M, Slamovich EB, Webster TJ. Enhanced osteoblast adhesion on hydrothermally treated hydroxyapatite/titania/poly(lactide-*co*-glycolide) sol–gel titanium coatings. Biomaterials 2005;26:1349–1347.

[11-190] Wang X-X, Hayakawa S, Tsuru K, Osaka A. Bioactive titania gel layers formed by chemical treatment of Ti substrate with a H_2O_2/HCl solution. Biomaterials 2002;23:1353–1357.

[11-191] Manjubala I, Kumar TSS. Effect of TiO_2-Ag_2O additives on the formation of calcium phosphate based functionally graded bioceramics. Biomaterials 2000;21:1995–2002.

[11-192] Nonami T, Taoda H, Hue NT, Watanabe E, Iseda K, Tazawa M, Fukaya M. Apatite formation on TiO_2 photocatalyst film in a pseudo body solution. Mater Res Bull 1998;33:125–131.

[11-193] Shibata Y, Kawai H, Yamamoto H, Igarashi T, Miyazaki T. Antibacterial titanium plate anodized by being discharged in NaCl solution exhibits cell compatibility. J Dent Res 2004;83:115–119.

[11-194] Wheeler KR, Karagianes MT, Sump KR. Porous titanium alloy for prosthesis attachment. In: Titanium alloys in surgical implants, ASTM Proceedings, Phoenix, AZ, 1983. pp. 241–254.

[11-195] Engelhard G, Zaharias R, Keller JC. Effects of CP Ti surface roughness on osteoblast mineralization. J Dent Res 1995;74:189 (Abstract No. 1424).

[11-196] Petronis S, Gretzer C, Kasemo B, Gold J. Model porous surfaces for systematic studies of material–cell interactions. J Biomed Mater Res 2003;66A:707–721.

[11-197] Xiaoxiong M, Shizhong H. The states of bromide on titanium surface prior to pit initiation. J Chin Soc Corr Protect 1989;9:29–35.

[11-198] Ungersboeck A, Pohler OEM, Perren SM. Evaluation of soft tissue reactions at the interface of titanium limited contact dynamic compression plate implants with different surface treatments: an experimental sheep study. Biomaterials 1996;17:797–806.

[11-199] Larsson C, Thomsen P, Aronsson BO, Rodahl M, Lausmaa J, Kasemo B, Ericson LE. Bone response to surface-modified titanium implants: studies on the early tissue response to machined and electropolished implants with different oxide thickness. Biomaterials 1996;17:605–616.

[11-200] Chung FH, McAlarney ME. Effects of variations in surface treatment on titanium topography. J Dent Res 1995;74:112 (Abstract No. 802).

[11-201] Thelen S, Barthelat F, Brinson LC. Mechanics considerations for microporous titanium as an orthopedic implant material. J Biomed Mater Res 2004;69A:601–610.

[11-202] Oshida Y. Requirements for successful, biofunctional implants. In: International advanced biomaterials. Yahia L'H, editor. Montréal, Canada: Ecole Polytechnique/GRBB. 2000. pp. 5–10.

[11-203] Takemoto M, Fujibayashi S, Neo M, Suzuki J, Kokubo T, Nakamura T. Mechanical properties and osteoconductivity of porous bioactive titanium. Biomaterials 2005;26:6014–6023.

[11-204] Kohn DH, Ducheyne P. A parametric study of the factors affecting the fatigue strength of porous coated Ti-6Al-4V implant alloy. J Biomed Mater Res 1990;24:1483–1501.

[11-205] Energy Research & Generation, Inc. Duocel foam metal: a new basic design material. 1998.

[11-206] Oshida Y, Hashem A, Nishihara T, Yapchulay MV. Fractal dimension analysis of mandibular bones: towards a morphological compatibility of implants. J Biomed Mater Eng 1994;4:397–407.

[11-207] Evans UR. Metallic corrosion passivity and protection. London: Edward Arnold. 1937. pp. 672–681.

[11-208] Oshida Y. Coloration of titanium materials. to be submitted to J Mater Res.

Chapter 12
Future Perspectives

12.1.	Titanium Industry and New Materials Development	385
12.2.	Amorphous Material	388
12.3.	Gradient Functional Material System	389
12.4.	Coating	390
12.5.	Fluoride Treatment	396
12.6.	Surface Texturing and Porous/Foamed Materials	397
12.7.	Laser Applications	398
12.8.	Near-Net Shape (NNS) Forming and Nanotechnology	400
12.9.	Tissue Engineering and Scaffold Structure and Materials	403
12.10.	Bioengineering and Biomaterial-Integrated Implant System	406
References		411

Chapter 12
Future Perspectives

Before we discuss the new and interesting things ahead of us, it is appropriate here to review what happened in the past and what is happening currently. Titanium materials science is a typical interdisciplinary science and engineering discipline. Figure 12-1 illustrates the complicated, yet nicely correlated interrelationship among different disciplines to establish a promising titanium materials science and engineering discipline for assisting not only industry but also health providers, as well as the receiver of such services. In Figure 12-1, in order for engineered titanium materials to serve as titanium biomaterials, we have been discussing and reviewing numerous articles to prove that appropriate surface modifications and characterizations should be properly preformed and reflected to fabrication technologies and methods. Then such titanium biomaterials are really ready to be used in different dental and medical applications. We have reviewed the ever-growing Ti materials research and development for meeting specific aims including V-free alloys, beta-Ti alloys, Ti materials having better properties of fatigue, as well as wear, amorphatizable materials, materials exhibiting better superplastic-formability, macro, micro, and nanoscale structure-controlled materials, etc. These new Ti materials, along with conventional Ti materials, are characterized to evaluate whether they meet specific required characteristics. All these activities, as seen in the figure, are nicely correlated to establish titanium biomaterials, from which various dental and medical applications can be realized. Furthermore, currently and continuously in the future, with tremendous valuable and supportive technologies (including newly developed surface modification, near-net shape forming, better understating of bone healing mechanisms, advanced tissue-engineering materials, and technologies, etc.) implant systems can be further developed to bring benefits to both patients and clinicians.

In the past, when tissues became diseased or damaged, a physician had little recourse but to remove the offending part, with obvious limitations. Removal of joints, vertebrae, teeth, or organs led to only a marginally improved quality of life. However, human survivability seldom exceeded the progressive decrease in quality of tissues, so the need for replacement parts was small. During the last century the situation changed greatly. The discovery of antiseptics, penicillin and other antibiotics, chemical treatment of water supplies, improved hygiene, and vaccination all contributed to a major increase in human survivability in developed countries. Life expectancy is now in the range of 80+ years. In the past, it was the major practice to remove the diseased tissues, whereas at the present, either transplants (using

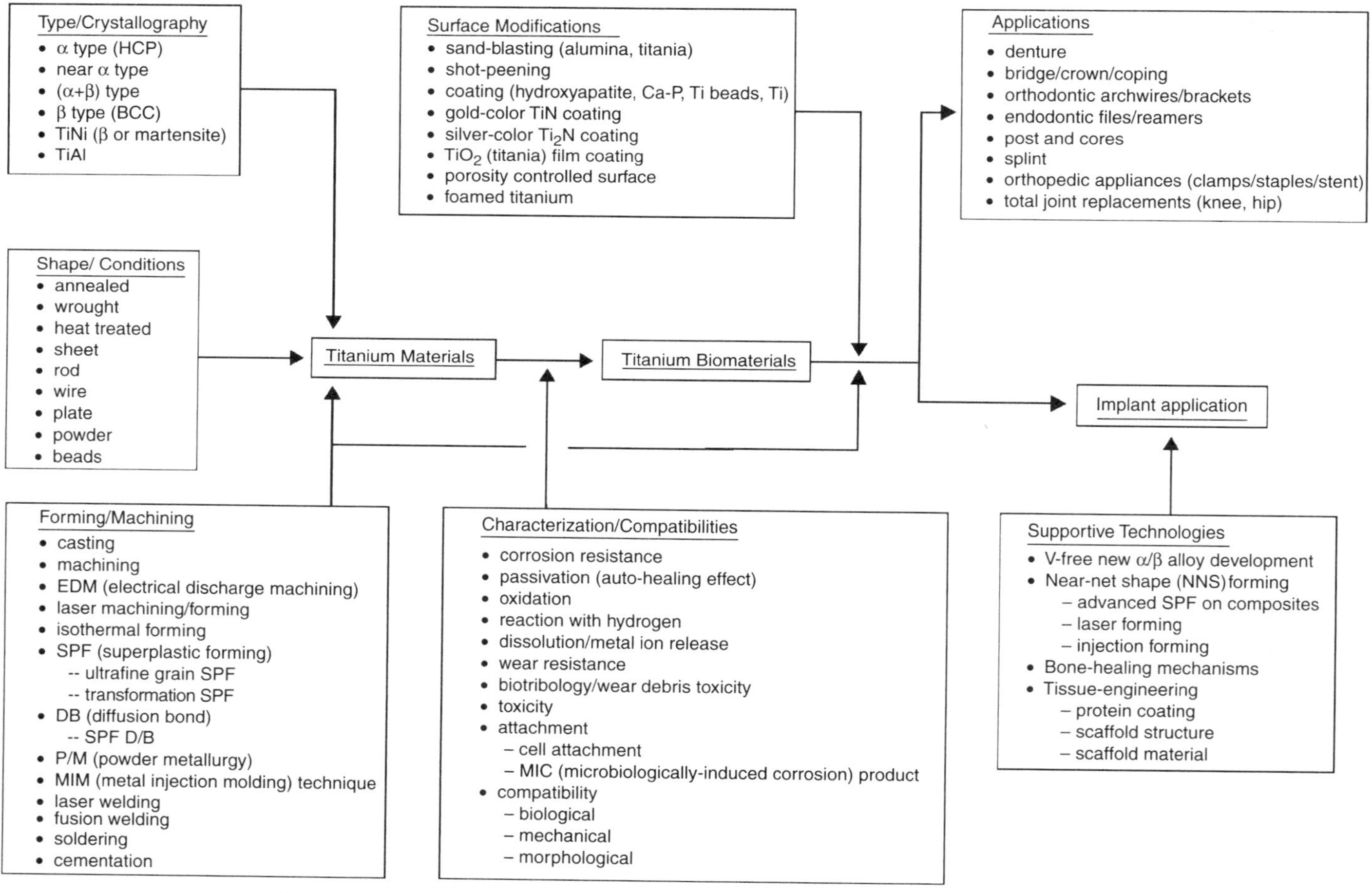

Figure 12-1. Interacting map of various disciplines involved in titanium materials.

autografts, heterografts, or homografts) or implants (using biological fixation, bioactive fixation, or cement fixation) are commonly utilized, and in the future, regeneration of tissues, based on engineered tissues and regenerative bioactive materials will become a major clinical impact. Regeneration of tissues means the following: (1) restoration of structure, (2) restoration of function, (3) restoration of metabolic and biochemical behavior, and (4) restoration of biomechanical performance. Our challenge for the future is to extend these findings to studies in compromised bones with osteopenia and osteoporosis, to apply the findings to larger animals, and especially humans, with aging bones, and to use the findings to design the 3-D architectures required for engineering of tissues [12-1]. On the basis of what we have seen so far, we can see further research and development in materials, technology, engineering, and clinical applications to provide better health services in the coming years, particularly characterized by an ever-aging society.

In this chapter, among many future perspectives associated with medical and dental, as well as industry, Ti biomaterials development and related technologies will be revealed.

12.1. TITANIUM INDUSTRY AND NEW MATERIALS DEVELOPMENT

The titanium industry is rebounding with projections that the worldwide mill product shipments by 2008 will be about 56,000 tons per year – up significantly from the 43,000 tons per year of 2002/2003. This is a result of the increase in commercial airplane production along with increased sales to the military, industrial, and consumer markets. As always, cost remains the major barrier to ore titanium use, especially in the industrial and consumer markets. However, there is light at the end of the tunnel with new reduction processes being developed and lower-cost fabrication processes ready for implementation (such as the powder-metallurgy approaches, including spraying concepts). A better characterization of the behavior of titanium will also lead to potential cost reductions [12-2].

Numerous attempts have been undertaken in the last 60 years to reduce the cost of producing titanium [12-3]. Primary titanium can be produced from a number of feed materials using different reduction reactions. Depending on the process, many physical forms of titanium can be produced. The different feed materials that can be used include slag or rutile (TiO_2) potentially acceptable for producing ferrotitanium alloys are high-purity $TiCl_4$, TiI_4, and Na_2TiF_6. Depending on the process used, the following forms of titanium are produced: titanium sponge, sintered electrode sponge, powder, molten titanium, electroplated titanium, hydride powder, and vapor-phase deposited titanium. The process of electrowinning molten titanium dioxide is conceptually the same as the Hall–Héroult process to produce aluminum from alumina. The oxide is dissolved in a molten salt electrolyte at a sufficiently high temperature so that when it is reduced, molten metal settles out of the electrolyte. The feed is introduced continuously, and the product is recovered periodically from the cell or pot [12-4]. The magnesium reduction reaction of $TiCl_4$-$MgCl_2$ has been shown to produce titanium metal where the $TiCl_4$–$MgCl_2$ was synthesized through the reaction of seed titanium in molten $MgCl_2$ and $TiCl_4$ gas. The macroscopic shape of the titanium product and its microscopic morphology depend on reduction temperature. Titanium metal thus obtained was then button-melted in an arc furnace. It appears that the titanium metal product obtained is suitable for its powder, as well as melted product applications [12-5].

With the advent of high-quality, lower-cost titanium powders, the emphasis in titanium powder metallurgy P/M technology has centered on production of near-net shapes with acceptable levels of mechanical properties. The pre-alloyed, blended elemental, and metal injection molding (MIM) approaches are all looking attractive, and should post-significant growth in the next few years [12-6].

Advantages and disadvantages of beta Ti alloys are listed as follows. (1) While beta alloys offer the highest strength-to-weight ratios of any Ti alloys, they are nonetheless higher in density compared to other alloys. Densities on the order of 4.70–4.98 are typical, with ultimate strength levels in the 1517–1586 MPa (or 1.52–1.59 GPa) range. (2) In the solution-treated condition, values of modulus of elasticity (MOE) from 69 to 76 GPa are typical, and in the aged condition, MOE values of 103 –110 GPa are common. In some cases, such as medical implants, matching the low modulus of bone is important (to achieve the mechanical compatibility). (3) Although beta alloys are expensive for formulate, the strip-rollable alloys can take advantage of low-cost/high-yield cold strip processing vs. hot-band mill operation. (4) Although segregation can be a problem during melting (solidification), the rapid cooling rate obtained in most weldments often prevents transformation and improves weldability. (5) Beta-phase alloys typically exhibit outstanding cold formability. However, their low elastic modulus and high strength can often result in a high springback, which must be accounted for in orthodontic archwires. Because of the high degree of stability of the beta phase, a water-quench is often not required for solution treating; air-cooling greatly reduces distortion, as does low temperature for solution treatment. However, beta alloys are notorious for surface oxygen contamination, even in supposed vacuum heat treatments, so care is needed to remove such contamination; usually the contamination is removed by pickling. (6) While beta alloys usually exhibit excellent fill characteristics during casting, segregation may limit the utility of some alloys. (7) Beta-phase alloys are generally not considered for elevated or cryogenic temperature service. Also, many beta alloys have outstanding corrosion resistance. Generally, corrosion resistance increases with molybdenum content, as the 316 (18Cr-8Ni-2Mo) stainless steel exhibits better resistance against the pitting corrosion than the 304 (18Cr-8Ni) stainless steel. Accordingly, new types of titanium, such as gamma alloys (e.g., 46Ti-48Al alloy with Ti$_3$Al and TiAl phases), were developed [12-7–12-10]. There are still strong areas in developing V-free Ti materials [12-11–12-13].

TiNi is not the only Ti-based alloy exhibiting the shape memory effect. TiPd-Co alloys [12-14] and Ti-V-Fe-Al alloys [12-15] are other examples showing the shape recovery capability. The electrochemical behavior of Ti, Pd, Co, and TiPd alloys was investigated to obtain the fundamental data for evaluating the corrosion resistance of TiPd-Co alloys. The potential-time curve and anodic polarization of the casting specimen were measured in 0.9% NaCl solution at 37°C. Pd showed the highest positive value of the rest potential and Co showed the lowest negative value. The potential of Ti increased with time, and reached a fairly high value of -0.03 V after 5 h. TiPd alloys exhibited a potential between the values of Ti and Pd. In anodic polarization, an abrupt increase in current density at 1.0 V was observed for Pd. From 0.0 to 1.2 V, Ti showed a constant, small amount of current density attributed to passivity. The trans-passivity (passivation breakdown) potential of the TiPd alloys was 0.55 V. On the basis of these findings, it was reported that TiPd-Co new shape memory alloys exhibited good corrosion resistance [12-14]. The application of TiNi for medical and dental implants has been attempted, but the toxicity of Ni is still doubtful. Shape recovery in the Ti-V-Fe-Al alloy was studied. Since this alloy contains more than 80% Ti without Ni, it seems to be biocompatible. Shape recovery in this alloy system occurred in Ti-(10.0–12.0)V-(1.5–3.0)Fe-(2.0–4.5)Al. By controlling the content ratio, shape recovery obtained in Ti-11.5V-1.7Fe-4.0Al was 92% at 60°C. A model implant with this alloy was prepared, and it was reported that an opening angle of 60° in the edge part was memorized. After treatment, an opening angle of 50° was recovered at 60°C. The corrosion resistance of Ti-11V-2Fe-3Al was evaluated by anodic polarization measurement in 1% NaCl solution at 37°C. It was found that stable passivation was maintained up to 2000 mV, while NiTi showed a transpassivity at about 1200 mV. These results suggest that this Ti-V-Fe-Al alloy system has applicability for dental implants [12-15].

Problems such as implant fracture and tarnish of the denture base have been on the rise due to their increased use in clinical applications. One cause of these problems is related to the presence of fluoride ion toothpaste [12-16, 12-17], mouth-rinses, and prophylactic agents, which are used to help prevent dental caries. Takemoto *et al.* [12-18] investigated the correlation between corrosion resistance and surface composition of an experimental Ti-20Cr casting alloy in saline solution fluoride. The alloy had a greater resistance to corrosion in fluoride-containing saline solution than did commercially pure titanium. However, with confirmed dissolution of titanium and chromium, it meant that the fluoride in the saline solution corroded the alloy slightly. It was reported that (i) the [Ti]/([Ti]+[Cr]) ratio in the surface oxide film decreased when immersed in fluoride-containing saline solution, that is, the surface oxide film became Cr-rich oxide, and (ii) the alloy obtained good corrosion resistance to fluoride due to formation of a chromium-rich oxide film [12-18].

He *et al.* [12-19] prepared a group of Ti-60Cu-14Ni-12Sn-4M (M: Nb, Ta, Mo) alloys using arc melting and copper mold casting. The as-prepared alloys have a composite microstructure containing a micrometer-sized dendrite β-Ti(M) phase dispersed in a nanocrystalline matrix. It was reported that (i) alloys exhibit a low MOE in the range of 50–103 GPa, and a high yield strength of 1,037–1,755 MPa, together with large plastic strains, (ii) with addition of refractory elements, Nb, Ta, and Mo, a micrometer-sized dendritic β-phase-nanostructured matrix composite has been achieved, and (iii) the dendritic β-Ti(M) solid solution acts as a reinforced phase stabilizing the deformation of the nanostructure, so as to achieve a very good balance between high strength and large plasticity of the alloys, indicating that new Ti alloys can be promising candidates as biomedical materials [12-19]. In addition to these developments, several Ti-Cu series were investigated, including Ti-10Cu-5Pd [12-20], and Ti-Cu(2–5w/o)-Si(0.2–2w/o) [12-21].

Three new Ti alloys (Ti-20Zr-3Nb-3Ta-0.2Pd-1In, Ti-20Zr-3Nb-3Ta-0.2Pd, Ti-15Zr-3Nb-3Ta-0.2Pd) with Zr, Nb, Ta, Pd, and In as alloying elements were developed and compared with currently used implant metals, namely, CpTi and Ti-6Al-4V, in terms of mechanical and corrosion properties, and cytotoxicity. It was shown that (i) new alloys exhibited comparable mechanical properties with that of Ti-6Al-4V, (ii) there were no significant differences in L-929 cell growth on the surface of the various metal specimens, indicating that the cells cannot differentiate between the passivated surfaces of the various Ti metals, and (iii) after regression analysis, among five alloying elements, Ta was the most effective in improving corrosion resistance, hot workability, and preventing the brittle-phase formation [12-22]. It was reported that Ti-30Zr [12-23] and Ti-30Hf [12-24, 12-25] improved mechanical properties as well as machinability.

12.2. AMORPHOUS MATERIAL

Amorphous materials are non-crystalline solids. For the last three decades, amorphous alloys have attracted great interest because of the results from their new alloy compositions and new atomic configurations. However, amorphous alloys found before 1988 did not have large glass-forming ability and their formation required high-cooling rates above 10^4 K/s for Fe-, Co-, and Ni-based amorphous alloys, and 10^3 K/s for Pd- and Pt-based amorphous alloys. Therefore, the resulting sample thickness for glass formation was usually below 100 μm for the former type amorphous alloys and less than several millimeters for the latter type amorphous alloys.

The critical cooling rates of these amorphous alloys are higher by several orders than those for oxide and fluoride glasses. The finding of an amorphous alloy with a much lower critical cooling rate has strongly been desired for the future development of basic science and applications of amorphous alloys [12-26], including Ti-Cu amorphous alloy [12-27].

TiAl amorphous alloy provides high strength, linear elastic behavior, and the infinite fatigue life necessary for high device reliability. This alloy was originally developed for material for the digital micromirror device (DMD) chip, and actually it is a $TiAl_3$ phase [12-28]. Bulk amorphous Ti-based alloys were found to be formed in the range up to 5 mm in diameter for the Ti-Ni-Cu-Sn and Ti-Ni-Cu-Si-B systems, which possess a high glass-forming ability [12-29]. There are newly reported Ti45-Ni20-Cu25-Sn5-Zr5 [12-30] and Ti50-Cu20-Ni24-Si4-B2 [12-31] that can be amorphatized, too.

12.3. GRADIENT FUNCTIONAL MATERIAL SYSTEM

Material is composed of multilayer, with each layer having unique characteristics, yet adjacent layers having some similarity is called gradient functional material (GFM). Although such functions can include various properties, it is limited to mechanical, physical, or thermal properties since other properties, such as chemical or electrochemical, are more likely important to the surface layer, and not related to bulky or semi-bulky behavior. For example, if hydroxyapatite is needed to spray-coat onto CpTi, this GFM concept can be applied. Instead of applying HA powder directly onto the CpTi surface, a multilayer of $HA/HA+Al_2O_3/Al_2O_3+TiO_2/TiO_2/CpTi$ can be prepared to enhance the bonding strength [12-32]. From the HA side to CpTi side, the mechanical properties (particularly, MOE) and thermal properties (such as linear coefficient of thermal expansion) are gradually changing, so that when this HA-coated CpTi is subjected to stressing, interfacial stress between each constituent layer can be minimized, resulting in that the degree of discreteness in the stress field can also be minimized.

The National Industrial Research Institute of Nagoya, of the Agency of Industrial Science and Technology, has established technology for forming a gradient functional titanium-oxide film on a titanium alloy (Ti-6Al-4V) by the reactive DC sputtering vapor deposition method [12-33]. The method was developed for fabricating denture bases and artificial implants, and the oxygen concentration was changed continuously during sputtering to provide a gradient in the film composition, by which adhesivity to the alloy, surface hardness, and biocompatibility was improved. Denture bases produced by superplastic forming (SPF) are cleaned with an organic solvent, the oxygen concentration is changed continuously during sputtering, and pure titanium is vapor-deposited by the reactive DC sputtering.

In the initial stage, intermetallic bonding is achieved by oxygen-free vapor deposition, so the adhesion is excellent and there is no fear of exfoliation, but farther away from the metal surface, the oxygen concentration is raised gradually to form a gradient film. At the surface some titanium oxides are formed. Titanium oxide features excellent biocompatibility, and since it is a hard material, resists damage. The overall film thickness in the experimental was 3 μm, and the Vickers hardness of the surface was 1500 (200–300 for pure titanium) [12-33].

Bogdanski *et al.* [12-34] fabricated the functionally graded material, which was prepared through powder metallurgical processing with thoroughly mixed powders of the elements. Ten mixtures were prepared ranging from Ni:Ti of 90:10 (by atomic %) through 80:20, to Ni:Ti = 10:90, and pure Ti (0:100). Each mixture was homogenized in a mixer for 24 h in a bottle without addition of grinding balls to keep impurities low. The compaction was done by hot isostatic pressing (HIP) at 1050°C and 195 MPa for 5 h. It was reported that, using cells (comprised of osteoblast-like osteosarcoma cells, primary human osteoblasts, and murine fibroblasts), good biocompatibility of Ni-Ti alloys has shown up to about Ni 50% [12-34].

12.4. COATING

Surface modifications have been applied to metallic biomaterials in order to improve their wear properties, corrosion resistance, and biocompatibility. Methods of applying calcium phosphate-based materials are being actively investigated with the aim of enhancing osteoinduction on titanium materials [12-35–12-37]. This work is necessary because a plasma sprayed calcium phosphate coating has disadvantages, such as the need for a critical thickness to ensure complete coverage of the implant surface [12-38]. Another approach to enhancing osteoinduction is to promote the formation of hydroxyapatite on titanium in the human body. Calcium ion implantation [12-39] and a $CaTiO_3$ coating [12-40] for titanium materials have been examined, and the improvement of biocompatibility with bone was confirmed.

Laser surface engineering (LSE) is advanced enough to modify the surface. A high power laser beam melts a ceramic powder precursor along with a thin layer of the substrate to produce a laser melt zone. High cooling rates produce metastable (or non-crystalline) phases by exceeding the solid solubility limit of the equilibrium-phase diagram. This leads to the possibility of a wide variety of microstructures having novel properties that cannot be produced by any conventional processing technique. The TiB_2 produced is a transition metal-based refractory ceramic that has the unique properties of high hardness, high melting point, wear resistance, corrosion resistance, electrical conductivity, and MOE of 477 GPa [12-41].

The development of a post-operative infection following the implantation, such as a Ti-6Al-4V alloy total joint prosthesis, is a severe complication in many orthopedic surgeries. Preventing these bacterial infections could theoretically be accomplished by administering therapeutic doses of antibiotics as close to the implant site as possible. Mixing antibiotics with polymethylmetaacrylate (PMMA) bone cements has been shown to provide adequate local antibiotic concentrations for extended periods of time [12-42–12-44]. Because metallic materials dominate orthopedic bioprosthetic devices, there exists a definite need for developing methods to attach antibiotics to metallic surfaces. Since the naturally forming passive surface oxide layer of Ti-6Al-4V is thought to be responsible for the excellent biocompatibility and corrosion resistance of this alloy, this oxide layer would be a natural choice for facilitating antibiotic attachment and retainment. By carefully controlling the surface chemistry of the oxide and utilizing the pH dependence of surface charge characteristics of the oxide, the attachment of charged antibiotics may be facilitated at suitable pH values. Such a concept has already been successfully tested with macroporous oxides (1–10 μm pores) formed in sulfuric acid solutions [12-45]. Dunn *et al.* [12-42] also studied the microporous (about 1.5 μm) anodized oxides formed on Ti-6Al-4V alloy to facilitate the attachment and sustained release of antibiotics for longer times. The degree of entamicin sulfate attachment and retainment to microporous oxide layers created on the surface of Ti-6Al-4V materials was determined to be a function of the oxide morphology and surface chemistry. Sulfuric (5–10%) anodized samples were observed to retain the electrostatically attached antibiotic for a period of 13 days when washed in saline at a pH of 7.4. It was found that a longer retention of gentamicin by potentiostatically anodized surfaces in phosphoric acid may be attributed to the lower isoelectric point and more negative zeta potential of these surfaces [12-42].

Similar studies were conducted by Kato *et al.* [12-46, 12-47] to evaluate the applicability of the titanium material as a carrier or a substratum. Spongy titanium adsorbed bone morphogenetic protein (BMP) was implanted in muscle pouches in the thighs of mice. It was found that the quantity of new bone induced was somewhat less than that of the control. In order to form a stable coating in passive state with the adsorption site on the material, small plates of pure titanium and Ti-6Al-4V alloy were anodized in the electrolytic solution of mixed sulfuric acid and phosphoric acid. The sparking voltage was about 160 V on the pure titanium and 150 V on the alloy. It was reported that on the treatment, below the sparking voltage, the film formed by anodic oxidation was thin and optically interferential; above the sparking voltage, the thick film had a hole or worm-eaten spot wad that formed [12-46]. With four electrolytic solutions (3% sulfuric acid, 1.5% phosphoric acid, 2% oxalic acid, and 2% oxalic acid – 3% sulfuric acid mixture), the anodizing of small plates of pure titanium and Ti-6Al-4V alloy were investigated to apply the materials to carriers for BMP. It was reported that (i) when pure titanium was anodized by applying a higher voltage than the sparking voltage in 3% sulfuric acid, an oxide film with many fine craters smaller than 0.5 μm in diameter was formed on the treated surface, while (ii) in 1.5% phosphoric acid, a thick but porous oxide film with open pores smaller than 1 μm in diameter was obtained, and (iii) when Ti-6Al-4V was anodized by applying higher voltage than the sparking voltage, an oxide film with craters about 1 μm was formed in any of the solutions [12-47].

The adsorption of bovine serum albumin (BSA) on titanium powder has been studied as a function of protein concentration and pH, and in the presence of calcium and phosphate ions. Isotherm data have shown that the adsorption process does not follow the Langmuir model (inflection points). For the pH dependence of adsorption, it was found that (i) the amount adsorbed increased with decreasing pH (at pH 5.15, 1.31 mg/m^2 at maximum), indicating that hydration effects are important, and (ii) adsorption increases and decreases in the presence of calcium and phosphate ions, indicating that electrostatic effects are important. The time dependence, isotherm, and desorption data provide indirect evidence of possible conformational changes in the BSA molecule [12-48]. Hence, protein adsorption is a dynamic event with proteins adsorbing and desorbing as a function of time. McAlarney *et al.* [12-49] investigated the role of complement C3 in the competitive adsorption of proteins from diluted human plasma (the Vroman effect) onto TiO$_2$ surfaces. Ti oxide surfaces were made: (1) four anatase surfaces (70, 140, 70 nm aged and solid anatase), (2) three rutile surfaces (70, 140 nm, and solid rutile), and (3) one electropolished Ti. It was found that (i) in both rutile and anatase surfaces, there was an increase in adsorption with increasing oxide film thickness and/or crystallinity, and (ii) anatase surfaces had greater C3 concentration than the equivalent rutile surfaces [12-49].

Titanium dental implants are widely used with success, but their rejection is not rare. One of the causes for implant failure may be due to biofilms created by interactions between the implant material and the surrounding tissues and fluids. The study described the selective adsorption of a specific salivary protein to Ti-oxide and the mechanism of adsorption. Klinger *et al.* [12-50] treated enamel powder, CpTi powder, as well as Ti powder by Ca, Mg, or K, which were suspended *in vitro* in human clarified whole saliva, or in various concentrations of purified salivary constituents, at pH 3.0 and 7.0. The powders were then suspended in EDTA solution in order to release proteins that may have adsorbed to their surfaces. It was found that (i) Ti powders adsorbed considerably less salivary proteins as compared with the enamel powder, (ii) human salivary albumin was identified by Western-immunoblot as the main protein that adsorbed to Ca-treated Ti powder, (iii) the Ca effect was not evident at pH 3.0 due to a neutral-basic shift of the protein at a pH level lower than its isoelectric point, and (iv) the *in vivo* investigation of salivary proteins adsorbing to Ti parts confirmed these results. On the basis of these findings, it was concluded that (i) albumin was shown to be the main salivary protein adsorbing to Ti via a selective calcium and pH-dependent mechanism, and (ii) these findings are important for the understanding of Ti biocompatibility properties, as well as patterns of bacterial dental plaque accumulation on Ti implants, and the consequent implant success [12-50].

A surface reforming technique for improving the biocompatibility of titanium materials and increasing the possibilities for producing artificial bones and artificial tooth roots was developed. A porous ceramic film (or oxide film) was formed by anodizing in an electrolyte containing BMP. The spark discharge will be instantaneous but shifts from one to another, so numerous minute pores with a diameter of about 3 μm will form in the film to permit BMP fixation on the porous substance. BMP is a bone derivative substance contained in animal bones, teeth, and tusks. It generates new bones even inside muscles, so it is highly promising for applications involving the production of artificial bones. It was reported that the yield of BMP is extremely low with 0.001% of bone, but by reforming the titanium surface, bones can be introduced efficiently with an extremely small amount of BMP [12-51].

Coating metallic biomaterials has been extensively investigated in order to improve surface properties. Commonly, calcium phosphate coatings (particularly hydroxyapatite) are dominantly produced by the plasma spraying technique and thick coatings (up to 200 μm) can be easily produced. However, adhesion and mechanical properties of HA coatings are not sufficient to permit a large use of them in the human body. Low energy (in the keV range) ion beam sputtering and derivative methods have been used in order to produce adherent. High-energy ion implantation, being used initially for semiconductor devices, has been used successfully for other applications. It was shown that (i) ion implantation of nitrogen markedly enhances the wear and fatigue resistance of various steels, and (ii) implantation of other light species such as carbon and boron, gives rise to similar effects in steels [12-52]. Hanawa *et al.* [12-53] have modified the surface of titanium by calcium ion implantation and investigated the surface formed film. They have reported the conversion of the TiO_2 surface film of titanium into $CaTiO_3$ and CaO when the dose of implanted calcium ions increases. Biological tests have been performed by Howlett *et al.* [12-54] on Mg, Mn, P, and N ion implanted silicone for dose rate of 10^{16} ions cm^{-2}. They have observed that there is no evidence that ion beam implantation produces increased adhesion of human bone derived cells. Later, Howlett *et al.* [12-55] reported that the attachment and spreading of cultured human bone derived cells into Mg ion-implanted Al_2O_3 is significantly enhanced in comparison to non-implanted alumina. An alternative to the direct implantation of metal ions is the bombardment with noble gas or reactive gases of a thin coating layer in order to produce interdiffusion by the action of radiation-induced defects. The ion energies must be great enough to reach the layer/substrate interface. Typically for ion energies ranging 100–200 keV, the thickness of the coating would not exceed 0.1–0.2 μm for light elements [12-55].

Hayakawa *et al.* [12-56] investigated to attach fibronectin directly to a titanium surface treated with tresyl chloride (2,2,2-trifluoroethanesulfonyl chloride) for the development of a strong connection of a dental implant to subepithelial connective tissues and/or peri-implant epithelia. Basic terminal OH groups of mirror polished titanium were allowed to react with tresyl chloride at 37°C for 2 days. After the reaction of fibronectin with titanium, the X-ray photoelectron spectroscopy revealed the remarkable effect of the activation of terminal OH groups with the tresyl chloride treatment. It was mentioned that fibronectin, a well-known cell-adhesive protein, could easily be attached to the titanium surface by use of the tresyl chloride activation technique [12-56]. Studies in developmental and cell biology have established the fact that responses of cells are influenced to a large degree by morphology and composition of the extracellular matrix. In order to use this basic principle for improving the biological acceptance of implants by modifying the surfaces with components of the extracellular matrix (ECM), Bierbaum *et al.* [12-57, 12-58] modified titanium surfaces with the collagen types I and III in combination with fibronectin. It was reported that (i) increasing the collagen type III amount resulted in a decrease of fibril diameter, while no significant changes in adsorption could be detected, (ii) the amount of fibronectin bound to the heterotypic fibrils depended on fibrillogenesis parameters, such as ionic strength or concentration of phosphate, and varied with the percentage of integrated type III collagen, and (iii) the initial adhesion mechanism of the cells depended on the substrate (titanium, collagen, fibronectin) [12-57, 12-58].

Collagen, as a major constituent of human connective tissues, has been regarded as one of the most important biomaterials. Kim *et al.* [12-59] investigated the fibrillar self-assembly of collagen by incubating acid-dissolved collagen in an ionic-buffered medium at 37°C. It was reported that (i) the degree of assembly was varied with the incubation time and monitored by the turbidity change, (ii) the partially assembled collagen contained fibrils with varying diameters, as well as non-fibrillar aggregates, while the fully assembled collagen showed the complete formation of fibrils with uniform diameters of approximately 100–200 nm with periodic stain patterns within the fibrils, which are typical of native collagen fibers, and (iii) without the assembly, the collagen layer on Ti adversely affected the cell attachment and proliferation [12-59].

A unique surface treatment on Ti was developed by Wang *et al.* [12-60]. Titanium screws and titanium flat sheets were implanted into the epithelial mantle pearl sacs of a fresh water bivalve by replacing the pearls. After 45 days of cultivation, the implant surfaces were deposited with a nacre coating with iridescent luster. The coating could conform, to some extent, to the thread topography of the screw implant, and was about 200–600 μm in thickness. It was found that (i) the coating was composed of a laminated nacreous layer and a transitional non-laminated layer that consisted mainly of vaterite and calcite polymorphs of calcium carbonate, and (ii) the transitional layer was around 2–10 μm thick in the convex and flat region of the implant surface, and could form close contact with titanium surface while the transitional layer was much thicker in the steep concave regions, and could not form close contact with the titanium surface. It was hence concluded that it was possible to fabricate a biologically active and degradable, and mechanically tough and strong nacre coating on titanium dental implants [12-60].

Frosch *et al.* [12-61] evaluated the partial surface replacement of a knee with stem cell-coated titanium implants for a successful treatment of large osteochondral defects. Mesenchymal stem cells (MSCs) were isolated from bone marrow aspirates of adult sheep. Round titanium implants were seeded with autologous MSC and inserted into an osteochondral defect in the medial femoral condyle. As controls, defects received either an uncoated implant or were left untreated. Nine animals with 18 defects were sacrificed after 6 months. It was reported that (i) the quality of regenerated cartilage was assessed by *in situ* hybridization of collagen type II and immunohistochemistry of collagen types I and II, (ii) in 50% of the cases, defects treated with MSC-coated implants showed a complete regeneration of the subchondral bone layer, (iii) a total of 50% of MSC-coated and uncoated implants failed to osseointegrate, and formation of fibrocartilage was observed, (iv) untreated defects, as well as defects treated with uncoated implants, demonstrated incomplete healing of subchondral bone and formation of fibrous cartilage. It was therefore concluded that (i) in a significant number of cases, a partial joint resurfacing of the knee with stem cell-coated titanium implants occurs, and (ii) a slow bone and cartilage regeneration and an incomplete healing in half of the MSC-coated implants are limitations of the presented method [12-61]. The osseointegration of four different kinds of bioactive ceramic-coated Ti screws were compared with uncoated Ti screws by biomechanical and histomorphometric analysis by Lee *et al.* [12-62]. Calcium pyrophosphate, 1:3 patite-wollastonite glass ceramic, 1:1 apatite-wollastonite glass ceramic, and bioactive CaO-SiO_2-B_2O_3 glass ceramic coatings were prepared and coated by the dipping method. Coated and uncoated titanium screws were inserted into the tibia of 18 adult mongrel male dogs for 2, 4, and 8 weeks. It was reported that (i) at 2 and 4 weeks after implantation, the extraction torque of ceramic-coated screws was significantly higher than that of uncoated screws, (ii) at 8 weeks, the extraction torques of calcium pyrophosphate coated and both apatite-wollastonite glass ceramics-coated screws were significantly higher than those of CaO-SiO_2-B_2O_3 glass-coated and uncoated screws, and (iii) the fixation strength was increased by the presence of osteoconductive coating materials, such as calcium pyrophosphate, and apatite-wollastonite glass ceramic, which enabled the achievement of higher fixation strength even as early as 2–8 weeks after the insertion [12-62].

Bigi *et al.* [12-63] performed a fast biomimetic deposition of hydroxyapatite (HA) coatings on Ti-6Al-4V substrates using a slightly supersaturated Ca/P solution, with an ionic composition simpler than that of simulated body fluid (SBF) to fabricate nanocrystalline HA. It was found that (i) soaking in supersaturated Ca/P solution results in the deposition of a uniform coating in a few hours, whereas SBF, or even 1.5 × SBF, requires 14 days to deposit a homogeneous coating on the same substrates, (ii) the coating consists of HA globular aggregates, which exhibit a finer lamellar structure than those deposited from SBF, and (iii) the extent of deposition increases on increasing the immersion time [12-63].

Although some types of TiO_2 powders and gel-derived films can exhibit bioactivity, plasma-sprayed TiO_2 coatings are always bioinert, thereby hampering wider applications in bone implants. Liu *et al.* [12-64] produced a bioactive nanostructured TiO_2 surface with grain size smaller than 50 nm using nanoparticle plasma spraying followed by hydrogen plasma immersion ion implantation (PI^3). It was reported that (i) the hydrogen PI^3 nanoTiO$_2$ coating can induce bone-like apatite formation on its surface after immersion in a SBF, but (ii) apatite cannot form on either the as-sprayed TiO_2 surfaces (both <50 nm grain size and >50 nm grain size) or hydrogen-implanted TiO_2 with grain size larger than 50 nm, and (iii) introduction of surface bioactivity to plasma-sprayed TiO_2 coatings, which are generally recognized to have excellent biocompatibility and corrosion resistance, as well as high bonding to titanium alloys, makes them more superior than many current biomedical coatings [12-64].

12.5. FLUORIDE TREATMENT

It is known in the literature that fluoride ions have osteopromoting capacity leading to increased calcification of the bone. Titanium fluoride is reported to form a stable layer on enamel surfaces consisting of titanium, which share the oxygen atoms of phosphate on the surface of HA. Ellingsen *et al.* [12-65] investigated as to whether a similar, or rather reverse, reaction would take place on fluoride pre-treated titanium after implantation in bone. Threaded TiO_2-blasted titanium implants were pre-conditioned with fluoride. The implants were operated into the tibia of Chinchilla rabbits and let to heal for 2 months before sacrificing the animals. The strength of the bonding between the implants and bones was tested by removing the implants from the bones by the use of an electronic removal torque gauge. It was reported that (i) the fluoride conditioned titanium implants had a significantly increased retention in bone (69.5 N cm) compared to non-treated blasted implants (56.0 N cm) and smooth surface implants (17.2 N cm), and (ii) the histological evaluation revealed that new bones formed on the surface of the test implants, as well as in the marrow or cancellous regions, which was not observed in the control groups, suggesting that fluoride conditioning of titanium has an osteopromoting effect after implantation [12-65]. Furthermore, push-out tests of fluoridated and control Ti implants placed in rabbits for up to 8 weeks were conducted [12-66]. It was found that the fluoridated implants sustained greater push-out forces than the controls, and substantial bone adhesion was observed in fluoridated implants, whereas the controls always failed at the interface between the bone and foreign materials. In the other rabbit test by Ellingsen *et al.* [12-67], it was reported that the fluoridated, blasted implants showed a significantly higher removal torque than the blasted test implant, again indicative of a bioactive reaction of fluoridated Ti implants. Johansson *et al.* [12-68] also reported significantly greater bone contact to fluoride-modified titanium implants at 1 and 3 months of follow-up, despite the fact that the blasted controls moderately roughened.

12.6. SURFACE TEXTURING AND POROUS/FOAMED MATERIALS

A process was developed for texturing the surfaces of shaped metal parts and included a sputter-etching step with masking the unwanted area with small ceramic particles. Through the choice of sputtering parameters and the size and distribution of masking particles, the width and depth of the textured features can be controlled over a wide range. Texturing improves bonding to other parts. The process has been used to texture metal fixtures to be anchored to composite-material (matrix/fiber) structural components. It could also be used to texture such prosthetic items as hip implants; textures could be sized and shaped to favor the ingrowth of adjacent bone to anchor the implants. The process has the following sequence: smooth surface to be textured → coated with adhesive → ceramic microspheres pressed into adhesive → adhesive charred by heating → sputter-etching in argon plasma → removal of char in plasma asher → ceramic particles brushed off → textured part in use as a bone implant [12-69, 12-70].

As osteoinductive biomaterials contain calcium and phosphorous, the important factors for osteoinduction are thought to be (i) the chemical composition of the biomaterial, (ii) the specific dissolution properties of the biomaterial, and (iii) the surface morphology of the biomaterial acting as a bone substitute should possess osteoconductive and osteoinductive ability, and it should have superior mechanical properties. Four types of Ti implants were prepared [12-71]. Bioactive Ti was prepared by chemical treatment (immersion in 5M aqueous NaOH solution at 60°C for 24 h), followed by a hot water treatment (at 40°C for 48 h), followed by a thermal treatment (heated at 600°C for 1 h, followed by furnace cooling). These materials were implanted as porous blocks. The Ti fiber mesh cylinders were used. The porous blocks were manufactured as follows. A macroporous Ti layer was formed on the Ti substrate by plasma spraying commercial pure Ti powder with a particle size of 50–200 μm or titanium powder, which can be prepared by sintering to control granular diameter size in a range from 420 to 500 μm [12-72]. Blocks were cut from the porous layer using an electric discharge. The tensile strength and the bending strength of porous Ti were 80.0 and 91.9 MPa. The Ti fiber mesh implants were manufactured by compacting a single 250 μm fiber into a die to a porosity of 50%, followed by vacuum sintering. It was shown that even a non-soluble metal that contains no calcium or phosphorous can be an osteoinductive material when treated to form an appropriate macrostructure and microstructure. This finding may elucidate the nature of osteoinduction, and lead to the advent of epochal osteoinductive biomaterials for tissue regeneration [12-71].

12.7. LASER APPLICATIONS

As mentioned before [12-41], laser technology and laser application have advanced remarkably. Lasers can be made to heat, melt, or vaporize materials, depending on laser power density. Materials absorb power more readily from Nd:Yag laser beams (λ=1.06 μm) than they do from CO_2 laser beams (λ=10.6 μm). By heating materials, materials can be annealed, or solid-state phase-transformation hardened. Using melting, materials can be alloyed, cladded, grain refined, amorphatized, and welded. Using vaporization, materials can be thin film deposited, cleaned, textured, and etched. Using shocking, materials can be shock hardened/peened [12-73]. Recent advances in the Nd:Yag and CO_2 lasers have made possible a wide range of applications and emerging technologies in the computer, microelectronics, and materials fields. In the realm of materials processing, surface treatments and surface processing, surface treatments and surface modification for metals and semiconductors are of particular interest. With appropriate manipulation of the processing conditions (e.g., laser power density or interaction time), a single laser can be used to perform several processes [12-74, 12-75].

Laser alloying is a material-processing method, which utilizes the high power density available from focused laser sources to melt metal coatings and a portion of the underlying substrate. Since the melting occurs in a very short time and only at the surface, the bulk of the material remains cool, thus serving as an infinite heat sink. Large temperature gradients exist across the boundary between the melted surface region and the underlying solid substrate. This results in rapid self-quenching (10^{11} k/s) and resolidification (velocities of 20 m/s). What makes laser surface alloying both attractive and interesting is the wide variety of chemical and microstructural states that can be retained because of the repaid quench from the liquid phase. The types of observed microstructures include extended solid solutions, metastable crystalline phases and metallic glasses as an amorphous metal [12-76, 12-77]. Alloy production with a wide variety of elements, as well as a wide range of compositional content, can also be accomplished by a mechanical alloying or powder metallurgy, both of which do not involve liquids.

The potential advantages of laser welding are the following: (1) light is inertialess, hence processing speeds with very rapid stopping and starting become possible, (2) focused laser light can have high energy density, (3) welding can be achieved at room temperature, (4) difficult materials (e.g., Ti, quartz, etc.) can be handled, (5) the workpiece need not be rigidly held, (6) no electrode or filler materials are required, (7) narrow welds can be made, (8) very accurate welds are possible, (9) welds with little or no contamination can be produced, (10) the heat affected zone (HAZ) adjacent to the cut or weld is very narrow, (11) intricate shapes can be cut or welded at high speed using automatically controlled light deflection techniques, and (12) time sharing of the laser beam can be achieved [12-78]. The development of high power lasers has made possible a variety of material removal techniques. Straight-line cutting and hole drilling can be done with lasers. Laser turning and milling can be also performed. Two approaches are considered: laser-assisted machining, in which a laser beam heats materials by a single point cutting tool, and laser machining, in which the laser forms a groove in the material by vaporization [12-79].

Direct laser forming (DLF) is a rapid prototyping technique, which enables prompt modeling of metal parts with high bulk density on the base of individual three-dimensional data, including computer tomography models of anatomical structures. Hollander *et al.* [12-80] investigated DLFed Ti-6Al-4V for its applicability as hard tissue biomaterial. It was reported that rotating bending tests revealed that the fatigue profile of post-DLF annealed Ti-6Al-4V was comparable to cast/hot isostatic pressed alloy. In an *in vitro* investigation, human osteoblasts were cultured on non-porous and porous-blasted DLFed Ti-6Al-4V specimens to study morphology, vitality, proliferation and differentiation of the cells. It was reported that (i) the cells spread and proliferated on DLFed Ti-6Al-4V over a culture time of 14 days, (ii) on porous specimens, osteoblasts grew along the rims of the pores and formed circle-shaped structures, as visualized by live/dead staining, as well as scanning electron microscopy, and (iii) overall, the DLFed Ti-6Al-4V approach proved to be efficient, and could be further advanced in the field of hard tissue biomaterials [12-80].

Recently, the femtosecond-laser-based tooth preparation technique has been developed [12-81, 12-82]. Any one of the existing laser technologies using a CO_2 laser, Er:Yag laser, Ho:Yag laser, excimer laser, frequency-doubled Alexandrite laser, superpulsed CO_2 laser, or picosecond Nd:Yag laser induce severe thermal adverse effects or do not supply sufficient ablation rates for completion of the mechanical drill [12-83]. Using the femtosecond laser technology for micromachining was successfully developed, for example, in machining tools for repairing photolithographic masks or fuel injector nozzles [12-84].

12.8. NEAR-NET SHAPE (NNS) FORMING AND NANOTECHNOLOGY

In order to achieve the better condition for fit, for example, superstructure for implants or denture base for prostheses, the accuracy of the final products are very crucial and need to be improved. Among various manufacturing technologies, NNS forming and nanotechnology are most promising and supportive for these specific aims. New classes of fabrication processes, such as direct-write (DW) technique [12-85] and solid freedom fabrication (SFF) [12-86], have established the capability to produce 3-dimensional parts faster, cheaper, and with added functionality. The choice of starting materials and the specific processing technique will produce unique microstructures that impact the final performance, especially of macroscopic structural and electronic parts. Furthermore, the ability to do point-wise deposition of one or more materials provides the opportunity for fabricating structures with novel microstructural and macrostructural features, such as microengineered porosity, graded interfaces, and complex multimaterial constructions. NNS processing offers cost reduction by minimizing machining, reducing part count, and avoiding part distortion from welding. NSS technologies such as flow-forming (FF), casting, forging, P/M methods, three-dimensional laser deposition, and plasma arc deposition (PAD) have been explored for potential use in tubular geometries and other shapes of varying complexity. It appears that the reduction in cost of a given titanium product will be maximized by achieving improvements in all of the manufacturing steps, from extraction to finishing [12-87].

Taylor [12-88] pointed out that nanotechnology offers the key to faster and remote diagnostic techniques – including new high throughput diagnostics, multiparameter, tunable diagnostic techniques, and biochips for a variety of assays. It also enables the development of tissue-engineered medical products and artificial organs, such as heart valves, veins and arteries, liver, and skin. These can be grown from the individual's own tissues as stem cells on a 3-D scaffold, 3-D tissue engineering ECM, or the expansion of other cell types on a suitable substrate. The applications which seem likely to be most immediately in place are external tissue grafts; dental and bone replacements; protein and gene analysis; internal tissue implants; and nanotechnology applications within *in vivo* testing devices and various other medical devices. Nanotechnology is applied in a variety of ways across this wide range of products. Artificial organs will demand nanoengineering to affect the chemical functionality presented at a membrane or artificial surface, and thus avoid rejection by the host. There has been much speculation and publicity about more futuristic developments such as nanorobot therapeutics, but these do not seem likely within our time horizon [12-88].

There are several studied done on nanotubes. Frosch *et al.* [12-89] investigated the effect of different diameters of cylindrical titanium channels on human osteoblasts. Titanium samples having continuous drill channels with diameters of 300, 400, 500, 600, and 1000 μm were put into osteoblast cell cultures that were isolated from 12 adult human trauma patients. It was reported that (i) within 20 days, cells grew an average of 838 μm into the drill channels with a diameter of 600 μm, and were significantly faster than in all other channels, (ii) cells produced significantly more osteocalcin messenger RNA (mRNA) in 600 μm channels than they did in 1000 μm channels, and demonstrated the highest osteogenic differentiation, (iii) the channel diameter did not influence collagen type I production, and (iv) the highest cell density was found in 300 μm channels, suggesting that the diameter of cylindrical titanium channels has a significant effect on migration, gene expression, and mineralization of human osteoblasts [12-89]. Macak *et al.* [12-90] reported on the fabrication of self-organized porous oxide-nanotube layers on the biomedical titanium alloys Ti-6Al-7Nb and Ti-6Al-4V by a simple electrochemical treatment. These two-phase alloys were anodized in 1 M $(NH_4)_2SO_4$ electrolytes containing 0.5 wt% of NH_4F. It was shown that (i) under specific anodization conditions, self-organized porous oxide structures can be grown on the alloy surface, (ii) SEM images revealed that the porous layers consist of arrays of single nanotubes with a diameter of 100 nm and a spacing of 150 nm, (iii) for the V-containing alloy, enhanced etching of the α-phase is observed, leading to selective dissolution and an inhomogeneous pore formation, and (iv) for the Nb-containing alloy an almost ideal coverage of both phases is obtained. According to XPS measurements, the tubes are a mixed oxide with an almost stoichiometric oxide composition, and can be grown to thicknesses of several hundreds of nanometers, suggesting that a simple surface treatment for Ti alloys has high potential for biomedical applications [12-90]. A vertically aligned nanotube array of titanium oxide was fabricated on the surface of titanium substrate by anodization. The nanotubes were then treated with NaOH solution to make them bioactive, and to induce growth of HA (bone-like calcium phosphate) in a SBF. It is found that (i) the presence of TiO_2 nanotubes induces the growth of a "nano-inspired nanostructure", i.e., extremely fine-scale (~8 nm feature) nanofibers of bioactive sodium titanate structure on the top edge of the ~15 nm thick nanotube wall, (ii) during the subsequent *in vitro* immersion in a SBF, the nanoscale sodium titanate, in turn, induced the nucleation and growth nanodimensioned HA phase, and (iii) such TiO_2 nanotube arrays and associated nanostructures can be useful as a well-adhered bioactive surface layer on Ti implant metals for orthopedic and dental implants, as well as for photocatalysts and other sensor applications [12-91].

Three types of bioactive polymethylmethacrylate (PMMA)-based bone cement containing nanosized titania (TiO_2) particles were prepared, and their mechanical properties and osteoconductivity are evaluated by Goto *et al.* [12-92]. The three types of bioactive bone cement were un-silanized TiO_2, 50 wt%, silanized TiO_2 50, and 60 wt% mixed to PMMA. Commercially available PMMA cement was used as a control. The cements were inserted into rat tibiae and allowed to solidify *in situ*. After 6 and 12 weeks, tibiae were removed for evaluation of osteoconductivity. It was reported that (i) bone cements using silanized TiO_2 were directly apposed to bone, while un-silanized TiO_2 cement and PMMA control were not, (ii) the osteoconduction of cement with 60 wt% of silanized TiO_2 was significantly better than that of the other cements at each time interval, and (iii) the compressive strength of cement with 60 wt% of silanized TiO_2 was equivalent to that of PMMA, indicating that cement with 60 wt% of TiO_2 was a promising material for use as a bone substitute [12-92]. Since it is essential for the gap between the HA-coated titanium and juxtaposed bone to be filled out with regenerated bone, promoting the functions of bone-forming cells is desired. In order to improve orthopedic implant performance, Sato *et al.* [12-93] synthesized nanocrystalline HA powders to coat titanium through a wet chemical process. The precipitated powders were either sintered at 1100°C for 1 h in order to produce microcrystalline size HA, or were treated hydrothermally at 200°C for 20 h to produce nanocrystalline HA. These powders were then deposited onto titanium by a room temperature process. It was reported that (i) the chemical and physical properties of the original HA powders were retained when coated on titanium by the room temperature process, (ii) osteoblast adhesion increased on the nanocrystalline HA coatings compared to traditionally used plasma-sprayed HA coatings, (iii) greater amounts of calcium deposition by osteoblasts cultured on Y-doped nanocrystalline HA coatings were observed [12-93].

With a wide variety of applications, nanotechnology has attracted the attention of researchers as well as regulators and industrialists, including nanodrugs and drug delivery, prostheses and implants, and diagnostics and screening technologies. We can take advantages of availability of ultra-fine nanomicrostructures of solid metals, alloys, powder, fibers, or ceramics to fabricate superplastically formed products.

It is indispensable here to mention the minimally invasive dentistry (MID) and minimally invasive surgery (MIS). The MIS concept has been created to allow new thinking and a new approach to dentistry where restoration of a tooth becomes the last treatment decision rather than first consideration as at present. It provides a practical approach to carry preventive measures based on the notion of demineralization and remineralization in a microphase in order to retain healthy teeth. The medical model of MID is characterized by (1) reduction in cariogenic bacteria, (2) preventive measures, (3) remineralization of early enamel lesions, (4) minimum surgical intervention of cavitated lesions, and (5) repair of defective restorations [12-94]. At the same time, it is mentioned that MIS has several advantages: (1) since the surgical area is narrower, damage on surrounding soft tissue can be minimized, (2) post-operation pain can be minimized, (3) hospital time can be shortened, and (4) early rehabilitation can be initiated [12-95]. These MIs in both dentistry and medicine inevitably require precisely manufactured prostheses in microscale or even nanoscale. It is anticipated that the MI-based technologies, as well as MI-oriented technologies, will be advanced in the near future.

12.9. TISSUE ENGINEERING AND SCAFFOLD STRUCTURE AND MATERIALS

Tissue engineering can perhaps be best defined as the use of a combination of cells, engineering materials, and suitable biochemical factors to improve or replace biological functions in an effort to effect the advancement of medicine, indicating an interdisciplinary field that applies the principles of engineering and life sciences toward the development of biological substitutes that restore, maintain, or improve tissue function for understanding the principles of tissue growth, and applying this to produce functional replacement tissue for clinical use. The term "regenerative medicine" is often used synonymously with "tissue engineering", although those involved in regenerative medicine place more emphasis on the use of stem cells to produce tissues [12-96]. Tissue engineering *in vitro* and *in vivo* involves the interaction of cells with a material surface. The nature of the surface can directly influence cellular response, ultimately affecting the rate and quality of new tissue formation. Initial events at the surface include the orientated adsorption of molecules from the surrounding fluid, creating a conditioned interface to which the cell responds. The gross morphology, as well as the microtopography and chemistry of the surface, determine which molecules can adsorb and how cells will attach and align themselves. The local attachments made of the cells with their substrate determine cell shape, which, when transduced via the cytoskeleton to the nucleus, result in expression of specific phenotypes. Osteoblasts and chondrocytes are sensitive to subtle differences in surface roughness and surface chemistry. Boyan *et al.* [12-97] investigated the chondrocyte response to TiO_2 of differing crystallinities, and showed that cells can discriminate between surfaces at this level as well. Cellular response also depends on the local environmental and state of maturation of the responding cells. It was mentioned that optimizing surface structure for site-specific tissue engineering is one option; modifying surfaces with biological means is another biological engineering [12-97].

One major determination of the suitability of various engineering materials for use in biological settings is the relative strength of adhesion obtained between those materials and their contacting viable phases. Maximal adhesive strength and immobility are desired for orthopedic and dental implants. For example, while minimal bio-adhesion is critical to preventing unwanted thrombus formation in cardiovascular devices, plaque buildup on dental prostheses, and bacterial fouling [12-98]. Attention should be directed to adhesive phenomena in the oral environment, examining new surface conditioning methods for the prevention of microorganism deposits, as well as the promotion of excellent tissue bonding to implanted prosthetic devices. Other bio-adhesive phenomena considered included those important to the safe and effective function of new cardiovascular devices.

Scaffold material has a two-fold function: artificial extracellular matrices (ECM) and as a spacer keeping a certain open space. Furthermore, scaffold material has to be dissolved completely into the living body after auto-cell is regenerated with artificial ECM [12-99]. There are several important biodegradable and/or biofunctional scaffold architectures, structures, and materials. They include blended-polymer scaffolds, collagen-based scaffolds, and composite scaffolds of polyhydroxybutyrate-polyhydroxyvalerate with bioactive wollastonite ($CaSiO_3$) [12-100]. Using an ink-injection technique [12-101], a thin film (with thickness of about 0.1 mm) of calcium phosphate and binding agent is injected onto the substrate to build 3-D bony-like structures [12-102]. Lee *et al.* [12-103] employed three-dimensional printing (3DP) technology to fabricate porous scaffolds by inkjet printing liquid binder droplets. Direct 3DP, where the final scaffold materials are utilized during the actual 3DP process, imposes several limitations on the final scaffold structure. An indirect 3DP protocol was developed, where molds are printed and the final materials are cast into the mold cavity to overcome the limitations of the direct technique. Results of SEM observations showed highly open, well interconnected, uniform pore architecture (about 100–150 μm) [12-103]. Scaffold materials for bone tissue engineering often are supplemented with BMPs. Walboomers *et al.* [12-104] investigated a bovine ECM product containing native BMPs. Hollow cylindrical implants were made from titanium fiber mesh, and were implanted subcutaneously into the back of Wistar rats. It was reported that (i) after 8 weeks, in two out of six loaded specimens, newly formed bone and bone marrow-like tissues could be observed, and (ii) after 12 weeks, this had increased to five out of six loaded samples. It was therefore concluded that the bovine ECM product loaded in a titanium fiber mesh tube showed bone-inducing properties [12-105].

Electrospinning [12-105] has recently emerged as a leading technique for generating biomimetic scaffolds made of synthetic and natural polymers for tissue engineering applications. Li *et al.* [12-106] compared collagen, gelatin (denatured collagen), solubilized alpha-elastin, and recombinant human tropoelastin as biopolymeric materials for fabricating tissue-engineered scaffolds by electrospinning. It was reported that (i) the average diameter of gelatin and collagen fibers could be scaled down to 200–500 nm without any beads, while the alpha-elastin and tropoelastin fibers were several microns in width, and (ii) cell culture studies confirmed that the electrospun engineered protein scaffolds support attachment and growth of human embryonic palatal mesenchymal cells [12-106]. For fabricating meshes of collagen and/or elastin by means of electrospinning from aqueous solutions, Buttafoco *et al.* [12-107] added polyethylene oxide and NaCl to spin continuous and homogeneous fibers. It was reported that (i) upon crosslinking, polyethylene oxide and NaCl were fully leached out, and (ii) smooth muscle cells grew as a confluent layer on top of the crosslinked meshes after 14 days of culture.

Surface properties of scaffolds play an important role in cell adhesion and growth. Biodegradable poly(α-hydroxy acids) have been widely used as scaffolding materials for tissue engineering; however, the lack of functional groups is a limitation. Liu *et al.* [12-108] mentioned in their studies that gelatin was successfully immobilized onto the surface of poly(α-hydroxy acids) films and porous scaffolds by an entrapment process. It was found that (i) the amount of entrapped gelatin increased with the ratio of dioxane/water in the solvent mixture used, (ii) chemical crosslinking after physical entrapment considerably increased the amount of retained gelatin on the surface of poly(α-hydroxy acids), (iii) osteoblasts were cultured on these films and scaffolds, (iv) cell numbers on the surface-modified films and scaffolds were significantly higher than those on controls 4 h and 1 day after cell seeding, (v) the osteoblasts showed higher proliferation on surface-modified scaffolds than on the control during 4 weeks of *in vitro* cultivation, and (vi) more collagen fibers and other cell secretions were deposited on the surface-modified scaffolds than on the control scaffolds [12-108].

There are still unique scaffold systems developed, such as the collagen-carbon nanotubes composite matrices [12-109], chitosan-based hyaluronan hybrid polymer fibers system [12-110], bioactive porous $CaSiO_3$ scaffold structure [12-111], or a three-dimensional porous scaffold composed of biodegradable polyesters [12-112].

12.10. BIOENGINEERING AND BIOMATERIAL-INTEGRATED IMPLANT SYSTEM

With the aforementioned supportive technologies, surfaces of dental and orthopedic implants have been remarkably advanced. These applications can include not only ordinal implant system but also miniaturized implants, as well as customized implants. Dental implant therapy has been one of the most significant advances in dentistry in the past 25 years. The computer and medical worlds are both working hard to develop smaller and smaller components. Using a precise, controlled, minimally invasive surgical (MIS) technique, the mini dental implants (MDI) are placed into the jawbone. The heads of the implants protrude from the gum tissue and provide a strong, solid foundation for securing the dentures. It is a one-step procedure that involves minimally invasive surgery, no sutures, and none of the typical months of healing.

Advantages associated with the MDI are (1) It can provide immediate stabilization of a dental prosthetic appliance after a minimally invasive procedure. (2) It can be used in cases where traditional implants are impractical, or when a different type of anchorage system is needed. (3) Healing time required for mini-implant placement is typically shorter than that associated with conventional 2-stage implant placement and the accompanying aggressive surgical procedure. According the clinical reports, a biometric analysis of 1029 MDI mini-implants, 5 months to 8 years *in vivo* showed that the MDI mini-implant system can be implemented for long-term prosthesis stabilization, and delivers a consistent level of implant success [12-113].

In addition to the aforementioned miniature implants, an immediate loading, as well as customized implants, have been receiving attention recently. Conventionally, a dental implant patient is required to have two-stages of treatment consisting of two dental appointments 5 to 6 months apart. Recently, a single-stage treatment has received attention. Placing an implant immediately or shortly after tooth extraction offers several advantages for the patient as well as for the clinician. These advantages include shorter treatment time, less bone sorption, fewer surgical sessions, easier definition of the implant position, and better opportunities for osseointegration because of the healing potential of the fresh extraction site [12-114–12-117]. Titanium bar (particularly the portions in direct contact to connective tissue and bony tissue) is machined to have the exact shape of the root portion of the extracted tooth of the patient. The expected outcome of this method is a perfect mechanical retention, and therefore an ideal osseointegration can be achieved. This is called a custom (or customized) implant, which is fabricated by the electro-discharge machining (EDM) technique. In addition to the immediate placement of dental implants, another concept has been introduced. Techniques such as stereoscopic lithography and computer-assisted design and manufacture

(CAD/CAM) have been successfully used with computer-numerized control milling to manufacture customized titanium implants for single-stage reconstruction of the maxilla, hemimandible, and dentition without the use of composite flap over after the removal of tumors [12-118]. Nishimura *et al.* [12-119] applied this concept to dental implants to fabricate the individual and splinted customized abutments for all restoration of implants in partially edentulous patients. It was claimed that complicated clinical problems such as angulation, alignment, and position can be solved. However, with this technique, the peri-implant soft tissues are allowed to heal 2–3 weeks, so that at least two dental appointments are required.

Many oral implant companies (about 20 companies are currently marketing 100 different dental implant systems) have recently launched new products with claimed unique, and sometimes bioactive surfaces. The focus has shifted from surface roughness to surface chemistry and a combination of chemical manipulations on the porous structure. To properly explain the claims for new surfaces, it is essential to summarize current opinions on bone anchorage, with emphasis on the potentials for biochemical bonding. There were two ways of implant anchorage or retention: mechanical and bioactive [12-120, 12-121]. Mechanical retention basically refers to the metallic substrate systems such as titanium materials. The retention is based on undercut forms such as vents, slots, dimples, screws, and so on, and involves direct contact between bone and implant with no chemical bonding. The osseointegration depends on biomechanical bonding. The potentially negative aspect with biomechanical bonding is that it is time consuming. Bioactive retention is achieved with bioactive materials such as HA or bioglass, which bond directly to bone, similar to ankylosis of natural teeth. The bioactivity is the characteristic of an important material, which allows it to form a bond with living tissues. It is important to understand that bioactive implants may, in addition to chemical bonding, show biomechanical anchorage, hence a given implant may be anchored through both mechanisms. Bone matrix is deposited on the HA layer because of some type of physiochemical interaction between bone collagen and the HA crystals of the implant [12-122].

Recent research has further redefined the retention means of dental implants into the terminology of osseointegration versus biointegration. When examining the interface at a higher magnification level, Sundgren *et al.* [12-123] showed that unimplanted Ti surfaces have a surface oxide (TiO_2) with a thickness of about 35 nm. During an implantation period of 8 years, the thickness of this layer was reported to increase by a factor of 10. Furthermore, calcium, phosphorous, and carbon were identified as components of the oxide layer, with the phosphorous strongly bound to oxygen, indicating the presence of phosphorous groups in the metal oxide layer. Many retrospective studies on retrieved implants, as well as clinical reports, confirm the aforementioned important evidence (1) surface titanium oxide film grows during the implantation period, and (2) calcium, phosphorous, carbon, hydroxyl ions, proteins, etc. are incorporated in an ever-growing surface oxide even inside the human biological environments [12-124, 12-125]. Numerous *in vitro* studies on treated or untreated titanium surfaces were covered and to some extent were incorporated with Ca and P ions when such surfaces were immersed in SBF. Additionally, we know that bone and blood cells are rugophilia, hence in order not only to accommodate for the bone growth, but also to facilitate such cells adhesion and spreading, titanium surfaces need to be textured to accomplish and show appropriate roughness. Furthermore, gradient functional concept (GFC) on materials and structures has been receiving special attention not only in industrial applications, but in dental as well as medical fields. Particularly, when such structures and concepts are about to be applied to implants, its importance becomes more clinically crucial. For example, the majority of implant mass should be strong and tough, so that occlusal force can be smoothly transferred from the placed implant to the receiving hard tissue. However, the surface needs to be engineered to exhibit some extent of roughness. From such macrostructural changes from bulk core to the porous case, again the structural integrity should be maintained. The GFC can also be applied for the purpose of having a chemical (compositional) gradient. Ca-, P-enrichment is not needed in the interior materials of the implants. Some other modifications related to chemical dressing or conditioning can also be utilized for achieving gradient functionality on chemical alternations on surfaces as well as near-surface zones.

Summarizing and ending this book, the author is proposing an ideal implant structure, which is integrated by bioengineering and biomaterials science. Oshida previously proposed the four important factors and requirements for successful and biofunctional implant systems: biological compatibility (or, biocompatibility) mechanical compatibility (or mechanocompatibility), morphological compatibility, and crystallographic compatibility (or micromorphological compatibility) [12-126].

Figure 12-2 illustrates a schematic and conceptual Ti implant which possesses a gradual function of mechanical and biological behaviors, so that mechanical compatibility and biological compatibility can be realized. Since microtextured Ti surfaces [7-71, 7-73, 7-74] and/or porous Ti surfaces [8-94–8-96] promote fibroblast apposition and bone ingrowth, the extreme left side representing the solid Ti implant body should have gradually increased internal porosities toward the right side which is in contact with vital hard/soft tissue. Accordingly, mechanical strength of this implant system decreases gradually from left to right, whereas biological activity increases from left to right. Therefore, the mechanical compatibility can be completely achieved, and it can be easily understood by referring to Figure 6-1. Porosity-controlled surface zones can be fabricated by an

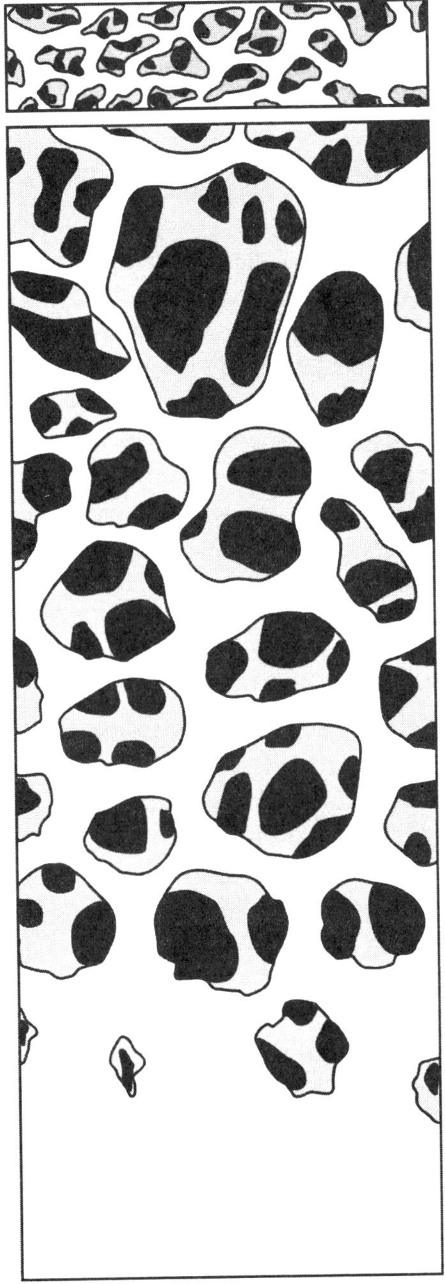

Figure 12-2. Schematic and conceptual Ti implant with gradient mechanical and biological functions.

electrochemical technique [3-108], polymeric sponge replication method [7-70], powder metallurgy technique, superplastic diffusion bonding method [10-87], or foamed metal structure technique [11-204].

Once the Ti implant is placed in hard tissue, TiO_2 grows and increases its thickness [3-135, 3-136, 4-76–4-82, 7-150], due to more oxygen availability inside the body fluid, as well as co-existence of superoxidant. It is very important to mention here that Ti is not in contact with the biological environment, but rather there is a gradual transition from the bulk Ti material, stoichiometric oxide (i.e., TiO_2), hydrated polarized oxide, adsorbed lipoproteins and glycolipids, proteoglycnas, collagen filaments and bundles to cells [7-64]. Such gradient functional structure was also fabricated in CpTi and microtextured polyethylene terephthalate (PET) system [7-81]. In addition, a gradient structural system of Ti and TiN was developed [7-87]. During HA coating, a gradient functional layer was successfully fabricated [11-123]. To promote these gradient functional (GF) and gradient structural (GS) transitions, there are many *in vivo*, as well as *in vitro*, evidences indicting that surface titanium oxide is incorporated with mineral ions, water, and other constituents of biofluids [4-35–4-38]. Since a surface layer of TiO_2 is negatively charged, the calcium ion attachment can be easily achieved [7-50, 7-51]. Retrieved Ti implants showed that surface TiO_2 was incorporated with Ca and P ions [7-149], while *in vitro* treatment of TiO_2 in extracellular fluids or SBF for prolonged periods of incubation time resulted in the incorporation of Ca, P, and S ions into TiO_2 [3-135, 3-136, 4-35–4-38, 7-150, 11-126, 11-127]. Without prolonged treatment, there are several methods proposed to relatively short-time incubation for incorporation of Ca and P ions. For example, TiO_2 can be electrochemically treated in an electrolyte of a mixture of calcium acetate monohydrate and calcium glycerophosphate [11-41]. As a result of incorporation of Ca and P ions, bone-like HA can be formed in macroscale [6-33] or nanodimension [12-93]. Again for reducing the incubation time, bone-like HA crystals can be formed by treating the TiO_2 surface with water and hydrogen plasma immersion ion implantation, followed by immersion in SBF [11-42], or by treating in hydrogen peroxide followed by SBF immersion [11-43], or immersion in SBF while treating the TiO_2 surface with microarc oxidation and irradiation with UV light [11-68]. It is also known that P ions can be incorporated into TiO_2 while it is immersed in the human serum [7-64].

Bony growth super-surface zones should have a same roughness as the roughness of receiving hard tissue through microporous texturing techniques. This area can be structured using nanotube concepts [12-89–12-91]. Because this zone responds strongly to osseointegration, the structure, as well as the chemistry, should accommodate favorable osteoinductive reactions. BMP [12-49–12-51], and nanoapatite can be coated [12-127]. The zone may be treated by femtosecond laser machining [12-80] to build a microscale 3D scaffold which is structured inside the macroporosities. Such scaffold can be made of biodegradable material (e.g., chitosan), which is incorporated with protein, Ca, P, apatite particles or other species possessing bone growth factors.

REFERENCES

[12-1] Hench LL. Biomaterials: a forecast for the future. Biomaterials 1998; 19: 1419–1423.

[12-2] Froes FH. The better characterization of titanium alloys. J Metals 2005; 57:41.

[12-3] Hartman AD, Gerdeman SJ, Hansen JS. Producing low-cost titanium for automotive applications. J Metals 1998;50:16–19.

[12-4] van Vuuren DS, Engelbrecht AD, Hadley TD. Opportunities in the electrowinning of molten titanium from titanium dioxide. J Metals 2005;57: 53–55.

[12-5] Fuwa A, Takaya S. Producing titanium by reducing $TiCl_4$-$MgCl_2$ mixed salt with magnesium in the molten state. J Metals 2005;57:56–60.

[12-6] Froes FH, Mashl SJ, Moxson VS, Hebeisen JC, Duz VA. The technologies of titanium powder metallurgy. J Metals 2004;56:46–48.

[12-7] Eylon D, Vassel A, Combres Y, Boyer RR, Bania PJ, Schutz RW. Issues in the development of beta titanium alloys. J Metals 1994;46:14–15.

[12-8] Bania PJ. Beta titanium alloys and their role in the titanium industry. J Metals 1994;46:16–19.

[12-9] Schutz RW. Environmental behavior of beta titanium alloys. J Metals 1994; 46:24–29.

[12-10] Kim Y-W. Ordered intermetallic alloys: Part III: gamma titanium aluminides. J Metals 1994;46:30–39.

[12-11] Lopez MF, Gutierrez A, Jimémez JA. *In vitro* corrosion behavior of titanium alloys without vanadium. Electrochim Acta 2002;47:1359–1364.

[12-12] Metikos M, Kwokal A. The influence of niobium and vanadium on passivity of titanium-based implants in physiological solution. Biomaterials 2003;24:3765–3775.

[12-13] Khan MA, Williams RL, Williams DF. The corrosion behavior of Ti-6Al-4V, Ti-6Al-7Nb and Ti-13Nb-13Zr in protein solutions. Biomaterials 1999;20: 631–637.

[12-14] Katakura N, Takada Y, Iijima K, Hosotani M, Honma H. Studies in the shape memory alloys for biomaterials (Part II) – Electrochemical behavior of Ti, Pd, Co and TiPd alloys. J Jpn Dent Mater 1991;10: 809–813.

[12-15] Sohmura T, Kimura HT. Shape recovery in Ti-V-Fe-Al alloy and its application to dental implant. Proceedings of the international conference on martensitic transformation ICOMAT-1986. Japan Inst Metals 1987. pp. 1065–1070.

[12-16] Lausmaa J, Kasemo B, Hansson S. Accelerated oxide grown on titanium implants during autoclaving caused by fluoride contamination. Biomaterials 1985;6:23–27.

[12-17] Pröbster L, Lin W, Hüttenmann H. Effect of fluoride prophylactic agents on titanium surfaces. Int J Oral Maxillofac Implants 1992;7:390–394.

[12-18] Takemoto S, Hattori M, Yoshinari M, Kawada E, Asami K, Oda Y. Corrosion behavior and surface characterization of Ti-20Cr alloy in a solution containing fluoride. Dent Mater J 2004;23:379–386.

[12-19] He G, Eckert J, Dai QL, Sui ML, Löser W, Hagiwara M, Ma E. Nanostructured Ti-based multicomponent alloys with potential for biomedical applications. Biomaterials 2003;24:5115–5120.

[12-20] Hattori M, Takemoto M, Yoshinari M, Kawada E, Oda U. Alloying effect of Pd to Ti-Cu alloy. Proceedings of 19th meeting, society of titanium alloys in dentistry, Tokyo. 2005, June. p. 45.

[12-21] Koike M, Okabe T. Mechanical properties of Ti-Cu-Si alloys. Proceedings of 19th meeting, society of titanium alloys in dentistry, Tokyo. 2005, June. p. 44.

[12-22] Kim T-I, Han J-H, Lee I-S, Lee K-H, Shin M-C, Choi B-B. New titanium alloys for biomaterials: a study of mechanical and corrosion properties and cytotoxicty. J Biomed Mater Eng 1997;7:253–263.

[12-23] Takahashi S, Kikuchi S, Okuno O. Machinability of Ti-Zr alloy. Proceedings of 19th meeting, society of titanium alloys in dentistry, Tokyo. 2005, June. p. 58.

[12-24] Kikuchi S, Takahashi M, Satoh H, Komatsu S, Okabe T, Okuno O. Machinability of Ti-Hf alloy. Proceedings of 19th meeting, society of titanium alloys in dentistry, Tokyo. 2005, June. p. 46.

[12-25] Satoh H, Kikuchi S, Komatsu S, Okuno O, Okabe T. Mechanical properties of Ti-Hf casts. Proceedings of 19th meeting, society of titanium alloys in dentistry, Tokyo (2005, June). p. 57.

[12-26] Inoue A. Recent progress of Zr-based bulk amorphous alloys. Sci. Rep. RITU A42 1996. pp. 1–11.

[12-27] Masumoto T. Amorphous Ti-Cu system alloys. Japan Patent Application Laid-Open No.7-54086 (1995).

[12-28] Tregilgas J. Amorphous Hinge Material. Adv Mater Processes 2005;163:46–49. [12-29] Inoue A. Synthesis and properties of Ti-based bulk amorphous alloys with a large supercooled liquid region. Mater Sci Forum 1999;313/314:307–314.

[12-30] Louzguine DV, Inoue A. Multicomponent metastable phase formed by crystallization of Ti-Ni-Cu-Sn-Zr amorphous alloy. J Mater Res 1999;14:4426–4430.

[12-31] Zhang T, Inoue K. Preparation of Ti-Cu-Ni-Si-B amorphous alloy with a large supercooled liquid region. Mater Trans JIM 1999;40:301–306.

[12-32] Oshida Y. Gradient functional implant system. to be submitted to Biomaterials.

[12-33] JETRO. Ti-O coating on Ti-6Al-4V alloy by DC reactive sputtering method, New Technology Japan. No. 94-06-001-01. 1994:22:18.

[12-34] Bogdanski D, Köller M, Müller D, Muhr G, Bram M, Buchkremer HP, Stöver D, Choi J, Epple M. Easy assessment of the biocompatibility of Ni-Ti alloy by *in vitro* cell culture experiments on a functionally graded Ni-NiTi-Ti material. Biomaterials 2002;23:4549–4555.

[12-35] Yoshinari M, Oda Y, Inoue T, Shimo M. Dry-process surface modification for titanium dental implants. Metall Mater Trans. 2002;33A: 511–519.

[12-36] Kasuga T, Mizuno T, Watanabe M, Nogami M, Niinomi M. Calcium phosphate invert glass-ceramic coating joined by self-development of compositionally gradient layers on a titanium alloy. Biomaterials 2001;22:577–582.

[12-37] Okido M, Kuroda K, Ishikawa M, Ichino R, Takai O. Hydroxyapatite coating on titanium solutions. Solid State Ionics 2002;151:47–52.

[12-38] van Dijk K, *et al*. Measurement and control of interface strength of RF mahnetron-sputtered Ca-PO coatings on Ti-6Al-4V substrates using a laser spallation technique. J Biomed Mater Res 1998;41:624–632.

[12-39] Takamura R. The bone response of titanium implant with calcium ion implantation J Dent Res 1997;76:1177. (Abstract No. 1123).

[12-40] Ohtsu N. Evaluation of degradability of $CaTiO_3$ thin films in simulated body fluids. Mater Trans. 2004;45:1778–1781.

[12-41] Agarwal A, Dahotre NB. Laser surface engineering of titanium diboride coatings. Adv Mater Process 2000;158:43–45.

[12-42] Dunn DS, Raghavan S, Volz RG. Gentamicin sulfate attachment and release from anodized Ti-6Al-4V orthopedic materials. J Biomed Mater Res 1993;27: 895–900.

[12-43] Hill J, Klenerman L, Trustey S, Blowers B. Diffusion of antibiotics from acrylic bone-cement *in vitro*. J Bone Joint Surg 1977;59-B:197–199.

[12-44] Petty W. Spanier S. Shuster JJ. Prevention of infection after total joint replacement. J Bone Joint Surg 1988;70-A:536–539.

[12-45] Dunn DS, Raghavan S, Volz RG. Anodized layers on titanium and titanium alloy orthopedic materials for antimicrobial activity applications. Mater Manuf Process 1992;7:123–137.

[12-46] Kato M, Naganuma K, Yamada M, Sonoda T, Kakimi H, Kotake S, Kawai T. On the dental application of titanium-base alloy, Part 5: Anodic oxidation aiming at the clinical use of the material as a carrier for bone morphogenetic protein (BMP). Gov Ind Res Inst 1991;40: 269–280.

[12-47] Kato M, Naganuma K, Yamada M, Sonoda T, Kotake S, Kakimi H. On the dental application of titanium-base alloy, Part 6: Effect of the composition of electrolytic solutions in anodizing the material. Gov Ind Res Inst 1991;40: 282–290.

[12-48] Wassell DTH, Embery G. Adsorption of bovine serum albumin on to titanium powder. Biomaterials 1996;17:859–864.

[12-49] McAlarney ME, Oshiro MA. Possible role of C3 in competitive protein adsorption onto TiO_2. J Dent Res 1994:73:401 (Abstract No. 2397).

[12-50] Klinger A, Steinberg D, Kohavi D, Sela MN. Adsorption of human salivary Albumin to titanium-oxide. J Dent Res 1996;75:273 (Abstract No. 2043).

[12-51] JETRO. Titanium surface reforming technology. New Technology Japan No. 92-02-100-08. 1992;19:18.

[12-52] Dearnaley G, Goode PD. Techniques and equipment for non-semiconductor applications of ion implantation. Nucl Instr Meth 1981;189: 117–132.

[12-53] Hanawa T, Ukai H, Murakami K. X-ray photoelectron spectroscopy study of calcium-ion implanted titanium. J Electron Spectr Relat Phenom 1993;63:347–354.

[12-54] Howlett L, Eavans MD, Wildish KL, Kelly JC, Fisher LR, Francis W, Best DJ. Effect of ion implantation on cellular adhesion. Clin Mater 1993;14:57–64.

[12-55] Howlett R, Zreiqat H, Odell R, Noorman N, Evans P, Dalton BA, McFarland C, Steels JG. The effect of magnesium ion implantation into alumina upon the adhesion of human bone derived cells. J Mater Sci: Mater Med 1994;5:715–722.

[12-56] Hayakawa T, Yoshinari M, Nemoto K. Direct attachment of fibronectin to tresyl chloride-activated titanium. J Biomed Mater Res 2003;67A: 684–688.

[12-57] Bierbaum S, Beutner R, Hanke T, Scharnweber D, Hempel U, Worch H. Modification of Ti6Al4V surfaces using collagen I, III, and fibronectin. I. Biochemical and morphological characteristics of the adsorbed matrix. J Biomed Mater Res 2003;67A: 421–430.

[12-58] Bierbaum S, Hempel U, Geißler U, Hanke T, Scharnweber D, Wenzel K-W, Worch H. Modification of Ti6AL4V surfaces using collagen I, III, and fibronectin. II. Influence on osteoblast responses. J Biomed Mater Res 2003;67A:431–438.

[12-59] Kim H-W, Li L-H, Lee E-J, Lee S-H, Kim H-E. Fibrillar assembly and stability of collagen coating on titanium for improved osteoblast responses. J Biomed Mater Res 2005;75A:629–638.

[12-60] Wang X-X, Xie L, Wang R. Biological fabrication of nacreous coating on titanium dental implant. Biomaterials 2005;26:6229–6232.

[12-61] Frosch K-H, Drengk A, Krause P, Viereck V, Miosge N, Werner C, Schild D, Stürmer EK, Stürmer KM. Stem cell-coated titanium implants for the partial joint resurfacing of the knee. Biomaterials 2006;27: 2542–2549.

[12-62] Lee JH, Ryu H-S, Lee D-S, Hong KS, Chang B-S, Lee C-K. Biomechanical and histomorphometric study on the bone–screw interface of bioactive ceramic-coated titanium screws. Biomaterials 2005;26: 3249–3257.

[12-63] Bigi A, Boanini E, Bracci B, Facchini A, Panzavolta S, Segatti F, Sturba L. Nanocrystalline hydroxyapatite coatings on titanium: a new fast biomimetic method. Biomaterials 2005;26:4085–4089.

[12-64] Liu X, Zhao X, Fu RKY, Ho JPY, Ding C, Chu PK. Plasma-treated nanostructured TiO_2 surface supporting biomimetic growth of apatite. Biomaterials 2005;26:6143–6150.

[12-65] Ellingsen JE, Birkeland G. The effect on bone healing of fluoride conditioned titanium implants. J Dent Res 1996;75:400 (Abstract No. 3060)

[12-66] Ellingsen JE. Pre-treatment of titanium implants with fluoride improves their retention in bone. J Mater Sci: Mater Med 1995;6:749–758.

[12-67] Ellingsen JE. Increasing biocompatibility by chemical modification of titanium surfaces. In: Ellingsen JE, Lyndstadaas SP ed. Bioimplant interface. Boca Raton: FL CRC 2003. pp. 323–340.

[12-68] Johansson C, Wennerberg A, Holmén A. Ellingsen JE. Enhanced fixation of bone to fluoride-modified implants. In: Transaction of the sixth world biomaterials conference. Sydney: Society for Biomaterials 2002. p. 601.

[12-69] NASA Lewis Research Center. Sputter texturing to produce surfaces for bonding. NASA Tech Briefs. 1997 May., p. 81. www.nasatech.com

[12-70] Banks BA. Improved texturing of surgical implants for soft tissues. LEW-15805. NASA Tech Briefs 1997;21:70.

[12-71] Fujibayashi S, Neo M, Kim H-M, Kokubo K, Nakamura T. Osteoinduction of porous bioactive titanium metal. Biomaterials 2004;25:443–450.

[12-72] Asaoka K, Kuwayama N, Okuno O, Miura I. Mechanical properties and biomechanical compatibility of porous titanium for dental implants. J Biomed Mater Res 1985;19:699–713.

[12-73] Dahotre NB. Laser surface engineering Adv Mater Process 2002;160: 35–39.

[12-74] Kopel A, Reitza W. Laser surface treatment. Adv Mater Process 1999;157: 39–41.

[12-75] Singh J. The constitution and microstructure of laser- surface-modified metals. J Metals 1992;44:8–14.

[12-76] Draper CW, Poate JM. Laser surface alloying. Int Metal Rev 1985;30: 85–108.

[12-77] Galerie A, Fasasi A, Pons M, Caillet M. Improved high temperature oxidation resistance of Ti-6Al-4V by superficial laser alloying. In: Surface Modification Technologies V. Sudarshan TS, Braza JF, editors. London, UK: The Institute of Materials. 1992. pp. 401–411.

[12-78] Mazumder J. Laser welding. In: M. Bass ed. Laser Materials Processing. North-Holland Pub. 1983. pp. 115–200.

[12-79] Bass M, Jau B, Wallace R. Shaping materials with lasers. In: M. Bass ed. Laser Materials Processing. North-Holland Pub. 1983. pp. 290–336.

[12-80] Hollander DA, von Walter M, Wirtz T, Sellei R, Schmidt-Rohlfing B, Paar O, Erli H-J. Structural, mechanical and *in vitro* characterization of individually structured Ti–6Al–4V produced by direct laser forming. Biomaterials 2006;27: 955–963.

[12-81] Weigl P, Kasenbacher A, Werelius K. Dental Applications. In: Femtosecond Technology for Technical and Medical Applications. Dausinger F, Lichtner F, Lubatschowski H, editors. Top Appl Phys. 2004;96:167–187.

[12-82] Gamaly EG, Rode AV, Luther-Davis B, Tikhonchuk VT. Ablation of solids by femtosecond lasers: Ablation mechanism and ablation thresholds for metals and dielectrics. Phys Plasmas 2002;9:949–953.

[12-83] Frentzen M, Winkelstraeter C, van Benthem H, Koort HJ. Bearbeitung der Schmelzorberflachen mit gepulster Laserstrahlung. Dtsch Zahnaertzl Z 1994;49:166–168.

[12-84] www.mdatechnology.net/techsearch.asp?articleid=191.

[12-85] Ringeisen BR, Chrisey DB, Piqué A, Krizman D, Bronks M, Spargo B. Direct write technology and a tool to rapidly prototype patterns of biological and electronic systems. Nanotechnology 2001;1:414–417.

[12-86] Finke S, Feenstra K. Solid freedom fabrication by extrusion and deposition of semi-solid alloys. J Mater Sci 2002;37:3101–3106.

[12-87] Klug KL, Ucok I, Gungor MN, Gulu M, Kramer LS, Tack WT, Nastac L, Martin NR, Dong H. The near-net-shape manufacturing of affordable titanium components for the M777 lightweight howitzer. J Metals 2004;56:35–42.

[12-88] Taylor J. Nanobiotechnology. http://www.dti.gov.uk.

[12-89] Frosch K-H, Barvencik F, Viereck V, Lohmann CH, Dresing K, Breme J, Brunner E, Stürmer KM. Growth behavior, matrix production, and gene expression of human osteoblasts in defined cylindrical titanium channels. J Biomed Mater Res 2004;68A:325–334.

[12-90] Macak JM, Tsuchiya H, Taveira L, Ghicov A, Schmuki P. Self-organized nanotubular oxide layers on Ti-6Al-7Nb and Ti-6Al-4V formed by anodization in NH_4F solutions. J Biomed Mater Res 2005;75A:928–933.

[12-91] Oh S-H, Finõnes RR, Daraio C, Chen L-H, Jin S. Growth of nano-scale hydroxyapatite using chemically treated titanium oxide nanotubes. Biomaterials 2005;26:4938–4943.

[12-92] Goto K, Tamura J, Shinzato S, Fujibayashi S, Hashimoto M, Kawashita M, Kokubo T, Nakamura T. Bioactive bone cements containing nano-sized titania particles for use as bone substitutes. Biomaterials 2005; 26:6496–6505.

[12-93] Sato M, Sambito MA, Aslani A, Kalkhoran NM, Slamovich EB, Webster TJ. Increased osteoblast functions on undoped and yttrium-doped nanocrystalline hydroxyapatite coatings on titanium. Biomaterials 2006; 27: 2358–2369.

[12-94] Tyas MJ. Minimum intervention dentistry – the cutting edge? http://www.mdhs.unimelb.edue.au.

[12-95] Minimally invasive surgery. http://www.wpahs.org/agh/mis.

[12-96] http://biomed.tamu.edubiomaterials/TissueENgineering.htm

[12-97] Boyan BD, Hummert TW, Dean DD, Schwartz Z. Role of materials surfaces in regulating bone and cartilage cell response. Biomaterials 1996;17: 137–146.

[12-98] Baier RE. Conditioning surfaces to suit the biomedical environment: recent Progress. J. Biomechanical Eng 1982;104:257–271.

[12-99] Taira M, Araki Y. Scaffolds for tissue engineering – kinds and clinical application-. J Dental Eng 2005;152:27–30.

[12-100] Li H, Chang J. Fabrication and characterization of bioactive wollastonite/PHBV composite scaffolds. Biomaterials 2004;25:5473–5480.

[12-101] http://www.wipo/cgi-pct/guest/getbyKy5?Key=02.

[12-102] Burgess DS. InkJet Technique precludes microvessels and microlenses. Technology World 2005;May: 213.

[12-103] Lee M, Dunn JCY, Wu BM. Scaffold fabrication by indirect three-dimensional printing. Biomaterials 2005;26:4281-4289.

[12-104] Walboomers XF, Jansen JA. Bone tissue induction, using a COLLOSS®-filled titanium fibre mesh-scaffolding material. Biomaterials 2005;26: 4779-4785.

[12-105] Hohman MM, Shin M, Rutledge G, Brenner MP. Electrospinning and electronically forced jets. Phys Fluids 2001;13:2201-2220.

[12-106] Li M, Mondrinos MJ, Gandhi MR, Ko FK, Weiss AS, Lelkes PI. Electrospun protein fibers as matrices for tissue engineering. Biomaterials 2005;26:5999-6008.

[12-107] Buttafoco L, Kolkman NG, Engbers-Buijtenhuijs P, Poot AA, Dijkstra PJ, Vermes I, Feijen J. Electrospinning of collagen and elastin for tissue engineering applications. Biomaterials 2006;27:724-734.

[12-108] Liu X, Won Y, Ma PX. Surface modification of interconnected porous scaffolds. J Biomed Mater Res 2005;74A:84-91.

[12-109] MacDonald RA, Laurenzi BF, Viswanathan G, Ajayan PM, Stegemann JP. Collagen-carbon nanotube composite materials as scaffolds in tissue engineering. J Biomed Mater Res 2005;74A:489-496.

[12-110] Funakoshi T, Majima T, Iwasaki N, Yamane S, Masuko T, Minami A, Harada K, Tamura H, Tokura S, Nishimura S-I. Novel chitosan-based hyaluronan hybrid polymer fibers as a scaffold in ligament tissue engineering. J Biomed Mater Res 2005;74A:338-346.

[12-111] Ni S, Chang J, Chou L. A novel bioactive porous $CaSiO_3$ scaffold for bone tissue engineering. J Biomed Mater Res 2006;76A:196-205.

[12-112] Wu L, Jing D, Ding J. A "room-temperature" injection molding/particulate leaching approach for fabrication of biodegradable three-dimensional porous scaffolds. Biomaterials 2006;27:185-191.

[12-113] Bulard RA, Vance JB. Multi-clinic evaluation using mini-dental implants for long-term denture stabilization: a preliminary biometric evaluation. Compend Condin Educ Dent 2005;26:892–897.

[12-114] Lazzarra RJ. Immediate implant placement into extraction sites: surgical and restorative advantages. Int J Periodontics Restorative Dent 1989;9:333–343.

[12-115] Parel SK, Triplett RG. Immediate fixture placement: a treatment planning alternative. Int J Oral Maxillofac Implants 1990;5:337-345.

[12-116] Werbitt MJ, Goldberg PV. The immediate implant. Bone preservation and bone regeneration. Int J Periodont Restor Dent 1992;12:207–217.

[12-117] Gruder U, Polizzi G, Goebe R, Hatano N, Henry P, Jackson WJ, Kawamura K, Koehler S, Renouard F, Rosenberg R, Triplett G, Werbitt MJ, Lithner B. A 3-year prospective multicenter follow-up report on the immediate and delayed-immediate placement of implants. Int J Oral Maxillofac Implants 1999;14:210–216.

[12-118] Peckitt NS. Stereoscopic lithography: customized titanium implants on orofacial reconstruction. Br J Oral Maxillofacial Surg 1999;37: 353–369.

[12-119] Nishimura RD, Chang TL, Perri GR, Beumer J. Restoration of partially edentulous patients using customized implant abutmens. Pract Period Aesthet Dent 1999;11:669–676.

[12-120] De Putter, de Lange GL, de Groot K. Permucosla dental implants of dense hydroxyapatite: fixation in alveolar bone. Proceedings of international congress on tissue integration in oral and maxillofacial reconstruction. May 1985, Brussels. Current Practice Series #29, Amsterdam: Excerta Media, 1986. pp. 389–394.

[12-121] Albrektsson T, Wennerberg A. Oral implant surfaces: Part 1 – review focusing on topographic and chemical properties of different surfaces and *in vitro* responses to them. Int J Prosthodont 2004;17:536–543.

[12-122] Denissen HW, Veldhuis AAH, van den Hooff A. Hydroxyapatite titanium implants. Proceedings of international congress on tissue integration in oral and maxillofacial reconstruction. May 1985, Brussels. Current Practice Series #29, Amsterdam: Excerta Media 1986. pp. 395–399.

[12-123] Sundgren JE, Bodo P, Lundström I. Auger electron spectroscopic studies of the interface between human tissue and implants of titanium and stainless steel. J Colloid Interface Sci 1986;11:9–20.

[12-124] Albrektsson T, Wennerberg A. Oral implant surfaces: Part 1 – review focusing on topographic and chemical properties of different surfaces and *in vivo* responses to them. Int J Prosthodont 2004;17:536–543.

[12-125] Ellingsen JE. A study on the mechanism of protein adsorption to TiO_2. Biomaterials 1991;12:593–596.

[12-126] Oshida Y. Requirements for successful biofunctional implants. In: Proceedings of international 2nd symposium on advanced biomaterials. Yahia L'H, editor. Montreal Canada: Ecole Polytechnique/GRBB. 2000. pp. 5–10.

[12-127] Pham MT, Reuther H, Maitz MF. Native extracellular matrix coating on Ti surfaces. J Biomed Mater Res 2003;66A:310–316.

Epilogue

It is said that about 2 million journals are published annually, in which about 20,000 journals have articles related, to some extent, to the bio-medical area. Furthermore, out of 20,000 journals, about 500 journals have at least one article, which is related to medicine/dentistry. Hence, we have to look over 500 journals to cover, at least superficially, the medical/dental articles. Suppose that we spend 1 h every day reading through such journals, and 15 min for each journal, we would need 125 days to cover all possible articles which might be useful to our expertise to avoid the delay for updated follow-up. Obviously, what this calculation indicates is that we need at least one out of every 3 days to spend time to go through all the journals. Today, journals such as *Biomaterials, Biomedical Materials Research, Biomedical Materials and Engineering, Prosthetic Dentistry, Oral Maxillofacial Implants*, and so on and so forth have at least three articles in each issue which are related to titanium materials.

I have also found while writing this book that in most European journals, as well as European authors (regardless of their publication in either European journals or American journals), there are few references cited from American journals or American authors, and vice versa. In addition to this unique phenomenon, there are quite a few excellent works done by Asian researchers published in obscure journals. I paid special attention to equalize this unbalanced appearance of many excellent articles written by internationally recognized researchers.

I have had an experience, during the course of writing this book, similar to a painter who must decide when and how to finish his work. Similarly, knowing that every month I will have about 20 good articles, I need to stop and read these before I will be able to finish this book. I have my last wish that I do not miss any crucial articles, and at the same time, I have a fear that I might miss some because I had to stop reading journal articles to analyze them from the June issue of the year 2006.

As society is aging with the increasing population of the elderly, special care and attention should be paid to managing diseases as well as treating these populations. For example, a geriatric dentist needs to be seriously changed to accommodate such social changes and challenges. Orthopedic and dental implants still have a window left to be further developed to facilitate the aged population. But I have a great hope, from analyzing numerous articles, that since the research and development in this sector has been so advanced and well-developed, sooner or later public health services will become much easier and cheaper.

Closing this small book, I admit to having a very long list of individual's names to be mentioned with my memories and appreciation from my teachers who inspired me, friends and colleagues who challenged me, and most importantly students (some of them are now my colleagues) who continue to both inspire and challenge me. To all of them, I owe my knowledge and capability to comprehend the cited articles presented in this book. These people, who were and are valuable to me and to this book, should at least include the late T. Nakayama, S. Iguchi, V. Weiss, H.W. Liu, the late J.A. Schwartz, the late T. Osaka, Y. Miwa, N. Saotome, T. Nishihara, M. Ito, F. Farzin-Nia, R. Sachdeva, S. Miyazaki, P-C. Chen, M. Okamura, G.K. Stookey, C.J. Andres, T.M. Barco, O. Okuno, B.K. Moore, A. Zuccari, A. Hashem, A. Wu, M. Yapchulay, S. Isikbay, C. Kuphasuk, C-M. Lin, I. Garcia, W. Panyayong, Y-J. Lim, M. Reyes, W.C. Lim, J-C. Chang, S. Al-Ali, B. Anbari, D. Bartovic, F. Hernandez, I.W. Koh, Z.H. Khabbaz, R. Xirouchaki, P. Agarwal, N. Al-Nasr, C.B. Sellers, J. Dunigan-Miler, S. Al-Johancy, C.S. Wang, E. Matsis, R. Rani, Y-C. Wu, K. Mirza, I. Katsilieri, and L.A. Oshida. Thanks should also go to M.A. Dirlam for excellent professional illustrations of all figures in this book, and J.D. Kummer for his elaborative time-consuming proof-reading of the text. Lastly, but not least important, I have no words to appreciate Elsevier individuals who support and assist me throughout the entire process of editing this book; they should include D.G. Sleeman, A. Kitson, D. Coleman, P. Wilkinson, and editorial team of MacMillan India. Thank you all.

Subject Index

316L stainless steel
    corrosion behavior 44
3DP (three-dimensional printing) technology
    404
3-Point bend flexural test 263–264
4-Point bending test 264
    cohesive fracture 264
    delamination of porcelain 264

A
ADA (American Dental Association) 138
Adhesion, *see under* Cell adhesion
AE (acoustic emission) 110, 235
AES (Auger electron spectroscopy) 355
Aerobic and anaerobic 34
AFM (Atomic force microscopy) 337
AIDS (acquired immune deficiency syndrome)
    133
Al_2O_3, as a blasting media 315–317
Allergic reaction 133–136
Allogenous grafts 325
Allografts 325
Alloplast 325
Alpha case 286–287
Amalgam xiii, 34
Amorphous materials 388–389
Anchorage mechanisms 193
Anodization, titanium surfaces modification by
    184
ANSI (American National Standards Institute)
    138
Apatite 359, *see also under* HA (hydroxyapatite)
APC (anodic plasma-chemical) process 358
APF (acidulated phosphate fluoride) 29
Apoptosis/cell death, at the implanted materials
    165
Artificial saliva 25, 49
Autologous grafts 325

B
Bacteria, causing corrosion 55, *see also*
    under MIC
    Actinobacillus actinomicetemcomitans
        56–57, 59

    Actinomyces viscosus 56–57
    on commercially pure titanium 60
    Porphyromonas gingivalis 56, 59
    Streptococcus epidermidis 59
    Streptococcus mutans 57–58, 60
    Streptococcus sanguis 56, 58–60
BAG (bioactive glass) 323
BCA (bone-like carbonate apatite) 336
BCP (biphasic calcium phosphate) 315
Beryllium 7
Beta Ti alloy
    advantages and disadvantages 386
    cytotoxicity 133
    immersion tests on 62
    surface roughness and static frictional force
        resistance of 115
Beta transus 11, 14, 262, 297
Biocompatibility 138–143
    definitions 138
    of biological materials 34
Bioengineering and biomaterial-integrated
        implant system 406–411
Biological reaction 127–148
    allergic reaction 133–136
    biocompatibility 138–143
    biological toxicity 129–130
    biomechanical compatibility 143–148, *see*
        also separate entry
    cytocompatibility (toxicity to cells) 130–133
    metabolism 136–138, *see also separate entry*
    toxicity 127–130, *see also separate entry*
Biomaterials
    as a 'foreign material' 6
    as scaffolds for bone tissue engineering 347
    biofunctioning materials, surface roughness
        181f
    biomaterials strength and rigidity of 144f
    compatibilities 7, 143–148
    metallic biomaterials, biocompatibility 129
    microroughness 180
    roughness and cellular response to
        179–182
    surface topography, importance 179–185
    surface topography, osteoblast response 185
Biomechanical issues 231
Biomimetic deposition 341

Bio-Oss® 194
Biotolerant biomaterial 163
Biotribological actions 40, 113–119
   friction 114–116
Blasting media 238
Bleaching treatments
   fluoride treatment 27
   for teeth 26–27
   hydrogen peroxide 27
   nightguard bleaching 27
BMP (bone morphogenetic protein) 163, 391, 393, 404
Bonding osteogenesis 324
Bone fusing to Ti 194
Bone healing 162–163
   de novo bone formation phase 162
   factors influencing 195
   healing phases 162
   osteoconduction phase 162
Bone implants
   fixation 146
   interface 237
   interface, topography influencing 176
   surface characteristics 174
Bone ingrowth 228, 243
   interstitial bone ingrowth 189
   into uncemented hemispherical canine
      acetabular components 189
   quantification 189
   to control (nontreated) and heat-treated
      Ti-6Al-4V implants 189
Bracket/archwire system 115
Brånemark implant 224, 234
   mechanical *in vitro* tests 233
Breakdown potential 37
BSA (bovine serum albumin) 172, 392
BSE (bovine spongiform encephalopathy) 324
BSEM (backscattering electron microscopy) 336
BSF (body simulated fluid) 38, *see also* SBF

C
Ca-P (calcium phosphate) coating 335–342
   biomimetic deposition and electrochemical
      deposition for 341
   CaP bioceramic, protein adsorption behavior
      of thin films of 169
   homogeneous deposition 184
   in vivo effect of 191
   RF magnetron sputter technique 339
   sol–gel method for 340

substrate straining method 340
Carbamide peroxide 26
Carbon, glass, ceramic coating 322
Casting 285–287
   evaluation 285
   melting methods 287
   reaction layer structures 287
CaTiO₃, at implant–bone interface 172
Cell adhesion, adsorption, spreading, and
      proliferation 166–179
   bio-adhesion 238
   inflammatory processes 166
   pain 166
   polymorphonuclear granulocytes 167
   redness 166
   repair 166
   resolution 167
   swelling 166
   tissue remodeling 166
Cemented implant restorations 218–219
Cementless fixation systems 195
CFS (chronic fatigue syndrome) 134
Chemical and electrochemical reactions 25–64
   amino-alcohol solution in 320
   and thermal surface modifications 319–322
   corrosion resistance under influences of
      environmental and mechanical-assisted
      actions 31–42
   corrosion, in media containing fluorine ion
      and bleaching agents 28–31
   crevice corrosion 62
   crevice–corrosion susceptibility 37
   discoloration 26–28
   electrochemical corrosion studies,
      interpretation 32
   internal discoloration 26
   metal ion release and dissolution 42–51,
      see also separate entry
   microbiology-induced corrosion 53–61, *see
      also under* MIC
   passivation breakdown potential 37
   reaction with hydrogen 61–64
   seawater embrittlement behavior 63
Chemical bonding 260–261
Chemicals, toxic entity 127
CIE-L*a*b* 27
Chondromalacia 219
Clamps and staples 273–274
Clasps 265–266
Coating, surface modification by 322–360
   Ca-P coating 335–342
   carbon, glass, ceramic coating 322

composite coating   342–347, *see also*
   separate entry
  fluoridated apatite coatings   327
  future perspectives   390–396
  HA coating   324–335
  post-spray coating heat-treatment effects
   328
  sol–gel method for   329, 358–359
  Ti coating   353–357, *see also separate entry*
Co-Cr alloy wires, surface roughness
   and static frictional force resistance
   of   115
Co-Cr-Mo
  abrasion   226
  corrosion behavior   44, 354
  surface oxides on   349
Collagen   394
Coloring   364
  anodic oxidation   364
  chemical-oxidation   364
  nitrizing   364
  thermal-oxidation   364
Composite coating   342–347
  "Ecopore"   346
  composite coatings   343
  HA/yttria-stabilizedzirconia(YSZ)/Ti-6Al-4V
  HTBC (HA–TiO$_2$ bond coating)   344
  wollastonite/TiO$_2$ composite coating   357
Contact guidance   236
Contact osteogenesis   324
Contamination   316
Core materials   266–267, *see also* Posts and
   cores
Corrosion behavior, *see also* Oxidation
  chemical corrosion   34
  corrosion–fatigue interactions   41
  corrosive effect, of bleaching agent   26
  in media containing fluorine ion and
   bleaching agents   28–31
  in vivo corrosion measurements   33
  microbiology-induced corrosion   53–61,
   see also under MIC
  of heavily alloyed Ti materials   38
  of NiTi materials   36
  of NiTi SME alloy in human body   38
Corrosion resistance   25
  methods for improving   38
CPP (calcium pyrophosphate)   342
CpTi (commercially pure titanium)/CpTi
   implant   13–16
  as fixtures for dental implantation   257
  bone formation around   240

bone-bonding ability of differently treated
  samples   244
clinical performance of machined/turned   187
corrosion behaviors   29, 35
CpTi (grade 1) samples, corrosion tests   15,
  48
CpTi (grade 2) Nd:Yag laser weldment,
  fatigue cycling   15, 109
CpTi (grade 3)   15
CpTi (grade 4)   15
CpTi (grade 7) samples, *in vitro* metal
  leaching tests   48
cytotoxicity   133
fibrin deposition and blood platelet adhesion
  study   165
in orthopedic internal fracture
  fixation   178
in vitro and *in vivo* electrochemical behavior
  39, 229
in vitro mineralization of osteoblast-like cells
  on   361
interface mechanical characteristics and
  histology   232
on bone ingrowth   190
pit initiation process on   360
preparation, laser welding and mechanical
  adhering method   52
surface modifications   321
surface treatment effects   170
Crowns and bridges   259–265
Crystalline structures, of Ti oxides   88–90
  anatase-type   87–88
  brookite-type   87–88
  characterization by AES   90
  characterization by EIS tests   90
  characterization by SIMS   90
  characterization by XPS   90
  formation conditions   88
  identification   88
  interfacial phenomena between   90
  rutile-type   87–88
Cullet method   343
Cyclic potentiodynamic tests   37
Cylindrical implants   190
Cytocompatibility (toxicity to
  cells)   130–133
  and mutagenicity of metal salts   131

D
De novo bone formation   162
Degenerative joint disease   161

Dental alloys
  for metal–ceramic restorations   265
  metallic ion release from   44
Dental hygiene practice   160
Dental implants, *see also* Implant-related
    biological reactions; Shape-memory dental
    implant
  and orthopedic implants, biological
      performance   241
  designing and design consideration in   146,
    191
  failure   160
  in vivo bone response   147
  perio root type   268
  plate form implants   159
  recent research   407
  retrieved screw system   234
  root form implants   159
  roughened titanium surfaces for   192
  subperiosteal implants   159
  surfaces cleanliness, importance   245
  surgical procedures   160
Dentistry
  bleaching treatments for teeth   26
  casting processes in   262
  dental materials, primary
      tests   138
  dental materials, secondary tests   138
  fluoride mouthwashes   28
  friction in   114
  hazardous substances in   7
  Ti application in   26
  viral bleaching   26
Denture bases   257–259
Direct bony interface   217
Discoloration   26–28, *see also* Bleaching
    treatments
  internal discoloration   26
Distance osteogenesis   324
DLF (direct laser forming)   399
DW (direct-write) technique   400

E
EBL (evidence-based learning)   xiii
ECM (artificial extracellular matrices)   404
EDM (electro discharge machining) and
    CAD/CAM   289–291, 406
  Procera process   290
  spark erosion   289
EDTA (ethylene diamine tetraacetic acid)   50,
    169

EIS (electrochemical impedance
    spectroscopy)   30, 39, 333
Electron Diffraction (ED)   88
Electrospinning   405
ELID (electrolytic in-process dressing)
    grinding technique   289
ELISA (enzyme-linked immunosorbent assay)
    176
Ellingen   168
EMPA (electron probe microanalyzer) analysis
    30
Endodontic files and reamers   271–273
Endosseous implants
  bioinert type   163
  bioreactive type   163
  biotolerant type   163
  therapy   224
ESCA (electron spectroscopy for chemical
    analysis)   28
ESEM (environmental scanning electron
    microscopy)   337
E_{pass} (passivation potential)   47
Epithelium   232

F
Fabrication technologies   285–303
  casting   285–287, *see also under* Casting
  DB (diffusion bonding)   294–295
  electro discharge machining (EDM) and CAD/
      CAM   289–291, *see also under* EDM
  HT (heat treatment)   302–303
  isothermal forging   291
  laser welding/forming   299–301
  machining   288–289, *see also under*
      Machining
  metal injection molding technology
      297–298, *see also under* MIM
  P/M (powder metallurgy)   295
  SPF (superplastic forming)   291–294, *see
      also under* SPF
Fatigue   107–111
  acoustic emission (AE) phenomenon to
      record   110
  affecting microstructural modification   110
  clasps causing   108
  corrosion-fatigue tests   109
  damage due to, measuring methods   273
  fatigue strength, ways to improve   111
  hip joint prostheses, fatigue failure of   109
  strain-controlled   108
  stress-controlled   108

FDA (Food and Drug Administration) 138
FEM (finite element analysis method) 233
Fenton reaction 95
FF (flow-forming) 400
Fibronectin 394
Fibro-osseo integration 146
Fluoride discoloration
  fluoride treatment 27
  temperature effect 28
Fluoride treatment 396–397
Fluorine ion and bleaching agents, corrosion in
  media containing 28–31
Foamed metal structure technique 363–364, 410
Fractal dimension analysis 316
Fracture and fracture toughness 111–113
  AE analysis 112
  Nanoindentation techniques for 113
Freestanding implant 233
Fretting wear 43
Friction 114–116
  coefficients of, determination 115
FTIR (Fourier transform infra red
  spectroscopy) 336

G
Galvanic corrosion
  factors influencing 52
  macro-cells 51–53
  micro-cells 51–53
  surface area ratio effect 52, 60
GFC (gradual functional concept) 408
GFM (gradient functional material) 389–390,
  408
Guinea-pig sensitization test 132

H
HA (hydroxyapatite) 140
  and Ti surfaces, topographic effects 184
  as a bone grafting material 325
  HA coating 324–335
  HA plasma-sprayed coating, porous 238
  HA/TCP (hydroxyapatite/ tricalcium
    phosphate) coating 197–198
  HA-coated CpTi implants 331
  HA-coated CpTi, interface mechanical
    characteristics and histology 232
  HA-coated implants, utilization 192
Hall–Héroult process 385
Hank's solution 46, 93, 327
Haptotaxis 236

Hardening rate (n) 288
HAZ (heat-affected zone) 300, *see also* Laser
  welding/forming
Hemocompatibility 164–166
  characterization 164
  systematic evaluation 164
HF (hydrofluoric acid) 30
High-energy ion implantation 393
Hip joint prostheses, fatigue failure of 109
HPF (human plasma fibronectin) 169
HT (heat treatment) 302–303
  PWHT (post-welding heat-treatment) 302
HTBC (HA–TiO$_2$ bond coating) 344
Human mandubular subperosteal dental
  implants 58
Hydrides 63
Hydrogen, reaction with 61–64
Hydrogen embrittlement (HE) 29, 63
Hydrogen peroxide 94–97, *see also* Ti-H$_2$O$_2$
  system 95
Hydroxyl radicals 95

I
ICP-MS (inductively coupled plasma mass
  spectroscopy) 354
Immediate loading 159
Implant application 217–246
  contraindicative factors 159–160
  clinical reports 222–226
  prostheses and fixtures, long-term outcome
    of 222
  reaction in biological environment 235–246
  reactions in chemical and mechanical
    environments 230–235
  single tooth replacement 233
  surface and interface characterization
    227–229
Implant-related biological reactions 159–200
  biomaterials, roughness and cellular response
    to 179–182, *see also separate entry*
  bone healing 162–163
  cell adhesion, adsorption, spreading, and
    proliferation 166–179, *see also separate*
    entry
  cell growth 182–185
  hemocompatibility 164–166, *see also*
    separate entry
  osseointegration and bone/implant interface
    193–199, *see also separate entry*
  tissue reaction and bone ingrowth 185–193,
    see also separate entry

Implants
  and bone close contact between   167
  and bone interface between   147
  and tissue, interlocking between   188
  biocompatibility   83
  biomechanical compatibility   143–148
  biomechanics in   231
  bone formation after   196
  by screw-tightening   218
  conceptual Ti implant with gradient
    mechanical and biological functions
    409f
  cylindrical coated   356
  healing around, evaluation   163
  implant instability, *see* Loosening implants
  implant materials, corrosion rates, *in vitro*
    laboratory tests   44
  implant prostheses, design and restoration of
    218
  implant surface preparation   180
  implant tooth and natural tooth, comparison
    232f
  in process of primary mineralization   188
  interactions of cells and   236
  metallic biomedical implants   175
  metallic implants, wound-healing responses
    around   237
  Mg-incorporated   197
  micromotion chamber for   190
  nonsubmerged ITI-Bonefit implant   190
  osseointegration of laser-textured implants
    199
  osteoblast adhesion on   174
  osteoblast response to   178
  osteoconductive apatite coatings   192
  physical and chemical treatments   229
  prosthetic and osteosynthetic
    implants   200
  required compatibilities   220
  restorations by cementation   218
  screw-shaped titanium plasma-sprayed
    implants   190
  single-crystal and polycrystalline titanium
    implants   112
  success of, reasons   228
  topography of   36
IMZ titanium implants   225
Infection   114
Interface   6–7
Intermetallic alloys   19
Intraoral environments, corrosion resistance
    under   31–42

Intraoral prostheses   52
Ion implantation onto NiTi   165
I_{pass} (passivation onset current density)   47
ISO (International Organization for
    Standardization)   138
Isothermal forging   291
ITI® dental implants   225

L
Laser applications   398–400
  laser alloying   398
  laser welding, advantages   399
Laser welding/forming   52, 299–301
  distortion, investigation   301
  in peening technology   301
  laser peening   301
Laser-peening method   313, 317–319
Laser-superfinished implants   198–199
LD50 (lethal dose 50%)   127
Loosening implants   114
  adverse factors for   199–200
LSE (laser surface engineering)   390

M
Machining   288–289
  buff-polishing   288
  chemical polishing   288
  plastic deformation due to   289
  springback   289
Macro-cells   51–53
MAR (mineral apposition rate)   196
Masticatory force   25
Materials classification   9–19
  medical/dental titanium and its alloys   9,
    13–19, *see also separate entry*
  near α type, (α+β) type   11
  α stabilizers   11
  α type (HCP)   11
  β stabilizers   11
  β type (BCC)   11
Maxillofacial implants   231
MDI (mini dental implants)   406
Mechanical and tribological behaviors
    107–119, *see also* Fatigue; Fracture and
    fracture toughness
  biotribological actions   113–119, *see also
    separate entry*
  wear and wear debris toxicity   116–119, *see
    also separate entry*
Mechanical surface treatment   111

Medical/dental titanium and its alloys   13–19
  CpTi (commercially pure titanium)   13–16,
    see also separate entry
  intermetallic alloys   19
  Ti-30Ta alloy   19
  Ti-3Al-2.5V   16
  Ti-5Al-3Mo-4Zr   16–17
  Ti-5Al-2.5Fe   17
  Ti-5Al-2Mo-2Fe (SP700) alloy   18–19
  Ti-6Al-4V   16
  Ti-6Al-7Nb   16
  Ti-Cu alloys   18
  Ti-Mo alloy   18
  TiNi alloy   17–18
MELISA® test   134
Metabolism   136–138
  TiO_2 transport   136
  nitroblue tetrazolium test   136
Metal ion release and dissolution   42–51
  biological significance   43
  fretting wear   43
  from 316L stainless steel   45
  from Co-Cr-Mo, CpTi (grade 2)   45
  from dental alloys   44
  from Ti-15Zr-4Nb-4Ta   45
  from Ti-6Al-4V   45, 50
  pitting corrosion   44
  polarization process   43
  repassivation kinetics   44
  Ti-6Al-7Nb   45
  transpassivation   43
Metal-ceramic systems   260
Metallic implants   40
MGCs (multinucleated giant cells)   315
MIC (microbiology-induced corrosion)   53–61
  anaerobic and aerobic bacteria in   59
  in vitro MIC-related corrosion tests   59
  peri-implant disease   58
  plaque development   57–58
  sulfate-reducing bacteria in   59
Micro-cells   51–53
Micrograin superplasticity   292
Micromotion, at the soft-tissue-implant
  interface   173
Microporous texturing techniques   410
MID (minimally invasive dentistry)   402–403,
  406
MIM (metal injection molding) approaches   6,
  220, 297–298, 386
  sequences   298
Mineralization   188
MIS (minimally invasive surgery)   402–403, 406

Mitek bone suture anchors   141
MOE (modulus of elasticity)   112, 144, 221, 233
Morphological compatibility   180–182
MS (multiple sclerosis)   134
MSCs (mesenchymal stem cells)   395
MTT method   133
Myelointegration   239

N
NaF-containing solution   29
Nanotechnology   400–403
  nanoindentation techniques   113
  nanophase materials   173
NiTi alloys
  alloy wires, surface roughness and static
    frictional force resistance of   115
  as implants in clinical surgery for fixation of
    spinal dysfunctions   142
  biocompatibility surface modifications on
    142
  biocompatibility   131
  characterization   142
  dental orthodontic archwires   37
  fatigue resistances of   109
  implants, oxide film coating   49
  in the medical and dental industries   141
  ion implantation onto   165
  ion release measurements on   142
  NiTi SME alloy   243
  NiTi superelastic endodontic, torsional
    properties   272
  orthodontic archwires, *in vitro* testing of   115
  orthodontic superelastic NiTi wires   49
  osteosynthesis clamps in   273
NITINOL (NIckel-TItanium Naval Ordnance
  Laboratory)   17, 141
Nitriding treatments   132
Nitrogen implantation   132
NNS (near-net shape) forming technique   6,
  295, 400–403, *see also* SPF/DB
NR method   133

O
Oral implants   228, 231
Orthodontic appliances   269–271
  curved canal, instrumentation   272
  HA coating with   197
  heat treatment process   269
  NiTi arch wires, acidic fluoride solution on
  270

NiTi wires, physio-mechanical properties 270
orthodontic archwires, *in vitro* testing of 115
orthodontic mechanotherapy 28, 63
orthodontic NiTi archwires, surface
  roughness and static frictional force
  resistance of 115
orthodontic tooth movement 115
orthopedic implants, advantages 161
orthopedic implants, disadvantages 161
recycling and re-usable 269
titanium in 26, *see also* Dentistry
Osseointegration 83, 146–147, 159–160,
  224
  and bone/implant interface 193–199
  definition, mechanism, understanding 194
Osteoarthritis 219
Osteoblast phenotype expression 132
Osteocalcin 237
Osteoclast activation 119
Osteoclastogenesis 119
Osteoconduction 162, 324
  osteoconductive apatite coatings 192
Osteogenesis 324
Osteoinduction 324
Osteolysis 118, 161
Oxidation and oxides 81–97, *see also* Passivity
  air-formed oxide 82
  anatase-type 87–88
  anodic oxide films on, electrochemical
    dissolution 94
  blood plasma protein interactions with,
    studies 176
  brookite-type 87–88
  cathodic reduction behavior 93
  chemical treatments causing 84
  crystalline structures 88–90, *see also under*
    Crystalline structures
  formation 81
  growth, stability, and breakdown 91–94
  in cellular behavior of osteoblasts,
    composition and characteristics 177
  influences on biological process 86
  nature and properties 84
  oxidized implants, osseointegration of 196
  pit generation on Ti 94
  reaction with hydrogen peroxide 94–97
  rutile-type 87–88
  structurally sensitive *in situ* methods for 85
  surface treatments effects on 169
  uniform film formation and dissolution,
    kinetics 93
  unique applications 91

P

P/M (powder metallurgy) 295
  advantages 296
  technique 410
  to modify the surface zone of implants 296
PACVD (plasma-assisted chemical vapor
  deposition) 131–132, 357
PAD (plasma arc deposition) 400
Parabolic rate constant 352
Passivity/Passivation 84–85
  at elevated temperatures 85–86
  definitions 82
  repassivation capability 85
  repassivation kinetics 44
  stability 46–47
  transpassivation 43
PBM (porcelain-bonded-to-metal) 259
PBS (phosphate-buffered saline) solution 39, 96
PC (polycarbonate) 257
Peening technology 301
Percutaneous and permucosal implants 186
Peri-implantitis 224, 240
  peri-implant bone healing 162
  peri-implant bone 147
  peri-implant disease 58
Periodontal membrane 232
PET (polyethylene terephthalate) 173
PFM (porcelain-fused-to-metal) 259
Phase 6
Physically toxic entities 130
Photo-catalytic function 91
PI³ (plasma immersion ion implantation or
  PIII) method 321
PI³&D (plasma immersion ion implantation and
  deposition) 174
Pitting corrosion, in Ti-based implant materials
  44–46
Placed implants, success rates of, factors
  influencing 160
Plaque biofilm removal 141
Plasma nitriding technique 132, 175
Plasma spray method 192, 329, 329, 331, 345
  corrosion behavior 326
  plasma-sprayed HA-coated devices 325
  porous 238
Plate form dental implants 159
Plunger-retained prosthesis 233
PMMA (polymethylmethacrylate) 257, 267,
  354, 402
PMN (polymorphonuclear neutrophil
  grannulocytes) 337
Polarization process 43

Polishing 111
Poly (α-hydroxy acids) 405
Polymeric sponge replication
    method 171, 410
Polymorphonuclear granulocytes, adhesion and
    activation 167
Polystylene 257
Polysulfone 257
Porcelain
    bonding agent application 261
    bonding mechanism 260
    chemical bonding 260–261
    compressive forces 261
    fired porcelain, bond strength
        evaluation 264
    mechanical retention through surface
        modifications 261
    physical bonding 261
    porcelain–titanium interface reactions,
        characterization 263
Posts and cores 266–267
Poubaix [E-pH] diagram 32
Powder metallurgy technique 410, *see also
    under* P/M
Procera process 290
Prosthetic restoration 225
Proteoglycans 168
PTM (porcelain-to-metal) 259
Pullout tests 195
PVD (Physical vapor deposition)
    coating 349
PWHT (post-welding heat-treatment) 302

R
RAP (regional acceleratory
    phenomenon) 240
Repassivation kinetics 44
Restorative dentistry devices 52
RGD peptide coating 193
Rigid fixation 194
Ringer's solution 33, 39, 45, 52, 333
Root form dental implants 159
Roughness 179–182, *see also under* Surface
    roughness
R_p (polarization resistance) 30
RPD (removable partial denture) framework
    cast 107, 265, 285
    fatigue resistance of the cast clasps 266
RT-PCR (Reverse Transcription Polymerase
    Chain Reaction) 178
Rugophilia 236

S
Sandblasting 313
    and surface texturing 314–317
    purposes 314
    using TiO_2 powder 316
SBF (simulated body fluid) 51, 140, 193,
    337, 345, 350, 396, 402
Scaffold material 404–405
SCE (saturated calomel electrode) 36
Screw-retained restorations 218–219
SE (superelasticity) 11, 36
Sealer material 268
Seawater embrittlement behavior 63
'Semiliquid transformation front'
    model 350
SFF (solid freedom fabrication) 400
SEM (scanning electron microscope) 171
Shear test 263
Shock-absorbing material 147–148
Shot peening 111, 313,
    317–319
Simon Nitinol Filter 141
Simplified mixed theory 52
SIMS (secondary ion mass spectroscopy) 28
Single tooth replacement 233
SiO_2, as a blasting media 315
SME (shape memory effect) 11, 36, 11
SME NiTi dental implants 228, 268–269
SnF_2-containing solution 29
Soldering 300, 301–302
    aluminum-based 302
    eutectic solders 302
    silver-based 302
Sol–gel method for coating 329–330
SP700 18–19
Sparks
    advantages 290
    spark erosion 289
SPF (superplastic forming) 4, 258, 291–294
    micrograin superplasticity 292
SPF/DB (Superplastic forming/diffusion
    bonding) 294–295
Spin trapping techniques 95
SPR (surface plasmon resonance) 229
Springback phenomenon 289
SSP (surface sol–gel processing) 176
α Stabilizers 11
β Stabilizers 11
Stainless steel
    endodontic files torsional properties 272
    implants, immuno-inflammatory responses
        187

Stainless steel (*Continued*)
  in orthopedic internal fracture fixation 178
  surface roughness and static frictional force
    resistance of 115
Steri-Oss osseointegrated implant 223–224
Stern's method 140
Strain-controlled fatigue test 108
Strain rate 292–293
Straumann titanium implants 224
Stress-controlled fatigue test 108
Stress ratio 44
Subperiosteal dental implants 159
Superplasticity
  diffusion bonding phenomenon associated
    with 297
  superplastic diffusion bonding method 410
  superplastic forging 293
Surface area ratio 52, 60
Surface concave texturing 313
Surface contact angle 133
Surface convex texturing 313
Surface modifications 313–364, *see also*
    chemical, electrochemical, and
    thermal surface modifications; coating;
    coloring; laser-peening method; PI^3;
    sandblasting; shot-peening; surface
    concave texturing; surface convex
    texturing
  bonding agent application 352
  bonding agent plus gold bonding agent
    application 352
  Cullet method 343
  foamed metal 363–364
  granule size effect on 325
  mono and triple-layered nitridation 352
  mono-layered chrome-doped nitridation 352
  porosity controlled surface and texturing
    360–363
  Procera porcelain application 352
Surface rolling 111
Surface roughness 170–172, 178, 314, 355,
  see also under Roughness
Surface texturing and porous/foamed materials
  397–398
Synthetic implants 325

T
TCP (tricalcium phosphate) 144, 342
TEM (transmission electron microscope) 339
Thermal expansion, linear coefficient of 260

Ti (Titanium) materials/alloys, *see also*
  individual entries; Materials classification
bone fusing to 194
Ca-deficient carbonate apatite coating 179
castability 286
characteristic properties 6
discovery 3
for medical and dental applications 25
for skeletal repair 183
in (CAD/CAM) process 4
in dental and medical fields 4, *see also*
  Dentistry
in industrial sectors 4
in vitro tests, containing chlorine ions 25
interacting map 384
oxygen content 15
properties 3
published articles on 5f
research and development 3
Ti copings, metallurgy-forming process to
  fabricate 265
Ti fiber and Ti oxide powder, as
  reinforcement for bone cement 267–268
Ti industry and new materials development
  385–388
Ti oxides, *see under* Oxidation and oxides
Ti thin film, by radio frequency sputter
  coating 49
tissue–bone compatibility of 83
various disciplines in, interacting map 384f
Ti coating 353–357
  anodization 355
  ion-beam-assisted sputtering deposition
    technique 357
  plasma-sprayed coatings 355
  Titania film and coating 357–360
Ti/porcelain system
  3-point bend flexural
    test 263–264
  4-point bending test 264, *see also separate*
    entry
  bonding mechanism 260, *see also under*
    Porcelain
  fracture processes 264, *see also under*
    4-point bending test
  problems in 262
  shear test 263
Ti-13Mo-7Zr-3Fe
  heat-treatment capabilities 112
  microstructural evolution and attendant
    strengthening mechanisms 302

Ti-20Cr casting alloy 387
Ti-29Nb-13Ta-4.6Zr, cytotoxicity of 133
Ti-34Nb-9Zr-8Ta
  heat-treatment capabilities 112
  microstructural evolution and attendant
    strengthening mechanisms 302
Ti-50Zr alloy 39
Ti-5Al-2.5Fe alloy
  cytocompatibility of 132
  friction behavior of hip prosthesis from 114
Ti-60Cu-14Ni-12Sn-4M (M: Nb, Ta, Mo)
  alloys 388
Ti-6Al-4V alloy
  abrasion 226
  alloy behavior, towards BSA 164
  as a biocompatible metal 258
  as an orthopedic implant material 135
  biocompatibility 177
  corrosion behavior 44
  corrosion behaviors 29, 36
  corrosion sensitivity of 93
  cytotoxicity of 133
  ELI (extra low level of interstitial content)
    13
  fatigue properties 297
  HA coating and macrotexturing on
    332–335
  in orthopedic internal fracture fixation 178
  in vitro metal leaching tests 48
  microstructure, chemical composition, and
    wettability of thermally and chemically
    modification studies 172–173
  microstructure, chemical composition, and
    wettability 320
  porous coating, high-cycle fatigue strength
    of 363
  silicate glass coating on 323
  SPF/DB processes 295
  SPFed denture bases of 294
  superplastic dentue base forming of 294
  surface oxides on 349
  surface roughness effects 170
  surfaces, osteoconduction of 192
  surgical implants, surface characterization
    242–243
  tribological properties 348
Ti-6Al-7Nb alloy 93
  anterior–posterior bar frameworks 258
  chemical and thermal surface modifications
    50–51
  corrosion behaviors 29

  fatigue studies on 110
  horseshoe-shaped bar frameworks 258
  palatal strap frameworks 258
  pin-on-disc wear tests on 117
Ti-H$_2$O$_2$ system 95
  cellular deterioration due to 95
  ellipsometry 95
  spectrophotometry 95
  spin trapping techniques for 95
TiN (Titanium nitride) 163
  antimicrobial characteristic 191
  coating 347–353
  onto CpTi 351–352
TiO$_2$
  crystal structures 83
  physical property that is
    unique 83
  preparation 322
  sandblasted using 316
TiPd-Co alloys 387
Tissue engineering and scaffold structure and
    materials 403–405
Tissue reaction and bone
    ingrowth 185–193
  peri-implant tissues 185–186
Titan 3
Ti-V-Fe-Al alloys 387
TMA (Ti-Mo alloy) archwires 116
Toxicity 127–130
  and corrosion rates 128
  biological toxicity 129–130
  chemicals 127
  cytocompatibility (toxicity to cells)
    130–133
  measurement 127
  physically toxic entities 130
TPS (titanium plasma-sprayed) 315
Transpassivation 43, 45
Tribo-chemical treatment 315
Tribology 113, *see also* Biotribological
    actions
Two-center effect 236
α Type (HCP) alloy 11
α+β Type alloy 11
Tyrode's solution 32

U
UHMWPE (ultrahigh molecular weight
    polyethylene) 117–118, 353
Ultrasonic fatigue tests 41–42

V
Vital hard/soft tissue   6
VMC (void metal composite)   360

W
Wear and wear debris toxicity   41, 114,
        116–119
    fraction/wear products,
        formation   118
    from orthopedic joint
        implants   119
    particulate wear debris   119
Wettability   133, *see also under* Surface
        contact angle

Wolff's law   189
Wollastonite (CaSiO$_3$) ceramic   357

X
Xenogenous grafts   325
XPS (x-ray photoelectron spectroscopy)   28,
        169
XSAM (x-ray scanning analytical microscope)
        186

Y
Yield strength   144, 289
YSZ (Yttria-stabilized-zirconia) coating   343